3rd International Symposium on Frontier of Applied Physics (ISFAP 2017)

Journal of Physics: Conference Series Volume 985

Jakarta, Indonesia
23 - 24 October 2017

Editor:

Deni S. Khaerudini

ISBN: 978-1-5108-6067-4
ISSN: 1742-6588

Printed from e-media with permission by:

Curran Associates, Inc.
57 Morehouse Lane
Red Hook, NY 12571

CURRAN ASSOCIATES INC.
proceedings
.com

Some format issues inherent in the e-media version may also appear in this print version.

Printed by Curran Associates, Inc. (2018)

For permission requests, please contact the Institute of Physics
at the address below.

Institute of Physics
Dirac House, Temple Back
Bristol BS1 6BE UK

Phone: 44 1 17 929 7481
Fax: 44 1 17 920 0979

techtracking@iop.org

Additional copies of this publication are available from:

Curran Associates, Inc.
57 Morehouse Lane
Red Hook, NY 12571 USA
Phone: 845-758-0400
Fax: 845-758-2633
Email: curran@proceedings.com
Web: www.proceedings.com

Table of contents

Volume 985

3rd International Symposium on Frontier of Applied Physics (ISFAP 2017)

23–24 October 2017, Jakarta, Indonesia

Accepted papers received: 1 March 2018
Published online: 19 March 2018

Preface

Papers

Optics, Photonics, and Optoelectronics

Geophysics

Advanced Characterization of Materials

Advanced Materials and Nanotechnology

High Temperature Materials and Coatings

Materials for Energy

IOP Conf. Series: Journal of Physics: Conf. Series **985** (2018) 011001 doi:10.1088/1742-6596/985/1/011001

Preface

International Symposium on Frontier of Applied Physics is organized annually by Research Center for Physics-Indonesian Institute of Sciences. The 3rd ISFAP 2017 conference is part of Indonesian Science Expo 2017 to commemorate 50th year anniversary of Indonesian Institute of Sciences. The conference theme is "Toward Innovation and Global Collaboration".

The third ISFAP cover basic and applied research in the fields of physics, such as optics, photonics, optoelectronics, magnetic materials, materials for energy, advanced materials, nanotechnology, high temperature materials, coatings, advanced characterization of materials, and earth physics.

ISFAP 2017 was held in Jakarta on October 23 - 24, 2017 and more than 60 researchers presented their work in the conference. The programme included leading talks from Prof. Ohashi Naoki from Research Center for Functional Materials-National Institute for Material Science (Japan), Prof. Nishimura Toshiyuki from Structural Non-Oxide Ceramics Group-National Institute for Material Science (Japan) and Dr. Bahman Shabani from School of Engineering-RMIT University (Australia).

IOP Conf. Series: Journal of Physics: Conf. Series **985** (2018) 011001 doi:10.1088/1742-6596/985/1/011001

Organising and Editor Committee

Chair:

Titi Anggono (Indonesian Institute of Sciences)

Editor:

Deni S Khaerudini
Indonesian Institute of Sciences, Indonesia

Febty Febriani
Indonesian Institute of Sciences, Indonesia

Ferensa Oemry
Indonesian Institute of Sciences, Indonesia

Iwan Sugihartono
Jakarta State University, Indonesia

Riza Setiawan
Gadjah Mada University, Indonesia

Syuhada
Indonesian Institute of Sciences, Indonesia

Wahyu Bambang Widayatno
Indonesian Institute of Sciences, Indonesia

Eni Sugiarti
Indonesian Institute of Sciences, Indonesia

Yufei Ma
Guangdong University of Technology, China

Isnaeni
Indonesian Institute of Sciences, Indonesia

Hubby Izzuddin
Indonesian Institute of Sciences, Indonesia

I Putu Susila
National Nuclear Energy Agency, Indonesia

Toto Sudiro
Indonesian Institute of Sciences, Indonesia

Yuliati Herbani
Indonesian Institute of Sciences, Indonesia

IOP Conf. Series: Journal of Physics: Conf. Series **985** (2018) 011001 doi:10.1088/1742-6596/985/1/011001

Reviewer Committee

Azwar Manaf, Indonesia University
Husin Alatas, Bogor Agricultural University
Muhammad Amal, Indonesian Institute of Sciences
Anawati, Indonesia University
Ahmad Ridwan, Tohoku University
Amanda, Indonesian Institute of Sciences
Widi Astuti, Indonesian Institute of Sciences
Vivi Fauzia, Indonesian Institute of Sciences
Lina Handayani, Indonesian Institute of Sciences
Ika Kartika, Indonesian Institute of Sciences
Sintia Niasari, Gadjah Mada University
Henri Uranus, Pelita Harapan University
Paulus Lobo Gareso, Hasanudin University
Lusi Safriani, Padjajaran University
Alfirano, Sultan Ageng Tirtayasa University

IOP Conf. Series: Journal of Physics: Conf. Series **985** (2018) 011002 doi:10.1088/1742-6596/985/1/011002

Peer review statement

All papers published in this volume of *Journal of Physics: Conference Series* have been peer reviewed through processes administered by the proceedings Editors. Reviews were conducted by expert referees to the professional and scientific standards expected of a proceedings journal published by IOP Publishing.

The effect of macro-bending on power confinement factor in single mode fibers

T B Waluyo, D Bayuwati and I Mulyanto

Research Centre for Physics – Indonesian Institute of Sciences
Kawasan PUSPIPTEK, Serpong 15314, Tangerang Selatan, Indonesia

E-mail: thom002@lipi.go.id

Abstract. One of the methods to determine the macro-bending effect in a single mode fiber is by calculating its power loss coefficient. We describe an alternative method by using the equation of fractional power in the fiber core. Knowing the fiber parameters such as its core radius, refractive indexes, and operating wavelength; we can calculate the V-number and the fractional power in the core. Because the value of the fiber refractive indexes and the propagation constant are affected by bending, we can calculate the value of the fractional power in the core as a function of the bending radius. We calculate the fractional power in the core of an SMF28 and SM600 fiber and, to verify our calculation, we measure its transmission loss using an optical spectrum analyzer. Our calculations and experimental results showed that for SMF28 fiber, there is about 4% power loss due to bending at 633 nm, about 8% at 1310 nm, about 20% at 1550 nm, and about 60% at 1064 nm. For SM600 fiber, there is about 6% power loss due to bending at 633 nm, about 11% at 850 nm, and this fiber is not suitable for operating wavelength beyond 1000 nm.

1. Introduction

Macro-bending loss in optical fibers has been realized since in the early years of its application in communication systems. Many researchers have investigated this phenomenon theoretically and experimentally. Gambling et al., for example, explained that when light propagates from a straight fiber to a curved section the field distribution changed considerably so that in addition to a pure bending loss there is also a transition loss due to mode conversion in the transition region [1]. Modeling of bend losses for a realistic single mode fiber with multiple cladding layers has been presented by Wang et al. [2] and Zendehnam et al. [3]. They showed that light reflection at the interface between the cladding layer and the coating layer of the fiber has an apparent effect so that the bend loss value of a standard single mode fiber was not a monotonic function of the bending radius but oscillating with both wavelength and bending radius.

Power loss occurred in a bent optical fiber was not desired in the application in communication systems. However, in contrast with the application in communication systems, bending loss phenomena could be useful for optical fibers intended as sensors. By tailoring the refractive index profile of the fiber, a bend-sensitive fiber can be produced with bending loss five times more than that of the ordinary G.652A single mode fiber [4]. One of the methods to calculate bending loss in an ordinary single mode fiber is by calculating its power loss coefficient, for example by using a classical formula introduced by Marcuse [5]. In this paper, we describe an alternative method to determine the

IOP Conf. Series: Journal of Physics: Conf. Series **985** (2018) 012001 doi:10.1088/1742-6596/985/1/012001

macro-bending effect by using the equation of fractional power propagating in the fiber core. Knowing the fiber parameters such as its core radius, the refractive index of the cladding, numerical aperture, and operating wavelength we can calculate the *V*-number, the normalized propagation constant, and the fractional power propagating in the core for the LP_{01} and LP_{11} modes. Because the value of the fiber refractive indexes and the normalized propagation constant are affected by bending, we can calculate the value of the fractional power propagating in the core as a function of the bending radius. As a case study, we calculate the optical power loss of an SMF28 and SM600 fiber. To verify our analysis, we measure its transmission spectra using an optical spectrum analyzer. This method could be useful to determine the combination of the wavelength of the light source with an ordinary single mode fiber (not a bend-sensitive fiber) that will be used as an intensity-based sensor based on macro-bending effect.

2. Methods

Most practical telecommunication-grade single mode fibers are classified as weakly guiding because the difference in refractive index of the core and the cladding is very small, less than 0.01 [6, 7]. Under this condition, the modes propagate in the fiber are assumed to be nearly transverse with an arbitrary set of polarization, called as Linearly Polarized (*LP*) modes. The equation for studying these modes is simply written as [7]

$$U\frac{J_{m+1}(U)}{J_m(U)} = W\frac{K_{m+1}(W)}{K_m(W)} \tag{1}$$

where U and W are fiber parameters that can be expressed in terms of the normalized propagation constant b and the normalized frequency V of the fiber, so that

$$W = \sqrt{b} \text{ and } U = \sqrt{1-b} \tag{2}$$

Using equation 2 in equation 1 we obtained that for $m = 0$ (fundamental mode)

$$V(1-b)^{1/2}\frac{J_1\left[V(1-b)^{1/2}\right]}{J_0\left[V(1-b)^{1/2}\right]} = Vb^{1/2}\frac{K_1\left[Vb^{1/2}\right]}{K_0\left[Vb^{1/2}\right]} \tag{3}$$

and for $m \geq 1$ (higher order modes)

$$V(1-b)^{1/2}\frac{J_{m-1}\left[V(1-b)^{1/2}\right]}{J_m\left[V(1-b)^{1/2}\right]} = -Vb^{1/2}\frac{K_{m-1}\left[Vb^{1/2}\right]}{K_m\left[Vb^{1/2}\right]} \tag{4}$$

For a given value of m, there will be a finite number of solutions and the nth solution ($n = 1, 2, 3, \ldots$) is referred as the LP_{mn} mode. The modal power carried in the core, as well as in the cladding, of a single mode optical fiber is one of the important parameters of the fiber. The fractional power propagating in the core is mathematically written as [7]

$$\eta = \frac{P_{core}}{P_{total}} = \left[\frac{W^2}{V^2} + \frac{U^2}{V^2}\frac{K_m^2(W)}{K_{m-1}(W)K_{m+1}(W)}\right] \tag{5}$$

The normalized propagation constant b is directly related to the effective refractive index (n_{eff})

$$b = \frac{n_{eff}^2 - n_{clad}^2}{NA^2} \tag{6}$$

where NA is the numerical aperture of the fiber

$$NA = \sqrt{n_{core}^2 - n_{clad}^2} \tag{7}$$

and the normalized frequency (V) of the fiber is

$$V = \frac{2\pi}{\lambda} \cdot a \cdot NA \tag{8}$$

where a is the fiber core radius and λ is the operating wavelength.

When the fiber is circularly bent with a bending radius of R, the refractive index of the fiber is changed due to stress-optic effects [8]

$$n'_{clad} = n_{clad}\left(1 + \frac{a}{R_{eff}}\right) \text{ and } n'_{core} = n_{core}\left(1 + \frac{a}{R_{eff}}\right) \tag{9}$$

where R_{eff} is the effective bend radius due to the stress-optic effects, for silica (SiO_2) fiber $R_{eff} = 1.28R$ [8], n_{clad} and n_{core} is, respectively, cladding and core refractive index of the unbent fiber. To determine the value of the refractive index at a different value of wavelength, we use the Sellmeier equation for SiO_2 as the cladding material [9]. The normalized propagation constant of the bent fiber (b_{bent}) is therefore [8]

$$b_{bent} = \frac{n_{eff}^2 - (n'_{clad})^2}{(n'_{core})^2 - (n'_{clad})^2} \tag{10}$$

Using this normalized propagation constant value into equation 2, the fractional power in the core of the bent fiber, η_{bent}, can be calculated using equation 5.

To verify our analysis, we then measure the transmission spectra of the fiber by using an optical spectrum analyzer (ANDO AQ-6312B). The measurement range is from 600 to 1600 nm with 10 nm resolution. The fiber samples are a two meters patchcord of SMF28 and SM600 from Thorlabs Inc. Firstly, we record the transmission of white light when the fiber patchcord is in a straight condition, and when it is circularly bent with a certain radius. As a white light source, we use a stabilized Xenon lamp (Thorlabs SLS201). To bend the fiber patchcord we use a concentric cylinder made from acrylic with varied diameter from 50 mm to 10 mm in 5 mm step. The block diagram of our measurement is shown in figure 1.

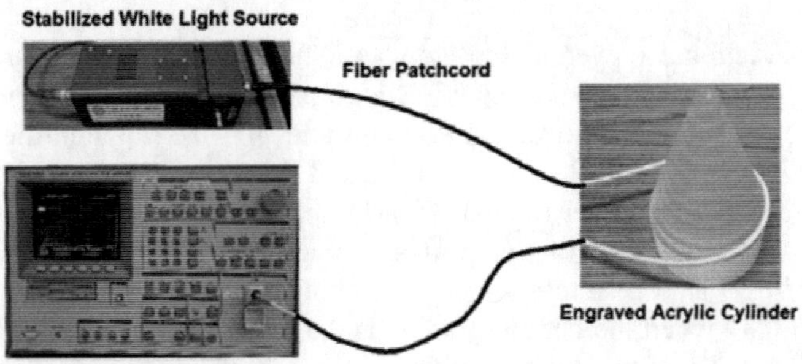

Figure 1. Block diagram of the transmission measurement.

3. Results and discussion

The graph of the fractional power in the core (P_{core}/P_{total}), calculated from equation 5, as a function of inverse bending radius ($1/R$, in cm^{-1}) for an SMF28 fiber ($a = 3.9$ μm, $NA = 0.12$) is shown in figure 2.

For light with 633 nm wavelength, the value of the *V*-number of this fiber is 4.65 with unbent propagation constant b = 0.82. This means the fiber is multi-moded in this wavelength and not sensitive to bending. As shown in the graph, when the fiber is straight (bending radius $R = \sim$ or $1/R = 0$) the fractional power in the core of the LP_{01} mode is 0.96 and decreases slightly to 0.92 when the bending radius is 0.5 cm ($1/R = 2$ cm^{-1}), indicating 4% or 0.18 dB loss of power [10]. At 1310 nm, the fiber is single-moded because the *V*-number is 2.24 with unbent propagation constant b = 0.47. At this wavelength, the power loss is 8% or 0.36 dB because the fractional power in the core of the LP_{01} mode for a straight fiber is 0.79 and decreases to 0.69 when the bending radius is 0.5 cm. At 1550 nm the fiber is more sensitive to bending than that at 1310 nm, because at this wavelength the unbent propagation constant b = 0.36 and the *V*-number is 1.9. The fractional power in the core of the LP_{01} mode for a straight fiber is 0.70 and decreases to 0.56 when the bending radius is 0.5 cm, indicating power loss of 20% or 0.97 dB. This SMF28 fiber is very sensitive to bending at 1064 nm wavelength. At this wavelength, the *V*-number is 2.76; this means that this fiber is few-moded with the unbent propagation constant for LP_{11} mode of 0.18. The fractional power in the core of the LP_{11} mode for a straight fiber is 0.64 and decreases to 0.24 when the bending radius is 0.5 cm so that the power loss at this wavelength is 60% or 3.98 dB.

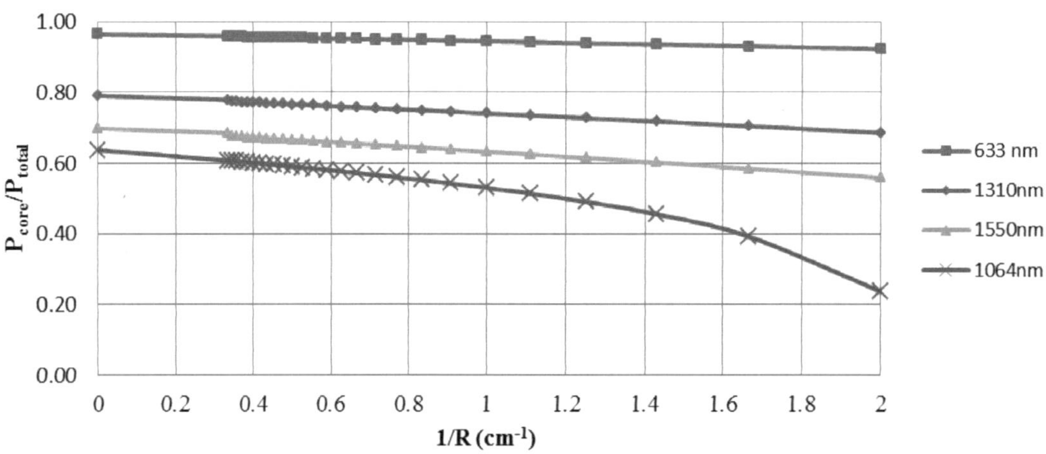

Figure 2. Fractional power in the core as a function of inverse bending radius for SMF28 fiber.

The graph of the fractional power in the core as a function of inversed bending radius for an SM600 fiber (a = 1.3 μm, NA = 0.14) is shown in figure 3. This fiber is single-moded at 633 nm with the *V*-number = 1.81 and the unbent propagation constant b = 0.33. When the fiber is straight, the fractional power in the core of the LP_{01} mode is 0.67 and decreases a little to 0.63 when the bending radius is 0.5 cm (6% or 0.27 dB power losses). Clearly, this fiber is not so sensitive to bending at this wavelength. At 850 nm, this fiber is single-moded with the *V*-number = 1.35 and the unbent propagation constant b = 0.17. The fractional power in the core of the LP_{01} mode for a straight fiber is 0.44 and decreases to 0.39 when the bending radius is 0.5 cm (power loss of 11% or 0.5 dB). At this wavelength, the fiber could be sensitive to bending if the bending radius is less than 0.5 cm. At 1310 nm, our calculation showed that the fractional power in the core of the LP_{01} mode for a straight fiber is 0.15 and decreases to 0.07 when the bending radius is 1.4 cm (53% or 3.31 dB power losses). This fiber, however, cannot be used at this wavelength since the *V*-number is 0.88 with propagation constant b = 0.04.

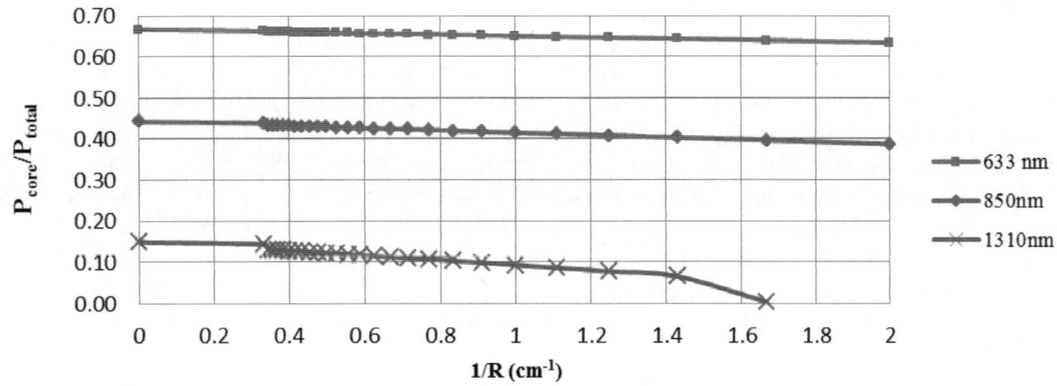

Figure 3. Fractional power in the core as a function of inverse bending radius for SM600 fiber.

The result of our measurements on transmission spectra of the SMF28 and SM600 patchcords at different bending radius is shown in figure 4. The setting of the optical spectrum analyzer was as follows: Start wavelength = 600 nm, Stop wavelength = 1600 nm, Resolution = 10 nm. For the SMF28 patchcord, the top trace is a straight fiber and then, respectively, for bending radius *R* of 1.5, 1, and 0.5 cm (lower trace). As can be seen, the fiber is sensitive to bending in the 1020 - 1120 nm wavelength range as well as in 700 – 720 nm range. There are about 3.5 dB losses when the fiber is bent with bending radius 0.5 cm at 1064 nm, while the bending loss at 633 nm, 1310 nm, and 1550 nm is negligible. For the SM600 patchcord, the top trace is a straight fiber and then for *R* of 1, 0.50, and 0.25 cm (lower trace), respectively. It can be seen that this fiber is not sensitive to bending at 633 nm; there is about 0.5 dB loss when the fiber is bent with bending radius 0.5 cm at 850 nm (and about 3 dB loss when the fiber is bent with bending radius 0.25 cm but not recommended in practice because it can damage the fiber), and no transmission at all beyond 1000 nm. Overall, the measurement results matched with the power confinement factor analysis.

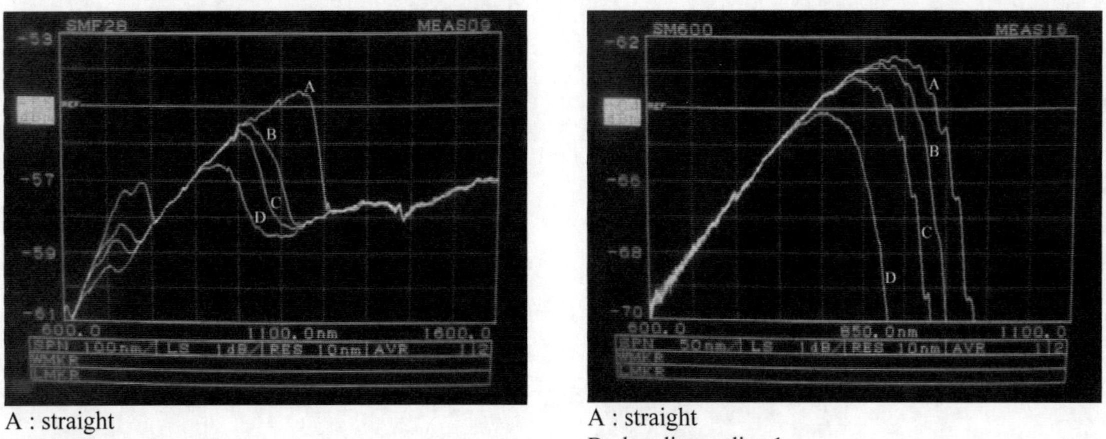

A : straight
B : bending radius 1.5 cm
C : bending radius 1 cm
D : bending radius 0.5 cm

A : straight
B : bending radius 1 cm
C : bending radius 0.5 cm
D : bending radius 0.25 cm

Figure 4. Transmission spectra of the SMF28 (*left*) and SM600 (*right*) patchcords at different bending radius.

4. Conclusions

The fractional power in the core (P_{core}/P_{total}) of a single mode optical fiber is one of the important parameters of the fiber. Since its value is affected by macro-bending, we can determine the bending characteristics for a certain fiber at a particular wavelength. This calculation does not take into account the presence of the fiber coating. The fiber is modeled as a core surrounded by virtually infinity cladding and the Fresnell reflection at the interface between cladding and coating is neglected. This calculation, however, could be used as a quick tool to determine the combination of the operational wavelength of the light source with an ordinary single mode fiber (not a bend-sensitive fiber) that will be used as an intensity-based sensor based on macro-bending effect. Our calculations and experimental results showed that, if the fiber is bent with a bending radius of 0.5 cm, about 3.5 dB losses occurred at 1064 nm for SMF28 fiber, and about 0.5 dB loss occurred at 850 nm for the SM600 fiber.

References

[1] Gambling W A, Matsumura H and Ragdale C M 1979 *Opt. Quant. Electron* **11** 43
[2] Wang Q, Farrel G and Freir T 2005 *Opt. Express* **13** 4476
[3] Zendehnam A, Mirzaei M, Farashiani A and Farahani L H 2010 *Pramana – J. Phys.* **74**, 591
[4] Lee D-H, Kwon K-H and Song J-W 2004 *J. Opt. Soc. Korea* **8** 83
[5] Marcuse D 1976 *J . Opt. Soc. Am.* **66** 216
[6] Gloge D 1971 *Appl. Opt.* **10** 2252
[7] Dutta A 2015 International Journal of Electronics and Communication Technology **6** 74
[8] Schermer RT and Cole J H 2007 *IEEE J. Quantum Electron.* **43** 899
[9] Agrawal G P 2002 *Fiber Optic Communication Systems 3rd Edition* (New York: John Wiley & Sons Inc.), p 39
[10] dB vs. optical power loss http://timbercon.com/dB-power-loss/ (last accessed July 2017)

Determination of the effective cut-off wavelength of several single-mode fiber patchcords

D Bayuwati, T B Waluyo and I Mulyanto

Research Center for Physics – Indonesian Institute of Sciences
Kawasan PUSPIPTEK Serpong, Tangerang Selatan 15314, Indonesia

E-mail: dwib001@lipi.go.id

Abstract. The effective cut-off wavelength is one of the important parameters in single-mode optical fiber. Nevertheless, the data sheet of an optical fiber patchcord generally does not specify the exact value of the cut-off wavelength. However, for certain applications, for example, to know the value of the numerical aperture, core radius and refractive index of the fiber at a certain wavelength, the exact value of the cut-off wavelength should be known. In this paper, we report the measurement of the cut-off wavelength of several commercially-available single-mode fiber patchcords, namely, the 780HP, SMF28, SM600, and SM2000. We used the multi-mode reference method in our measurement because it is simpler than the bend reference counterpart. From our measurements, the effective cut-off wavelengths for the above-mentioned fiber patchcords are (0.731±0.006) μm, (1.225±0.005) μm, (0.485±0.010) μm, and (1.665±0.018) μm, respectively, and its comparison with catalogue values has been discussed. The usefulness of the cut-off wavelength determination for calculating other fiber parameters has also been described.

1. Introduction

The cut-off wavelength (λ_c) is the wavelength above which a fiber behaves as a single-mode fiber [1, 2]. The theoretical cut-off wavelength of an optical fiber depends on the radius and refractive index of the fiber core, the normalized frequency, the relative refractive index between core and cladding, and the fiber numerical aperture (*NA*) [1, 2]. Nevertheless, the data sheet of an optical fiber generally does not specify the exact value of the cut-off wavelength whereas for certain application, for example, in an intensity-based sensor which utilized bending loss effect, cut-off wavelength highly determines the bending loss. The differences in core radius or *NA* of about 5% can result in bending losses that differ by more than one order of magnitude [3], while according to Schermer and Cole [4], the numerical aperture of SMF28 in the data sheet varies between 0.12 and 0.14. The cut-off wavelength value can also be used to calculate several fiber parameters such as core and refractive index of the fiber core and the fiber's *NA* [4, 5]. Thus, an exact value of cut-off wavelength in optical fiber applications is important.

There are many techniques for cut-off wavelength measurement [1, 6 – 9], among other is bend reference method where the long wavelength edge of the bend-induced loss is greater than the long-wavelength baseline by 0.1 dB and the effective cut-off wavelength can be determined from the point of intersection of a straight line parallel to the λ-axis and displaced by 0.1 dB from the long-wavelength baseline. Another technique is the multi-mode reference technique where transmission powers in a single-mode and a reference multi-mode fiber are compared to determine the effective cut-off wavelength.

In this paper, we report the measurement of the cut-off wavelength of several commercially-available single-mode fiber patchcords, namely, the 780HP, SMF28, SM600, and SM2000 patchcords. We use the multi-mode reference method in our measurement because it is simpler and no need to introduce the bending to the fiber.

2. Method

Fiber cut-off wavelength has been defined by the ITU-T G650 [10] which is obtained from measurement with bending reference technique using 2 m length of fiber with 14 cm single loop radius. As stated before, there are many techniques for measuring the cut-off wavelength and basically, it involves the comparison of transmitted power from the test fiber to the reference fiber at different wavelengths [6 – 9]. In the bend reference technique, the reference fiber is the same sample of single-mode fiber which is coiled to several diameter values. In the multi-mode reference technique, a multi-mode fiber is used as a reference fiber.

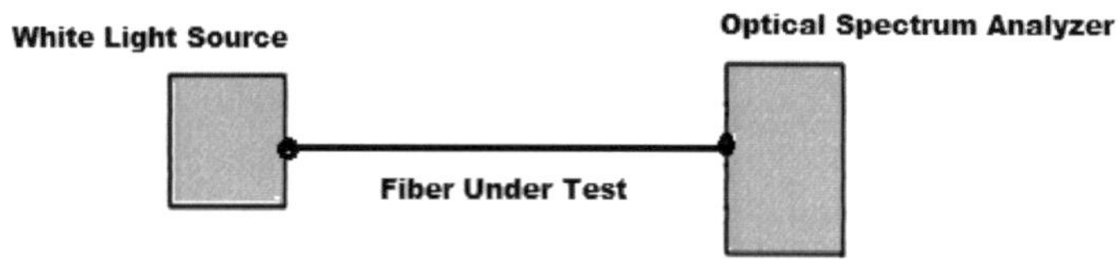

Figure 1. The multi-mode reference technique set-up.

The multi-mode reference technique is presented by the experimental set-up shown in figure 1. The white light source was launched into the test single-mode fiber and the spectral variation of its output power, $P_s(\lambda)$ is recorded by an optical spectrum analyzer. Then, the single-mode test fiber is replaced by a multi-mode fiber and the spectral dependence of power $P_m(\lambda)$ is again recorded. The variation of loss $R_m(\lambda)$ is the comparison output of both spectrum, the $P_s(\lambda)$ and $P_n(\lambda)$, which is given as below [1].

$$R_m(\lambda) = 10 \log\left[\frac{P_s(\lambda)}{P_m(\lambda)}\right] \quad (1)$$

To find the cut-off wavelength, firstly the R_m is plotted as a function of λ. A straight line is then fitted to the long wavelength region of $R_m(\lambda)$ and shifted up to 0.1 dB. The wavelength value at the point of intersection of this line with the $R_m(\lambda)$ curve gives the effective cut-off wavelength (figure 2).

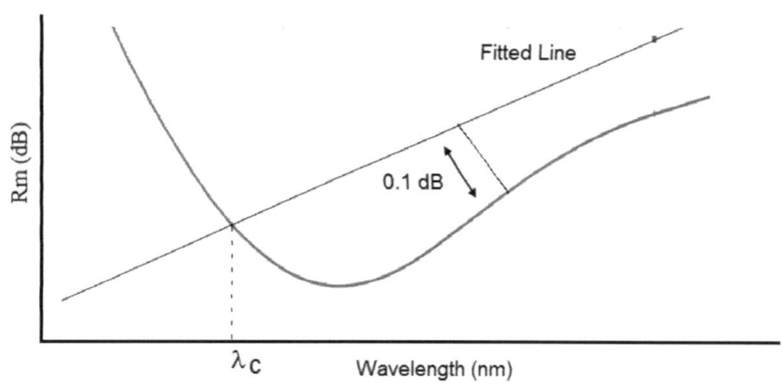

Figure 2. Determination of the cut-off wavelength with the multi-mode reference technique.

As mentioned before, the test single-mode fibers used here are the 780HP, SMF28, SM600, and SM2000 fiber patchcords from Thorlabs Inc. [11] with a length of 2 m each. The multi-mode fiber used for the reference is the M31L02 from Thorlabs Inc. which has a core diameter of 62.5 μm and *NA* of 0.275 [11]. The white light source is a stabilized broad band light source type SLS201/M (Thorlabs) while the optical spectrum analyzer/OSA is the AQ6310B model from ANDO.

Interestingly, the cut-off wavelength (λ_c) value can be used to determine other fiber parameters such as core radius (a) and numerical aperture (*NA*) of a single-mode fiber. The fiber data sheet rarely informs the value of the core radius but, instead, its mode field diameter (*MFD*) value at a certain operating wavelength (λ). To determine the value of a, we use an empirical formula derived by Marcuse [5].

$$MFD = 2a\left(0.65 + 0.434\left(\lambda/\lambda_c\right)^{3/2} + 0.015\left(\lambda/\lambda_c\right)^{6}\right) \tag{2}$$

Furthermore, the value of *NA* can be calculated using the cut-off wavelength expression which is given by [1, 2]:

$$\lambda_C = \frac{2\pi a}{V_c}\sqrt{n_1^2 - n_2^2} \tag{3}$$

or

$$\lambda_C = \frac{2\pi a}{2.405} NA \tag{4}$$

because $NA = \sqrt{n_1^2 - n_2^2}$, a is the radius of the core, n_1 is the refractive index of the core, and n_2 is the refractive index of the cladding. Equation 3 and 4 are derived from the equation for the normalized frequency V of the fiber written as:

$$\begin{aligned} V &= \frac{2\pi}{\lambda} aNA \\ &= \frac{2\pi}{\lambda} n_1 (2\Delta)^{1/2} \end{aligned} \tag{5}$$

Where $\Delta = \frac{1}{2}\frac{n_1^2 - n_2^2}{n_2^2}$ is the relative index difference, and $V = V_c = 2.405$ (the first root of the J_0 Bessel function) is the normalized frequency for the cut-off condition for a single-mode fiber.

3. Results and discussion

Figure 3a shows the typical result of the variation loss curve, $R_m(\lambda)$, for the 780HP fiber patchcord at a wavelength range from 680 nm to 760 nm. Figure 3b shows the detailed graph for the cut-off wavelength determination which results in the effective cut-off wavelength value of 730 nm from the intersection point. Similarly, the variation losses with their determinations of effective cut-off wavelengths for the SMF28, SM600 and SM2000 fiber patchcords, respectively, are presented in figure 4 – 6. Table 1 presents the results of our cut-off wavelength measurement, the calculated of the core radius, the numerical aperture of each fiber patchcords, and the catalogue value for comparison.

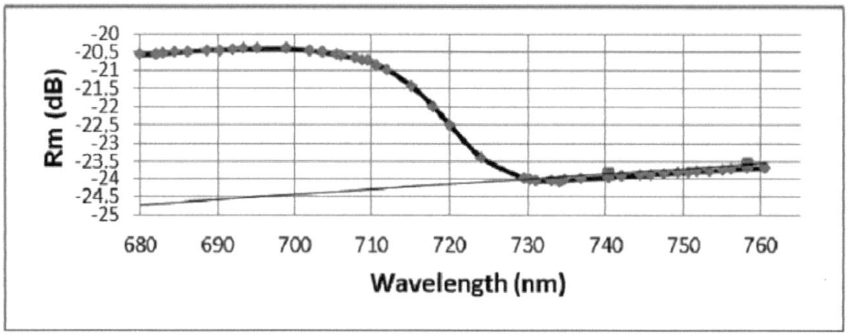

(a)

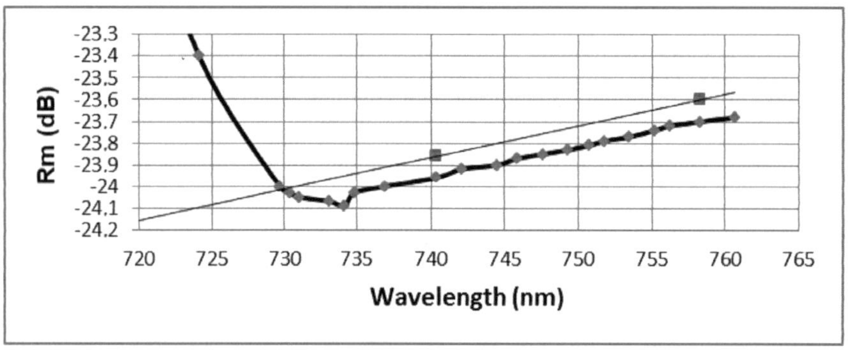

(b)

Figure 3(a). The plot of variation loss and the determination of cut-off wavelength for the 780HP fiber patchcord. **(b).** Details of its cut-off wavelength determination result in intersection point at wavelength 730 nm (λ_c).

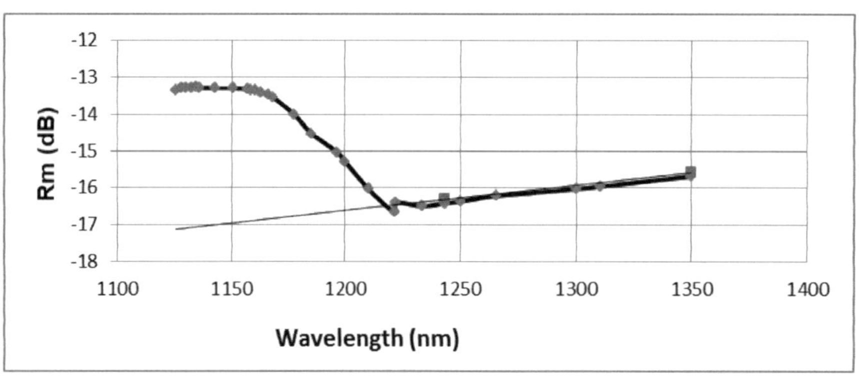

Figure 4. The plot of variation loss as a function of wavelength for the SMF28 fiber patchcord with its cut-off wavelength determination.

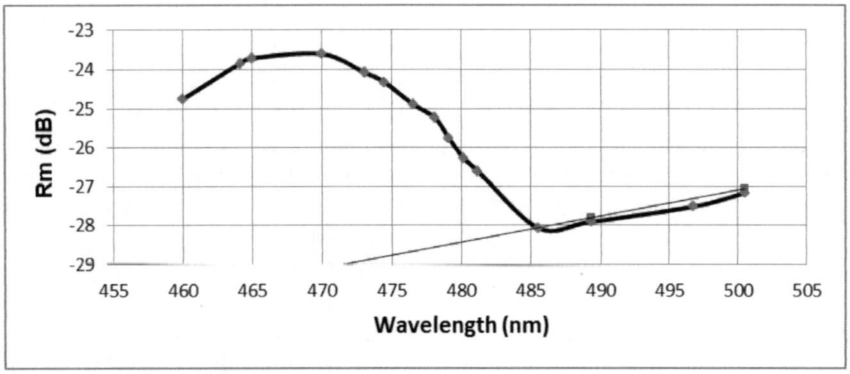

Figure 5. The plot of variation loss as a function of wavelength for the SM600 fiber patchcord with its cut-off wavelength determination.

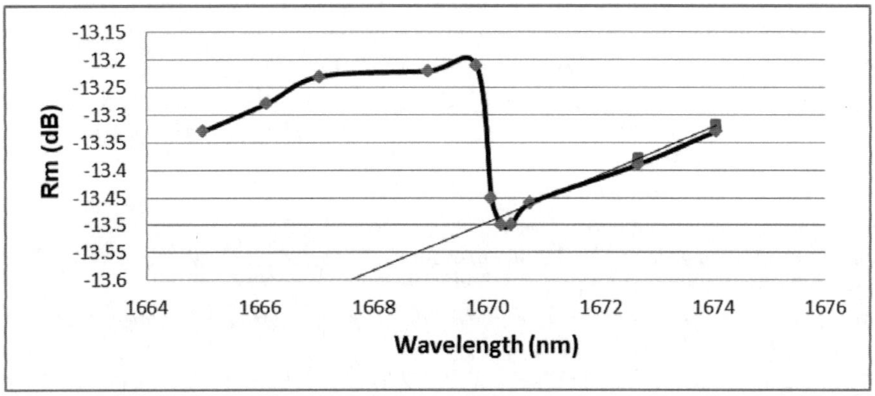

Figure 6. The plot of variation loss as a function of wavelength for the SM2000 fiber patchcord with its cut-off wavelength determination.

Table 1. Comparison of the measured cut-off wavelength, the calculated core radius and numerical aperture with its catalogue values.

Type of patchcord	Operating wavelength [11] (μm)	Cut off wavelength, λ_c (μm)		Calculated core radius[a] (μm)	*NA*	
		Measured	Catalogue value [11]		Calculated[b]	Catalogue value [11]
780HP	0.780 – 0.970	(0.731±0.006)	(0.730±0.030)	2.03	0.14	0.13
SMF28	1.260 – 1.625	(1.225±0.005)	<1.260	3.91	0.12	0.14
SM600	0.635 – 0.780	(0.485±0.010)	0.500 – 0.600	1.93	0.09	0.1 – 0.14
SM2000	1.700 – 2.300	(1.665±0.018)	1.700	5.14	0.12	0.11

[a] Calculated using equation 2.
[b] Calculated using equation 4.

The cut-off wavelength measurement shows that the experimental values for all the fiber patchcords are comparable or still within the range of its catalogue values. By using this obtained effective cut-off wavelength, we can then calculate the core radius and the numerical aperture of the fiber, as presented in table 1. This calculation is important because relying merely on the catalogue value, would give confusing results. For example, as pointed out by Schermer and Cole [4], the catalogue values for SMF28 fiber are as follows: core diameter 8.2 μm, refractive index difference 0.36%, and the numerical aperture 0.14. Using these values, we obtained $NA = 0.123$, calculated from equation 5, and $\lambda_c = 1.49$ μm, calculated from equation 4, that would imply multi-mode behavior at 1.31 μm, which is not true because this type of single-mode fiber is designed to operate at 1.3 – 1.5 μm. Caution should be given because the catalogue value of *NA* is typically based on beam output measurement which is different with the index-based *NA*; and the refractive index difference is usually measured in the fiber preform, rather than in the fiber itself [4].

4. Conclusions

By using the multi-mode reference method, we have determined the effective cut-off wavelength of four commercially available fiber patchcords, namely, the 780HP, SMF28, SM600, and SM2000. The data sheet of each fiber does not mention the specific value of its cut-off wavelength. The average value of the effective cut-off wavelength for each fiber patchcords is 0.731, 1.225, 0.485, and 1.665 μm, respectively. Furthermore, from the calculation of the fiber core radius, we obtained that the core radius of the SM2000 > SMF28 > 780HP > SM600, as expected from its operating wavelength.

Acknowledgments

The authors would like to acknowledge the Management of the Research Center for Physics, Indonesian Institute of Sciences/Pusat Penelitian Fisika-LIPI. The research was funded by Program Tematik Pusat Penelitian Fisika-LIPI 2017.

References

[1] Ghatak A and Thyagarajan K 1998 *Introduction to Fiber Optics* (New York: Cambridge University Press) pp 430-434

[2] Senior J M and Jamro M Y 2009 *Optical Fiber Communications: Principles and Practices* 3rd ed. (London: Prentice Hall) pp 54-61

[3] Schulze C, Lorenz A, Flamm D, Hartung A, Schröter S, Bartelt H and Duparré M 2013 *Opt. Express* **21** 3170

[4] Schermer R T and Cole J H 2007 *IEEE J. Quantum Electron.* **43** 899

[5] Yang B, Duan J, Xie Z and Xiao H 2013 *J. Cent. South Univ.* **20** 1832

[6] Krahn F, Sange B, Neumann E G, Schwierz H, Streckert J and Wilczewski F 1989 *Fiber and Integrated Optics* **8** 203

[7] Kato Y and Miyauchi M 1985 *Appl. Opt.* **24** 2351

[8] Brinkmeyer E, Heckmann S 1984 *Opt. Lett.* **9** 28

[9] Cut-off wavelength, http://www.tpub.com/neets/tm/109-4.htm (last accessed on July 31st, 2017)

[10] ITU-T Study Group 15 1997 Recommendation ITU-T G.650 (04/97), Series G: Transmission Systems and Media, Digital Systems and Networks, Transmission Media Characteristics - Optical Fiber Cables : Definition and Test Methods for the Relevant Mode Fibres (Switzerland: International Telecommunication Union) pp 22-30

[11] Single Mode FC/PC Fiber Optic Patch Cables, https://www.thorlabs.com/newgrouppage9.cfm?objectgroup_id=1362

Tuning sputtered gold thickness to enhance absorption and emission in core-shell type erbium doped upconversion nanoparticles

R V Manurung[1,3], C T Wu[2] and S Chattopadhyay[3]

[1]Research Center for Electronics and Telecommunication, Indonesian Institute of Sciences, Jl. Cisitu No.21/154 D, Bandung, Indonesia
[2]Nano Device Materials Characterization Division, National Nano Device Laboratories, Science Park, Hsinchu, Taiwan
[3]Institute of Biophotonics, National Yang Ming University, Li Nong Street No.155 – Taipei City, Taiwan

E-mail: robe007@lipi.go.id

Abstract. Upconversion nanoparticles (UCNPs) converts near-infrared excitation to visible emission with advantages e.g. photostable, non-blinking, and background-free probes for bioimaging and biosensor. However, low quantum yield and low efficiency (~1%) as drawback need to be enhanced. A plasmonic gold nano-structured surface was designed and fabricated to couple with the 980 nm radiation and produce plasmonic enhancement of the upconversion luminescence. The synthesis of the UCNPs was done by thermal decomposition and SiO_2 coating prepared by the reverse microemulsion process. Here, we report a novel tunable plasmon-enhanced fluorescence by modulating the thickness and surface roughness of gold island film on Si. The localized surface plasmon resonance (LSPR) at 980 nm was obtained, matched with the native excitation of UCNPs resulting in maximum enhancement of 10-fold of green emission band at 540 nm for the Er-doped UCNPs.

1. Introduction

Lanthanide-doped upconversion nanoparticles (UCNPs) can use infra-red (IR) excitation for its visible emission, which is fit for deep tissue imaging [1]. The upconversion process employs sequential absorption process of multiple low energy (IR) photons, the use of long lifetime, and through ladder-like energy levels of trivalent lanthanide ions embedded in a suitable inorganic host lattice to produce higher energy (visible) anti-Stokes luminescence [1]. Nevertheless, the major drawback of UCNPs is low (<1%) quantum yield due to the small absorption cross-sections. The low quantum yield/efficiency in these UCNPs can be enhanced by different techniques, and among various strategies metal-enhanced fluorescence (MEF) is a focus in this study. Normally, adjacent metallic structures quench fluorescence from fluorophores [2]. However, if placed at a certain distance away from them, these metallic structures can enhance fluorescence. In this paper, we have adopted the plasmonic enhancement technique for the luminescence enhancement. A novel sputtered gold substrate has been used whose preparation is experimentally simple. We have used a silica spacer on the lanthanide doped UCNPs to create the separation [3], and dispersed these particles on the gold-coated substrate to optimize the plasmonic enhancement of the emission. The objective of enhancing the luminescence by surface plasmons in a gold film is schematically shown in figure 1.

2. Experimental

2.1. Materials

Ytterbium(III) chloride ($YbCl_3$, 99.99%), yttrium trichloride (YCl_3, 99.99%), erbium(III) chloride ($ErCl_3$, 99.99%), polyoxyethylene(5) nonylphenylether (IGEPAL CO-520), sodium hydroxide (NaOH, 98%), ammonium fluoride (NH_4F, 98%), silver nitrate ($AgNO_3$, 98%), ethylene glycol (99%), and ethanolamine (99%) were purchased from Sigma-Aldrich (USA). Oleic acid (90%), octadecene (90%), and tetraethyl orthosilicate (TEOS, >99%) were purchased from Acros (Belgium). All the chemicals were used as received from the vendors without any further purification.

2.2. Synthesis of $NaYF_4$:Yb,Er core nanoparticles

Upconverting lanthanide $NaYF_4$:Yb,Er nanoparticles were synthesized as follows [4]: YCl_3 (0.8 mmol), $YbCl_3$ (0.18 mmol), and $ErCl_3$ (0.02 mmol) were mixed with 6 ml oleic acid and 15 ml octadecene in a 50 ml flask. The solution was heated to 160 °C under vigorous stirring to form a homogeneous solution and then cooled to room temperature. Subsequently, 10 ml methanol solution of NaOH (2.5 mmol) and NH_4F (4 mmol) were slowly added to the flask and vigorously stirred for 30 minutes. The solution was heated slowly to remove methanol, degassed at 100 °C for 10 minutes, followed by heating at 300 °C for 1 h under nitrogen flow. The solution was remained undisturbed and cooled naturally; afterwards, the nanoparticles were precipitated from the solution with acetone and washed with ethanol/water (1:1 v/v) for three times.

2.3. Synthesis of the SiO_2 shell as spacer on the $NaYF_4$:Yb,Er nanoparticles

The coating procedure of the $NaYF_4$:Yb,Er nanoparticles with a SiO_2 layer as a spacer was reported previously [4] as follows: 0.1 ml of CO-520, 6 ml of cyclohexane, and 4 ml of 10 mM $NaYF_4$:Yb,Er nanoparticles solution in cyclohexane were mixed and stirred for 10 minutes. 0.4 ml of CO-520 and 0.08 ml of 30 wt.% ammonia were then added to the mixture solution and the container was sealed and sonicated for 20 minutes until a transparent emulsion was formed. Subsequently, 0.04 ml of TEOS was added to the solution. The thickness of silica coated outside UNCPs could be adjusted by controlling the concentration of TEOS. The solution was stirred for 48 hours at a speed of 600 rpm. Finally, $NaYF_4$:Yb,Er/SiO_2 core/shell nanoparticles were precipitated by adding acetone, and washed with ethanol/water (1:1 v/v) twice and re-dispersed in water.

Synthesis and preparation of gold nanostructured film layer on crystalline silicon (c-Si) substrates for MEF was done as follows: first, Si substrates (2 cm × 3 cm) were cleaned and sonicated in deionized water and acetone, separately, for 15 and 30 minutes, respectively, followed by drying under a nitrogen stream. Gold was coated on substrates by a sputter coater (Emitech K550X, United Kingdom) for 2, 4, and 8 minutes. Subsequently, 2 ml of 10 mM $NaYF_4$:Yb,Er/SiO_2 core/shell nanoparticles were dropped on each of gold coated c-Si substrates and dried in ambient air. The proposed schematic to study metal-enhanced fluorescence effect between UCNPs and different thickness of gold sputtered on a silicon substrate (2, 4, and 8 minutes) has shown in figure 1.

2.4. Characterization of morphology, structure, and emission

Transmission electron microscope (TEM) and scanning electron microscopy (SEM) imaging for morphology and surface roughness of gold thin film coated silicon wafer was performed using TEM JEM-2000EX (JEOL, Japan) and Field Emission SEM 6700F (JEOL, Japan), respectively. UV-Vis absorption measurement was performed using UV-Visible-Near Infrared spectrophotometer (V-770, JASCO, Japan). Upconversion and metal-enhanced fluorescence spectra were measured by Fluorolog-3 fluorescence spectrofluorometer (Jobin Yvon, USA) using 980 nm laser head (SDL-5000T, Shanghai, China) with 60° angle of incident and power up to 5 W as the excitation source.

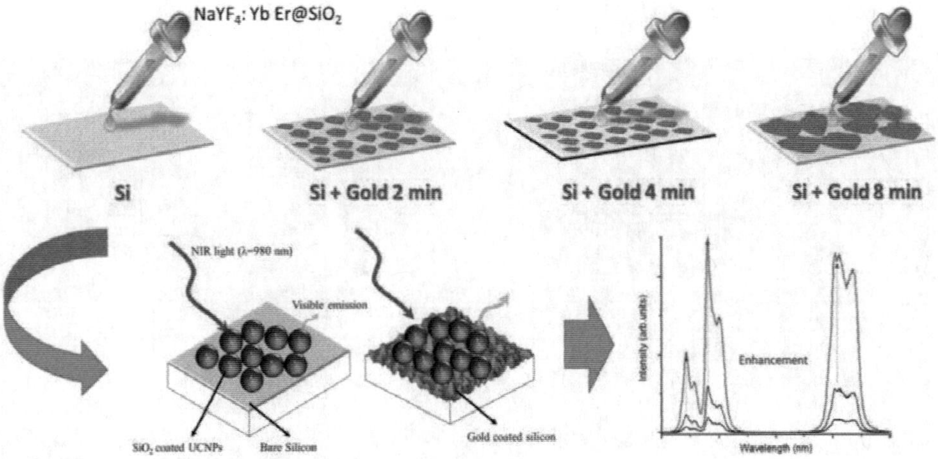

Figure 1. Schematic of the proposed setup gold island on Si for metal-enhanced fluorescence in upconversion nanoparticles.

3. Results and discussion

Figure 2a and b shows typical TEM images of the $NaYF_4$:Yb,Er core and core-shell upconversion nanoparticles. The core UCNPs have a lattice spacing of 0.3 nm, and diameter of 8±2 nm (figure 2a). The use of the dense SiO_2 coating on the core UCNP is important for stabilizing the core and enabling dispersion in aqueous media instead of cyclohexane which cannot be used for cellular or tissue imaging. Another reason to use the SiO_2 layer as a gap between UCNPs and metal structured for MEF. The core-shell UCNPs has shown in figure 2b and obtained 7 nm of silica thickness.

Figure 2c described energy transfer upconversion (ETU) mechanism. ETU is a process which employs two types of luminescent centers, which are usually called the sensitizer (Yb^{3+}) and activator (Er^{3+}), embedded in the crystal matrix. Under the excitation of photons with lower energy, both the sensitizer and the activator would be excited. With the non-radiative energy transfer from the sensitizer to the adjacent activator, sequential absorption of the activators to the higher excited state takes place. Radiative emission could be observed as the activator returned to the ground state. In Yb^{3+}–Er^{3+} dual ions activated systems, green emissions at 520 and 540 nm together with red emissions at 650 nm are frequently observed characteristics. Moreover, another peak, particularly in blue emission bands such as 380, 410, and 487 nm, also appeared in weak intensity (figure 2c).

Figure 2d exhibit the up-conversion emission spectra of the core (black line), and core-shell (red line) type nanoparticles under 2 W excitation. The fluorescence spectra indicated signal reduction up to 25% after SiO_2 coating has been applied on the surface of UCNPs core. Similarly, the optical photograph in the inset of figure 2d display the intensity of core-shell that was weaker than core, which indicates SiO_2 coating as a spacer for MEF purpose reduced the intensity of UCNPs.

Metallic (Au or Ag) nanostructures including nanofilms and nanoparticles have been widely used in metal-enhanced fluorescence including the UCNPs [5]. The optical properties of metal films can be tailored by manipulating the organization of the constituent nanostructures, columns [5, 6], islands, or layers. Specifically, ultrathin gold or silver island films shows frequency tunable surface plasmon (SP) absorption bands in the visible and near-infrared (400 – 1000 nm) [7, 8]. The localized surface plasmon resonance (LSPR) of the metallic nanostructures then increases the radiative rate and/or the excitation intensity producing the luminescence enhancement. The luminescence quenching that stems from resonance energy transfer from the fluorophore or UCNPs to the metallic structure can be prevented by introducing a spacer layer to delineate the fluorophore from the metal [9]. MEF in

UCNPs has been reported before by using a variety of gold or silver nano-morphologies including film [10], nanoparticles [11], and nanorods [12].

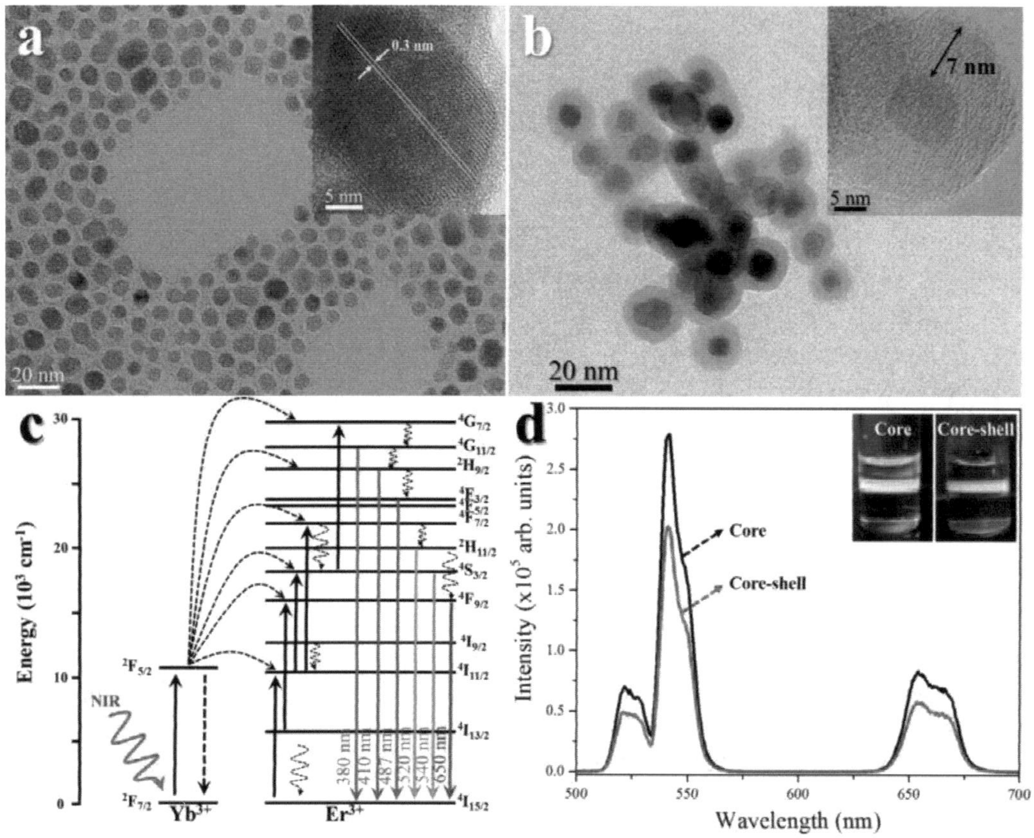

Figure 2. TEM image of $NaYF_4$:Yb,Er (a) core and (b) core-shell. The inset indicated high magnification TEM image of core and core-shell. (c) Upconversion mechanism of the Yb,Er. Under excitation at 980 nm, an electron of Yb^{3+} is excited from the $^2F_{7/2}$ to the $^2F_{5/2}$ level. The energy may be transferred to Er^{3+} non-radiatively to excite it to the corresponding excited level. For Yb,Er co-doped UCNPs, the emission bands at 520, 540, and 650 nm assigned to $^2H_{11/2}$ - $^4I_{15/2}$, $^4S_{3/2}$ - $^4I_{15/2}$, and $^4F_{9/2}$ - $^4I_{15/2}$ transitions of Er^{3+}. (d) The up-conversion emission spectra of the core (black line), and core–shell (red line) type nanoparticles under 2 W excitation. The inset shows an optical photograph of the tube showing green luminescence from core ($NaYF_4$:Yb,Er) and core-shell ($NaYF_4$:Yb,Er/SiO_2).

In this paper, we introduce our design of thin film metallic nanostructures for the observation of MEF in the UCNPs (figure 1). The gold layer would be tunable in surface coverage, morphology, and thickness by adjusting the sputtering time. With a sputtering machine at our disposal this would mean low density gold nano-islands, coalesced gold islands, and thick layers of a gold island with gold nano-clusters, with increasing sputtering time. Fixed volume core-shell structure ($NaYF_4$:Yb,Er/SiO_2) UCNPs dispersed on clean silicon served as the control and gold coated silicon with different time deposition (figure 1). Gold-coated silicon was optimized for the maximum luminescence from the same volume of UCNPs dispersed on them. We have carefully prepared gold-coated silicon substrates by sputtering and optimized for the best MEF.

Figure 3 shows the SEM and AFM images of the gold layer on c-Si, displaying morphology for sputtering times of 2, 4, and 8 minutes. The SEM images clearly reveal discontinuous gold island film

IOP Conf. Series: Journal of Physics: Conf. Series **985** (2018) 012003 doi:10.1088/1742-6596/985/1/012003

morphology for 2 minutes coating time (figure 3a), which becomes a coalesced quasi-continuous film by 4 minutes (figure 3b) coating time. Further increase in the sputtering time yielded fully coalesced, continuous, and comparatively thicker gold films (figure 3c). Figure 3e – g shows a representative AFM image of the gold nanostructure for different sputtering times. The surface roughness of the gold nanostructures for 8 minutes sputtering time, as shown in figure 3g, is higher than 2 and 4 minutes and the data has been consistent with the SEM images. Coupling of the UCNPs with these metallic surfaces should lead to enhanced fluorescence going by earlier reports [9, 13]. Tuning the optical reflection of the gold nanostructures may efficiently couple more light into, and concentrate plasmonic electric field at the UCNP site to facilitate enhanced fluorescence.

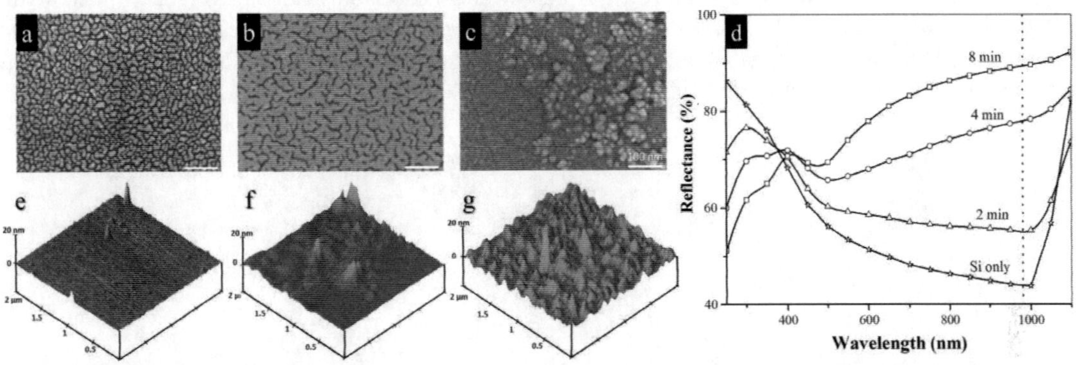

Figure 3. Top view SEM and AFM images showing the morphology of the gold island films on Si substrate produced by sputtering for (a, e) 2, (b, f) 4, and (c, g) 8 minutes, respectively. (d) Reflectance spectra of silicon only (star), and silicon substrates coated with 2 (triangle), 4 (circle), and 8 (square) minutes of gold by sputtering. Short dash line shows $NaYF_4$:Yb,Er/SiO_2 absorption band at 980 nm.

To verify our hypothesis mentioned above, fixed volume of UCNPs were dispersed on these gold coated c-Si and measured for their luminescence. To characterize the gold nanostructures, number identification tags would be used to represent sputtering time (in minutes) of gold coating. Therefore, (0) would represent luminescence from UCNPs on bare c-Si or control, (2) would represent a sample with a bottom gold layer of 2 minutes, and (8) would imply a gold thin film layer of 8 minutes. All samples were prepared and measured under identical conditions.

Figure 4a exhibits MEF from the core-shell UCNPs for samples (2), (4), and (8) compared to (control), where (8) showed the best fluorescence signal which we would identify as the optimized bottom gold layer coating time. Maximum enhancements of ~8, ~9, and ~10 fold were observed for the 520, 540, and 650 nm emissions, respectively, for the (8) compared to the (control) configuration. The spectral emissions were similar to those in solution (figure 4b) with signature peaks at 540 and 650 nm, and a comparatively weaker emission peak at 520 nm. The intensity variation of the main spectral features of the UCNP emission at 520, 540, and 650 nm, as a function of excitation power density, is shown in figure 4b – d, respectively, for all configuration. The slopes of the linear fits of the intensity variation (figure 4b – d) are between 2.6 – 3, signifying approximately 2 – 3 participating photons (n) in the luminescence process which is consistent with earlier reports [14]. This presentation would reflect the efficacy of the bottom gold layer, in the emission process with respect to those without any use of gold. The gold thin film nanostructures (8) yields an enhancement factors of ~ 8, 9, and 10 for the emissions at 520, 540, and 650 nm, respectively, against the no-gold (control) configuration. The 8 minutes gold sputtered on c-Si proved to be the most efficient enhancer of luminescence compared to other configurations.

To understand the effect of the gold nanostructured layer, we have to study their optical properties by considering the reflectance spectra of a gold nanostructured layer on silicon as shown in figure 3d. The bare Si substrate, as control, shows typical silicon reflectance spectra with a decreased absorption below the band gap 1100 nm energy. The data shows that the reflectance near 980 nm increase significantly (absorption decreasing) with increasing thickness (2 – 8 minutes of sputtering) of the gold layer. The result indicates a mirror-like gold coating (8 minutes gold sputtering on Si) which reflects 90% of the 980 nm excitation which is also the absorption band of the UCNPs (shown by the dashed band around 1000 nm). The elevated reflection of the 980 nm radiation by the gold nanostructured layer would ensure increased optical path length and superior absorption in the UCNP layer. The enhancement was also observed on the 2 and 4 minutes gold coated Si substrates but in lesser proportions (figure 3d). The efficiency of MEF will be weaker on the discontinuous island films, as shown on 2 or 4 minutes gold coated Si substrates case, because a significant number of the UCNPs would be sitting on bare Si surface which would absorb instead of reflecting most of the available 980 nm radiation. The observed enhancement for the 2 and 4 cases, is similar to the one reported by Zhang et al [10] who obtained 5-fold enhancement just by a single coating the core UCNPs without a SiO_2 shell on them with gold NPs.

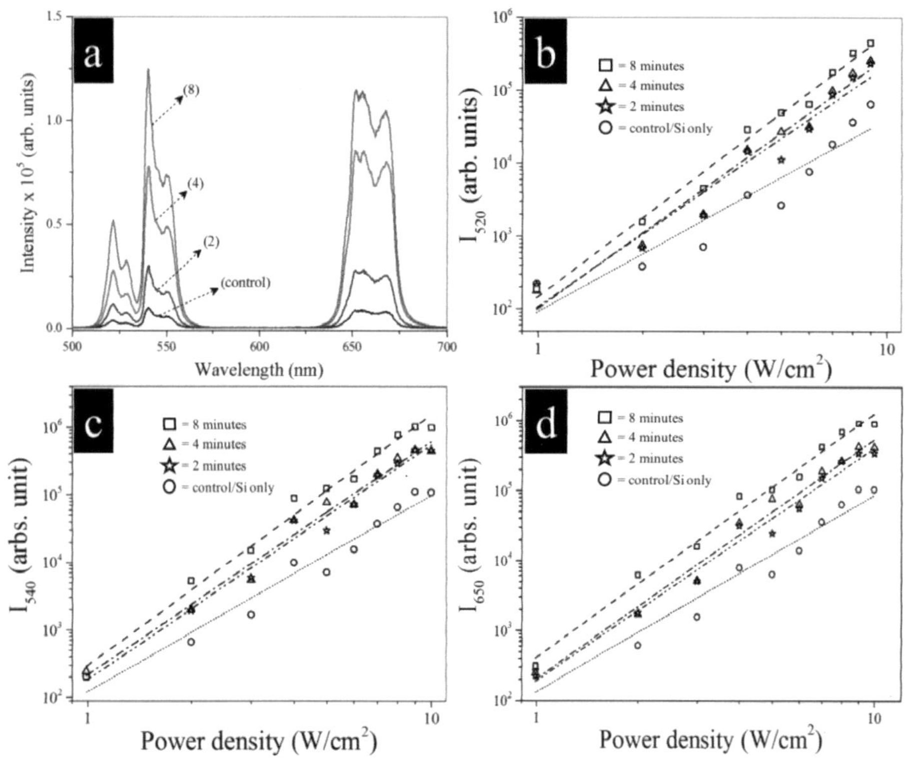

Figure 4. (a) Upconversion luminescence spectra, and power dependence of the (b) 520, (c) 540, and (d) 650 nm peak emission of $NaYF_4$:Yb,Er/SiO_2 on Si only (control), and Si coated with 2 (2), 4 (4), and 8 (8) minutes of gold. Laser power of 2 W was used in (a). The line joining the data points in (b – d) represents linear fits to the data points (symbols). The spectra in (a), and the symbols in (b – d) represent average measurements done on five different spots on the samples.

4. Conclusion

In summary, we have proposed and successfully demonstrated the use of a gold nanostructured layer for metal-enhanced fluorescence from $NaYF_4$:Yb,Er/SiO_2 core-shell nanoparticles (in solid state) with core sizes 8 nm in diameter prepared by thermal decomposition and ~7 nm thick SiO_2 shell prepared by the microemulsion technique. We have optimized a thick gold layer, with higher reflectance in the infrared region, to work best for the fluorescence enhancement by increasing the optical path length and therefore efficient multiphoton absorption of the excitation in the UCNP layer. A maximum enhancement factor of 10 was observed when using the gold nanostructured layer compared to the bare nanocrystals without the use of any gold surface.

Acknowledgments
This work was supported by the Ministry of Science and Technology, Taiwan, Grant Nos. MOST-101-2112-M-010-003-MY3 and 104-2112-M-010-002-MY3. The author also thanks Research Center for Electronics & Telecommunication, Indonesian Institute of Sciences and the Functional Nanomaterial Laboratory, Institute of Biophotonics, National Yang-Ming University, Taiwan, for the support.

References

[1] Kumar R, Nyk M, Ohulchanskyy T Y, Flask C A and Prasad P N 2009 *Adv. Funct. Mater.* **19** 853
[2] Luther J M, Jain P K, Ewers T and Alivisatos A P 2011 *Nat. Mater.* **10** 361
[3] Li Z, Zhang Y and Jiang S 2008 *Adv. Mater.* **20** 4765
[4] Manurung R V, Wu C T, Roy P K and Chattopadhyay S 2016 *RSC. Adv.* **6** 87088
[5] Jensen T R, Schatz G C and Van Duyne R P 1999 *J. Phys. Chem. B* **103** 2394
[6] Yang B, Lu N, Qi D, Ma R, Wu Q, Hao J, Liu X, Mu Y, Reboud V, Kehagias N, Torres C M, Boey F Y, Chen X and Chi L 2010 *Small* **6** 1038
[7] Shumaker-Parry J S, Rochholz H and Kreiter M 2005 *Adv. Mater.* **17** 2131
[8] Myroshnychenko V, Rodríguez-Fernández J, Pastoriza-Santos I, Funston A M, Novo C, Mulvaney P, Liz-Marzán L M and García de Abajo F J 2008 *Chem. Soc. Rev.* **37** 1792
[9] Lakowicz J R 2001 *Anal. Biochem.* **298** 1
[10] Zhang H, Xu D, Huang Y and Duan X F 2011 *Chem. Commun. (Camb.)* **47** 979
[11] Saboktakin M, Ye X C, Oh S J, Hong S H, Fafarman A T, Chettiar U K, Engheta N, Murray C B and Kagan C R 2012 *ACS Nano* **6** 8758
[12] Feng A L, You M L, Tian L, Singamaneni S, Liu M, Duan Z, Lu T J, Xu F and Lin M 2015 *Sci. Rep.* **5** 7779
[13] Schietinger S, Aichele T, Wang H Q, Nann T and Benson O 2010 *Nano Lett.* **10** 134
[14] Pollnau M, Gamelin D R, Lüthi S R, Güdel H U and Hehlen M P 2000 *Phys. Rev. B* **61** 3337

IOP Conf. Series: Journal of Physics: Conf. Series **985** (2018) 012004 doi:10.1088/1742-6596/985/1/012004

Photoluminescence study of carbon dots from ginger and galangal herbs using microwave technique

Isnaeni[1], I Rahmawati[2], R Intan[3] and M Zakaria[2]

[1]Research Center for Physics, Indonesian Institute of Sciences, Riset Rd. Building 442, Puspiptek, South Tangerang, Banten, 15314, Indonesia
[2] Department of Physics, State University of Semarang, Sekaran Street, Gunung Pati, Semarang, Central Java, 50229, Indonesia

E-mail: isnaeni@lipi.go.id

Abstract. Carbon dots are new type of fluorescent nanoparticle that can be synthesis easily from natural sources. We have synthesized carbon dots from ginger and galangal herbs using microwave technique and studied their optical properties. We synthesized colloidal carbon dots in water solvent by varying microwave processing time. UV-Vis absorbance, photoluminescence, time-resolved photoluminescence, and transmission electron microscope were utilized to study properties of carbon dots. We found that microwave processing time significantly affect optical properties of synthesized carbon dots. UV-Vis absorbance spectra and time-resolved photoluminescence results show that luminescent of carbon dots is dominated by recombination process from n-π* surface energy level. With further development, these carbon dots are potential for several applications.

1. Introduction

Carbon dots are newly discovered carbon-based nanostructure in last decade [1, 2]. Carbon dots differ from other carbon-based nanostructure, such as graphene dots, carbon nanotubes, etc. Carbon dots are solid spherical carbon nanostructures, which have good optical properties [2 – 4]. Carbon dots have good optical stability and moderate luminescent efficiency so that they can be utilized for various optical devices such as light emitting diodes and solar cells [3 – 5]. Carbon dots are also can be applied as photo-catalyst and bio and chemical sensors. Synthesis of carbon dots is also relatively simple. Several simple techniques, such as hydrothermal, solvothermal, ultrasonic, and microwave techniques have been demonstrated to produce carbon dots [1, 6]. Those techniques have shown simple and fast synthesis process of carbon dots. The interesting fact regarding carbon dot synthesis is that most of natural resources can be utilized as the carbon sources. Many researchers have shown that carbon dots can be synthesized from sugar, leaves, fruit peels, beverages, and many more [7, 8]. In addition, some researchers also employed high purity chemicals, such as oleic acid, to fabricate carbon dots [9]. However the optical properties of carbon dots prepared either from natural resources or high purity chemicals are comparable. Therefore, the utilization of natural resources for synthesizing carbon dots becomes an advantage for carbon dots research. The utilization of natural resources to produces luminescent carbon dots can reduce pollution in air, water, and soil; therefore, this technology becomes environmental friendly.

In this work, we focused on synthesis of carbon dot from herbs, i.e., ginger and galangal. Ginger and galangal are herbs that are commonly found in tropical countries. They have strong flavor and have been utilized as spices for cooking or healthy drinks. In this work, synthesis of carbon dots from ginger and galangal was done using microwave technique which was chosen due to its simplicity. Many researches previously have shown that microwave technique can easily synthesize carbon dots [1, 5]. However, in their articles, there was no information regarding the influence of microwave processing time on optical properties of synthesized carbon dots. The purpose of this work is to synthesize and optically characterize carbon dots made of ginger and galangal using microwave technique with different microwave processing time.

2. Sample preparation and measurements

The synthesis procedure of carbon dots from of ginger and galangal herbs using microwave technique is shown in figure 1. Each herb was crushed using domestic electrical blender in water. The blended herb was then filtered to get extract solution and then placed in a 450 W domestic microwave oven. Microwave processing time was varied from 5, 10, 20, and 40 minutes until brown dried sample was obtained. After that, the sample was dissolved and centrifuged to separate the brown liquid and precipitate. Final brown liquid which contains colloidal carbon dots, was then characterized using transmission electron microscope (TEM FEI Tecnai 200 kV), UV-Vis absorbance, photoluminescence (PL) and time-resolved photoluminescence (TRPL). We measured absorbance using deuterium and halogen lamps as light source and MAYAPRO2000 from Ocean Optics as photospectrometer. PL and TRPL were performed using picosecond diode laser at wavelength 420 nm as excitation source.

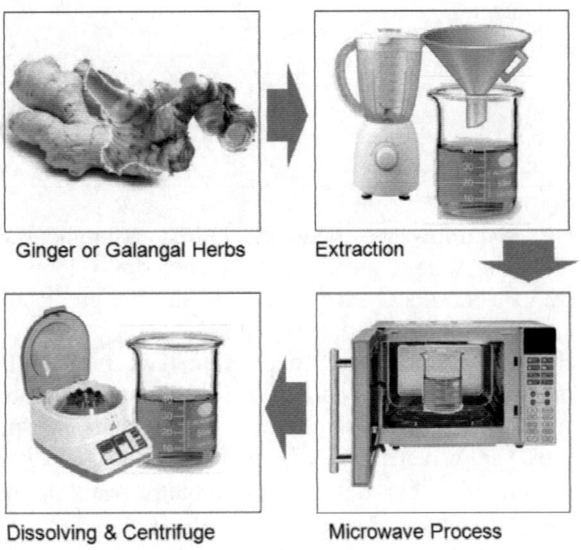

Figure 1. Schematic of carbon dots synthesis from ginger and galangal herbs using microwave technique.

3. Results and discussion

In general, we have successfully synthesized carbon dots from ginger and galangal herbs. There are two easy hints indicating the presence of carbon dots, i.e. brown color solution and light emitted from solution upon blue laser excitation as shown in figure 2. The brown color solution indicates that carbon dots perfectly dissolved in water solvent. After few weeks, we did not notice any sedimentation from sample indicated that carbon dots and water solvent are a stable colloid. The light emission coming from carbon dots colloid solution is mostly cyan color or light-green. However, not all of the samples give good indication of carbon dots existence. As shown in table 1, sample with 5 minutes

IOP Conf. Series: Journal of Physics: Conf. Series **985** (2018) 012004 doi:10.1088/1742-6596/985/1/012004

microwave processing time from galangal herb does not show brown color and light-green emission color. We notice that increasing microwave processing time will increase number of synthesized carbon dots indicated by darker brown color.

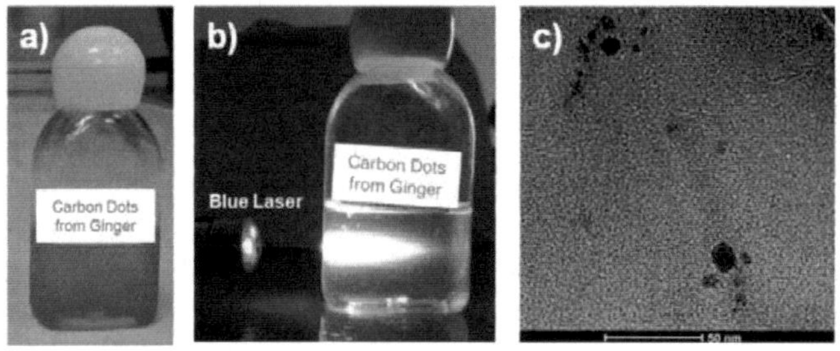

Figure 2. Appearance of colloidal carbon dots under (a) ambient illumination, (b) blue laser illumination, and (c) its TEM images.

Table 1. A slightly more complex table with a narrow caption.

Sample number	Original source	Microwave processing time (minutes)	Appearance color
1	Ginger	5	Brownish white
2	Ginger	10	Light Brown
3	Ginger	20	Brown
4	Ginger	40	Dark brown
5	Ginger	5	Milky white
6	Ginger	10	Brownish white
7	Ginger	20	Light brown
8	Ginger	40	Brown

In order to confirm our carbon dots, we conducted TEM measurement as show in figure 2c. We could not obtain clear TEM image due to small size of carbon dots. However, from TEM image, we can confirm that small size of carbon dots exist in the sample. We predicted that size of carbon dots below 10 nm, which is typical size of carbon dots [1, 10,11].

In order to study optical properties of carbon dots, first we measured absorbance of colloidal carbon dots. We only measured 7 samples from table 1. Sample number 5 did not show any absorbance peak. For all measured samples, we observed that there are two clear absorbance peak, i.e., at 270 nm wavelength and 370 nm wavelength. Absorbance peak around 270 nm wavelength is related to π-π^* energy level from core of carbon dots, and absorbance peak around 370 nm wavelength is related to n-π^* energy level from surface of carbon dots [10,11]. For further analysis in this work, we focused on energy level from surface of carbon dots. After normalization of absorbance spectra (as shown in figure 3), we notice that absorbance value at wavelength around 370 nm wavelength increases as microwave processing time increases (as shown in the inset figures of figure 3). For carbon dots made of ginger, the absorbance value at 370 nm wavelength linearly increases as microwave time increases. However, a little bit different trend was observed for carbon dots made of galangal. The increase of absorbance value at 370 nm wavelength indicates that the role of surface energy level increases in carbon dots.

Luminescent of carbon dots were obtained using photoluminescence experiment as shown in figure 4. For colloidal carbon dots made of ginger (figure 4a), we found that carbon dots with microwave processing time up to 20 minutes show similar luminescent spectra and peak position. Carbon dots with 40 minutes processing time shows different luminescent spectrum. The spectrum is red shift, which is shifted to longer wavelength. In addition, the luminescent spectra of carbon dots made of galangal show similar trend (figure 4b). Carbon dots with 5 minutes microwave processing time did not give any luminescent upon blue laser excitation. The emission occurs from carbon dots with 10 minutes microwave processing time. Luminescent spectra from carbon dots with longer microwave shift to longer wavelength. Comparison of carbon dots luminescent spectra from ginger and galangal (figure 4c and d) shows the luminescent spectra become identical after 40 minutes microwave processing time. We believe that 40 minutes microwave processing time to be the optimum condition for synthesizing carbon dots.

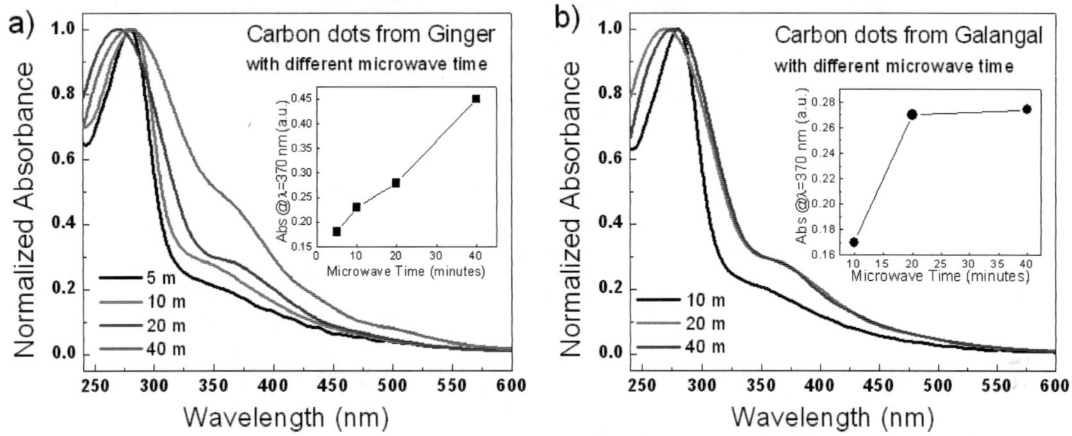

Figure 3. Absorbance spectra of colloidal carbon dots made of (a) ginger and (b) galangal.

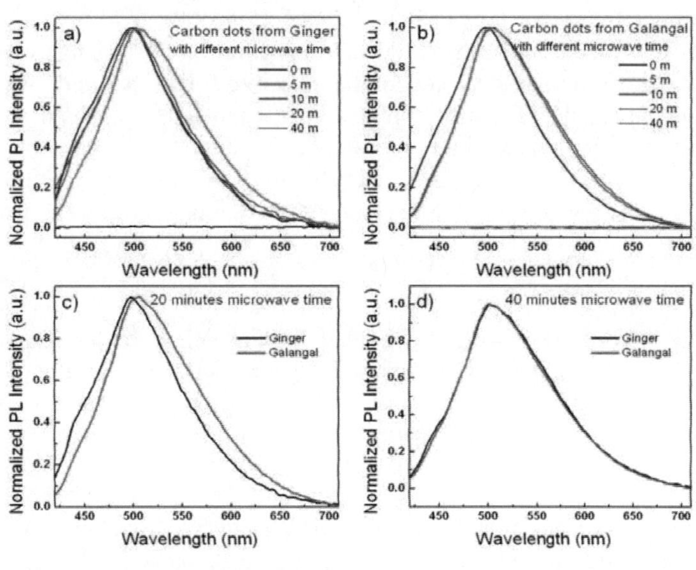

Figure 4. Photoluminescence spectra of colloidal carbon dots made from (a) ginger and (b) galangal and (c, d) their comparison.

In addition to photoluminescence, we also observed exciton decay time from every sample of carbon dots using TRPL measurement as shown in figure 5. For carbon dots made from ginger (figure 5a) and carbon dots made from galangal (figure 5b), we found that microwave processing time affects exciton decay time. Longer microwave processing time gives shorter exaction decay time (as shown in figure 4.d). This indicates that luminescent of carbon dots is dominated from n-π^* (surface) energy level as predicted by earlier. We also found that, although luminescent spectra of carbon dot made from both ginger and galangal (figure 4d) are similar, but the exciton decay time of carbon dots made from ginger is much shorter than those from galangal.

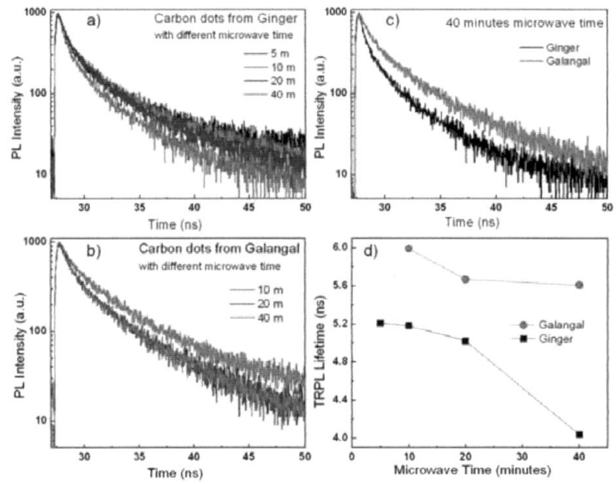

Figure 5. Time-resolved photoluminescence spectra of colloidal carbon dots made from (a) ginger and (b) galangal, and (c, d) their comparison.

4. Summary

We have successfully synthesized carbon dots from ginger and galangal herbs. Green color luminescent was obtained from every carbon dots sample. We found that microwave processing time significantly affected optical properties of synthesized carbon dots. For both samples (ginger and galangal), 40 minutes microwave time gives almost similar optical properties, especially photoluminescence spectra. UV-Vis absorbance spectra and time-resolved photoluminescence results show that luminescent of carbon dots is dominated by recombination process from n-π^* (surface) energy level.

Acknowledgments

The authors would like to thank to Kegiatan Unggulan LIPI and Kegiatan Penguatan Komptensi from Research Center for Physics, Indonesian Institute of Sciences for financial support of this work. The authors also would like to thank to researchers in the Group of Laser, Research Center for Physics for valuable discussion and support during this work.

References

[1] Reyes D, Camacho M, Camacho M, Mayorga M, Weathers D, Salamo G, Wang Z and Neogi A 2016 *Nanoscale Res. Lett.* **11** 424
[2] Puvvada N, Kumar B N P, Konar S, Kalita H, Mandal M and Pathak A 2012 *Sci. Technol. Adv. Mater.* **13** 045008
[3] Li H, He X., Kang Z, Huang H, Liu Y, Liu J, Lian S, Tsang C H A, Yang X and Lee S-T 2010 *Angew. Chem Int. Ed.* **49** 4430
[4] Jiang K, Zhang L, Lu J, Xu C, Cai C and Lin H 2016 *Angew. Chem Int. Ed.* **55** 7231
[5] Hola K, Zhang Y, Wang Y, Giannelis E P, Zboril R and Rogach A L 2014 *Nano Today* **9** 590
[6] Sciortino A, Marino E, van Dam B, Schall P, Cannas M and Messina F 2016 *J. Phys. Chem. Lett.* **7** 3419
[7] Sarswat P K and Free M L 2015 *Phys. Chem. Chem. Phys.* **17** 27642

[8] Pires N R, Santos C M W, Sousa R R, de Paula R C M, Cunha P L R and Feitosa P A 2015 *J. Braz. Chem Soc.***26** 1274
[9] Zhang F, Feng X, Zhang Y, Yan L, Yang Y and Liu X 2016 *Nanoscale* **8** 8618
[10] Sun C, Zhang Y, Kalytchuk S, Wang Y, Zhang X, Gao W, Zhao J, Cepe K, Zboril R, Yu W W and Rogach A L 2015 *J. Mater. Chem. C* **3** 6613
[11] Hsu Y-F, Chen Y-H and Chang C-W 2016 *Phys. Chem. Chem. Phys.* **18** 30086

IOP Conf. Series: Journal of Physics: Conf. Series **985** (2018) 012005 doi:10.1088/1742-6596/985/1/012005

Pulse laser ablation of Au, Ag, and Cu metal targets in liquid for nanoparticle production

Y Herbani[1], Irmaniar[2], R S Nasution[3], F Mujtahid[4] and S Masse[4]

[1]Research Center for Physics, Indonesian Institute of Sciences, Bldg 442 Kawasan Puspiptek Serpong, Banten, Indonesia 15314
[2]Department of Physics, Gadjah Mada University, Sekip Utara, Bulaksumur, Yogyakarta, Indonesia 55281
[3]Department of Physics, Universitas Sumatera Utara, Jl. Bioteknologi 1, Kampus Padang Bulan, Medan, Sumatera Utara, Indonesia 20155
[4]Department of Physics, Hasanuddin University, Kampus Unhas Tamalanrea Km. 10 Makassar, Sulawesi Selatan, Indonesia 90245

E-mail: yuliati.herbani@lipi.go.id

Abstract. We have fabricated metal and oxide nanoparticles using pulse laser ablation of Au, Ag, and Cu metal targets immersed in water. While laser ablation of Au and Ag targets in water produced metal nanoparticles which were stable for a month even without any dispersant, we found CuO nanoparticles for Cu target due to rapid oxidation of Cu in water resulted in its poor stability. Au, Ag, and CuO nanoparticles production were barely identified by naked eyes for their distinctive colour of red, yellow, and dark green colloidal suspensions, respectively. It was also verified using UV-Vis spectrometer that Au, Ag, and CuO colloidal nanoparticles have their respective surface plasmon resonance at 520, 400, and 620 nm. TEM observation showed that particle sizes for all the fabricated nanoparticles were in the range of 20 – 40 nm with crystalline structures.

1. Introduction

In our modern society, the nanosciences have as of late developed as a noteworthy research course, resulting from continuous endeavours to make anything in a nanometer size using various tools and processes. Particularly in the nanoparticle synthesis, it is well admitted that the physical approach should now complimented with the classic chemical one because the former seems more preferable over the later, based on the fact that the chemical approach involves various types of chemicals which are toxic and difficult to be cleaned afterward [1]. Among the physical approaches available so far, evaporation-condensation, arc-discharge, and pulse laser ablation in liquid (PLAL) are the most important since they are lack of solvent contaminant and can produce a relatively uniform distribution of nanoparticles. Physical synthesis of nanoparticle using a tube furnace in atmospheric pressure, however, has some disadvantages e.g. consume a great amount of energy (more than several kilowatts) and requires a lot of time to preheat and achieve thermal stability (several tens of minutes) [2, 3]. On the other hand, the arc-discharge was also claimed to be 'green' because it utilized no added surfactant on the nanoparticle synthesis [4]. In this technique where metal wires were submerged in the deionized water and used electrodes, a small amount of metal ions (in ppm concentration) can be released into the surrounding liquids and should be cleaned further, especially for the applications involving biological materials. Based on these facts, PLAL seems to be appropriate choice a 'green' physical approach amongst others.

Here, we explored a utilization of nanosecond laser for nanoparticle production, called as pulse laser ablation of bulk target in liquid (PLAL). Differs from laser-induced photochemical reduction [5 – 7], here metal nanoparticles were obtained by fragmenting metallic bulk materials immersed in solution [8 – 10]. Using pure metal plates of Au, Ag, and Cu, we successfully synthesized Au, Ag, and CuO nanoparticles in water using a nanosecond laser. We selected Au, Ag, and Cu metals as targets due to their widely use in the field of electronics and their interesting plasmonic properties. We also explored cetyl trimethylammonium bromide (CTAB) surfactant for stabilizing the obtained nanoparticles. It was found that CTAB can preserve the stabilization of Au and Ag nanoparticles for a month, but not for CuO nanoparticles. The results were expected to open more exploration on types of nanoparticles that can be produced using this technique.

2. Methods

Laser ablation was carried out with a nanosecond NdYAG laser (Q-Smart 850 mJ, Quantel, Mountain View, CA), which provided less than 6 ns full width at half maximum (FWHM) pulses (fundamental wavelength 1064 nm, maximum energy 850 mJ/pulse, repetition rate of 10 Hz). The radiation was focused by an objective, with focal length of 10.0 cm, onto a metal target that was placed on the bottom of a 20 ml beaker glass filled with the 5 ml aqueous solution. The laser beam with energy of 10 – 40 mJ was directed to the metal surface and the height of the liquid layer above the metal plate (99.99%, PT. Antam Gold) was 3 mm. The vessel was placed on a horizontal platform, which executed linear motions every 1 second to change the ablated position. Please see Fig 1 for detail setup. All solutions were prepared from high-purity deionized water (16.7 MΩ.cm). CTAB ($C_{16}H_{33}N(CH_3)_3Br$, 99+%, Sigma-Aldrich) solutions were prepared as stock solutions in 0.01wt.% concentrations immediately prior to their use. Characterization of the obtained colloids were carried out by a modular UV-Vis spectrometer (HR4000/MayaPro2000, Ocean Optics) to identify the surface plasmon resonance (SPR), and by transmission electron microscopy (TEM FEI Tecnai G20 S-Twin 200kV) to analyze the nanoparticle morphology and size.

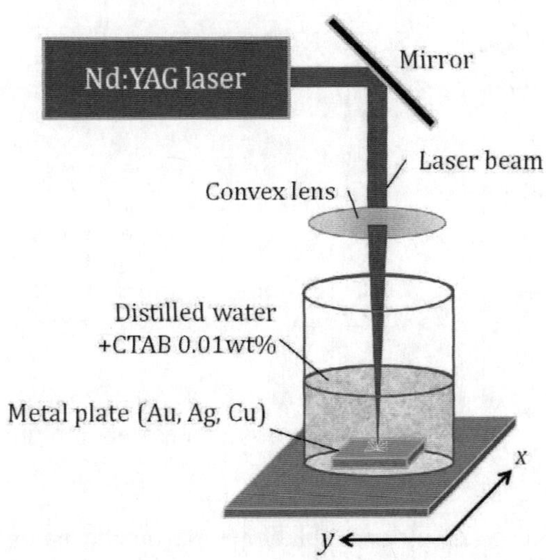

Figure 1. Setup of pulse laser ablation in liquid.

IOP Conf. Series: Journal of Physics: Conf. Series **985** (2018) 012005 doi:10.1088/1742-6596/985/1/012005

3. Results and discussion

The ablation of bulk Au, Ag, and Cu metal targets in aqueous phase by the radiation of 1064 nm, 532 nm, and 355 nm laser sources as shown in figure 2 resulted in the reddish, yellowish, and dark green colour of solutions, respectively. We examined the effect of these laser wavelengths for the ablation of Au, Ag, and Cu metal plates in water on the production of nanoparticle. The laser energy used was 20 mJ and the ablation duration was 60 minutes. Figure 2 shows the UV-Vis spectra of the obtained nanoparticles at the different laser wavelength. The produced nanoparticles can be indentify by the appearance of surface plasmon peaks in their respective spectra which were around 520, 420, and 634 nm for Au, Ag, and CuO nanoparticles, respectively. These obtained SPR peaks were in a good agreement with the ones reported in literatures [8 – 10]. From this figures, it was found that for the different metal target, the effective laser wavelength for its correlated nanoparticle's production (identified by the highest absorbance) is also different, e.g. for Au and Ag metal targets, the effective laser wavelength is 532 nm and for Cu metal target is 1064 nm. It might be due to the absorption coefficient of bulk Au and Ag is higher at 532 nm, so the intensity of absorbance peak of 532 nm higher than 1064 nm [10], whereas the opposite was true for bulk Cu. For all metal targets under investigation, the rest of experiment in this paper will be conducted at their respective laser wavelength.

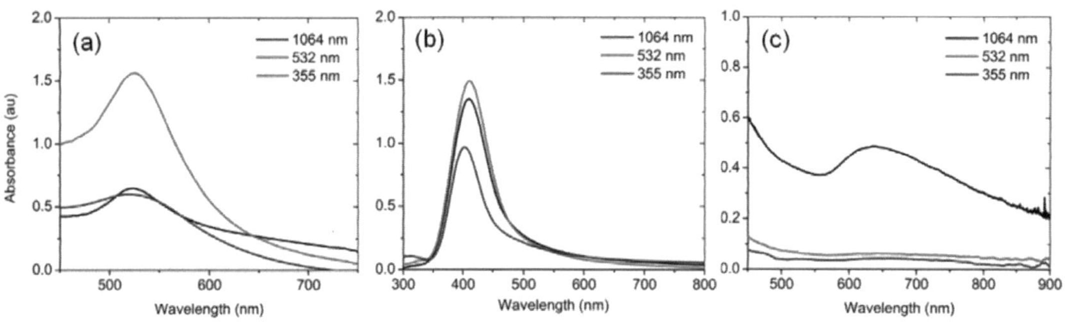

Figure 2. UV-visible absorption spectra of (a) Au, (b) Ag, and (c) CuO nanoparticles obtained by PLAL at the laser energy of 20 mJ and the ablation time of 60 minutes.

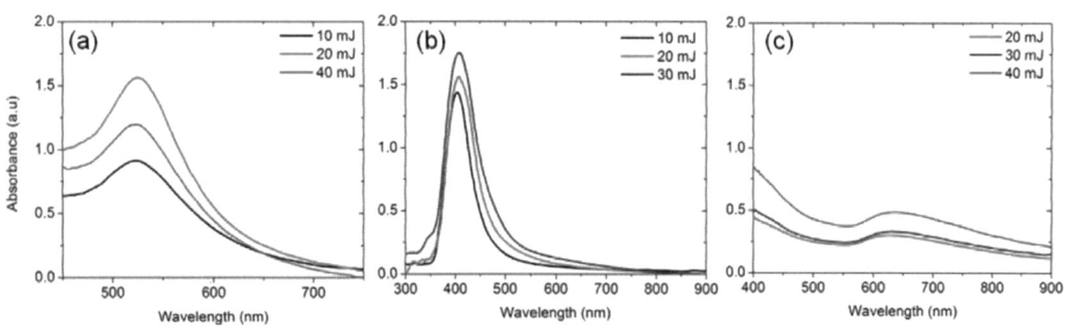

Figure 3. UV-visible absorption spectra of (a) Au, (b) Ag and (c) CuO nanoparticles obtained by PLAL at different laser energy of 10 – 40 mJ for 60 minutes ablation time. Each color identifies the same level of laser energy.

The influence of laser pulse energy and ablation time on the nanoparticle properties was also investigated. Figure 3 shows the absorbance spectra of the colloidal nanoparticles prepared in the different laser pulse energies. We analysed the dependence of basic features of the surface plasmon band namely its center wavelength, full width half-maximum, and the maximum intensity on the laser pulse energy. For all nanoparticle obtained within the studied pulse energy interval, the absorption plasmon band intensity increased more than 30% when the laser pulse energy is increased from 10 to 40 mJ. It can also be notified that CuO nanoparticles were produced at the pulse energy more than 20 mJ for its plasmon band intensity can be recognized clearly. This result may due to the fact that bulk Cu is the transition metal with less electrons (electron configuration of Au: [Xe] $4f^{14}$ $5d^{10}$ $6s^{1}$, Ag: [Kr] $4d^{10}$ $5s^{1}$ and Cu: [Ar] $3d^{10}$ $4s^{1}$ [11]) than the other two bulk metals of Au and Ag, and hence it needs more energy for the laser to ablate the Cu metal target compared to Au or Ag bulk metal. On the other hand, both center wavelength and the FWHM exhibit no significant dependence on the laser pulse energy, which suggest that both size and size distribution of nanoparticles were similar for each nanoparticle.

For the ablation time, the same features were observed. As shown in figure 4a, the plasmon band intensity increased as the ablation duration increased, except for Ag nanoparticles with its peak intensity tended to decrease for the ablation time more than 40 minutes. Time to time observation of UV-Vis absorption suggested that at longer ablation time, the absorption tail was observed at the higher absorption wavelength (e.g. at 60 minutes for the ablation of Ag metal target), indicating that the obtained nanoparticle started to aggregate due to the absorption of the laser energy by previously produced nanoparticles ejected to the surrounding liquid [8]. Using a dispersant dissolved in the surrounding liquid may prevent such aggregation. From figure 4b, it can be observed that there was no significant dependence of SPR peak on the ablation time, suggested that the nanoparticles were kept its size and size distribution.

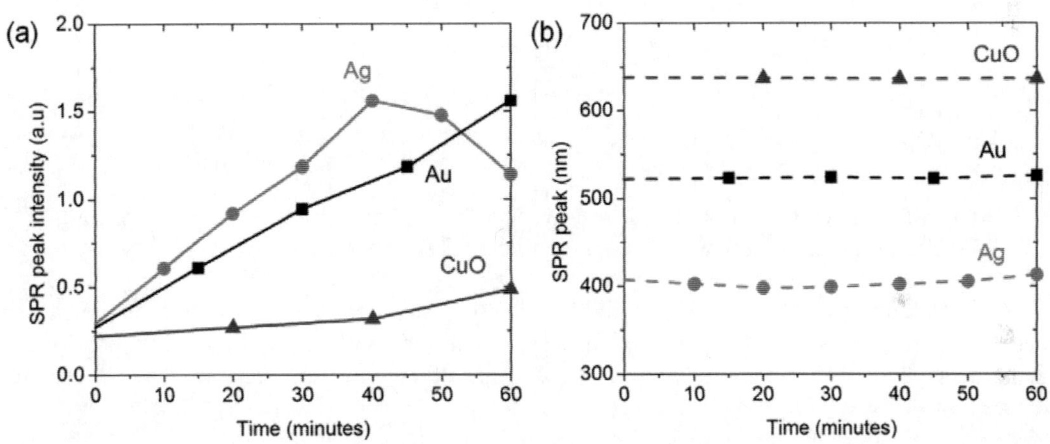

Figure 4. (a) SPR peak intensity and (b) SPR peak intensity dependence on the ablation time for all nanoparticles prepared by laser ablation in water.

We used CTAB to disperse the produced nanoparticle in water because it has a long chain of polymeric CTA+ ion when it was dissolved in water. Addition of CTAB in the fabrication of Au nanoparticle by laser ablation in liquid has been reported recently where CTAB was effectively inhibits the chain-like aggregation occurred in colloidal Au nanoparticles right after fabrication and promote a relatively good stabilization of the colloids for a few weeks [12]. Here, we reported the CTAB effect on the production of Ag and CuO nanoparticles using the same method. Figure 5 shows the evolution of the absorption spectra after the laser was ceased. For Ag nanoparticles (figure 5a and 5b), the nanoparticle stabilization can be preserved for a few weeks when CTAB was added into the solution (figure 5b). The Ag colloids without CTAB showed aggregation at the bottom of the storage vessel after 5 weeks (see figure 5a). While Au and Ag nanoparticles were still stable for weeks, the stability of CuO nanoparticles was only few days even for the ones with CTAB. As shown in figure 5c and 5d, the average stability of CuO nanoparticles was only 9 days before the aggregation was observed. It means that a more suitable dispersant needed for the stabilization of CuO nanoparticles.

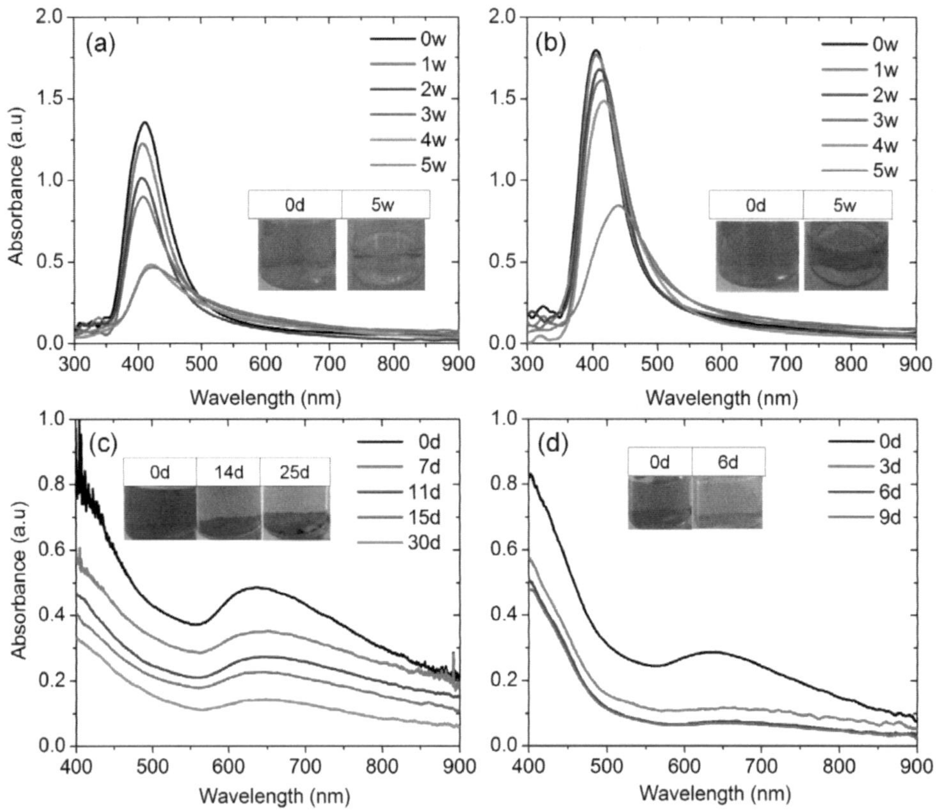

Figure 5. The evolution of the absorption spectra of Ag (a, b) and Cu (c, d) colloids after the laser was ceased, with (a,c) and without CTAB (b, d).

For the colloids with CTAB, we use TEM to analyze size and size distribution as well as their structure. Figure 6 shows the TEM images for all nanoparticles obtained by this technique. We found that the average size of nanoparticles was in the range of 20 – 40 nm and they have a crystalline structure.

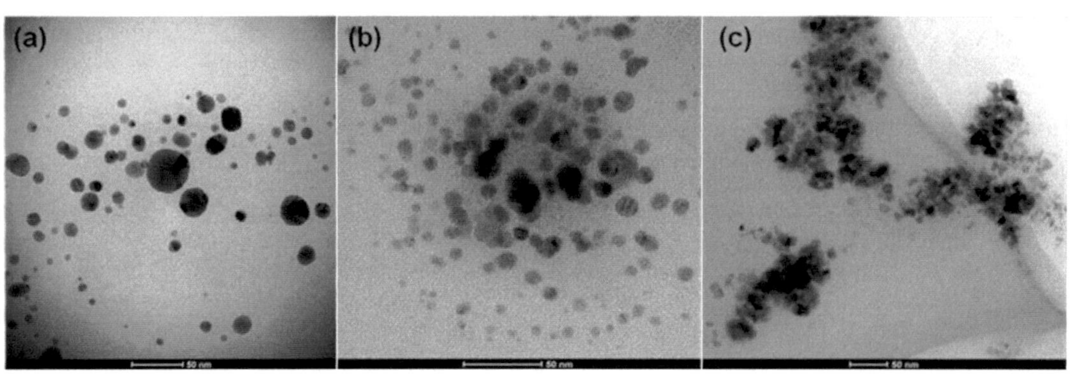

Figure 6. Typical TEM images for (a) Au, (b) Ag and (c) CuO nanoparticles prepared by laser ablation in water with CTAB.

4. Conclusion

In conclusion, we have demonstrated the synthesis of Au, Ag, and CuO nanoparticles using pulse nanosecond laser ablation in aqueous phase. For Au and Ag metal targets, the technique can produce Au and Ag nanoparticles identified by their characteristic surface plasmon at around 400 and 520 nm, while Cu targets produced a metallic oxide of CuO nanoparticles identified at 620 nm SPR band. The parameters such as laser wavelength, energy, and ablation time were also investigated, leading to a comprehensive understanding that every bulk metal target responds to a specific laser wavelength, energy, and ablation time differently. Dispersant used in this research has a positive effect to disperse Au and Ag nanoparticles and thus stabilized them at least for a month. However, it was not true for CuO nanoparticles due to a rapid aggregation occurred during the ablation. Nevertheless, TEM observation showed that particle size for all the fabricated nanoparticles were in the range of 20 – 40 nm with a crystalline structure. This work needs further exploration for both laser and bulk target properties so that a complete feature of the technique can be seen clearly and will provide more insight on how this technique can be improved as an alternative for the chemical synthesis.

Acknowledgments

The authors would like to deliver their gratitude to the Indonesian Institute of Science as well as the Ministry of Research Technology and Higher Education for providing financial support for this research.

References

[1] Zhang J, Malicka J, Gryczynski I and Lakowicz J R 2005 *J. Phys. Chem. B* **109** 7643
[2] Iravani S, Korbekandi H, Mirmohammadi S V and Zolfaghari B 2014 *Res. Pharm. Sci.* **9** 385
[3] Jung J H, Oh H C, Noh H S, Ji J H and Kim S S 2006 *J. Aerosol Sci.* **37** 1662
[4] Tien D-C, Tseng K-H, Liao C-Y, Huang J-C, Tsung T-T 2008 *J. Alloy Compd.* **463** 408
[5] Nakamura T, Mochidzuki Y and Sato S 2008 *J. Mater. Res.* **23** 968
[6] Nakamura T, Magara H, Herbani Y and Sato S 2011 *Appl. Phys. A* **104** 1021
[7] Nakamura T, Herbani Y and Sato S 2011 *Proc. Conf.on Lasers & Electro-Optics Europe & 12th European Quantum Electronic Conf.* (Munich) (New York: IEEE Xplore) 12237041
[8] Tomko J, Naddeo J J, Jimenez R, Tan Y, Steiner M, Fitz-Gerald J M, Bubb D M and O'Malley S M 2015 *Phys. Chem. Chem. Phys* **17** 16327
[9] Chen Y-H and Yeh C-S 2002 *Colloids Surf. A* **197** 133
[10] Kazakevich P V, Voronov V V, Simakin A V and Shafeev G A 2004 *Quantum Electronics* **34** 951
[11] Jacobs R L 1968 *J. Phys. C* **1** 1296
[12] Isnaeni, Irmaniar and Herbani Y 2017 *J. Phys.: Conf. Ser.* **817** (Bristol: IOP Publishing) 012039

IOP Conf. Series: Journal of Physics: Conf. Series **985** (2018) 012006 doi:10.1088/1742-6596/985/1/012006

Detection of colloidal silver chloride near solubility limit

K Y Putri[1] and R Adawiah[2]

[1]Research Center for Physics, Indonesian Institute of Sciences, Bd. 442 Kawasan Puspiptek Serpong, South Tangerang 15314, Banten, Indonesia
[2]Department of Physics, Faculty of Mathematics and Natural Sciences, Hasanuddin University, Jl. Perintis Kemerdekaan km. 10, Makassar 90245, South Sulawesi, Indonesia

E-mail: kirana.yuniati.putri@lipi.go.id

Abstract. Detection of nanoparticles in solution has been made possible by several means; one of them is laser-induced breakdown detection (LIBD). LIBD is able to distinguish colloids of various sizes and concentrations. This technique has been used in several solubility studies. In this study, the formation of colloids in a mixed system of silver nitrate and sodium chloride was observed by acoustic LIBD. Silver chloride has low solubility limit, therefore LIBD measurement is appropriate. Silver and chloride solutions with equal concentrations, set at below and above the solubility of silver chloride as the expected solid product, were mixed and the resulting colloids were observed. The result of LIBD measurement showed that larger particles were present as more silver and chloride introduced. However, once the concentrations exceeded the solubility limit of silver chloride, the detected particle size seemed to be decreasing, hence suggested the occurrence of coprecipitation process. This phenomenon indicated that the ability of LIBD to detect even small changes in colloid amounts might be a useful tool in study on formation and stability of colloids, i.e. to confirm whether nanoparticles synthesis has been successfully performed and whether the system is stable or not.

1. Introduction

Detection of nanoparticles presences in liquid medium can be achieved by applying laser-induced breakdown detection (LIBD). LIBD is a sensitive characterization technique developed especially for detecting colloids with small size and low concentration [1, 2]. The liquid samples subjected to LIBD measurement need no specific treatment for preparation, i.e. it can be measured as it is [3]. Previous study has shown that LIBD can detect nanoparticles as small as few nanometers and down to several part per trillion (ppt) [4], while also can be applied both in situ or online monitoring system [5 – 7].

Some studies related to solubility had applied LIBD to detect solid formation more precisely [8 – 17]. Most of the samples were oxides and optical LIBD was often employed [8 – 10, 12, 13]. Applications of acoustic LIBD in solubility study are somewhat less often. Meanwhile, solubility of metal salt is also interesting to be studied, since it can be an unwanted product in a chemical process. Some metal salts might be formed due to excess amount of precursors added during metal nanoparticle synthesis, e.g. formation of silver chloride (AgCl) precipitate when silver nitrate ($AgNO_3$) is introduced to chloroauric acid ($HAuCl_4$) at concentrations beyond the solubility limit of AgCl [18].

This research aimed to study the use of acoustic LIBD in solubility study, particularly in detecting the formation of colloidal AgCl as a reaction product of $AgNO_3$ and NaCl while observing its evolution near solubility limit. AgCl has low solubility, i.e. colloids are form at low concentration of Ag and Cl, and therefore it is appropriate for LIBD measurement.

2. Experimental method

Silver chloride (AgCl) samples were prepared by introducing sodium chloride (NaCl) solution into silver nitrate ($AgNO_3$) solution. The concentrations of Ag and Cl after mixing were set equally at 5×10^{-6} mol/L, 10^{-5} mol/L, 1.2×10^{-5} mol/L, 1.5×10^{-5} mol/L, and 2×10^{-5} mol/L. As stock solutions, 0.5 mol/L $AgNO_3$ and 0.5 mol/L NaCl were prepared by proper dissolution of $AgNO_3$ powder and NaCl powder in distilled water. Each sample was placed in a 10 ml glass container. Two other samples containing only $AgNO_3$ and only NaCl were also prepared to provide data comparisons.

The experimental setup for acoustic LIBD used in this study is shown in figure 1. Pulsed laser beam from nanosecond Nd:YAG laser (Q-smart 850, Quantel) at 532 nm wavelength and 10 Hz frequency was focused into the aqueous sample in glass container. Upon interaction between laser beam and colloid passing by the focused area, a breakdown was occurred following plasma formation. The acoustic signal was then detected by a microphone-recorder system and saved as an audio file. Each file, which contained data for one measurement, was processed and analyzed using a homemade processing program [19].

In this experiment, each sample was measured by LIBD technique at increasing laser energies, which were set from 0 to 5 mJ. The pulsed laser energy setting were carried out with considering the attenuation by a glass cover which was put on top of the glass container to minimize contamination. For statistical purpose, three measurements were done for each condition. Breakdown occurrences from 1000 laser shots were observed in every measurement.

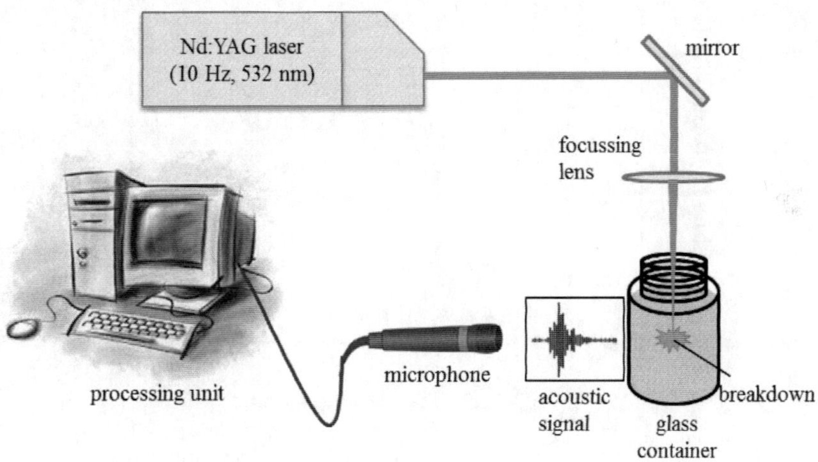

Figure 1. Experimental setup for acoustic LIBD.

3. Result and discussion

AgCl was expected to be formed from an exchange reaction between $AgNO_3$ and NaCl, as shown in equation 1 [20].

$$AgNO_3(aq) + NaCl(aq) \rightarrow AgCl(s) + NaNO_3(aq) \tag{1}$$

When $AgNO_3$ was dissolved in distilled water below its solubility, it dissociated into Ag^+ and NO_3^- ions. In the presence of Cl^- from NaCl, Ag^+ formed the insoluble AgCl. Such reaction is called coprecipitation, where an ion brings down another to precipitate. Na^+ and NO_3^- were present as spectator ions, because they do not participate in the reaction [20]. Therefore the mixed sample was simply mentioned as AgCl sample.

Figure 2 shows the LIBD measurement result of $AgNO_3$, NaCl, and AgCl samples. The homemade processing program processed the recorded signal by dividing it into blocks of signal with equal width [19]. The block width was set so that it represent the time gap between two consecutive laser shots. A breakdown was expected to generate within this gap, if any. To count the breakdown, the processing system detected any acoustic signal which exceeded the predetermined intensity threshold. The number of breakdowns generated for a certain number of laser shots was called breakdown probability. The breakdown probabilities increased with increasing pulsed laser energy, hence the so-called S-curve. The S-curve of AgCl was positioned more to the left compared to those of $AgNO_3$ and NaCl. This might indicate the formation of AgCl colloids as reaction product of $AgNO_3$ and NaCl. The dotted lines are precursor solutions before mixing and the solid line signifies the mixed solution. A closer look to the breakdown threshold energies suggested the similar colloidal size of $AgNO_3$ and AgCl, however the steeper slope of the mixed solution hinted that more colloids were present. The breakdown threshold energy was the pulsed laser energy needed to induce breakdown probability of 0.01. In LIBD analysis, breakdown threshold energy corresponds to colloidal size, while S-curve slope correlates to colloidal concentration. The lower the breakdown threshold energy, the larger the colloidal size; while the steeper the slope, the more concentrated the colloids [4].

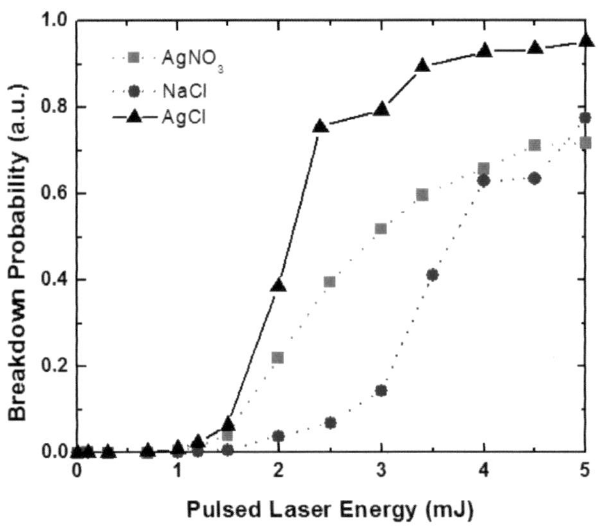

Figure 2. The shift in S-curves implied the formation of AgCl through coprecipation process.

Further experiment was carried out to see the evolution of colloidal AgCl near the solubility limit. The solubility product constant (Ksp) of AgCl is 1.66×10^{-10}, thus giving the solubility value of AgCl as low as 1.33×10^{-5} mol/L. Three of the AgCl samples prepared had Ag and Cl concentrations below AgCl solubility limit, whilst the others were above it. Figure 3a shows the LIBD results for AgCl with various concentrations of Ag and Cl. Curves with solid lines represent mixed solutions in which the Ag and Cl concentrations were below AgCl solubility limit, whereas the dotted lines represent mixed solution with Ag and Cl concentrations above AgCl solubility limit. As Ag and Cl concentrations increased, the S-curve shifted to the left; yet it was back to the right when the concentrations exceeded the solubility limit of AgCl.

A clearer detail on breakdown threshold energy is shown by figure 3b. For the three samples with Ag and Cl concentrations below AgCl solubility limit, the breakdown threshold energy values decreased as concentrations increased. It indicated that larger particles were formed as Ag and Cl concentrations increased. However, once Ag and Cl concentrations in the mixed sample exceeded AgCl solubility value, i.e. at 1.5×10^{-5} mol/L, the breakdown threshold energy rose again, which implied the presence of smaller particles in the sample. Further increase in Ag and Cl concentrations, i.e. at 2×10^{-5} mol/L, caused the breakdown threshold energy to decrease, suggesting that larger particles had been formed again. It was the same trend with those of below AgCl solubility. From this result, it implied that LIBD might be useful to detect when solubility limit has been exceeded.

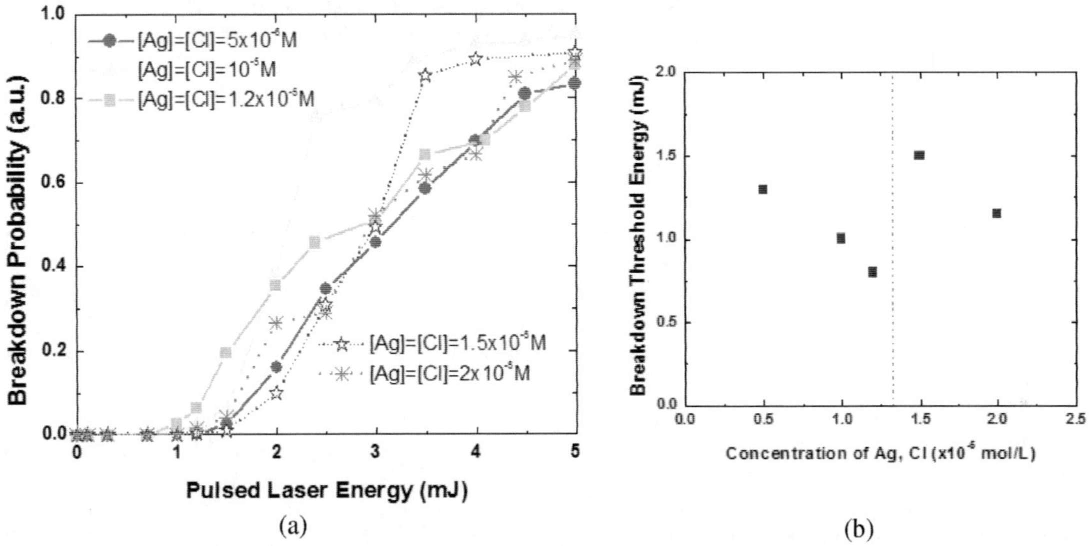

Figure 3. An increase in Ag and Cl concentration affected the (a) S-curves and (b) breakdown threshold energy. The dotted red line in the right figure represents the solubility value of AgCl.

4. Conclusion

Formation of colloidal AgCl and its evolution near solubility limit was observed by means of acoustic LIBD. Introduction of NaCl into Ag solution resulted in higher concentration of colloids observed due to the formation of AgCl colloid. For Ag and Cl concentrations below AgCl solubility limit, larger particles were detected when higher concentrations of Ag and Cl were given. Once the solubility limit was exceeded, the colloidal size decreased. It rose again when more Ag and Cl were added. This phenomenon indicated that the ability of acoustic LIBD to detect small changes in colloid amount might be useful in studying solubility of a substance, including metal salt. It can signify the formation of reaction product in a mixed sample and detect the solubility limit. It might also be useful for study on colloidal stability. However, to get more pronounced trend and clear solubility value, more detailed research with more variation of Ag and Cl concentration near solubility limit is needed.

Acknowledgments

This research was supported by Indonesian Institute of Sciences through the 2017 Excellent Research Program.

References

[1] Thang N M, Knopp R, Geckeis H, Kim J I and Beck P 2000 *Anal. Chem.* **72** 1
[2] Bundschuh T, Yun J-I and Knopp R 2001 *Fresenius J. Anal. Chem.* **371** 1063
[3] Walther C, Bitea C, Hauser W, Kim J I and Scherbaum F J 2002 *Nucl. Instr. and Meth. in Phys. Res. B* **195** 374

[4] Kim J I and Walther C 2006 Laser-induced breakdown detection *Environmental Colloids and Particles: Behaviour, Separation, and Characterisation* vol 10 eds K J Wilkinson and J R Lead (Chichester: John Wiley & Sons) chapter 12 pp 555-612

[5] Hauser W, Geckeis H, Kim J I and Fierz Th 2002 *Colloids Surf. A* **203** 37

[6] Zanker H 2010 In situ measurements on suspended nanoparticles with visible laser light, infrared light, and x-rays *Nanoparticles in the Water Cycle* eds F H Frimmel and R Nieβer chapter 8 pp 117-138

[7] Jung E C, Cho H R, Park M R and Baik M H 2011 *J. Kor. Rad. Waste Soc.* **9** 41

[8] Bundschuh T, Knopp R, Müller R, Kim J I, Neck V and Fanghänel Th 2000 *Radiochim. Acta* **88** 625

[9] Neck V, Müller R, Bouby M, Altmaier M, Denecke M A and Kim J I 2002 *Radiochim. Acta* **90** 485

[10] Bitea C, Walther C, Kim J I, Geckeis H, Rabung T, Scherbaum F J and Cacuci D G 2003 *Colloids Surf. A* **215** 55

[11] Bitea C, Müller R, Neck V, Walther C and Kim J I 2003 *Colloids Surf. A* **217** 63

[12] Walther C 2003 *Colloids Surf. A* **217** 81

[13] Bouby M, Geckeis H, Manh T N, Yun J-I, Dardenne K, Schäfer T, Walther C and Kim J-I 2004 *J. Chromatogr. A* **1040** 97

[14] Cho H R, Walther C, Rothe J, Neck V, Denecke M A, Dardenne K and Fanghänel Th 2005 *Anal. Bioanal. Chem.* **383** 28

[15] Cho H R, Park K K, Jung E C and Jee K Y 2007 *J. Kor. Rad. Waste Soc.* **5** 189

[16] Rothe J, Walther C, Brendebach B, Büchner S, Fuss M, Denecke M A and Geckeis H 2009 *J. Phys.: Conf. Ser.* **190** 012188

[17] Shin Y, Bae S and Lee W 2013 *Adv. Env. Res.* **2** 167

[18] Mallin M P and Murphy C J 2002 *Nano Lett.* **2** 1235

[19] Putri K Y, Zahar I and Isnaeni 2016 *Prosiding LINOF 2015* (Tangerang Selatan: Pusat Penelitian Fisika LIPI) 102

[20] Kotz J C, Treichel P M and Townsend J R 2009 *Chemistry & Chemical Reactivity*, 7th ed (Belmont: Thomson Brooks/Cole) pp 121-130

IOP Conf. Series: Journal of Physics: Conf. Series **985** (2018) 012007 doi:10.1088/1742-6596/985/1/012007

High resolution extensometer based on optical encoder for measurement of small landslide displacements

M I Afandi, H Adinanta, A Setiono, Qomaruddin and B Widiyatmoko

Research Center for Physics, Indonesian Institute of Sciences, Komplek PUSPIPTEK Building 442, Serpong, South Tangerang, Banten 15314 Indonesia

E-mail: afandiimam@gmail.com

Abstract. There are many ways to measure landslide displacement using sensors such as multi-turn potentiometer, fiber optic strain sensor, GPS, geodetic measurement, ground penetrating radar, etc. The proposed way is to use an optical encoder that produces pulse signal with high stability of measurement resolution despite voltage source instability. The landslide measurement using extensometer based on optical encoder has the ability of high resolution for wide range measurement and for a long period of time. The type of incremental optical encoder provides information about the pulse and direction of a rotating shaft by producing quadrature square wave cycle per increment of shaft movement. The result of measurement using 2,000 pulses per resolution of optical encoder has been obtained. Resolution of extensometer is 36 μm with speed limit of about 3.6 cm/s. System test in hazard landslide area has been carried out with good reliability for small landslide displacement monitoring.

1. Introduction

Landslides are among the most common and damaging natural hazards in terms of human lives and infrastructure damage. A landslide is a phenomenon of mass movement, which moves small displacement per day in wide area. The observation of landslide together with geophysical, hydrological, and environmental parameters helps to reveal the relationships between movements and different parameters, which is necessary to develop sensors used in landslide detection.

An extensometer is a sensor for landslide detection that measures extended or contracted distance of interest in a material when it is elongated and compressed such as in a tension test. A contact type of the extensometer measures the strain of material by directly attaching a sensor to the specimen of a material [1]. It is easy to measure the overall elongation or compression. However, when the material to be measured is small and not enough to fit to the clamp of a loading device, or the deformed rate is not uniform along the elongated (or compressed) direction, it is hard and even impossible to find out the strain using contact type extensometer [2].

The other monitoring techniques for landslide displacement detection, such as using multi turn potentiometer [1 – 3], space-borne or ground-based synthetic aperture radar (SAR) interferometry [4 – 6], fiber optic strain sensors [7, 8], and geodetic measurements (GPS, EDM, etc.) [9, 10] have been shown to be an effective complementary tool for landslide monitoring. However, highly sensitive extensometer is essential to measure small landslide displacement for a long period observation.

In this work, an optical encoder has been used to produces pulse signal with high stability of measurement resolution despite voltage source instability. It also has the ability of high resolution for wide range measurement. Developed extensometer using optical encoder has been tested and evaluated to meet the requirements for measurement of small landslide displacement. The extensometer also utilizes wireless communication, so that it can be used as a landslide early warning system [11].

2. Methodology

An optical rotary encoder uses optical sensing technology that has a rotation code and a pattern on it. The incremental optical encoder is the most widely used of all rotary encoders due to its low cost and ability to provide signals that can be easily interpreted to provide motion related information such as velocity or change in position. Figure 1a shows how incremental optical encoder provides information on instantaneous position of a rotating shaft by producing two output square wave cycles per increment of shaft movement. The two output wave forms are 90 degrees out of phase that usually called quadrature signals.

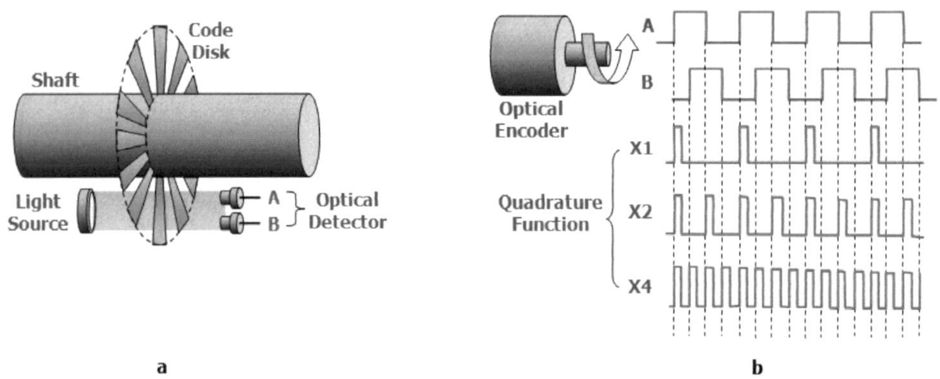

Figure 1. Working principle of incremental optical encoder.

In figure 1b, the quadrature signals are digital signals that produce two channels, channel A and channel B. When optical encoder rotates in a clockwise direction, channel A leads channel B. If optical encoder rotates in counterclockwise direction, channel B leads channel A. When more resolution is needed, it is possible for the counter to count the leading and trailing edges of the quadrature encoder's pulse train from one channel, which doubles (×2) the number of pulses per revolution. Counting both leading and trailing edges of both channels, a quadrature encoder will quadruple (×4) the number of pulses per revolution. For example, if we have 2,000 pulses per revolution (ppr) quadrature encoder, it can be increased to a maximum of 8,000 pulses per revolution (ppr) counter data.

3. Results and discussion

3.1. Design of extensometer system

In order to make optical encoder as extensometer, we used optical encoder with 2,000 pulses per revolution (ppr) and the microcontroller circuit that contains an 8-bit ATMega8A microcontroller to detect quadrature signal. It is then calculated as counter/position measurement. For monitoring purposes, wireless communication using WI-FI module was integrated to a 3G/GPRS gateway system and connected to a cloud system. We also used dual voltage regulators to supply the component system from variable range voltage of external power sources. The block diagram of the system is shown in figure 2a.

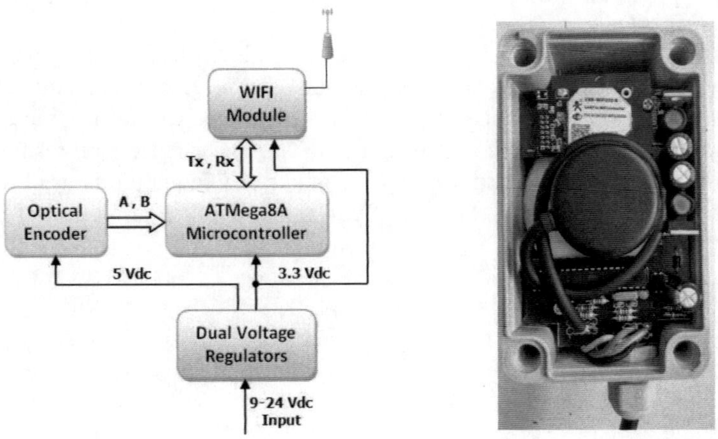

a b

Figure 2. (a) Block diagram and (b) hardware realization of extensometer system.

A system is placed inside a weatherproof box as shown in figure 2b. A spring disc merges to wire drum, and it is also attached to optical encoder. In case of displacement, a still wire will roll up the drum and rotate the spring disc. An optical encoder measures the rotation of spring disc (in degree), where it can be converted to the actual displacement (in cm). The steel wire from extensometer can be pulled up to reach a length of 1.5 meters. Figure 3 shows the works flow of developed extensometer system. The system starts by initializing an interrupt. Microcontroller reads every change from the optical encoder and calculates the displacement from the pulse signal. Microcontroller updates the system data, and sends the data to the gateway system. If there is no power loss, the system will repeat the reading for every minute.

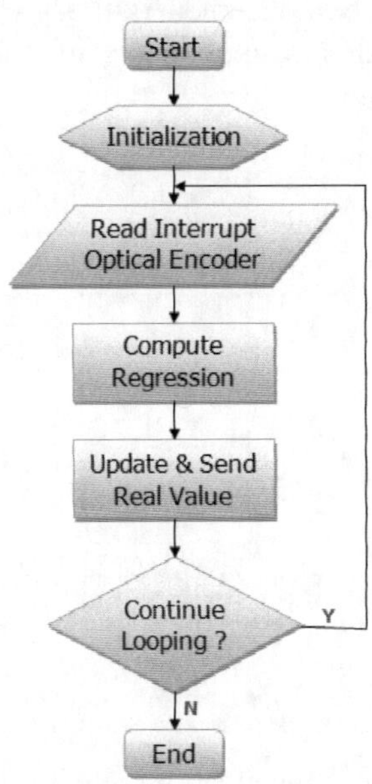

Figure 3. Flowchart of extensometer system.

IOP Conf. Series: Journal of Physics: Conf. Series **985** (2018) 012007 doi:10.1088/1742-6596/985/1/012007

3.2. System characterization

To obtain relationship between optical pulse counter signals with the actual position, we have to characterize the system by comparing the measurement distance to a standard length. By doing a comparison measurement every distance from 0.5 cm to 20 cm, we get the measurement result of optical encoder as linear equation as shown in figure 4. The linear regression results is $y = 0.0036x + 0.11354$, in which y is length in cm and x is pulse counter. The equation is used for microcontroller system to calculate the distance of each change of pulse counter. The resolution of equation is 0.0036 cm.

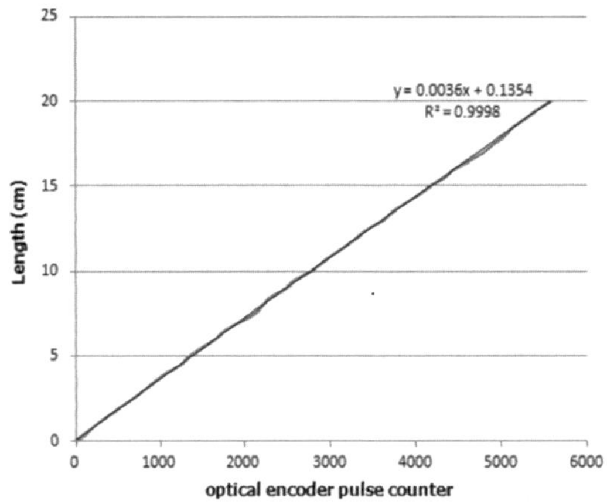

Figure 4. The system characterization of extensometer.

Figure 5 shows errors as the difference between the values obtained from the regression calculations and the actual measurements. Errors from the calculation tend to increase in proportion to its pulse counter value. In the measurement range of 20 cm, the largest absolute measurement error is 0.36 cm. Another factor that should be concerned is the speed of pulse counter from the optical encoder. If the speed counter of 1 kHz is used in the microcontroller system, the speed limit of displacement should not exceed 3.6 cm/s.

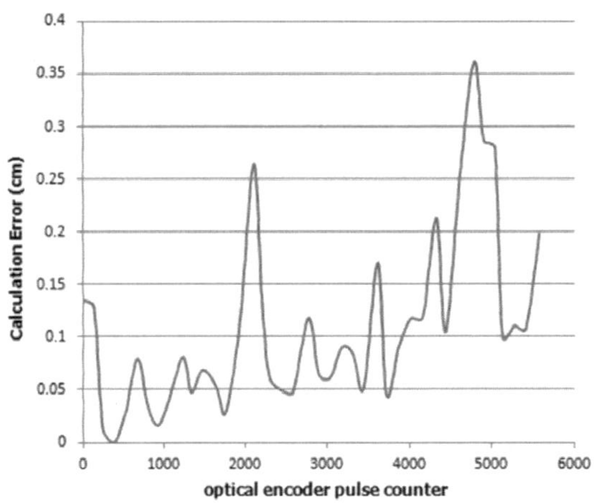

Figure 5. Absolute error of regression calculation.

3.3. Testing system in real condition

The extensometer system based on optical encoder was tested and evaluated in hazard landslide area starting from November 10, 2016 to January 25, 2017. Location with high risk of landslide occurrence, which is km. 107 of railway track in Purwakarta, West Java, was selected as the evaluation site. Figure 6 shows the evaluation site with extensometer is placed on high stand rather than on the ground.

Figure 6. Testing site for extensometer in a hazard landslide condition.

The extensometer system measured and recorded activities not only from landslide movement but also surrounding conditions. For examples, wind and vegetation around the extensometer system will cause changes in to the system, e.g. steel wire being pulled from the extensometer due to vegetation growth. These changes may cause incorrect recorded data in the system. By providing a sheath in the surrounding system will reduce some disturbances to the system and correct the recording data (figure 7).

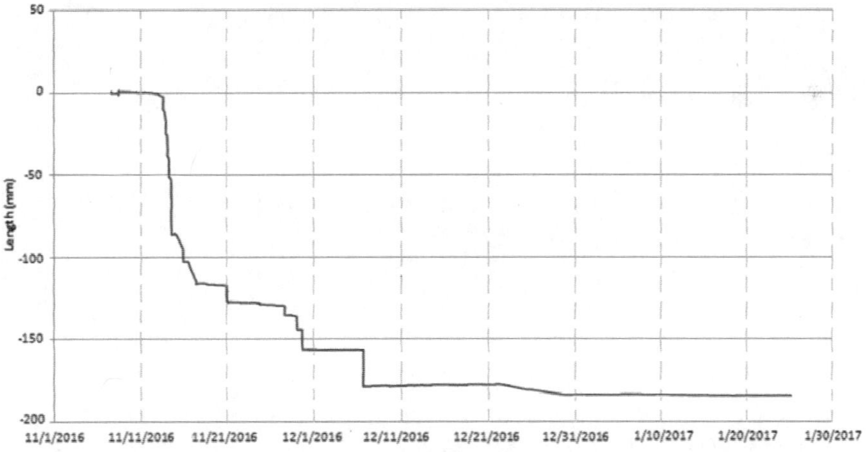

Figure 7. Monitoring landslide displacement using extensometer.

4. Conclusion

The extensometer system has been designed and tested using 2,000 pulses per resolution of optical encoder. The system has measurement resolution until 0.0036 cm with speed limit of displacement not exceeding 3.6 cm/s. In the measurement range of 20 cm, the largest absolute measurement error is 0.36 cm. The testing system in hazard landslide area has been reported in good reliability for monitoring small landslide displacement.

Acknowledgments
Authors wishing to acknowledge assistance or encouragement from colleagues, special work by technical staff or financial support from organizations should do so in an unnumbered Acknowledgments section immediately following the last numbered section of the paper.

References

[1] Corominas J, Moya J, Lloreta A, Gili J A, Angeli M G, Pasuto A and Silvano S 2000 *Eng. Geol.* **55** 149
[2] Mentes G 2015 *Procedia Earth and Planetary Sci.* **15** 421
[3] Intrieri E, Gigli G, Mugnai F, Fanti R and Casagli N 2012 *Eng. Geol.* **147-148** 124
[4] Casagli N, Farina P, Leva D, Nico G and Tarchi D 2003 *Proc. SPIE* **4886** 322
[5] Martire D D, Tessitore S, Brancato D, Ciminelli M G, Costabile S, Costantini M, Graziano G V, Minati F, Ramondini M and Calcaterra D 2015 CATENA **137** 406
[6] Uhlemann S, Smith A, Chambers J, N Dixon, Dijkstra T, Haslam E, Meldrum P, Merritt A, Gunn D and Mackay J 2016 *Geomorphology* **253** 438
[7] Dai Z-Y, Liu Y, Zhang L-X, Ou Z-H, Zhou C and and Liu Y-Z 2008 *Journal of Electronic Science and Technology of China* **6** 416
[8] Bi W, Wu G and L Feng 2008 *Proc. SPIE* **7278** 727808
[9] Gili J A, Corominas J and Rius J 2000 *Eng. Geol.* **55** 167
[10] Bányai L, Mentes Gy, Újvári G, Kovács M, Czap Z, Gribovszki K and Papp G 2014 *Geomorphology* **210** 1
[11] Suryadi, Puranto P, Adinanta H, Tohari A and Priambodo P S 2017 *J. Phys.: Conf. Ser.* **853** 012010

Effect of ablation time on femtosecond laser synthesis of Au-Ag colloidal nanoalloys

A N Hidayah[1,2], D Triyono[1], Y Herbani[2], Isnaeni[2] and M M Suliyanti[2]

[1]Department of Physics, Universitas Indonesia, Kampus UI Depok, Indonesia
[2]Research Center for Physics, Indonesian Institute of Sciences, Kawasan PUSPIPTEK Building 442, Serpong, Tangerang Selatan, Banten, Indonesia

E-mail: affi001@lipi.go.id

Abstract. Au-Ag nanoalloys have been synthesized by laser irradiation technique. First, Au and Ag nanoparticles were prepared from Au and Ag pure metal (99.9%) ablated using an 800 nm femtosecond laser in distilled water. Using the same laser, Au and Ag nanoparticle with 1: 1 ratio were subsequently mixed and irradiated with various irradiation time, i.e. 0, 5, 20, and 35 minutes. We varied the ablation time for each metal nanoparticles, i.e. 25 minutes and 1 hour to see its effect on the production of nanoalloys in the subsequent irradiation. Au-Ag nanoalloys were characterized and analyzed using transmission electron microscope and UV-Vis spectrophotometry. The result shows that Au-Ag nanoalloys were already formed in 20 minutes irradiation, either for the sample ablated for 25 minutes or 1 hour. The result of TEM shows that the size of Au-Ag nanoalloys prepared from 1 hour ablation was around 15.03 nm.

1. Introduction

Metallic nanoparticles (NPs) have been studied for more than a century and in the last ten years because of their special optical, electronic and catalytic properties, which may be very useful in chemistry, engineering, or biomedical application and those properties depend on plasmonic oscillation [1 – 3]. The combination of desirable features of mono-metallic and bimetallic NPs, either as alloy or as core-shell structures, is an attractive material. They have a diverse range of application related to photonics, catalysis, information storage, chemical/biological sensing, and surface-enhanced Raman scattering (SERS) [4 – 7].

Amongst the metallic alloy nanoparticles, Au-Ag bimetallic systems have unique surface plasmon resonance (SPR) band, which can absorb and scatter in the visible spectral region [8 – 9]. The Au-Ag system is interesting because the SPR band is tunable between ~520 nm for Au [10] and ~410 nm for Ag [11]. One important observation of Au-Ag bimetallic systems is that alloy and core-shell nanoparticles often show different optical properties, although the composition of Au and Ag within the nanostructures is same [12]. Core/shell and alloy bimetallic NPs are especially interesting because they provide opportunities to tune the NPs' optical and catalytic properties [13 – 14].

Pulse laser ablation of solid target in liquids (PLAL) has been known for a method to obtain monometallic Au and Ag nanoparticles [15]. Theory and experiment demonstrate that if NPs posses a strong absorption band whose energy coincides with the photon energy of a laser, the NPs can selectively be heated above their melting point into related liquid [16]. Therefore, the subsequent laser irradiation of the mixture of Au and Ag nanoparticles synthesized by a laser ablation can be expected to form their respective AuAg alloy nanoparticles in the medium. In this work, we will investigate effect of irradiation time on Au-Ag nanoalloys synthesis.

2. Experiment

Au-Ag nanoalloys were synthesized from the mixture of the colloidal Au and Ag NPs irradiated by a femtosecond laser. The colloidal Au and Ag NPs were previously prepared separately by the same laser using PLAL method. We use a Ti/Sapphire laser (Mai Tai, Spectra Physics), with 100 fs full-width-half maximum (FWHM) pulses at wavelength of 800 nm for ablation and irradiation source. Experimental setup of laser ablation for monometallic nanoparticle synthesis is shown in figure 1.

Au and Ag NPs were prepared from pure gold and silver targets with 99.9% purity. The target was placed in the glass vessel with 5 ml distilled water ablated using a femtosecond laser with a power of 2.1 Watt, which was focused with a focusing lens (focal length of 10.0 cm) on the target surface. All of Au and Ag NPs were prepared for both 25 minutes and 1 hour ablation time. Then, they were mixed with 1:1 ratio (2.5 ml Au: 2.5 ml Ag), placed in the glass vessel again and irradiated with the same laser with irradiation time of 0, 5, 20, and 35 minutes. Absorption spectra of the mixture were measured before and after irradiation.

The absorption spectra of Au NPs, Ag NPs, and Au-Ag nanoalloys were measured using UV-Vis spectrophotometry. Size and morphology of Au-Ag nanoalloys in 35 minutes irradiation time were investigated using transmission electron microscope (TEM). The size distribution of Au-Ag NPs was measured with software ImageJ and plotted with Origin. From the characterization, we can study optical properties and morphology of Au-Ag nanoalloys.

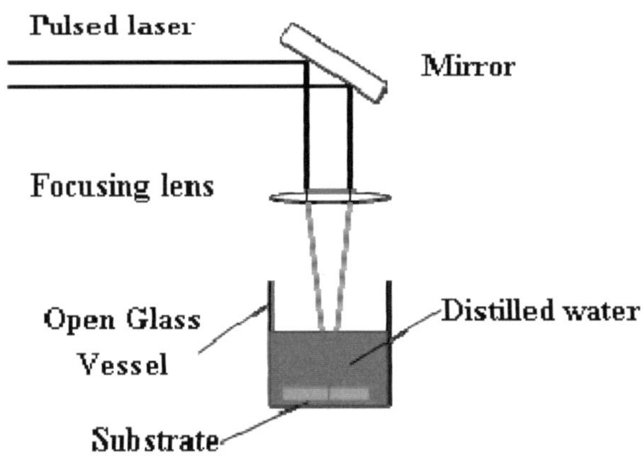

Figure 1. Schematic diagram of the experimental setup.

3. Results and discussion

Our research purpose is to analyze optical properties, morphology and the influence of irradiation time of femtosecond laser to the surface plasmon resonance, formation, and size distribution of Au-Ag nanoalloys.

3.1. Preparation Au NPs and Ag NPs

Au and Ag NPs have been prepared by femtosecond laser ablation with different ablation time of 25 minutes and 1 hour. The absorption spectra of colloidal Au and Ag NPs were shown in figure 2. UV-Vis absorption spectra relate to surface plasmon resonance (SPR) of Au and Ag NPs. The absorption spectra of Au NPs show a characteristic SPR at around 500 – 530 nm, while SPR of Ag NPs is at around 318 – 405 nm. We can see from figure 2a that the different ablation time of Au target resulted in a similar characteristic SPR for both 25 minutes and 1 hour, i.e. 517 nm. However, the absorption spectra of Ag NPs show different results or characteristics, where peaks of 25 minutes and 1 hour ablation times are at 403 nm and 408 nm, respectively. We can also observe form figure 2 that longer ablation time of pulsed femtosecond laser results in higher absorbance because ablation time is about how long the laser is interacting with matter. When laser interacts with matter (Au or Ag plate) in distilled water, there are phenomena occurred such as heating, melting, and evaporation of the matter. Longer ablation time means a longer interaction time between the laser and the matter, resulted in more material ablated from the bulk matter (Au and Ag) forming nanoparticles in distilled water.

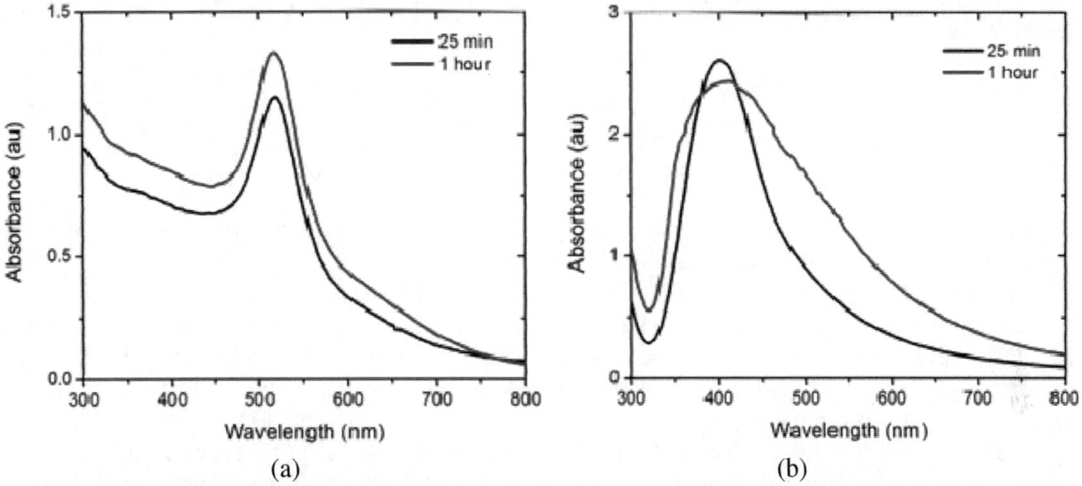

Figure 2. UV-Vis absorption spectra of colloidal (a) Au NPs and (b) Ag NPs.

3.2. Au-Ag nanoalloys: optical properties and morphology

After synthesizing Au and Ag NPs, the next step is to mix Au NPs and Ag NPs with 1:1 ratio (2.5 ml: 2.5 ml). Mixture of Au and Ag NPs was then irradiated again with different irradiation time of 0 minutes, 5 minutes, 20 minutes, and 35 minutes. Results from different irradiation time are shown in figure 3. Figure 3a shows the UV-Vis absorption spectra of Au-Ag NPs originated from the mixture of the respective monomettalic nanoparticles which were prepared with an ablation time of 25 minutes. UV-Vis absorption relates to surface plasmon resonance (SPR) of Au-Ag NPs. Before irradiation time (0 minutes), two distinct SPR were observed at 397 nm and 518 nm, indicating isolated Ag and Au NPs in the mixture. At 5 minutes irradiation time, the SPR absorption peak of Ag NPs shifts to longer wavelength at 409 nm, whereas the SPR absorption of Au NPs shifts to shorter wavelength at 515 nm. After 20 minutes, the SPR shows only one absorption peak at 445 nm. At the last 35 minutes, the SPR is unchanged and still at 445 nm, but it shows decreasing absorbance intensity.

The same behavior is also observed for the mixture nanoparticles which are prepared for 1 hour ablation time as shown in figure 3b. The figure shows the SPR of Au-Ag NPs in varying different irradiation time in 1 hour ablation time. When Ag and Au NPs were only mixed and without irradiation (0 minutes), there are two SPR absorption peaks with maximum at 398 nm and 515 nm, corresponding to SPR absorption peak of Ag and Au NPs, respectively. After Au and Ag NPs were mixed and irradiated, the SPR changed. At 5 minutes irradiation time, two SPR peaks still remain but Ag NPs peak shifts to longer wavelength at 417 nm and Au NPs peak shifts to shorter wavelength at 498 nm. With 20 minutes irradiation time, only one peak of SPR absorption appears at 444 nm. However, at 35 minutes irradiation time, the SPR absorption shifts to longer wavelength at 455 nm. This condition differs with the result in figure 3a, where SPR at 35 minutes irradiation time has same characteristics with SPR of 20 minutes irradiation time.

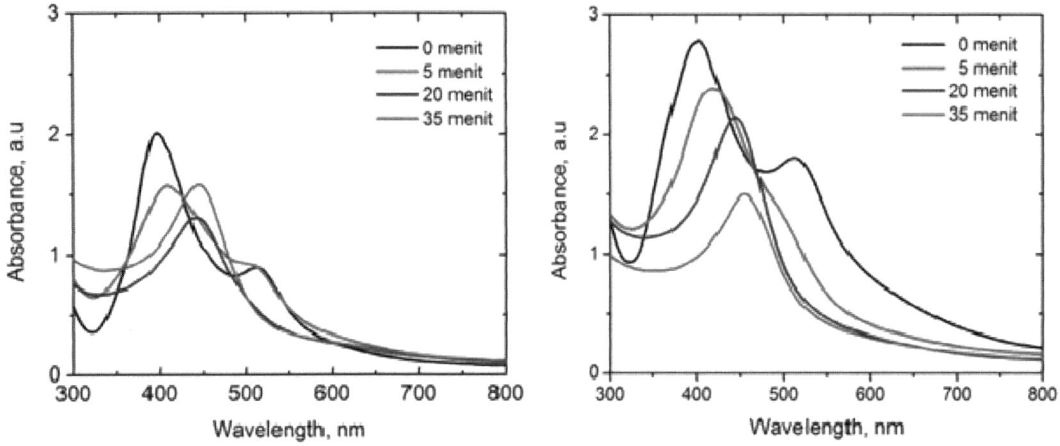

Figure 3. Optical absorption spectra of a mixture Au and Ag NPs in distilled water with different irradiation time in (a) 25 minutes and (b) 1 hour ablation time.

Briefly, for both prepared samples using PLAL (25 minutes and 1 hour ablation time), Au-Ag nanoalloys have been formed in 20 minutes irradiation time (figure 3) that only one surface plasmon resonance was observed. Alloying process of Au and Ag nanoparticle is easy since it is a thermodynamically favorable process as indicated by their phase diagram [17]. Moreover, Au and Ag have almost identical lattice constants, so that alloying process may take place easily to form homogeneous Au-Ag alloy particles. A red shift of SPR peak to a longer wavelength for the mixture irradiated for a longer time (35 minutes) could result in the formation of coalescence structure, as in agreement with the result by Haiss et al. for Au nanoparticles [18]. We predicted that this might be due to a longer fragmentation time by the laser (1 hour ablation time). This result should be confirmed by transmission electron microscopy.

TEM image of Au-Ag nanoalloys after irradiation time of 35 minutes for both monometallic nanoparticles prepared with 25 minutes and 1 hour ablation time are shown in figure 4. Figure 4 indicates the morphology image and the histogram of alloy Au-Ag NPs distribution size. Distribution of alloy Au-Ag NPs prepared by PLAL for 1 hour was more dispersed and bigger than those for 25 minutes. Because the morphology of Au-Ag NPs is homogeneously spherical without any visible phase separation, we could say that the colloidal Au-Ag NPs prepared in this work contained alloy rather than core shell. Despite their chained structure, distribution size of Au-Ag NPs prepared from a 25 minutes ablation time shows the highest number of particles in 8 – 12 nm range, whereas those prepared from 1 hour ablation time shows the highest number of particles in 5 – 10 nm range. This result shows that particle size of sample prepared for 1 hour ablation time is bigger than those for 25 minutes, but the structure were more dispersed or unchained.

In short, we have successfully fabricated bimetallic alloy Au-Ag nanoparticles rather than core-shell nanoparticles as suggested by a single SPR peaks in UV-Vis absorption spectra and supported by TEM image where no phase observation was observed. We suggest similar study can be carried out with different mixing ratio of Au/Ag NPs to have a greater possibility to tune the composition of alloy nanoparticle.

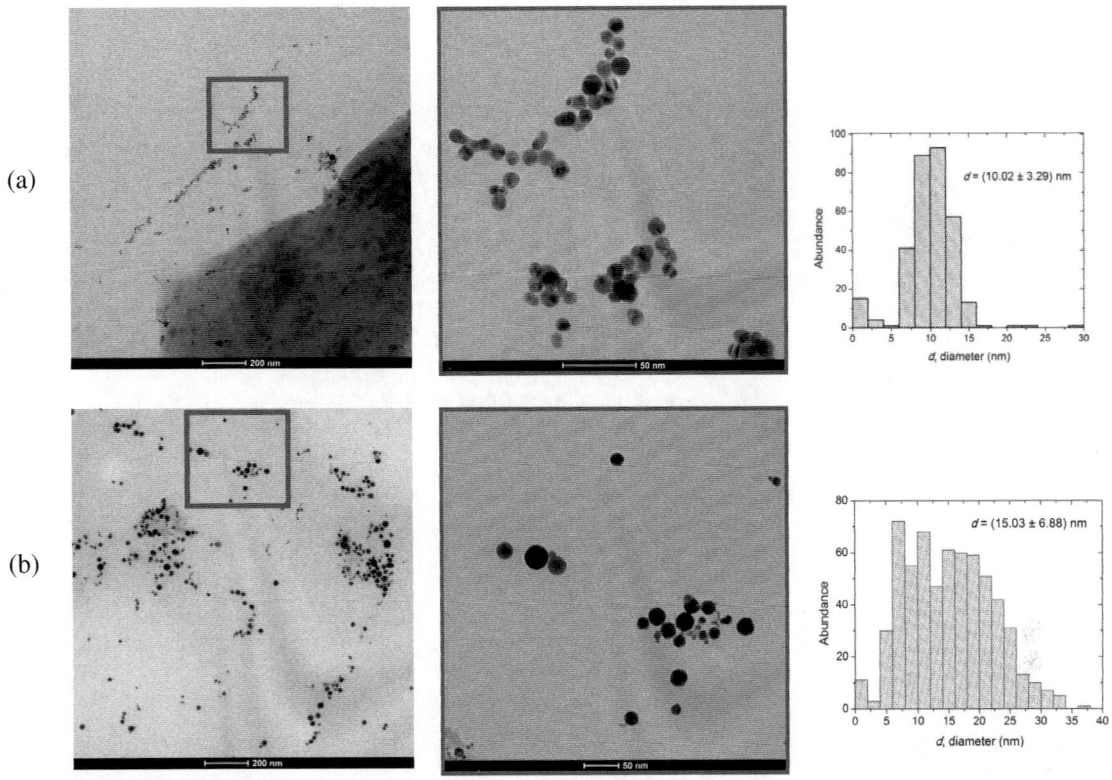

Figure 4. TEM image of Au-Ag Nanoalloys in distilled water in different ablation time (a) 25 minutes and (b) 1 hour.

4. Conclusions

Au-Ag nanoalloys have been synthesized from mixture of Au and Ag NPs prepared with different ablation time of 25 minutes and 1 hour, respectively. Each mixing Au and Ag NPs in 25 minutes and 1 hour ablation time was then irradiated with different irradiation time of 0, 5, 20, and 35 minutes. After 20 minutes irradiation, absorption spectra of mixing nanoparticles shows different surface plasmon resonance (SPR) from their monometallic counterpart suggesting Au-Ag nanoalloys have been formed. TEM image shows the size of Au-Ag nanoalloys of around 15.03 nm for a longer ablation time. This result indicates that different ablation time gives different optical and structural properties. Further parameter investigation is needed to synthesize Au-Ag nanoalloys such as wavelength of laser, mixing ratio of Au and Ag NP, and power laser. Therefore, we can learn more about optical properties of Au-Ag nanoalloys.

Acknowledgments

This work was supported by National Innovation System Research Incentive Program (Insinas 2017) from Ministry of Research, Technology and Higher Education of the Republic of Indonesia.

References

[1] Kreibig U and Vollmer M 1995 *Opical Properties of Metal Clusters* (Berlin: Springer-Verlag)
[2] Zang J, Worley J, Dénommée S, Kingston C, Jakubek Z J, Deslandes Y, Post M and Simard B 2003 *J. Phys. Chem. B* **107** 6920
[3] Murphy C J, Sau T K, Gole A M, Orendorff C J, Gao J, Gou L, Hunyadi S E and Li T 2005 *J. Phys. Chem. B* **109** 13857

[4] Jha R and Sharma A K 2009 *J. Opt. A* **11** 045502
[5] Sharma A K and Gupta B D 2006 *Nanotechnology* **17** 124
[6] Alvarez-Puebla R A, Bravo-Vasquez J P, Cheben P, Xu D X, Waldron P and Fenniri H 2009 *J. Colloid Interface Sci.* **333** 237
[7] Hao E, Li S, Bailey R C, Zou S, Schatz G C and Hupp J T 2004 *J. Phys. Chem. B* **108** 1224
[8] Feldheim D L, Foss Jr. C A 2002 *Metal Nanoparticles: Synthesis, Characterization, and Applications* (New York: Marcel Dekker)
[9] Jain P K, Lee K S, El-Sayed I H and El-Sayed M A 2006 *J. Phys. Chem. B* **110** 7238
[10] Daniel M C and Astruc D 2004 *Chem. Rev.* **104** 293
[11] Bright R M, Musick M D and Natan M J 1998 *Langmuir* **14** 5695
[12] Hodak J H, Henglein A, Giersig M and Hartland G V 2000 *J. Phys. Chem. B* **104** 11708
[13] Wang A Q, Chang C M and Mou C Y 2005 *J. Phys. Chem. B* **109** 18860
[14] Liu X, Wang A, Wang X, Mou C Y and Zhang T 2008 *Chem. Commun. (Camb.)* 3187
[15] Han H, Fang Y, Li Z and Xu H 2008 *Appl. Phys. Lett.* **92** 023116
[16] Kamat P V 2002 *J. Phys. Chem. B* **106** 7729
[17] Park J, Lee J 2008 *Calphad* **32** 135
[18] Haiss W, Thanh N T, Aveyard J and Fernig DG 2007 *Anal. Chem.* **79** 4215

Analyzing silver concentration in soil using laser-induced breakdown spectroscopy

S Prasetyo[1,2], Isnaeni[2], Zaitun[2], K Mitchell[2,3], M M Suliyanti[2] and Y Herbani[2]

[1]Department of Physics, Diponegoro University, Prof.Soedarto SH Street, Tembalang Semarang, Central Java, 502275, Indonesia
[2]Research Centre for Physics, Indonesian Institute of Sciences, Riset Rd. Building 442, Puspiptek, South Tangerang, Banten, 15314, Indonesia
[3]Department of Physics, The University of Melbourne, Grattan Street, Parkville, Victoria, 3010, Australia

E-mail: siswoyoprasetyo@st.fisika.undip.ac.id

Abstract. Determination of concentration of heavy metal ions in soil, such as silver, is very important to study soil pollution levels. Several techniques have been developed to determine silver ion concentration in soil. In this paper, we utilized laser-induced breakdown spectroscopy (LIBS) to study silver concentration in soil. We used four different data analysis methods to calculate silver concentration. In this case, we prepared soil samples with different silver ion concentrations from 400 ppm to 1000 ppm. Our analysis was focused on the 843.15 nm silver atomic absorption line. We found that plasma intensity increased as silver concentration increased. Our findings were based on our analysis using four different analysis methods. We believe that these analysis methods are able to calculate silver concentration in soil using LIBS.

1. Introduction

Laser-induced breakdown spectroscopy (LIBS) is a fast technique for studying elements in materials. The plasma spectrum, which is generated from laser-matter or photon-atomic interaction, provides essential information such as excitation and de-excitation processes of electrons in materials and elemental information. In principle, materials produce photons when electrons undergo de-excitation. Photons produced from de-excitation process are unique; therefore the energy of the produced photon provides information regarding specific elements in the material [1].

Rapid development of LIBS was significantly achieved due to its great potential applications. In comparison to other spectroscopy technique such as X-ray fluorescence spectroscopy (XRF), LIBS can detect elements with low atomic numbers. The plasma spectrum generated from photon-atomic interaction is mostly in the ultraviolet (UV), visible, and near infrared (NIR) regions, which aids detection. In addition, LIBS can be applied to various types of materials, including solids, liquids, and gases. Generally, LIBS measurements do not require extensive sample preparation. Most samples only require one preparation step before LIBS measurement. Low preparation time makes LIBS a powerful and fast spectroscopy technique.

LIBS can be utilized to detect almost all elements in nature, e.g., heavy metal. Detection of heavy metal elements in nature, such as soil, is crucial for determining contamination levels. Several techniques have been developed to detect heavy metals content in soil. Chemical analysis requires complicated and time consuming procedures. XRF and X-ray diffraction (XRD) techniques require an X-ray source and detector, which are expensive. Detection of elements in UV-visible-NIR range is one of the easiest detection techniques. LIBS offers fast detection of heavy metals contained in soil since detection is done in UV-visible-NIR range.

However, analysis of low heavy metal concentration in soil using LIBS is not an easy task. Therefore, there are not many reports related to the determination of heavy metal concentrations in soil. A calibration curve constructed from known concentrations is required to quantitatively analyse LIBS data. A problem that often occurs in LIBS measurement is that plasma formed from natural materials is unstable due to the heterogeneity of the sample. One way to analyse plasma spectra is by using a normalization technique [2]. Gornushkin et al. reported that normalization of surface density by considering mechanical effects of ablated mass provides a simple solution for analyzing plasma spectra of low concentration elements in a matrix [3].

For a specific case, silver element causes pollution in the environment and severe negative effects. Silver pollutants are known to accumulate in plants and animals and can cause serious illness in humans at the top in the food chain [4 – 6]. In this work, we will focus on analyzing plasma spectrum of low concentration of silver (Ag) ion in soil matrix using several data analysis techniques. The purpose of this work is to find the optimum analysis method of plasma spectra, which was obtained using LIBS measurement, for determining Ag ion concentration in soil.

2. Sample preparation and measurements

We utilized ordinary soil from local tropical farm fields as the sample matrix. The soil was ground and filtered to obtain soil particles of uniform size at 1 mm. 15 g of prepared soil was then mixed with 10 ml of various $AgNO_3$ concentrations, enabling calculation of Ag ion concentration in soil in unit of ppm. The mixture was then blended to make homogeneous Ag-embedded soil samples. The sample was then heated in oven at 100 °C for 1 hour to reduce water content. In order to ease plasma generation and analysis, the Ag-embedded soil was transformed into a pellet using pellet press with a pressure of 25 bar. The surface of pellet was flat, smooth, and ready for LIBS measurements. Low water content and flat pellet surface improved the quality of LIBS measurements [7 – 8].

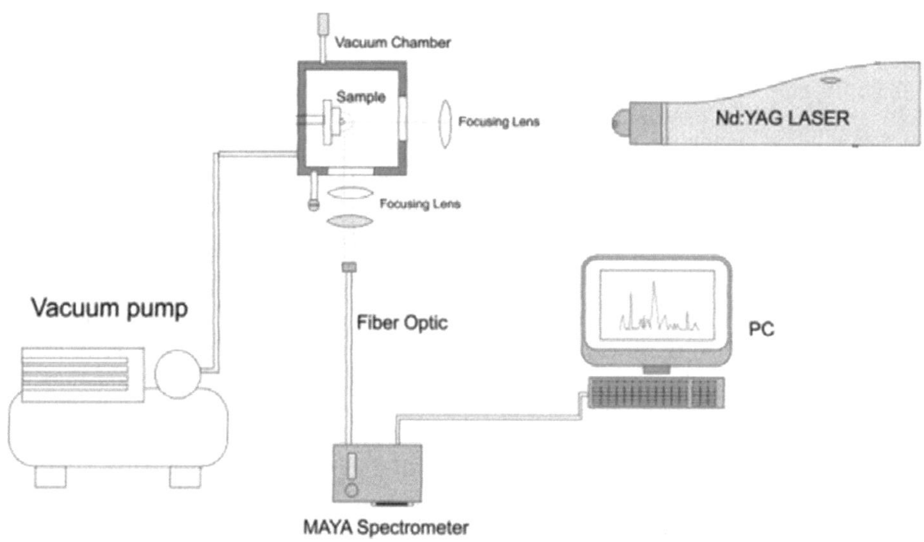

Figure 1. LIBS experimental setup.

IOP Conf. Series: Journal of Physics: Conf. Series **985** (2018) 012009 doi:10.1088/1742-6596/985/1/012009

LIBS measurement was done using setup shown in figure 1. A Quantel Nd:YAG laser operating with a 10 Hz repetition rate, a 532 nm wavelength, and a 24 mJ energy, was focused onto sample pellets using a 100 mm focusing lens. The sample was placed in a vacuum chamber and rotated during LIBS measurement to ensure that laser is incident on a new surface of the sample for each laser pulse. A spectrophotometer with integration time of 100 ms was used to detect and record emission spectra of generated plasma. It was assumed that for every 100 ms there was only one laser pulse to generate plasma. For analysis purposes, 250 plasma spectra were recorded for one sample. We also conducted LIBS measurements of a 99.99% pure silver plate to obtain silver lines in plasma spectrum.

3. Results and discussion

First, we analyzed the plasma spectrum of the Ag plate, shown in figure 2. We obtained seven detectable Ag atomic lines from plasma spectrum, i.e., Ag (I) 405.54 nm, Ag (II) 418.54 nm, Ag (II) 519.81 nm, Ag (II) 544.09 nm, Ag (I) 768.77 nm, Ag (II) 825.47 nm, and Ag (I) 843.15 nm. We also noticed that these Ag ion lines were detectable in low laser fluence.

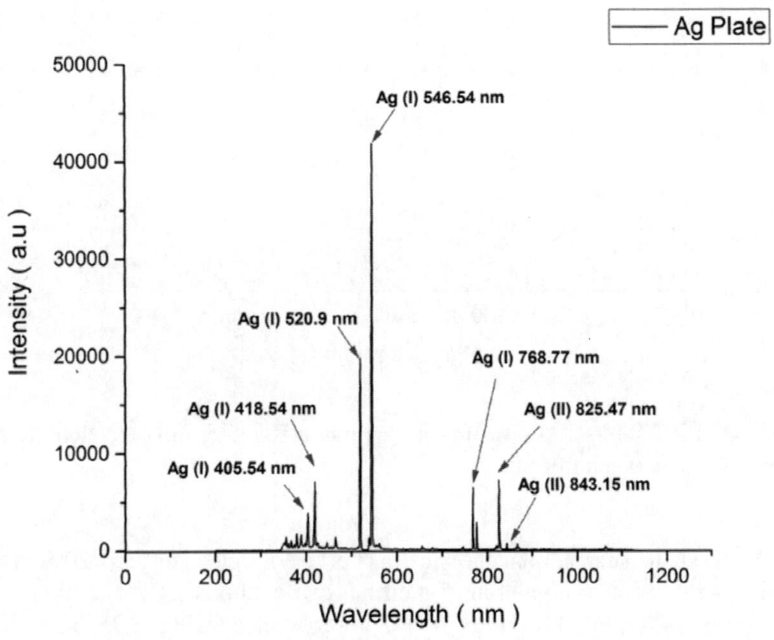

Figure 2. LIBS emission spectra from Ag plate.

Every recorded plasma spectrum was then analyzed to study Ag ion concentration in soil. Plasma peaks were then confirmed and corrected using the National Institute of Standards and Technology (NIST) atomic line database. We also confirmed the recorded plasma peaks using the silver plate spectrum. This confirmation was required since there was random shot-to-shot variation of plasma peak intensity due to the heterogeneity of the sample. Therefore, we calculated and normalized 250 spectra to get a single spectrum for every soil sample.

Plasma spectra obtained from soil samples were quite different compared to those from the Ag plate. We assumed that this was because matrix effects influenced generated plasma. Soil and other elements in the matrix could form new and different plasma lines [9]. A lot of spectral lines that are not linearly dependent on Ag concentration could occur in a complex sample such as soil. This complicated spectrum is difficult to understand and analyze. The set-up and spectrophotometer position will also affect recorded spectrum. This is the reason that 250 spectra were collected for each sample. By confirming soil sample spectra with the silver plate spectrum and NIST atomic database, we selected the Ag plasma line at 843.15 nm wavelength for further analysis. This line was chosen because intensity increased with increasing concentration as can be seen in figure 3. This means that it was an appropriate peak for analyzing silver ion concentration. We selected this peak wavelength since this peak was always detectable in all measurements and there was no overlapping with peaks from other elements. Therefore, it made analysis easier.

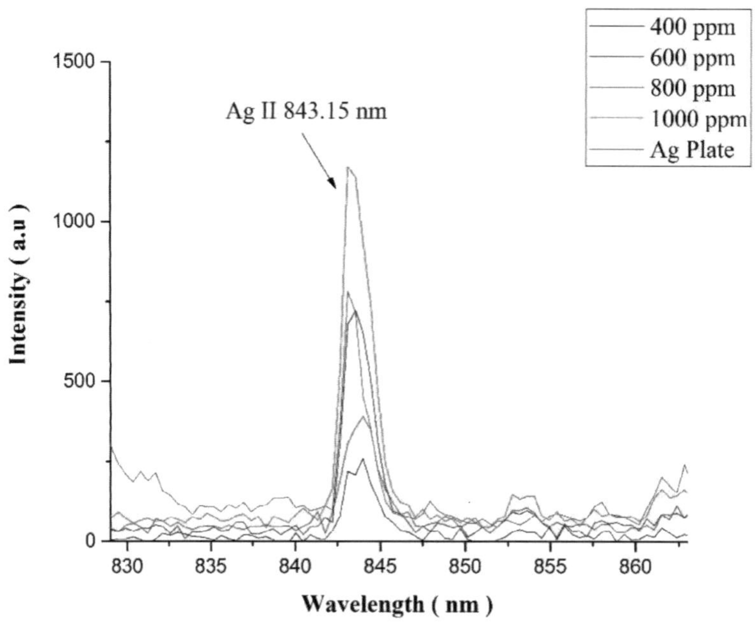

Figure 3. LIBS emission spectra of Ag ion at 843.15 nm wavelength for various Ag ion concentrations.

We used four methods to analyse plasma spectra, i.e. uncorrected, discard 20% of outlying data, peak-to-background ratio, and a combination of the final two methods [10]. The number of scans was chosen to be 250 as this was reported by Body [10] to improve data analysis. 250 scans was an optimal number as more scans could potentially damage the sample and affect the results. In uncorrected method, we calculated area under Ag ion line using trapezium formula for all 250 spectra and took the average as the peak area. In the discard 20% method, we discarded 10% of highest data and 10% of lowest data from calculation results of uncorrected method. It is assumed that the highest and lowest 10% data from the median contains potential outliers that can introduce error into the calculation of the mean. In the ratio method, we compared the peak area of Ag ion line to total plasma spectrum area. This method is to minimize laser power and matrix effect. In the last method we applied discard 20% method to results obtained from the ratio method.

The precision of analysis is given by the relative standard deviation (RSD) value as shown in figure 4. The accuracy of calculations of the 843.15 nm Ag (II) ion line in soil is shown in figure 5. The uncorrected method gives the worst calculation result, as expected. After applying the discard 20% method, the accuracy improves. The ratio and combined ratio and discard 20% methods also give good results.

Figure 5 shows analysis results of silver concentration in soil using discard 20% and ratio with discard 20% methods. There is a weak linear correlation between plasma intensity increasing as silver concentration is increased. However, the correlation is weak and non-linear due to many factors, such as the uncontrolled heterogeneity of silver concentration in soil.

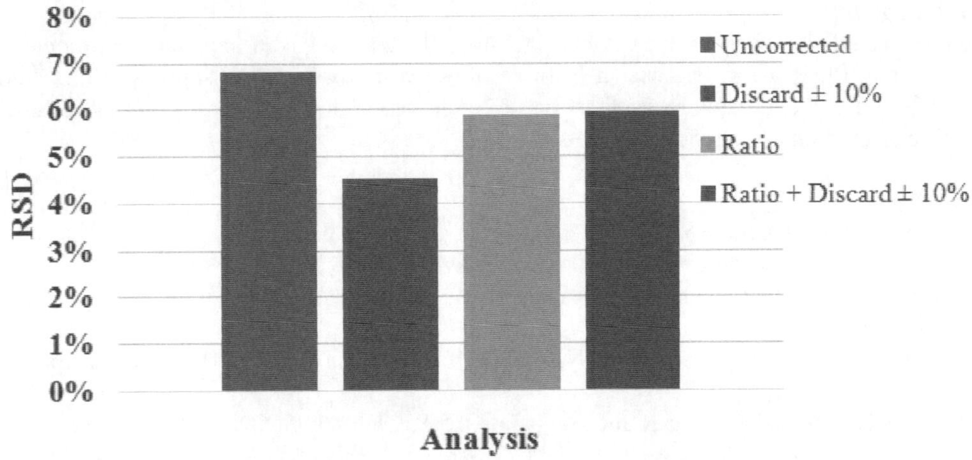

Figure 4. Precision calculation of Ag atomic line using four different methods.

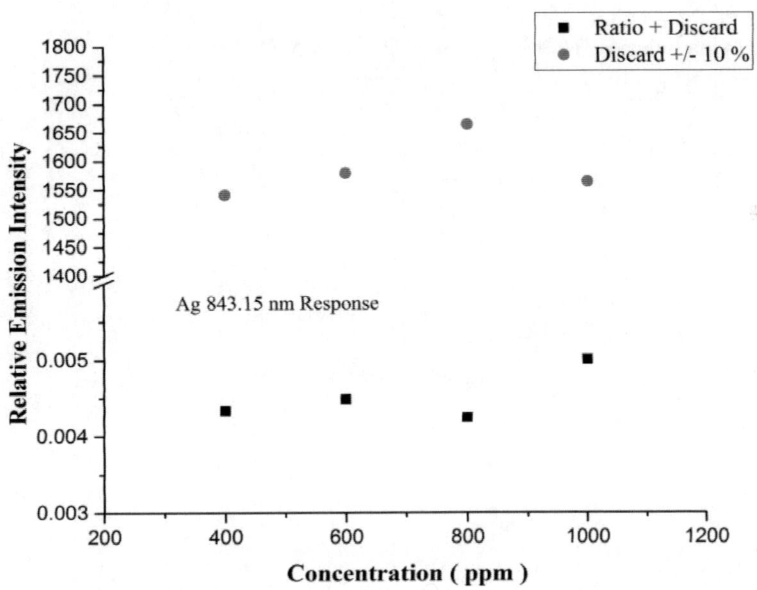

Figure 5. Calibration curves for a 843.15 nm Ag atomic line present in various Ag ion concentrations.

4. Conclusion

Ag emission lines were detected in soil samples using a 532 nm wavelength laser. Using Ag atomic lines from Ag plate and NIST database, we have confirmed and analyzed plasma spectra from soil to determine Ag content from 400 ppm until 1000 ppm. Analysis of 250 plasma spectra Ag at 843.15 nm was carried out using four different methods. We found that the discard 10% method gives better analysis than other methods. It requires more experiments by varying more Ag concentration to show linearity between Ag concentration and plasma intensity.

Acknowledgments
The authors would like to thank Kegiatan Unggulan LIPI and Kegiatan Penguatan Kompetensi from Research Centre for Physics, Indonesian Institute of Sciences for financial support of this work. The authors also would like to thank researchers in the Laser Research Group, Research Centre for Physics for valuable discussion and support during this work.

References

[1] Baudelet M and Smith B W 2013 *J.Anal. At. Spectrom.* **28** 624
[2] Huang J-S, Ke C-B and Lin K-C 2004 *Spectrochim. Acta B* **59** 321
[3] Gornushkin S I ,Gornushkin I B, Anzano J M, Smith B W and Winefordner J D 2002 *Appl. Spectrosc.* **56** 433
[4] Tomacheski D, Pittol M, Simões D N, Ribeiro V F, Santana R M C 2017 *Global J. Environ. Sci. Manage.* **3** 341
[5] Wakshlak R. B-K, Pedahzur R and Avnir D 2015 *Sci. Rep.* **5** 9555
[6] Nowack B, Krug H F and Height M 2011 *Environ. Sci. Technol.* **45** 1177
[7] Bousquet B et al. 2008 *Spectrochim. Acta B* **63** 1085
[8] Rusak D A, Clara M, Austin E E, Visser K, Niessner R, Smith B W and Winefordner J D 1997 *Appl. Spectrosc.* **51** 1628
[9] Ismail M A, Imam H, Elhassan A, Youniss W T and Harith M A 2004 *J. Anal. At. Spectrom.* **19** 489
[10] Body D and Chadwick B L 2001 *Spectrochim. Acta B* **56** 725

IOP Conf. Series: Journal of Physics: Conf. Series **985** (2018) 012010 doi:10.1088/1742-6596/985/1/012010

Quantitative analysis of titanium concentration using calibration-free laser-induced breakdown spectroscopy (LIBS)

Zaitun, S Prasetyo, M M Suliyanti, Isnaeni and Y Herbani

Research Centre for Physics, Indonesian Institute of Sciences, Riset Rd. Building 442, Puspiptek, South Tangerang, Banten, 15314, Indonesia

E-mail: zaitunbasyid@gmail.com

Abstract. Laser-induced breakdown spectroscopy (LIBS) can be used for quantitative and qualitative analysis. Calibration-free LIBS (CF-LIBS) is a method to quantitatively analyze concentration of elements in a sample in local thermodynamic equilibrium conditions without using available matrix-matched calibration. In this study, we apply CF-LIBS for quantitative analysis of Ti in TiO_2 sample. TiO_2 powder sample was mixed with polyvinyl alcohol and formed into pellets. An Nd:YAG pulsed laser at a wavelength of 1064 nm was focused onto the sample to generate plasma. The spectrum of plasma was recorded using spectrophotometer then compared to NIST spectral line to determine energy levels and other parameters. The value of plasma temperature obtained using Boltzmann plot is 8127.29 K and electron density from calculation is 2.49×10^{16} cm^{-3}. Finally, the concentration of Ti in TiO_2 sample from this study is 97% that is in proximity with the sample certificate.

1. Introduction

Laser-induced breakdown spectroscopy (LIBS) is one of the methods that can be used for quantitative and qualitative analysis. The advantages of LIBS are less destructive and more efficient. For qualitative analysis using LIBS, further preparation sample is not required. Until now LIBS has been successfully used for testing several types of samples such as gases, liquids, and solids.

For quantitative analysis LIBS, standardization is needed, such as using calibration curves. The commonly used calibration curve is matrix-matched standards [1]. When matrix-matched is used in LIBS analysis then the advantages of LIBS are reduced because it is no longer efficient. The use of calibration standards should meet several attempts of plasma spectral emission, such as the state of plasma temperature. To overcome these issues, a calibration-free LIBS (CF-LIBS) method can be used for solution. CF-LIBS was proposed by Ciucci et al. in 1999 [2]. Until now, quantitative analysis using CF-LIBS has been applied in a number of studies and some analytical methods for CF-LIBS have been proposed [3]. Dong et al. [4] in 2015 developed a procedure using determined plasma temperature by genetic algorithm (GA). Bulajic et al. [5] in 2002 proposed a procedure for correcting self-absorption in CF-LIBS. Self-calibrated LIBS was proposed by De Giacomo et al. [6] in 2007.

Using CF-LIBS for quantitative analysis must meet several circumstances, such as the laser induced plasma must be in local thermodynamic equilibrium (LTE) condition [7]. LTE condition is characterized with the achievement of the value of electron density (Ne) $\geq 1.6\times10^{16}$ cm^{-3} [8]. The concentration of a specific element of interest is given by the sum of the concentrations of neutral and singly ionized species.

In this paper, we used CF-LIBS for analyzing concentration of Ti in TiO_2 sample. The purpose of this work is to find the successive CF-LIBS as quantitative analysis without calibration curve. The standard sample used in this work was high purity TiO_2 powder from MERCK (99.99%).

2. Experimental methods

TiO_2 powder was mixed with polyvinyl alcohol (PVA) as a binding material for better compactness and then pressed into pellet by hydraulic pressing machine. The sample was pressed for 10 minutes at 48.78 MPa at room temperature. Diameter and thickness of the pellet are 29 mm and 3 mm, respectively.

LIBS setup scheme is shown in figure 1. Nd: YAG Q-Smart 850 laser from Quantel was used in this work, with laser wavelength of 1064 nm and repetition rate of 10 Hz. Laser beam was focused onto sample using 100 mm focusing lens. The energy of laser was 45 mJ and the pellet sample was placed in a vacuum chamber. Ocean Optics spectrometer was used to detect and record emission spectrum of generated plasma with integration time 100 ms.

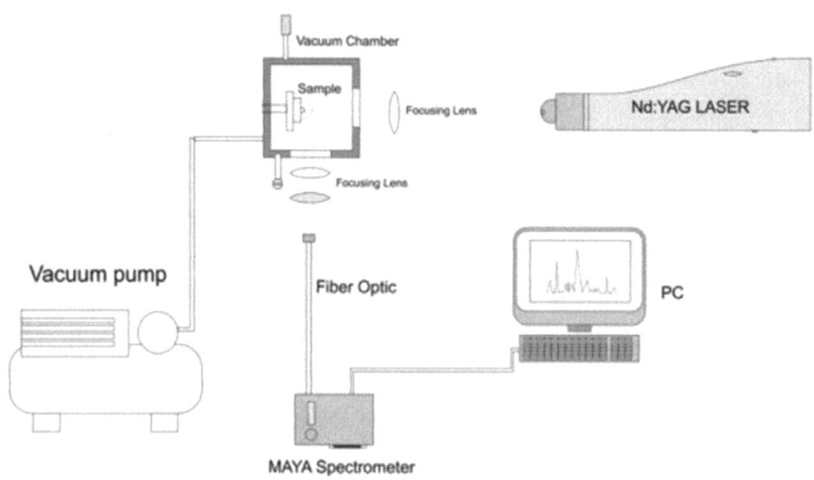

Figure 1. LIBS setup.

The CF-LIBS method as quantitative analysis is based on the assumption that laser induced plasma is optically thin and under LTE condition [7]. When the conditions are accomplished, the integrated intensity of spectral emission line can be expressed by Saha-Boltzmann Equation [9]:

$$I = FCsA_{ki}\frac{g_k}{U(t)}\exp\left(-\frac{E_k}{kT}\right) \qquad (1)$$

Where I is the measured integrated spectral line intensity, F is constant, C_s is the number density of an atomic or ionic species s, A_{ki} is transition probability, g_k is the statistical weight for the upper level, $U(T)$ is the partition function, E_k is energy of the upper energy, k is Boltzmann constant, and T is the plasma temperature.

For a plasma in LTE conditions, plasma temperature is calculated according to:

$$ln\frac{I}{A_{ki}} = -\frac{1}{kT}E_k + ln\frac{FCs}{U(T)} \qquad (2)$$

In a linear form $y = mx + bs$, plasma temperature can be obtained from the slope of the Boltzmann plot with $m = -1/kT$ [10]. The concentration of a specific element in plasma can be calculated following the steps prescribed by equations 3 – 5:

$$U(T) = \sum_i g_i e^{-E_i/kT} \tag{3}$$

$$Cs = \frac{1}{Fn^p} U(T) e^{bs} \tag{4}$$

$$Cs = \frac{U(T)e^{bs}}{\sum_s U'(T)e^{bs'}} \tag{5}$$

3. Result and discussion

First, pellet samples of TiO_2 were shot with a laser beam from Nd: YAG with wavelength 1064 nm and energy of laser beam 45 mJ. Emission spectrum of generated plasma from the sample was detected and recorded by spectrophotometer. Emission spectrum was converted to graphic using Ocean Optics spectrophotometer program. Graphic of emission spectrum is shown in figure 2.

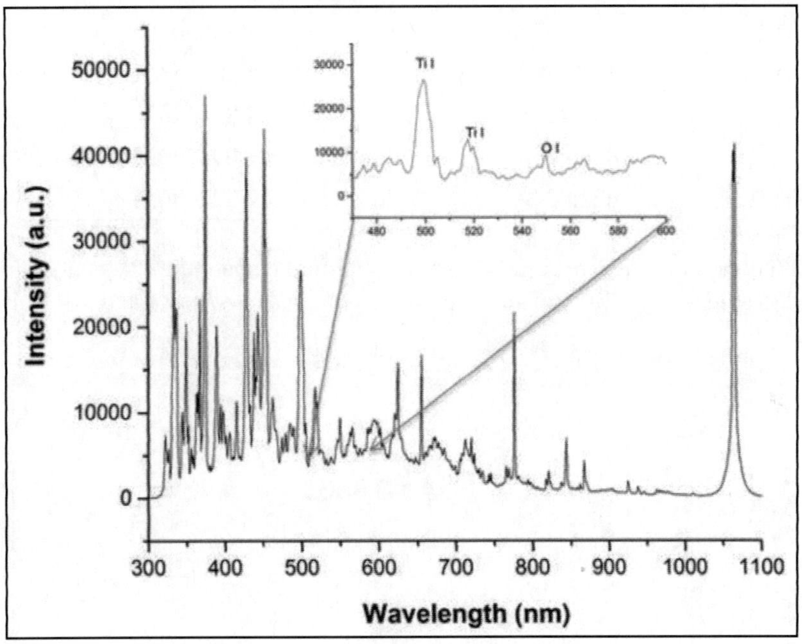

Figure 2. Graphic of emission spectrum.

Graphic of emission spectrum was analyzed to obtain spectroscopic constants using NIST database as a reference. Required spectroscopic constants are transition probability, energy in upper level, and degeneracies. The analysis result is shown in table 1.

Based on the literature, CF-LIBS can be used as a method for quantitative analysis with some conditions. One of the conditions is that the LTE state of plasma is fulfilled. Plasma can be considered in LTE state if the value of $Ne \geq 1.6\times10^{16}$ cm^{-3}, and Ne can be calculated according to equation 6 [8]:

$$Ne \geq 1.6 \times 10^{12} T_e^{1/2} (\Delta E_{nm})^3 \tag{6}$$

where Ne is electron density, T_e is electron temperature, and ΔE_{nm} is the largest gap between adjacent energetic levels.

Table 1. Spectral lines considered for the analysis.

Species	Wavelength (nm)	A_{ki} $(s^{-1}).10^6$	E_k (eV)	g_k
Ti I	320.90219	16.3	5.8310053	3
Ti I	363.79650	0.93	3.4070939	3
Ti I	366.06292	3.0	3.4070939	3
Ti I	387.32011	50.5	5.19706946	4
Ti I	451.2733	9.86	3.5826547	5
Ti I	517.3740	4.23	2.3957447	2
Ti I	549.0148	3.69	3.7178148	3
Ti I	842.6504	1.20	2.2968146	2
Ti II	332.33879	1.02	4.89451479	$^5/_2$
Ti II	335.20693	15.2	4.91904347	$^1/_2$
Ti II	348.36246	97	7.86630466	$^7/_2$
Ti II	373.9599	1.71	5.90475341	$^5/_2$
Ti II	442.19384	1.85	4.86432351	$^3/_2$
Ti II	498.279	4.20e-7	3.09479323	$^3/_2$
Ti II	622.724	2.7e-8	2.59768654	$^5/_2$

To fulfill LTE condition, the most crucial factor is plasma temperature. Plasma temperature can be obtained from the slope of the Boltzmann, a plot between $ln\ (I\ \lambda/A_{ki}\ g_k)$ and E_k. Boltzmann plot is shown in figure 3.

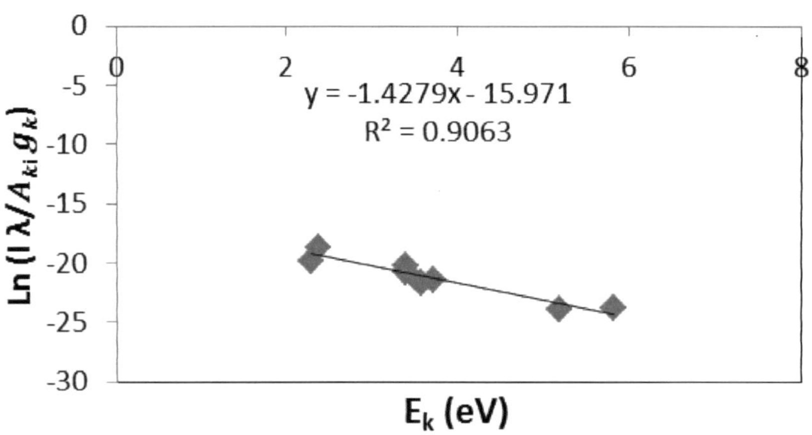

Figure 3. Boltzmann Plot.

The result of plasma temperature in this work is 8127.29 K. Using value of plasma temperature, the obtained electron density is 2.49×10^{16} cm^{-3}. The *Ne* result is greater than 1.6×10^{16} cm^{-3}, which means LTE condition of plasma in this work is fulfilled. Furthermore, CF-LIBS can be used as quantitative analysis in this work.

Concentration of the species *Cs* can be calculated using the intersection value *bs* of the linear regression in the Boltzmann in equations 3 – 4. From calculation result, concentration of Ti in TiO_2

sample in this work is up to 97%, which is very close from known concentration of certified samples, i.e. 99%.

4. Conclusion

In this work, the best linear fit is obtained in calculating temperature plasma. The value of *Ne* is 2.49×10^{16} cm^{-3}, which means that LTE condition of plasma in this work is fulfilled. In this work, quantitative analysis for concentration of Ti in TiO_2 sample using CF-LIBS is successful. The results of concentration Ti is obtained 97% which is very close from known concentration of certified samples, i.e. 99%.

Acknowledgments

The authors are grateful to Unggulan Program and Penguatan Kompetensi Program Research Center of Physics, Indonesian Institute of Science (LIPI) for financial support of this research.

References

[1] Eppler A S, Cremers D A, Hickmott D D, Ferris M J and Koskelo A C 1996 *Appl. Spectrosc.* **50** 1175
[2] Ciucci A, Corsi M, Palleschi V, Rastelli S, Salvetti A and Tognoni E 1999 *Appl. Spectrosc.* **53** 960
[3] Grifoni E, Legnaioli S, Lorenzetti G, Pagnotta S, Poggialini F and Palleschi V 2016 *Spectrochim. Acta B* **124** 40
[4] Dong J, Liang L, Wei J, Tang H, Zhang T, Yang X, Wang K and Li H 2015 *J. Anal. At. Spectrom.* **30** 1336
[5] Bulajic D, Corsi M, Cristoforetti G, Legnaioli S, Palleschi V, Salvetti A and Tognoni E 2002 *Spectrochim. Acta B* **57** 339
[6] De Giacomo A, Dell'Aglio M, Gaudiuso R, Santagata A, Senesi G S, Rossi M, Ghiara M R, Capitelli F and De Pascale O 2012 *Chemical Physics* **398** 233
[7] Tognoni E, Cristoforetti G, Legnaioli S and Palleschi V 2010 *Spectrochim. Acta B* **65** 1
[8] Horňáčková M, Horňáček M, Rakovský J, Hudec P and Veis P 2013 *Spectrochim. Acta B* **88** 69
[9] Davari S A, Hu S and Mukherjee D 2017 *Talanta* **164** 330
[10] Tognoni E, Cristoforetti G, Legnaioli S, Palleschi V, Salvetti A, Mueller M, Panne U and Gornushkin I 2007 *Spectrochim. Acta B* **62** 1287

IOP Conf. Series: Journal of Physics: Conf. Series **985** (2018) 012011 doi:10.1088/1742-6596/985/1/012011

The 1064 nm laser-induced breakdown spectroscopy (LIBS) inspection to detect the nutrient elements in freshly cut carrot samples

N Yudasari, S Prasetyo and M M Suliyanti

Research Center for Physics - Indonesian Institute of Sciences
Kawasan PUSPIPTEK Serpong, 442 Building, South Tangerang 15314
Banten, Indonesia

E-mail: nurfina.yudasari@lipi.go.id

Abstract. The laser-induced breakdown spectroscopy (LIBS) technique was applied to detect the nutrient elements contained in fresh carrot. Nd:YAG laser the wavelength of 1064 nm was employed in the experiments for ablation. Employing simple set-up of LIBS and preparing the sample with less step method, we are able to detect 18 chemical elements including some fundamental element of carrot, i.e Mg, Al, Fe, Mn, Ti, Ca, and Mn. By applying normalized profiles calculation on some of the element, we are able to compare the concentration level of each element of the outer and inner part of carrot.

1. Introduction

Elemental detection and analysis has played very important role for agricultural product examination. Hence we will be able to determine the nutritional and harmful element for human body. Detecting nutritional element is very useful to ensuring that our body will not be lacked of nutrient consisted on agricultural product or food that we consume [1]. For the plant itself, nutritional detection is also good to observe whether the plant gains sufficient inputs of macro-nutrients and/or micronutrients. It was reported that the deficiency of some elements could adversely give the bad effect of growth process. For example, the plant which is lack of Mg can lead to chlorosis and impair shoot and root growth. Another case, Ca has role to deliver nutrients into the plant, therefore Ca deficiency gives effect in deformation and chlorosis of leave [2]. In the other hand harmful elements such as toxic or heavy metals inside the food and agricultural product has also become urgent need to be detected. The accumulated amount of armful element may lead to intoxication for human health [3].

A lot research has been developed such that the examination of the food and agricultural product sample could be conducted with fast analysis and less preparation step. Some worked has been done by employing chemical method which need multistep of sample preparation, whilst the data acquisition and analysis took a long time [4]. The laser-induced breakdown spectroscopy (LIBS) offers more promising technique to detect the nutritional and harmful element in the agricultural product. This technique has basic rule on laser-induced material ablation (vaporization and ionization of a minute amount of sample materials) and on spectral analysis of the plasma emission [3]. The plasma emission is captured by fiber optic and translated into emission spectra data. The spectra

IOP Conf. Series: Journal of Physics: Conf. Series **985** (2018) 012011 doi:10.1088/1742-6596/985/1/012011

consist of some lines and they were identified as element consisted in the sample. This measurement technique is relatively simple and fast, and we do not need much sample preparation step.

In this study, we use carrot as our experimental sample. We choose carrot because this vegetable grew as root in the ground. This will be good sample to detect if there are some contaminant elements inside the carrot, because the root is the main part of the plant to absorb water and element in soil. Once the soil was contaminated by harmful elements, it is highly possible that the carrot was polluted as well. Some of the work has been done by employing LIBS technique to determine the chemical element inside of the agricultural product [1, 2, 4]. On 2008, a research has been done on using the LIBS technique to determine chemical element on freshly cut potato and carrot, this study used 266 nm wavelength and Echelle spectrometer, hence it was able to detect 27 elements inside the potato [5]. In this work, we have demonstrated a simple LIBS set-up and less preparation step of sample. We are able to detect 18 chemical elements including some fundamental element of carrot, i.e Mg, Al, Fe, Mn, Ti, Ca, and Mn.

2. Methodology

The schematic diagram of this experiment is shown in figure 1a. The LIBS system consists of Nd-YAG laser (Q-Smart 850mJ, Quantel) with the wavelength of 1064 nm, energy of 80 mJ and frequency of 10 Hz. The laser beam was focused to sample by a 150 mm focal length plano-convex lens. The sample was attached into a rotation holder inside the vacuum chamber which was operated on the pressure of 6.9 torr. The generated plasma was then collected by using collimating lens and focused into fiber optics. The fiber optic was connected to Ocean Optics Spectra Suite MAYA2000 Pro measuring emission lines from 200 to 1200 nm with a spectral resolution of 0.25 nm.

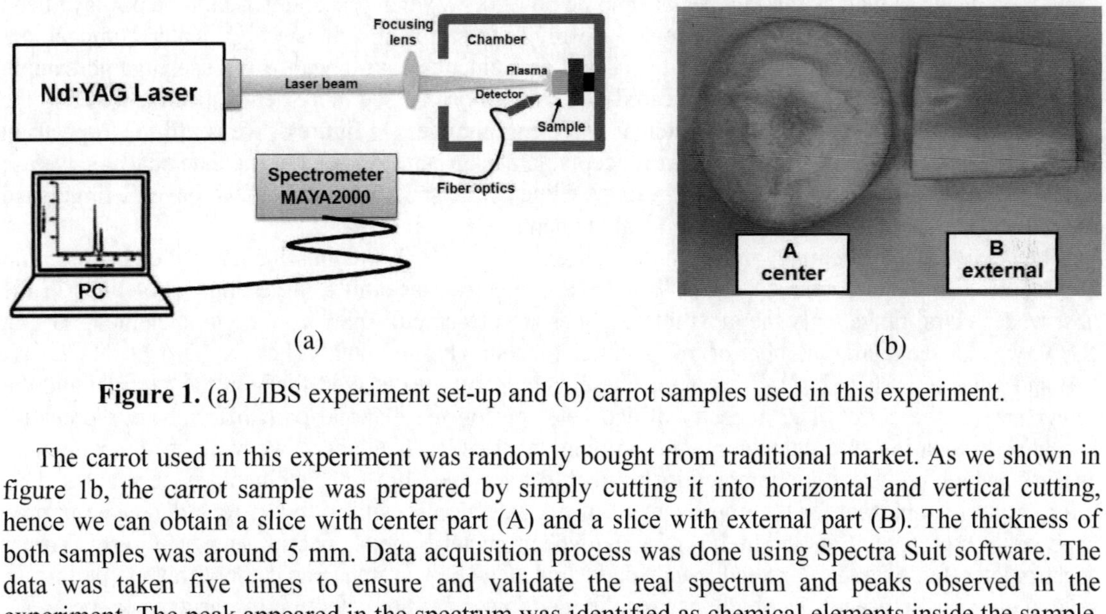

Figure 1. (a) LIBS experiment set-up and (b) carrot samples used in this experiment.

The carrot used in this experiment was randomly bought from traditional market. As we shown in figure 1b, the carrot sample was prepared by simply cutting it into horizontal and vertical cutting, hence we can obtain a slice with center part (A) and a slice with external part (B). The thickness of both samples was around 5 mm. Data acquisition process was done using Spectra Suit software. The data was taken five times to ensure and validate the real spectrum and peaks observed in the experiment. The peak appeared in the spectrum was identified as chemical elements inside the sample. The NIST database was used in this study to identify the chemical element of the carrot sample.

3. Results and discussion

Figure 2 gives an overview of the typical LIBS spectra of the carrot sample, the black and red data was respectively the center and external part of the carrot sample. The plasma was induced on the carrot sample. The spectra from the plasma was captured using spectrometer fiber in the dark room and by eliminating the background spectra.

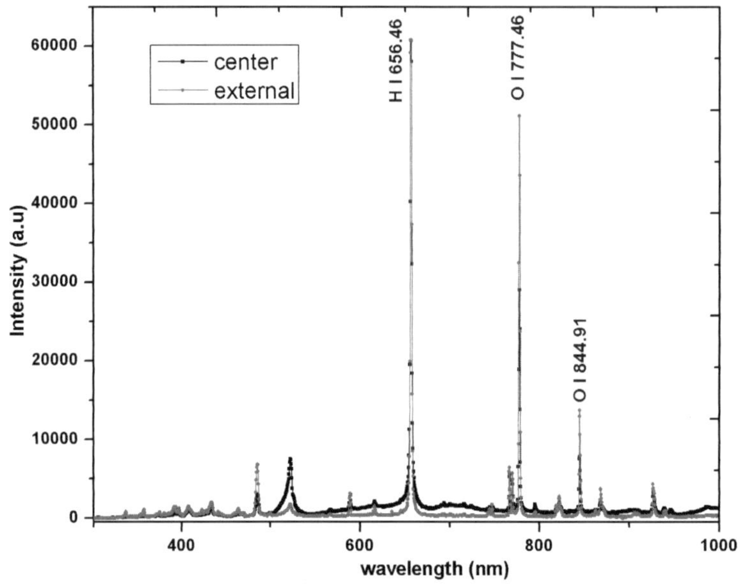

Figure 2. Typical LIBS spectra of the carrot sample

Figure 2 shows the LIBS spectra of the carrot sample. According to the spectra, we could identify some high peaks as well as the low peaks at some point of wavelength. The two high peaks have been identified as H and O. Because we use the freshly cute carrot, and it has high water composition. Therefore, it is most likely dominated by H and O element. Also, because we use organic sample another element that expected to be appeared is C, or carbon. To see more detail spectra, we chop the graph on figure 3 into several some particular wavelength range. On figure 3, we could see a spectrum with quite large number of lines with quite acceptable background signal. For the data analysis, we use the NIST spectral database [6] to identify line by line of the spectrum. In total we have distinguished around 35 lines, which consist of 17 chemical elements.

The detail spectra of carrot sample could be seen in figure 3. There are 18 chemical elements could be identified in the spectrum of carrot sample. For species that emit a large number of lines in the observed spectral range, only the most intense lines are presented. Apart from organic elements (H, C, N, O), we can see a large number of inorganic elements including metals (Fe, Cu, Ti, Mn, Th, Zr, W, Ne) and nonmetals (Mg, K, P, S, Cl, Ca). These elements were detected for both the external and the center part of the carrot. The intensity of each element on the external part differs from the central part. For comparing and calculating the amount level of each element we will need quantitative analysis, which require more complex instrument to trace and ultra-trace elements in vegetables. This information is important for the improvement of the detection sensitivity by using a dedicated narrow bandwidth, high resolution and high throughput spectrometer Echelle spectrometer or (Czerny-Turner spectrometer for example) instead. Such dedicated detection system can be optimized for one or several elements in order to get higher sensitivity and lower limit of detection for targeted elements.

In this study we might still be able to see rough calculation to compare the element intensity on the central and external part of the carrot. For this investigation, we chose some elements as examples, which are K, Mn, Fe, and Cu. The normalization was made for a given spectrum by dividing the line intensity of each element by the highest among the two data, external and central element in intensity. Figure 4 shows the normalized profile of K, Fe, Mn, and Cu element level inside the carrot sample. By applying this method the concentration level of each element could be compared between the outer and inner part of carrot. In the center part Fe level was higher than in the external part, while K, Mn, and Cu were more dominant in the external part than in center one.

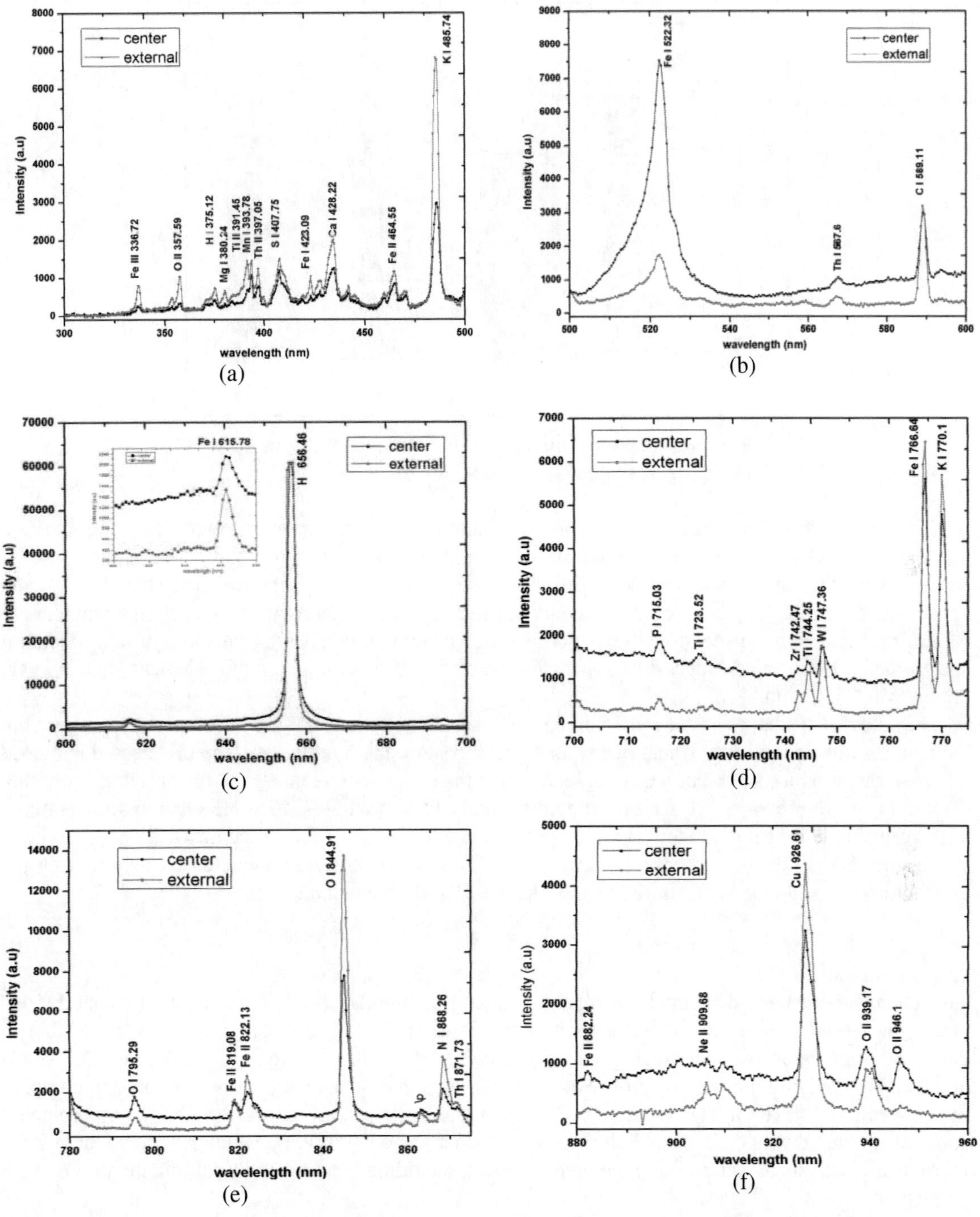

Figure 3. LIBS spectra of carrot samples, the black and red data was for the center and external part respectively.

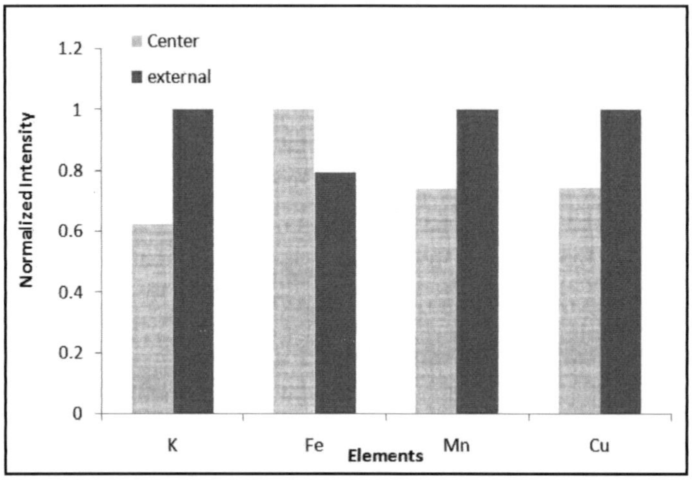

Figure 4. Normalized profiles of relative concentrations of K, Fe, Mn and Cu for fresh carrot

Some other works have been done in the same field to detect the chemical elements of vegetables. Some of the works have demonstrated the measurement with more complex sample preparation [1, 7], while some other did experiment with different laser type [3, 11], laser wavelength [5, 8, 9], and spectrometer tool [5, 10]. Juve et al. [5] conduct the same experiment with UV laser (the wavelength of 266 nm) and Echelle spectrometer which has much more sophisticated technology than common spectrometer used in this experiment, which is able to detect some of the major element such as Mg, Al, Fe, Mn, Ti, Ca, and Mn.

The study in this paper employed a simple LIBS set-up and the least step of sample preparation, but the result is quite promising. Using our simple LIBS system, the number of element detected is indeed less than the previous work but we are able to detect the same main elements as the previous work did. Hence, in the future work we would be able to apply these methods on some other organic sample with very simple preparation step. By adding and replacing some of the instruments, i.e using Echelle spectrometer and delay stage generator, more detail analysis could be conducted on determining the element level, including determining limit of detection of each elements.

4. Conclussion

The chemical element detection on freshly cut carrot sample has been done in this study. By employing simple set-up of LIBS and preparing the sample with less step method, we are able to detect 18 chemical elements including some fundamental element of carrot, i.e Mg, Al, Fe, Mn, Ti, Ca and Mn. By applying normalized profiles calculation on some of the element, we are able to compare the concentration level of each element of the outer and inner part of carrot. By adding and replacing some of the instruments, i.e using Echelle spectrometer and delay stage generator, we expect to have more detail analysis on determining the element level, including determining limit of detection of each elements.

References

[1] Kim G, Kwak J, Choi J and Park K 2012 *J. Agric. Food Chem.* **60** 718
[2] Brownell P F and Bielig L M 1996 *Funct. Plant. Biol.* **23** 171
[3] Bossu M, Hao Z-Q, Baudelet M, Jin Y, Zhang Z and Zhang J 2007 *Chin. Phys. Lett.* **24** 3466
[4] Sharma K D, Karki S, Thakur N S and Attri S 2012 *J. Food Sci. Technol.* **49** 22
[5] Juvé V, Portelli R, Boueri M, Baudelet M and Jin Y 2008 *Spectrochim. Acta B* **63** 1047

[6] NIST atomic spectra database lines form, https://physics.nist.gov/PhysRefData/ASD/lines_form.html
[7] Shukla P, Kumar R and Kumar Raib A 2016 *J. Appl. Spectrosc.* **83** 872
[8] Huang H, Yang L M and Liu J 2012 *SPIE Defense, Security, and Sensing* (Washington: SPIE Publishing) 8358
[9] Agnes N, Tao H-Y, Hao Z-Q, Sun C-K, Gao X and Lin J-Q 2013 *Chin. Phys. B* **22** 014209
[10] Peruchi L C, Nunes L C, de Carvalho G G A, Guerra M B B, de Almeida E, Rufini I A, Santos Jr. D and Krug F J 2014 *Spectrochim. Acta B* **100** 129
[11] Gurevich E L and Hergenröder R 2007 *Appl. Spectrosc.* **61** 233A

IOP Conf. Series: Journal of Physics: Conf. Series **985** (2018) 012012 doi:10.1088/1742-6596/985/1/012012

Nondestructive evaluation of Bakwan paddy grains moisture content by means of spectrophotometry

M Makky[1], R E Putry[1], K Nakano[2] and Santosa[1]

[1]Dept. of Agricultural Engineering, Andalas University, Kampus UNAND Limau Manis Padang 25163 West Sumatera, Indonesia
[2]United Graduate School of Agricultural Science, Gifu University, Yanagido 501-1193, Gifu Prefecture, Japan

E-mail: muh_makky@yahoo.com

Abstract. Paddy grains moisture content (*MC*) strongly correlated to the physical properties of rice after being milled. Incorrect *MC* will cause higher percentage of broken rice and prompts the grains to be more fragile. In general, paddy grains with 13 – 14% *MC* are ideal for post-harvest processing. The objective of this study is to measure the *MC* of intact paddy grain from CV. Bakwan by means of non-destructive evaluation using NIR spectral assessment. Paddy grains samples with identical *MC* were put into 30 mm tube glass and measured using NIR spectrophotometer. The electromagnetic radiation absorbance under consideration upon spectral measurement fell between 1000 and 2500 nm. The grains' actual *MC* was then measured by primary method, based on weight measurement i.e. oven method. In this study, the spectral data of the grains was then processed by means of Principal Component Analysis (PCA) before correlated with its *MC*s by Partial Least Square (PLS) method. The model calibration obtained correlation (r) of 0.983 and *RMSEC* of 1.684. Moreover, model validation produced correlation (r) of 0.973, *RMSEP* of 2.095, and bias of 0.2, indicating that the *MC* of paddy grains can be precisely identified by non-destructive evaluation using spectral analysis.

1. Introduction

Rice is the main staple food and commodities in Indonesia. It is consumed mainly in the middle and western part of the country with more than 45 million tons of rice annual consumption nationwide [1]. As the main crops and food, rice become the crucial factor for ensuring food security and independence of this young republic.

Directly after being harvested, the paddy were subsequently reaped and passed through several processes including stacking, handling, threshing, cleaning, and hauling [2] before stored or consumed. Before stored, the grain's moisture content (*MC*) should be reduced to a safe level, viz. 9 to 14% wet based (wb) [3] through natural or mechanical drying. The final edible white or brown kernel depends on the phases of the milling processes [2]. High *MC* will increase the product loss upon milling due to kernel break up or damage [4]. Normally, the *MC* of grains should be maintained at less than 14% before being milled into rice [3]. Beside losses, the *MC* of grain dictates the quality of rice produced after milling [5]. The percentage of kernel with cracks or damage correlates with its *MC* [6]. Higher *MC* increase yield losses and consequently reduces its retail price [7, 8].

Accurate measurement of the *MC* of the grains is commonly performed by means of weight methods [9, 10]. This method uses random samples from the bulk, and due to the huge resources required it only considered as representative of the overall grains. On the other hand, many non-destructive evaluation methods for food and other agricultural products had been established. These methods offer better accuracy, lower operating costs, more rapid measurements, and repeatability when used for determining the internal properties [11] of the samples. Various methods had successfully employed to determine the features of oil palm fruits [12 – 21], tomato leaf [22], citrus [23], star fruits [24], pomegranate [25], and other agricultural commodities [26 – 28], including grains [29]. Grains properties such as starch [30], protein [31], mineral [32], and *MC* [29] can be determined nondestructively, by means of machine vision and spectroscopy. The *MC* influences the rice spectrum properties when observed by near-infrared (NIR) electromagnetic radiation [33]. The radiation is most notable when the reflection of incandescent light from the probe falls between 400 nm and 1050 nm. At a certain frequency, the *MC* causes variation in the absorption [33]. This phenomenon can be used to correlate the *MC* of rice kernel with its optical properties.

Water absorbs wide range of electromagnetic radiation and the nature is heavily influenced by the condition and the state of the water. In liquid form, water absorbs electromagnetic radiation at wavelength of 698, 718, 810, and 935 nm [34 – 38], which is observable when measured using VIS/NIR spectrophotometer. According to these facts, the water (*MC*) inside the paddy grain is possible to be evaluated nondestructively using Vis/NIR spectroscopy.

The rate of electromagnetic radiation absorption by grains can be modeled to its *MC* using statistical analysis. Different established model such as the Multiple Linear Regressions (MLR) [29] and Partial Least Square Regression (PLSR) [33] had been used for *MC* predictions in food products. In addition, Principal Component Analysis (PCA) [19] can be used to reduce the variables for explaining the correlation. While each model individually may be used for model creation, the combination of two methods, in particular the PCA and PLS, offer better accuracy and more acceptable results. Employing these methods for evaluating *MC* of the paddy grain will explain its internal properties and constituent without necessarily damaged the samples.

The objective of this study is to establish model for *MC* determination of the paddy grains from CV. Bakwan. The model was created based on the spectral data of the samples when measured using NIR spectrophotometer. The absorption data were then subsequently correlated with the samples' *MC*, as measured by primary method. The model was established using PCA and PLS statistical analyses. The calibrated model was then verified for predicting the *MC* of the remaining grains samples. The results will offer non-destructive evaluation nature, while the performance of the model will be evaluated by means of Receiver Operating Characteristic (ROC).

2. Materials and methods

In this study, the samples' *MC* is determined through electromagnetic radiation absorbance, measured using NIR spectrophotometer. 10 kg paddy grains of CV. Bakwan were harvested by manual picking from the stem. The unhusked grains were cleaned and placed into 30 ml glass tubes. The samples were compacted by placing the tubes on vortex for 10 seconds, at 80% speed, to minimized space between grains [29]. More grains were added and procedure repeated until the tubes were fully filled.

The sample tubes were then placed into NIR spectrophotometer and their spectral absorbances (1000 – 2500 nm) were recorded at normal scan speed using average of 10 scans method. The absorbance measurements were taken at two points by rotating the tube to 180° and the average of the two data were calculated. Recorded signals were pre-processed with spectroscopy software (Unscrambler® X, Camo, USA), and all measurements were replicated five times to minimize errors.

The actual samples' *MC* was determined by primary method (oven) at 130 °C. The *MC* was calculated on wet basis. The *MC* measurements were replicated five times to minimize error. All measurements were following Makky et al. [29].

The absorbance data was then analyzed using engineering statistical software (Unscrambler® X, Camo, USA) to correlate with its *MC*. PCA method was used to identify significant correlation between certain electromagnetic radiation wavelengths with sample's *MC*. Selected principal component(s) (*PC*s) from the analyses were then considered as determining factor. A model to determine *MC* of the samples according to its electromagnetic radiation absorbance was developed using PLS. The electromagnetic radiation absorbance was considered as covariates in the analyses, the samples *MC* as dependent variables, and the *PC*s from PCA analysis as factors. Data was partitioned into two groups; the first 2/3 of data was used to train the model, and the rest of 1/3 data was used to validate the model [19]. Model performance was analyzed using Sum of Squares Error (SSE), Relative Error (RE), and ROC Curve [16, 17, 19].

3. Results and discussion

In this study, determination of samples' *MC* was done by primary (oven) method. The sample initial *MC* was measured at 11.8, 15.7, 19.5, 21.9, 24, 26.5, and 27.5% wet based (wb). The difference of *MC* is caused by the variation on the weather and harvest condition upon samples collection, as well as post-harvest handling. Raw spectral measurements data showed that samples with different *MC* produce different absorbance value when measured by the spectrophotometer (figure 1). In correlation with the *MC*, the absorbance value of the samples increased along with the *MC* level. The absorbance is influenced by sample *MC* throughout the electromagnetic radiation recording, between 1000 – 2500 nm. The differences were most notably in NIR spectrum of 1150 – 1800 nm (figure 1). In other spectrum range, the difference of absorbance of the electromagnetic radiation of the samples is observable, though the differences were minor.

The absorbance data of the sample were adjusted-normalized to set the value scaled from 0 to 1. PCA analysis was done to distinguish which wavelengths significantly differ according to sample *MC*. Using KMO and Bartlett's test of sphericity, the principal components (*PC*) were extracted by determining the Eigen values greater than 1 through limited number of iterations until all spectral data reached convergence and all data (100%) were explained. To better distinguish each principal component, a rotation technique was applied based on Varimax method [39], in which each principal component will be less dependent towards each other. Three *PC*s were obtained from the extraction which explained more than 99% of absorbance data. Of the three *PC*s obtained from Varimax extraction method using Eigen values greater than 1, the first two *PC*s, which explained the majority of the electromagnetic radiation absorbance data from the samples, are presented in figure 2.

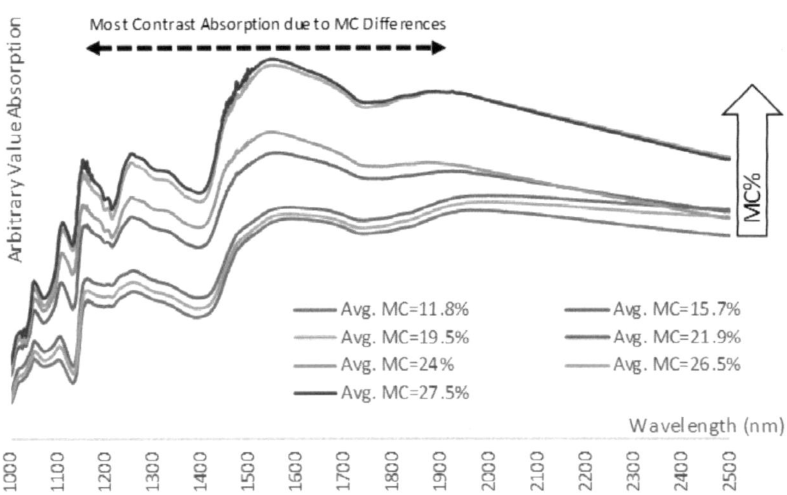

Figure 1. Electromagnetic radiation absorbance of CV. Bakwan paddy grains as measured by NIR spectrophotometer.

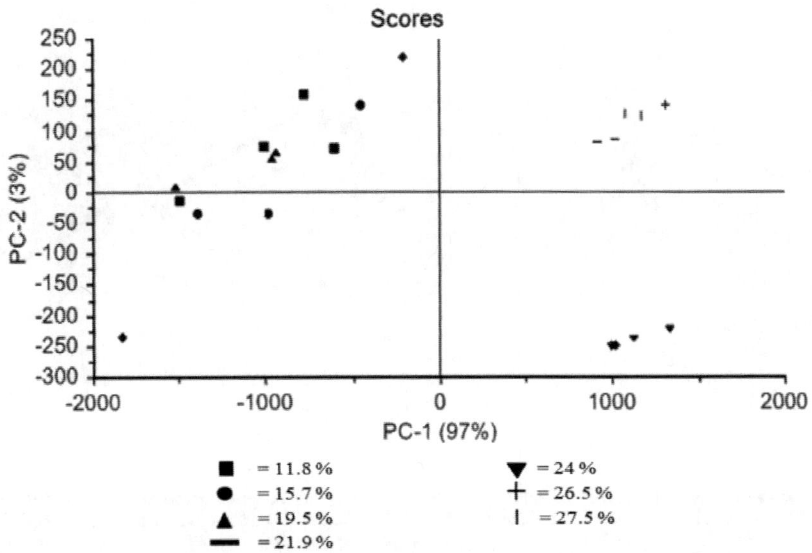

Figure 2. Two principal components from of the samples' electromagnetic radiation absorption as plotted in rotated space.

In figure 2, the majority of the first two *PC*, as plotted, placed in separated quadrants. The differences were corresponded to the samples' *MC* level. All two *PC*s explained 97% of variance from the extraction sums of squared loadings of the absorbance data. Samples with 11.8, 15.7, and 19.5% *MC* occupied similar region, while samples with *MC* of 21.9, 26.5, and 27.5% occupied similar quadrant on other section. Only samples with *MC* of 24% are located separately from the others in the diagram. Although the *PC*s show distinction between samples group, the PCA alone cannot be used to explain which wavelength corresponds the most to the sample *MC*. The first *PC* according to its coefficient value for every wavelength under consideration is presented in figure 3. From the figure 3, eight distinguished wavelengths were determined from the analysis. The selected peaks are 1013 nm, 1050 nm, 1115 nm, 1147 nm, 1257 nm, 1457 nm, 1523 nm, and 1859 nm. Nonetheless, the correlation to the *MC* of the samples cannot be established directly. It is the reason why while PCA results suggest eight absorption peaks of electromagnetic radiation wavelengths for explaining sample's *MC*, the absorbance data of the corresponding wavelengths cannot be directly calculated to predict the *MC* in the samples.

In order to perform the prediction, a PLS analysis was employed to develop the model based on the absorbance properties of the samples. Using statistical engineering software, the model was developed by assigning the spectral absorbance data of the samples as covariate, and rescaled the data using adjusted normalization. Samples *MC* was assigned as dependent variable for training the model. A hyperbolic tangent was used to activate the artificial grouping in this analysis, which is not limited to several groups. Using batch training type in the software to scale conjugated gradient of optimization algorithm, the initial calibration and validation model results are explained in the figure 4a and 4b respectively.

Figure 4 shows that the samples' *MC* can be accurately predicted by evaluating the absorbance of the selected wavelengths. For model calibration and validation, the r^2 is 0.966 and 0.950 respectively, while the *RMSEC* and *RMSEP* is 1.684 and 2.095 respectively. The results also showed that samples with lower *MC* can be predicted more accurately by model, as compared to the samples with higher *MC*. Higher residual of samples with more *MC* indicated that while the model result is still within acceptable value, more errors will occur if the samples contain more moisture above normal.

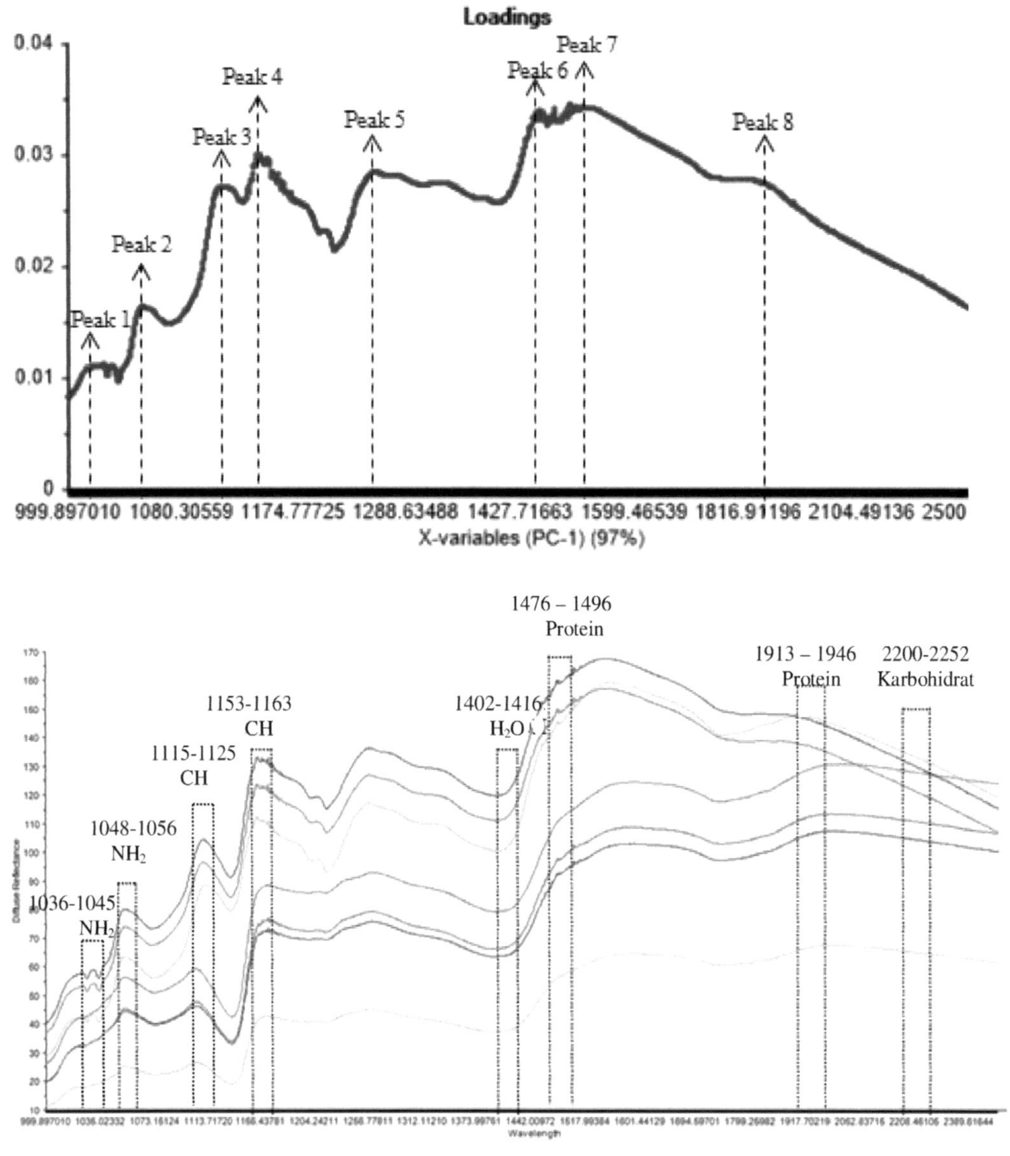

● = MC of 11.8% ● = MC of 24%
● = MC of 15.7% ○ = MC of 26.5%
● = MC of 19.5% ● = MC of 27.5%
● = MC of 21.9%

Figure 3. Selected electromagnetic radiation wavelengths according to the PCA analysis of the samples' absorbance.

This phenomenon may relate to the amount and phase of moisture in the samples, and the properties of the electromagnetic radiation itself. The absorption of electromagnetic radiation by the water can occur in the wide range of regions, and is heavily influenced by the state of the water itself, especially in the liquid phase [35 – 38]. In the UV and visible band, the absorption of electromagnetic radiation by water is very weak, in particular when the water present in liquid phase and no scattering effect produced during measurements by spectrophotometer [35]. In contrast, water absorption in organic material is stronger in the NIR band [33, 40].

Furthermore, the reason behind the bias of the models may correlate to the variance of physical properties of the grains, in which the model cannot exactly explain. Chemical properties, such as total protein content and composition of sugars or carbohydrates in the grains could influence this analysis. Therefore, the regression analysis of the model from the samples should incorporate these features as well. Nonetheless, by computing the relative error of the model, the deviation of prediction can be reduced. This study produces better prediction model for *MC* determination compared to previous finding [29, 41].

In previous studies [40, 42, 43], the chemical contents of paddy grains influence certain wavelength absorption spectrum. Furthermore, wavelengths selection from the developed model suggests that the *MC* of paddy grain may influence the electromagnetic radiation absorption by protein in the samples. Nonetheless, no previous results indicated how strong the influence was [40, 41]. In the developed model, the preference of electromagnetic radiation within the NIR range for *MC* determination in paddy grains samples was in agreement to other studies. Chang et al. [41] considered three wavelengths for the absorption observation: 785, 910, and 1020 nm, for third harmonic generation, C – H stretching vibration, and second octave N – H stretching vibration respectively. Other finding [44] proposed 990 nm wavelength to observe O – H stretching vibration at the second octave. Nevertheless, both results were used to identify the protein and amylase content of rice grain while in this study, only *MC* of samples is considered, regardless of the amylase and proteins content of the samples.

Beside the coefficient of correlation, the model performance was evaluated by ROC. The ROC curve explained the trade-off between sensitivity and specificity. Higher sensitivity will at the same time lower the specificity of the model. The model performance as plotted in ROC curve (figure 5) is significantly higher than by chance (diagonal line), indicating high accuracy from the test results. The plot shows that the developed model obtained high true positive rate against the false positive rate for the different possible cut-points. The area under the ROC curve indicates the test accuracy for the model. The developed model, according to ROC curve, resulted in high acceptability results with high accuracy with area under the curve of 0.847.

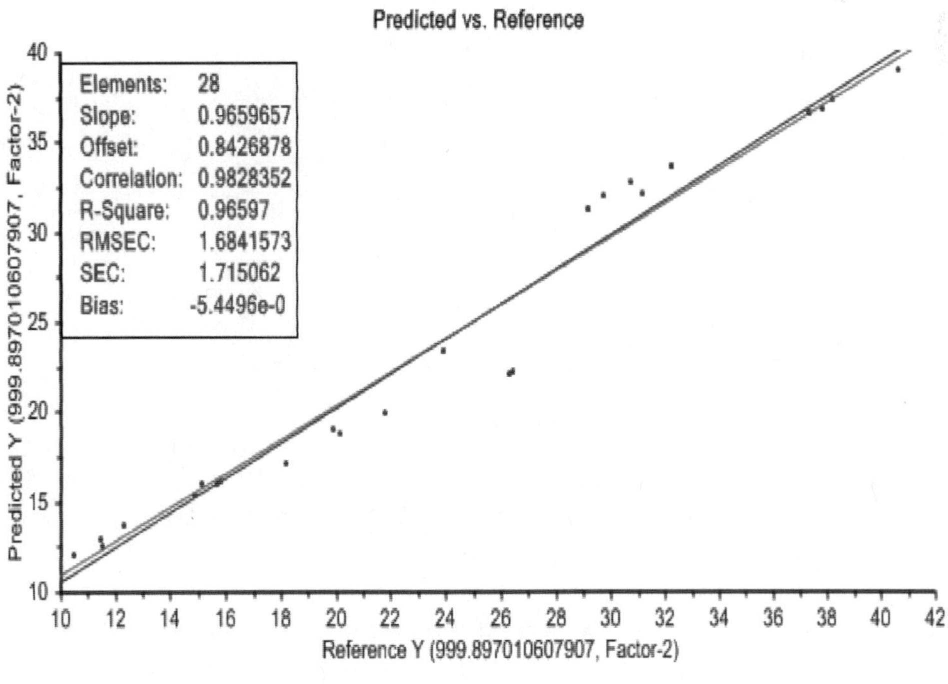

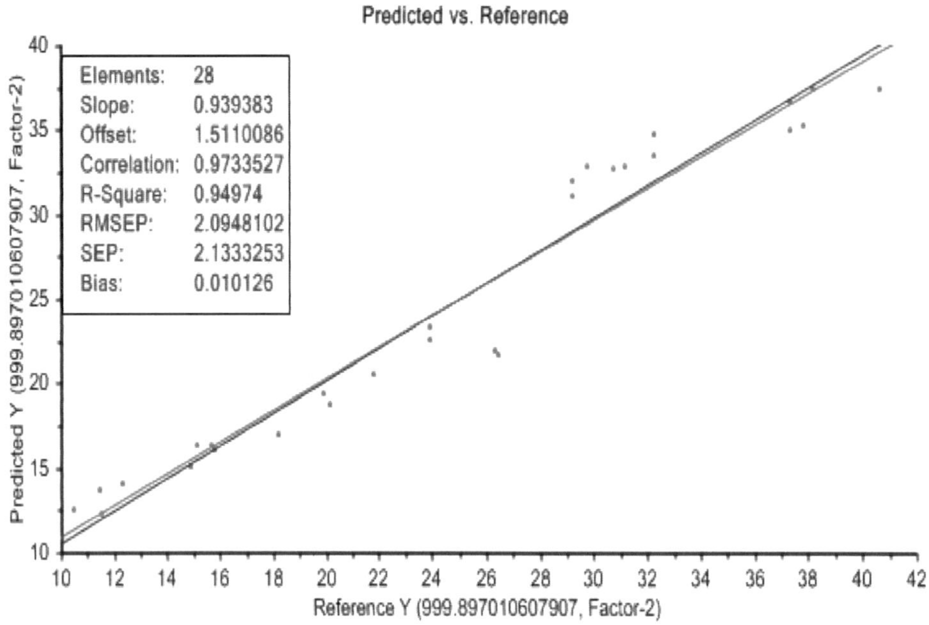

Figure 4. (Top) PLS model calibration for samples' *MC* determination according to the absorption of electromagnetic radiation in NIR region. (Below) PLS model validation for samples' *MC* determination according to the absorption of electromagnetic radiation in NIR region.

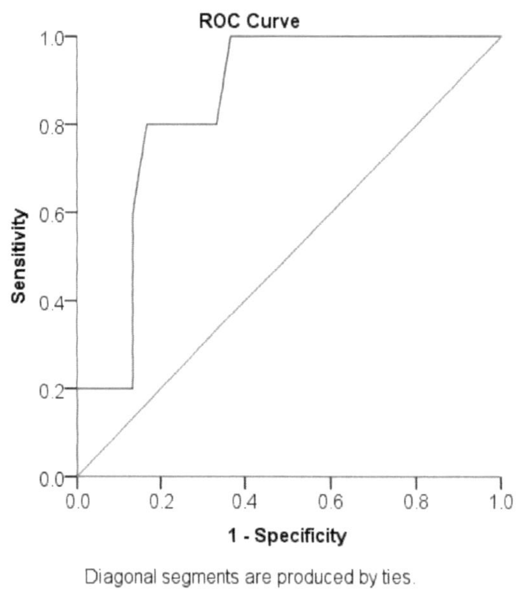

Figure 5. Model performance evaluation using ROC curve.

4. Conclusions

In this study, a nondestructive evaluation method was developed to model the *MC* of paddy grains from CV. Bakwan. The model calculates the absorbance value of the electromagnetic radiation from the samples and correlated the data with samples' *MC* using PLS analysis. The electromagnetic radiation for the measurements is within the NIR spectrum (1000 – 2500 nm). Upon validation, the model obtained r^2 of 0.950 with *RMSEP* of 2.095. In conclusion, the model produces high accuracy and sensitivity, although some residuals were observed in samples with higher *MC*. This study offers a more accurate, reliable and practical solution for nondestructive evaluation of *MC* in paddy grains, a crucial information to prevent losses upon milling.

Acknowledgments
The authors express their gratitude to Directorate General of Higher Education (DIKTI) and Universitas Andalas for providing the fund for this research through "Klaster Riset Guru Besar" Scheme, contract number. 65/UN.16.17/PP.HGB/LPPM/2017 signed on 21 March 2017.

References

[1] BPS 2015 *Official Statistical data: Paddy production in Indonesia in every province (1993-2015)* (Jakarta: Statistics Indonesia)
[2] Milling http://www.knowledgebank.irri.org/step-by-step-production/postharvest/milling
[3] Measuring moisture content in harvesting http://www.knowledgebank.irri.org/step-by-step-production/postharvest/harvesting/measuring-moisture-content
[4] Thompson J F and Mutters R G 2006 *Trans. ASABE* **49** 435
[5] Siebenmorgen T J, Bautista R C and Counce P A 2007 *Appl. Eng. Agric.* **23** 517
[6] Grain moisture content and grain quality http://www.knowledgebank.irri.org/step-by-step-production/postharvest/drying/drying-basics/grain-moisture-content-and-grain-quality
[7] Bautista R C and Siebenmorgen T J 2005 *Appl. Eng. Agric.* **21** 637
[8] Bautista R C, Siebenmorgen T J and Mauromoustakos A 2009 *Trans. ASABE* **52** 1611
[9] Methods for measuring moisture content http://www.knowledgebank.irri.org/step-by-step-production/postharvest/milling/milling-and-quality/measuring-moisture-content-in-milling
[10] IRRI moisture meter http://www.knowledgebank.irri.org/step-by-step-production/postharvest/drying/drying-basics/methods-in-measuring-moisture-content-for-drying/measuring-moisture-content-in-drying/irri-moisture-meter
[11] Manickavasagan A and Jayasuriya H 2014 *Imaging with Electromagnetic Spectrum: Applications in Food and Agriculture* 1st ed (London: Springer-Verlag)
[12] Cherie D, Herodian S, Ahmad U, Mandang T and Makky M 2015 *International Journal on Advanced Science, Engineering and Information Technology* **5** 255
[13] Cherie D, Herodian S, Mandang T and Ahmad U 2015 *International Journal on Advanced Science, Engineering, and Information Technology* **5** 314
[14] Makky M, Herodian S, Cherie D, Ahmad U and Mandang T. 2012 *Proc. Bogor Agricultural University Research Dissemination Seminar (Bogor)* (Bogor: Bogor Agricultural University) p 1
[15] Makky M, Soni P and Salokhe V M 2012 *Proc. Asian Forum of 2012 CSAM (Chinese Society for Agricultural Machinery) International Academic Annual Meeting (Hangzhou)* (Hangzhou: Chinese Society for Agricultural Machinery) p 1
[16] Makky M, Soni P and Salokhe V M 2014 *Int. Agrophys.* **28** 319
[17] Makky M and Soni P 2013 *Comput. Electron. Agric.* **93** 129
[18] Makky M and Soni P 2013 *International Journal on Advanced Science, Engineering and Information Technology* **3** 1
[19] Makky M and Soni P 2014 *J. Food Eng.* **120** 248
[20] Makky M 2016 *Agric. Agric. Sci. Procedia* **9** 230
[21] Makky M 2016 *International Food Research Journal* **23** S81
[22] Makky M 2016 *International Journal on Advanced Science, Engineering and Information Technology* **6** 349
[23] Aleixos N, Blasco J, Navarrón F and Moltó E 2002 *Comput. Electron. Agric.* **33** 121

[24] Siti Mazlina M K, Abdul Ghani L A, Nur 'Aliaa A R, Siti Aslina H and Rozita O 2008 *Pertanika J. Sci. & Technol.* **16** 1
[25] Blasco J, Cubero S, Gómez-Sanchís J, Mira P, Moltó E 2009 *J. Food Eng.* **90** 27
[26] Kondo N. 2009 *Sens. Instrum. Food Qual. Saf.* **3** 81
[27] Jin H, Li L and Cheng J 2015 *J. Food Anal. Methods* **8** 25242
[28] Williams E J and Drexler J S 1981 *Peanut Science* **8** 134
[29] Makky M, Santosa, Putri R E and Nakano K 2017 *AIP Conf. Proc.* **1813** 020002
[30] Ali A, Wani T A, Wani I A, Masoodi F A 2016 *Journal of the Saudi Society of Agricultural Sciences* **15** 75Nehete J Y, Bhambar R S, Narkhede M R and Gawali S R 2013 *Pharmacogn Rev.* **7** 107
[31] Shih F F and Daigle K W 2000 *J. Am. Oil Chem. Soc.* **77** 885
[32] Heman A and Hsieh C-L 2016 *Engineering in Agriculture, Enviro nment and Food* **9** 280–290
[33] Duarte F J 2016 *Tunable Laser Applications* 3rd ed. (New York: CRC Press)
[34] Pope R M and Fry E S 1997 *Appl. Opt.* **36** 8710
[35] Kou L, Labrie D and Chylek P 1993 *Appl. Opt.* **32** 3531
[36] Jonasz M 2007 Absorption coefficient of water: data sources http://www.tpdsci.com/Tpc/AbsCfOfWaterDat.php
[37] Wozniak B and Dera J 2007 *Light Absorption in Sea Water. Atmospheric and Oceanographic Sciences Library* (Ney York: Springer-Verlag)
[38] Kaiser H F 1958 *Psychometrika* **23** 187
[39] Natsuga M and Kawamura S 2006 *Trans. ASABE* **49** 1069
[40] Chang R K, Zhang W Y, Cui J, Wang Y H, Wei Y, and Liu Y 2009 *Computer and Computing Technologies in Agriculture* **III** 523-531
[41] Layne E 1957 Spectrophotometric and turbidimetric methods for measuring proteins *Methods in Enzymology* 3 eds P S Colowick and N O Kaplan (New York: Acadmic Press, Inc.) pp. 447-455
[42] Simonian M H and Smith J A 2006 Spectrophotometric and colorimetric determination of protein concentration *Current Protocols in Molecular Biology* (New York: John Wiley & Sons, Inc.) 76:I:10.1A:10.1.1
[43] McClure W F 2006 Introduction *Near-Infrared Spectroscopy in Food Science and Technology* eds Y Ozaki, W F McClure and A A Christy (New Jersey: John Wiley & Sons, Inc.)

S-velocity structure in Cimandiri fault zone derived from neighbourhood inversion of teleseismic receiver functions

Syuhada [1], T Anggono[1], F Febriani[1] and M Ramdhan[2]

[1]Research Centre for Physics - Indonesian Institute of Sciences (LIPI), Tangerang, Indonesia
[2]Indonesian Agency for Meteorology Climatology and Geophysics (BMKG), Jakarta, Indonesia

E-mail: hadda9@gmail.com

Abstract. The availability information about realistic velocity earth model in the fault zone is crucial in order to quantify seismic hazard analysis, such as ground motion modelling, determination of earthquake locations and focal mechanism. In this report, we use teleseismic receiver function to invert the S-velocity model beneath a seismic station located in the Cimandiri fault zone using neighbourhood algorithm inversion method. The result suggests the crustal thickness beneath the station is about 32-38 km. Furthermore, low velocity layers with high Vp/Vs exists in the lower crust, which may indicate the presence of hot material ascending from the subducted slab.

1. Introduction

The study area is located in the Cimandiri fault zone. The fault is located in the area where there is high population cities and infrastructure exposed to the seismic hazard (figure 1). The Cimandiri fault has produced some moderate and shallow earthquakes causing serious damage on the surrounding area such as the Rajamandala earthquake in 1910, the Tanjungsari earthquake in 1972 and the Sukabumi earthquake in 2001 [1, 2]. Thus, knowing realistic earth models in the area located near this fault zone is an essential tool for assessing the seismic hazards in this area. However, the information about velocity crustal structure in this region is limited and mostly focused on the shallow structure derived from surface geological surveys (see e.g. [1, 2]) and geophysical investigations using audio-frequency magnetotelluric and gravity method (see e.g. [3, 4]). In recent years extensive receiver functions studies [5, 6] have been conducted to identify the structure and the geometry of the fault. In addition to these studies and to give better constraint about crustal structure information around this area, we investigate the detailed S-wave velocity structure by applying further analysis on receiver functions of teleseismic events recorded at a single three-component seismic station near the Cimandiri fault zone.

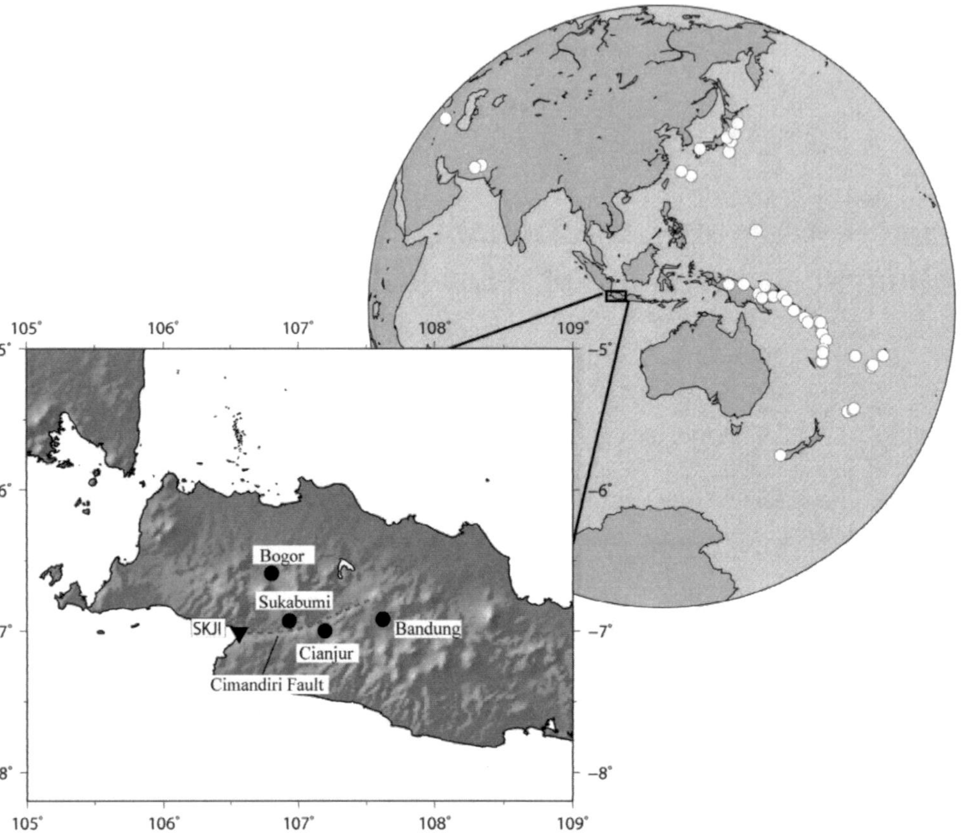

Figure 1. Map showing the location of the study area. The globe on the right figure shows the distribution of teleseismic events used in this study (white circles). A black box in the globe marks the study area enlarged on the left figure.

2. Data and method

In this study, we investigate S-wave velocity structure using earthquake data recorded in the period from 2008 to 2015 by single station located near the Cimandiri fault zone [5, 6]. This seismic station is part of the IA-Network operated by BMKG and GFZ-Postdam. For computing receiver functions, we use over 49 seismograms with clear P-arrivals from earthquakes having magnitude larger than 6 and epicentre distance between 30° and 90°. This selection provides seismic records that may have good signal to noise ratios [7]. It is important to note that some of these records may have low signal to noise ratios and thus, are removed from the analysis if the computed receiver functions do not meet the misfit criteria. Before computing receiver functions, the seismograms are inspected manually to choose the best quality records as well as to discard traces without clear P arrivals. The seismograms are initially baseline-corrected by removing the linear trend and eliminating the effect of instrument response. The horizontal components of the seismogram are then rotated into radial and transverse components. The time windows of the seismogram are finally chosen 10 s before and 50 s after the P-wave arrival.

We applied the time domain iterative deconvolution technique [8] to compute the receiver functions. To control the high frequency noise of the receiver function, we perform the Gaussian filter with bandwidths of 1.5. The selected Gaussian width eliminates the frequency contents greater than approximately 0.75 Hz. This value is chosen by trial and error, considering the balance between the detailed waveform of receiver function and the effect of the high frequency noise to the receiver function. The quality of computed receiver function are then evaluated using a least-square misfit

IOP Conf. Series: Journal of Physics: Conf. Series **985** (2018) 012013 doi:10.1088/1742-6596/985/1/012013

criterion, which calculates the difference between the observed radial component and the convolution of the calculated receiver function with the observed vertical component. In this study, we only use radial receiver functions that achieve at least a 90% fit for further analysis. Once determined, the selected receiver functions are then stacked according to distance and back azimuth of the events and the waveform similarities to increase the signal to noise ratio ,as well as to assure that the individual receiver function are passing through similar structure [7]. The staking process resulted in four receiver function stacks for this seismic station as shown by figure 2.

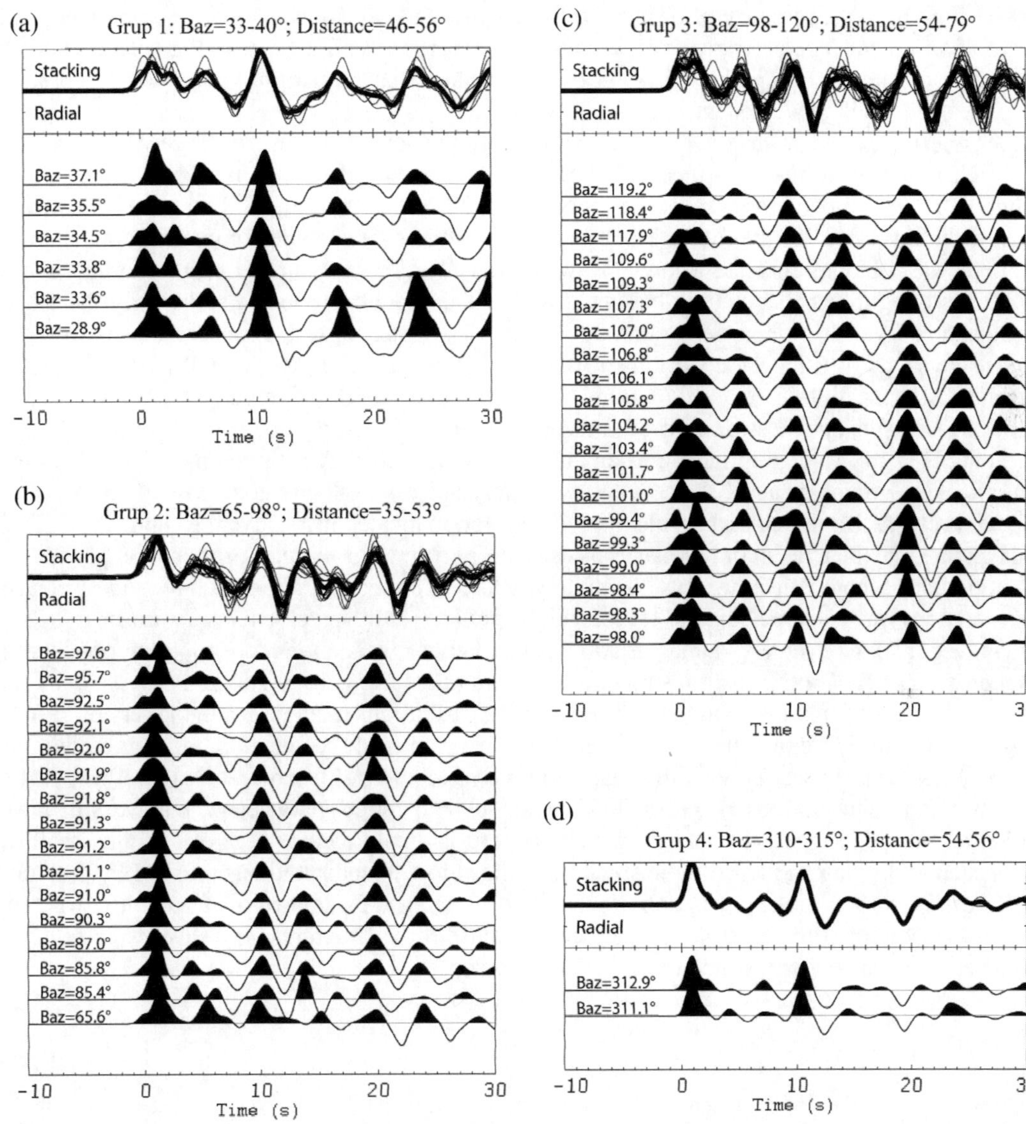

Figure 2. Plot of radial receiver functions at station SKJI arranged by group: (a) Group 1, (b) Group 2, (c) Group 3 and (d) Group 4. The top trace in each panel displays the stacked radial receiver functions.

To derive an S wave velocity profile beneath the seismic station, we apply the non-linear neighbourhood algorithm (NA) [9, 10] based on random choice of model samples to find the best model using the misfits of the previous models. We divide the crustal structure into 6 layers (sediment, basement, upper crust, middle crust, lower crust and upper mantle) consisting of 4 parameters in each layer. These parameters describe the layer thickness, Vp/Vs, the S wave velocity at the top and bottom of each layer. After a number of trials, the tuneable parameters are set for each inversion involving 5000 iterations, producing 250050 velocity models.

3. Result and discussion

All four stacks for this station (figure 2) are characterized by relatively complicated waveform in the first arrivals indicating the presence of shallow complex structure [5, 6], then followed by P-s conversion wave arrival at around 5.5-6 seconds. A large amplitude reflection is also identified at around 10 s indicating the presence of deep reflector. As the study area located in the southern edge of the Eurasian continent, characterized by the subduction process of Indo-Australian plane beneath Eurasian plate at rate 67 mm/yr [11, 12], thus we speculate that this strong amplitude is possibly related to the reflection in the subducted slab.

In order to estimate the S-wave velocity model beneath the station, we invert the receiver function stack in each group using the non-linear neighbourhood algorithm described by Sambridge [9] and Shibutani [10]. The inversion models for the four groups are shown by figure 3. In general, the model solutions of the inversion for the four stacked receiver functions have some common characteristics. For example, the solutions indicate a low S-wave velocity layer at the surface corresponding to the soft sediments lying directly above the basement rocks. The model solutions also identify the crust-mantle transition layer at a depth of around 32-38 km. This result is in a good agreement with the previous receiver study using H-k stacking method [5]. Syuhada and Anggono [5] found that the crustal thickness beneath this station is around 33.6 km. Other studies also suggest that West Java generally have crustal thickness around 30 km derived from modelling of local and global crustal structure data [13, 14], reflecting the Eurasian continental character [15]. Furthermore, we also note that some stacking groups (group 1, 2 and 4) show slightly negative velocity gradient with high Vp/Vs in the lower crust. As discussed earlier that the seismic station located just above the subduction zone, thus, this feature may be a result of hot materials ascending from the slab during subduction process.

It is important to note that in the case of the raypaths sampling the same structure, some stacked receiver functions on the same seismic station might produce similar velocity structure in the final models (see e.g. [7]). Conversely, some variation in the output inversion models might be obtained if there is a prominent variation in crustal structure with azimuth (see e.g. [6]). This can only be detected by using receiver function with good enough azimuthal distribution. In all cases, the inversion results of this study produced reasonable fitting between observed and synthetic model, although the amplitude of the first arrivals is not fully matched. We have tried to improve this waveform fit by running inversion using various range of incident angles and model parameters, but we still have unsatisfied results. We suggest that this poor waveform fit is due to the presence of complex shallow structures such as dipping and anisotropic structures related to Cimandiri fault as inferred by previous studies [5, 6]. Therefore, further inversion process incorporating the parameters of complex structure (e.g. percent anisotropy, dip and strike of dipping and anisotropic structures) is necessary in order to provide a better crustal velocity structure beneath this station.

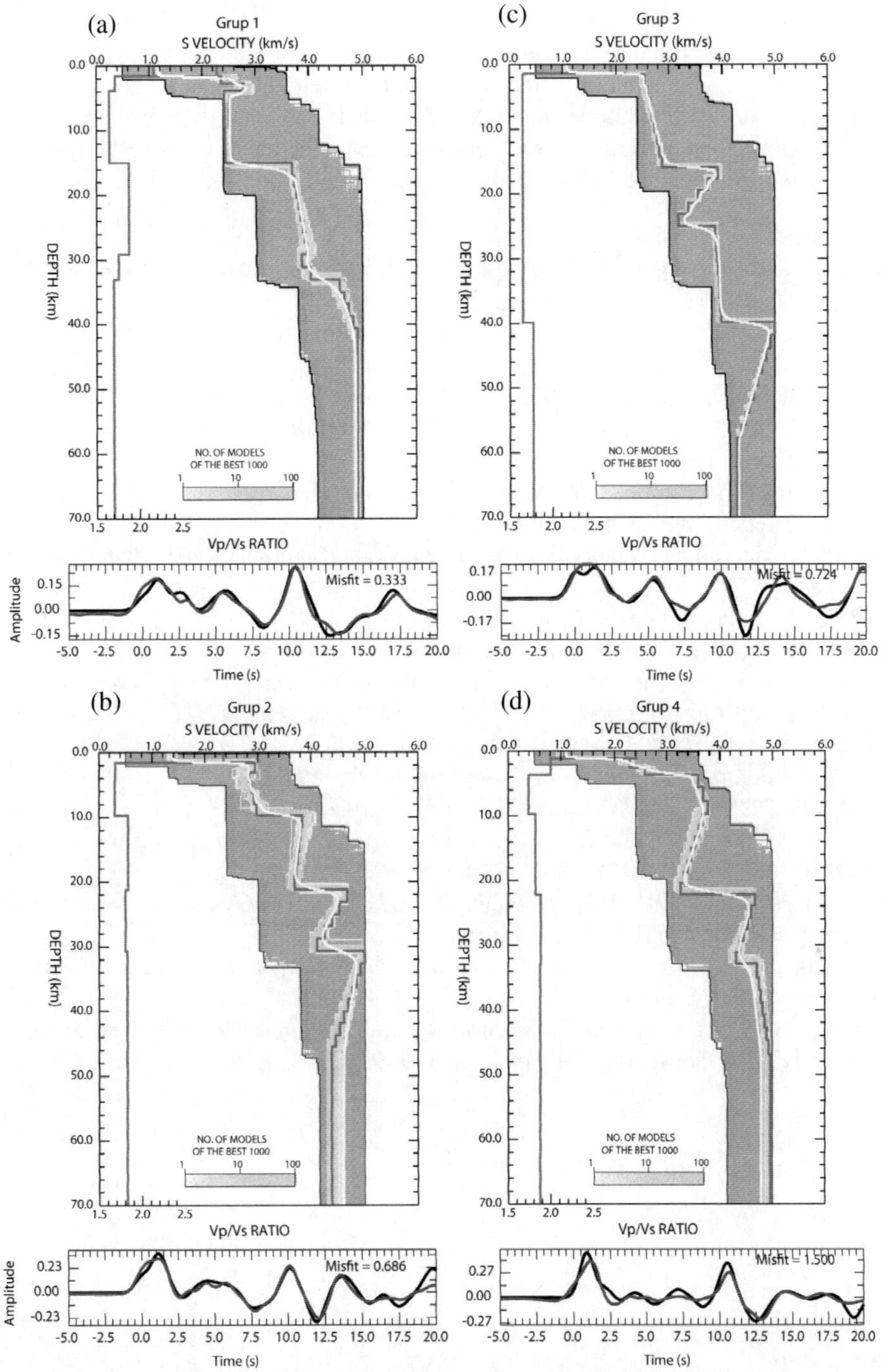

Figure 3. Plot of NA-inversion results obtained for station SKJI arranged by group: (a) Group 1, (b) Group 2, (c) Group 3 and (d) Group 4. The solid red and white lines represent the best and the average fitting model, respectively. The grey shaded area are all models (250050 models) searched to find the best model. The red line in the left panel is the best fitting Vp/Vs ratio model. The bottom trace in each panel displays the predicted and observed staked receiver functions indicated by blue and black lines, respectively.

4. Conclusions

We investigate the velocity structure beneath a single seismic station located near the Cimandiri fault zone by inverting the receiver function using the non-linear neighbourhood algorithm. The obtained velocity profiles indicate a crust-mantle transition at 32-38 km depth. Another main feature in these crustal velocity models is the presence of low velocity layer in the lower crust with high Vp/Vs which can be interpreted as ascending hot materials from the subducted slab.

Acknowledgments

The authors are grateful to Indonesian Agency for Meteorology, Climatology, and Geophysics (BMKG) and GFZ Potsdam for their seismogram data.

References

[1] Dardji N, Villemni T and Rampnoux J P 1994 *J. Southeast Asian Earth Sci.* **9** 3
[2] Martodjojo S 2003 *The evolution of Bogor, West Java basin* Doctoral Thesis (Bandung: ITB Bandung Publishing)
[3] Febriani F 2014 *Subsurface structure of the Cimandiri fault zone, West Java, Indonesia* Doctoral Thesis (Chiba: Chiba University)
[4] Handayani L, Maryati M, Kamtono K, Mukti M M and Sudrajat Y 2017 *Indonesian Journal on Geoscience* **4** 39
[5] Syuhada and Anggono T 2016 *Proc. Int. Sym. on Frontier of Applied Physics* (Bandung) vol 1711 (New York: AIP Publishing) p 070001
[6] Syuhada and Titi Anggono 2017 *Proc. Int. Sym. on Frontier of Applied Physics* (Serpong) Series 817 (Bristol: IOP Conf. Series: Journal of Physics) 012074-1
[7] Macpherson K A, Hidayat D and Goh S H 2012 *J. Asian Earth Sci.* **46** 161
[8] Ligorría J P and Ammon C J 1999 *Bull. Seism. Soc. Am.* **89** 1395
[9] Sambridge M 1999 *Geophysical Journal International* **138** 479
[10] Shibutani T, Sambridge M and Kennett B 1996 *Geophysical Research Letter* **23** 1829
[11] Hamilton W 1988 *Geological Society of America Bulletin* **100** 1503
[12] Kopp H, Klaeschen D, Flueh E R and Bialas J 2002 *Journal of Geophysical Research* **107** 2034
[13] Nishimura S and Harjono H 1992 *Geojournal* **28** 87
[14] Heine C 2007 *Formation and evolution of intracontinental basins* Doctoral Thesis (Sydney: The University of Sydney)
[15] Hamilton W B 1979 *Tectonics of Indonesian Region* (Virginia: US Geological Survey)
[16] Midzi V and Ottemöller L 2001 *Tectonophysics* **339** 443

Investigation of sediment thickness effect to the receiver function through forward modelling

T Anggono[1], Syuhada[1], F Febriani[1], B Soedjatmiko[1] and A Amran[2]

[1] Research Center for Physics - LIPI, Kawasan PUSPIPTEK Serpong, Indonesia
[2] Indonesian Agency for Meteorology, Climatology and Geophysics, Jl. Angkasa I, No. 2, Kemayoran, Jakarta

E-mail: titi.anggono@lipi.go.id

Abstract. In this study, we investigated sediment thickness effect to the receiver function waveform. We conducted forward modelling of receiver function using different thickness of low velocity sediment layers. We varied low velocity layer from 0 - 4 km thickness. We applied Gaussian width parameter of 1.5 and 2.5, which corresponds to low pass filter with a corner frequency of 0.75 and 1.25 Hz, respectively. In this numerical analysis, we applied halfspace model with low velocity layer near the surface representing sedimentary layer. We generated synthetic receiver function using Thomson-Haskell matrix method. From our modelling, changes occur in the first few seconds of calculated synthetic receiver functions. Such changes occur due to reverberation in the sedimentary layer near the surface. Such changes differ with different sedimentary thickness used in the model. We applied *H-k* stacking method to the calculated synthetic receiver functions to see the effect of sedimentary layers to the results of *H-k* stacking.

1. Introduction

Receiver function is one of a few methods that can be used to characterize detailed structure of the Earth's crust and upper mantle. Receiver function may also obtain information about discontinuities that occur in the crust and the upper mantle. The method uses teleseismic event waveforms from a record of three-component seismic station. The recorded teleseismic waveforms may contain information of source time function, propagation effect between source and receiver and structure beneath a seismic station or receiver. Receiver function is obtained by removing the effect and source time function and propagation effects.

Assumption used in this method is that incident of P wave signals from teleseismic events to a discontinuity in the crust or upper mantle will result in the converted S waves (Ps). This converted phase (Ps) will arrive at a station after a direct P phase. The resulted Ps converted phase is sensitive to the shear velocity contrast beneath a station. Several studies have been carried out to estimate shear wave velocity model from inversion of calculated receiver functions [1-3]. However, a unique inversion results will be difficult to obtain. Hence, it is necessary to minimize the difference between observed and synthetic receiver functions using a priori information available in the target area. The converted phases Ps and crustal multiples PpPs and PpSs+PsPs may contain information on the average crustal properties such Moho depth and Vp/Vs ratio [4]. The delay time of the direct Ps conversion from Moho and crustal multiples can be used to estimate the crustal thickness and Vp/Vs ratio [3-5].

Distribution of stations may be beneficial to estimate structure using receiver functions. However, if the site is filled by sediments, the sediment-bedrock interface may become a velocity discontinuity. Hence, the calculated receiver functions can be dominantly affected by the sedimentary structure in the first few seconds after the direct P arrival [6]. From these effect, information from the deeper structure might be hidden by the multiples or phases from the near surface discontinuity. Several studies have tried to estimate the sediment thickness from the inversion of the receiver functions, such as in New Madrid Seismic Zone and Bohai Bay Basin in Eastern China [7, 8]. Bannister et al. [9] observed complex receiver functions in the first 3 s in the radial component from the analysis of receiver functions Antarctica [9]. They suggested that the complex shape of the waveform might be the effect of reverberation in inside sedimentary layers. Several studies also observed similar characteristics of complex receiver functions above sallow sedimentary basins [1, 10]. Strong sediment reverberations offer significant challenges to the analysis based on calculated receiver functions, such as H-k stacking analysis [4] and common inversion point (CCP) stacking method [11, 12]. In this study, we examined effect of sediment thickness to the waveform shape of receiver functions. We carried out forward modelling of receiver functions by changing sediment thickness and its influences to waveform shape of the receiver functions.

2. Methodology

To examine the effect of a sedimentary layer, we applied forward modeling to calculate synthetic receiver functions. We calculated several synthetic receiver functions of radial component using one dimensional velocity model involving sedimentary layers (figure 1). Homogeneous medium consist of crust and mantle with Moho depth at 40 km was used as a model. Sedimentary layer thickness was varied from 0 km to 4.0 km thick at the top of the crust. To calculate synthetic receiver functions, we used the following parameters: for the sedimentary layers, V_p, V_s, and density are 1.8 km/s, 0.6 km/s, 1850 kg/m^3, respectively; for the crustal layer, the values are 6.1 km/s, 3.52 km/s, and 2730 kg/m^3 with Moho depth at 40 km; and for the mantle, the values are 8.15 km/s, 4.7 km/s and 3364 kg/m^3. Gaussian width parameters were set at 1.5 and 2.5, which corresponds to low pas filter with a corner frequency of 0.75 and 1.25 Hz. Synthetic receiver functions were computed using the program hrftn96 implemented in Computer Programs in Seismology [13]. Receiver function computation in the code hrftn96 is based on the Thomson-Haskell matrix method.

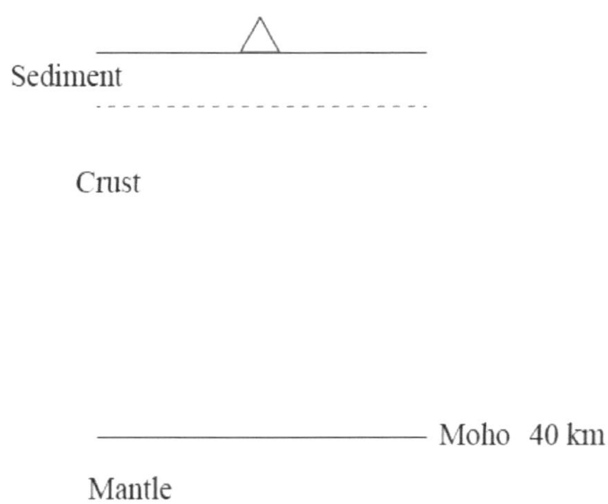

Figure 1. Schematic diagram of one dimensional structure to calculate synthetic receiver functions.

IOP Conf. Series: Journal of Physics: Conf. Series **985** (2018) 012014 doi:10.1088/1742-6596/985/1/012014

In the propagator matrix method, heterogeneous medium is represented by a stack of homogeneous layers overlying a homogeneous half-space. Velocity and stress are introduced as motion-stress vectors. v_r and v_z represent the velocity in the radial and vertical directions, τ_{zr} and τ_{zz} are shear and normal stress on a horizontal surface. Elastic wavefield at each layer can be divided into four types of waves, upgoing *P*, downgoing *P*, upgoing *S* and downgoing *S*. In the calculation of receiver function, teleseismic *P* wave can be approximated as *P* wave coming at a bottom of a vertical heterogeneous structure beneath a receiver. A 1-D velocity structure can be parameterized by a stack of homogeneous layers above a half space layer. Detailed explanation of Thomson-Haskell matrix method can be found in Aki and Richards [14].

3. Results and discussions

Figure 2 shows synthetic radial receiver functions with different sediment thickness with Gaussian parameter of 1.5 and 2.5. The synthetic receiver functions show direct P phase at 0 s, Ps phase at 5 s and PpPs at about 17 s for sediment thickness of 0 km. With sediment thickness of 0.5 km, large amplitude appears after direct P phase that obstruct the arrival Ps phase. With sediment thickness of 1 km thick, the synthetic receiver function becomes more complex and the pulse appears a slight delay. As we increase the sediment thickness, the effect of sedimentary layers on the receiver functions becomes more apparent. We observe strong pulse at around 0 - 10 s that cause the first few seconds of receiver functions become more complex. From the synthetic receiver functions, the conversion phase and reverberations due to the presence of sedimentary layers may obstruct the conversion phases, such as Ps, PpPs and PpSs+PsPs, from the Moho discontinuity.

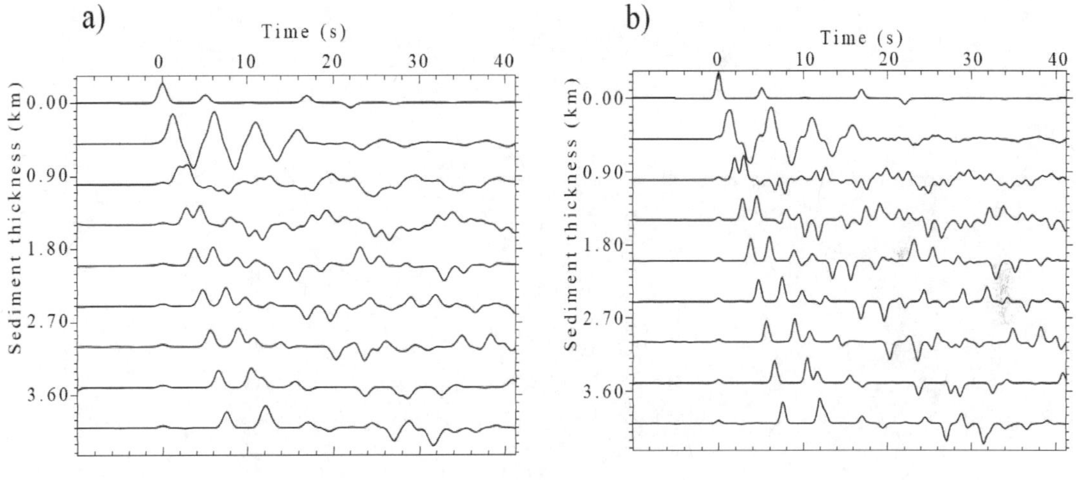

Figure 2. Synthetic receiver functions with different thickness of sedimentary layers. Synthetic receiver functions with Gaussian width parameter of a) 1.5 and b) 2.5.

H-k stacking analysis is very powerful tool to estimate crustal thickness and Vp/Vs ratio [4]. In this method, we stack the calculated receiver functions by summation of Ps conversion phase and its reverberation such as PpPs and PpSS + PsPS. We calculated synthetic receiver function using 1 km-thick of sedimentary layer to see how the sediment thickness might affect the result of *H-k* stacking analysis. We set the Moho depth at 40 km. Synthetic receiver functions with 0 km and 1 km thick of sediment layer with the results obtained from H-k stacking analysis are shown in figure 3. Figure 3a shows synthetic radial receiver functions no sedimentary layer. For the calculation of H-k stacking analysis, we applied weighting parameters of 0.7 for Ps phase, 0.2 for PpPs and 0.1 for PpPs + PsPs multiples. We observed that although the Moho depth is obtained at 40.1 km, which is small different from our model of 40 km depth. We obtained same value for the Vp/Vs ratio, which is 1.73. The slight difference of Moho depth from the H-k stacking analysis might be due to the coverage of ray parameter. Stacking process, which is a summation of the conversion phases, may result in slight different shape of the stacked phases and the slight changes in the arrival time. The differences of stacked phases and its individual receiver functions can be seen in right side of figure 2a. Figure 2b shows H-k stacking analysis from the synthetic radial receiver function with 1 km-thick sediment layer. The search grid for the synthetic receiver functions is set to be same with the H-k analysis of sedimentary layer model of synthetic receiver functions. We obtained that the Moho depth is about 20.0 km, which is significantly different with our model of 40 km. The obtained Vp/Vs ratio is 1.60. It is clear from the analysis that sediment layers may affect the result from the H-k stacking analysis.

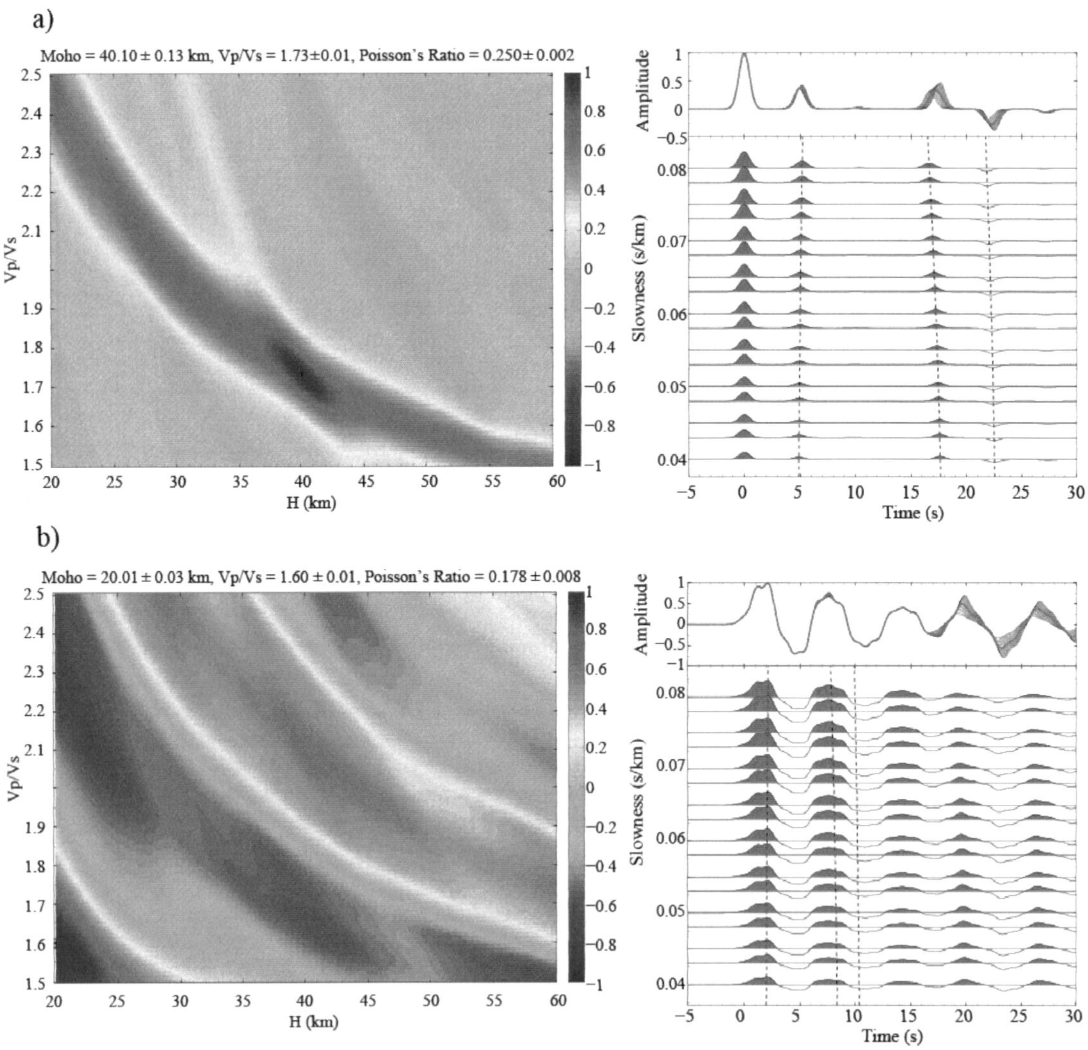

Figure 3. H-k stacking analysis from synthetic receiver functions by considering sediment layers of a) 0 km thick and b) 1 km thick. Left panel shows the result of H-k stacking analysis and right panel shows synthetic radial receiver functions with Gaussian width parameter of 1.5. Dashed lines show estimated arrival time of Ps, PpPs and PpSs+PsPs conversion phases, respectively.

Reverberations in the observed receiver function due the sedimentary layers are obvious in the first few seconds of receiver functions that cause the waveform is complex. However, it is suggested that reverberation might not occur if sedimentary layer is thinner than about one fourth of the Ps phase wavelength [15]. If the sedimentary layer is too thick, the reverberation decays. The first large amplitude, which might be misinterpreted as direct P phase, also shows slight delay. From the forward modelling of receiver functions by taking consideration of sedimentary layers, we suggest that cautions should be taken when calculating H-k stacking from receiver functions. Figure 4 shows examples of calculated radial receiver functions from two seismic station of IA network. We interpreted the observed receiver functions as examples of receiver functions influenced or affected by sedimentary layer at station JCJI (figure 4a) and little or not affected by sedimentary layer at station LEM (figure 4b). We also calculated the H-k stacking analysis for the observed receiver functions. We also compare the theoretical arrival time of converted phases from the results of H-k stacking analysis. At station JCJI we observed that the estimated arrival time of converted Ps from the result of H-k stacking is close to the direct P. However, at the station LEM the estimated arrival time of converted Ps is close to the Ps phase we have expected, which is ~5 s.

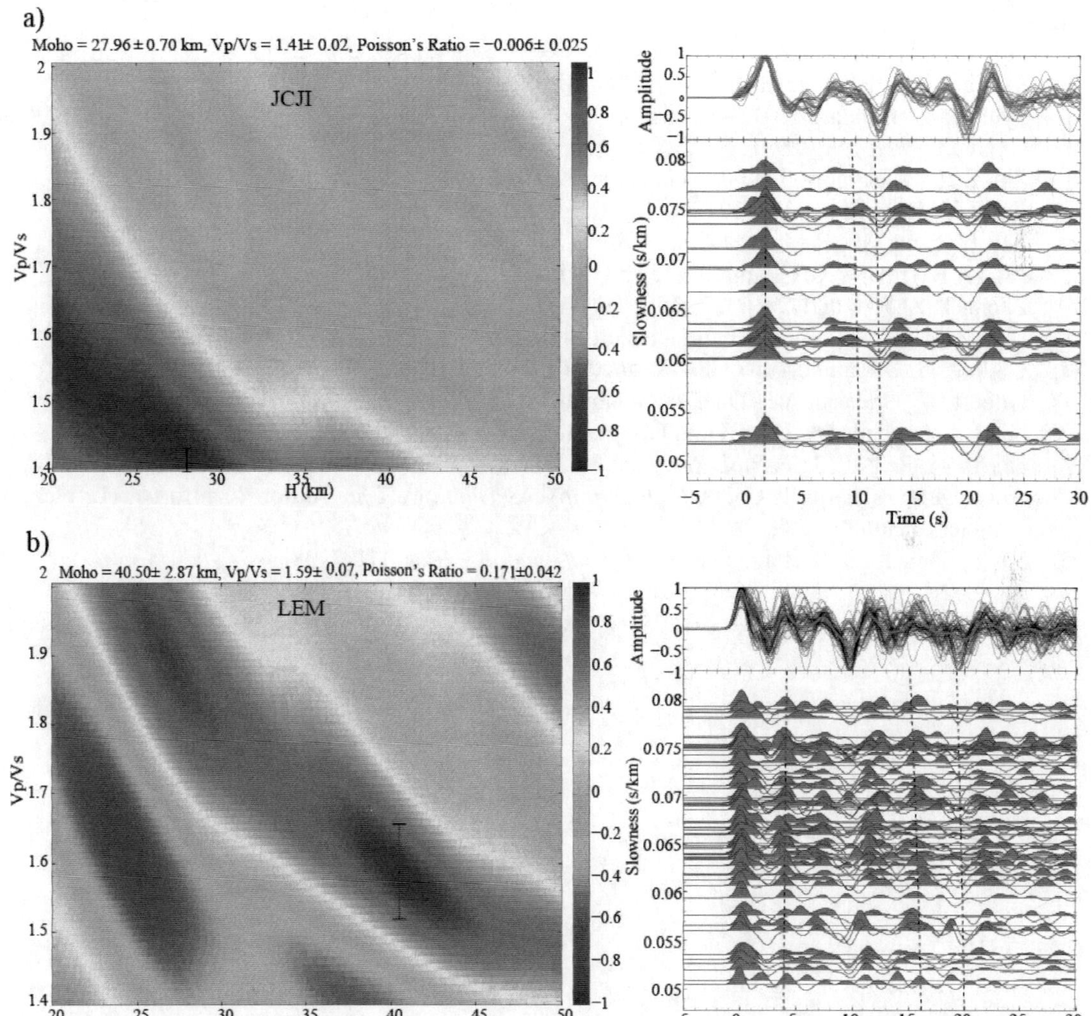

Figure 4. Examples of observed receiver functions from two seismic stations of IA network using Gaussian width parameter of 1.5. Result of H-k stacking analysis at stations a) JCJI and b) LEM.

4. Conclusions

We have calculated synthetic receiver functions with different thickness of sedimentary layers from 0 - 4 km thick with different Gaussian width parameter. We observed different shape of waveforms in the first few seconds of receiver functions with different thickness of sedimentary layer. With increasing sedimentary layer thickness, reverberation seems to decrease. We also applied H-k stacking analysis to the synthetic receiver functions, where we found that the H-k stacking results is influenced by the sedimentary structure. Cautions must be taken when analysing the receiver functions from stations near sedimentary basins or with significant reverberation in the observed receiver functions.

Acknowledgments

This study was supported by LIPI. Seismic data were obtained from the GEOFON data centre of the GFZ German Research Centre for Geosciences.

References

[1] Shibutani T, Sambridge M and Kennett B 1996 *Geophys. Res.Lett.* **23** 1829
[2] Sambridge M 1999 *Geophys. J. Int.* **138** 727
[3] Syuhada S, Hananto N D, Abdullah C I, Puspito N T, Anggono T and Yudistira T 2016 *Acta Geophysica* **64** 2020
[4] Zhu L and Kanamori H 2000 *J. Geophys. Res.* **105** 2969
[5] Eagar K C and Fouch M J 2012 *Seismol. Res. Lett.* **83** 596
[6] Zelt B C and Ellis R M 1999 *Can. J. Earth Sci.* **36** 585
[7] Julia J, Herrmann R B, Ammon C J and Akinci A 2004 *Bull. Seismol. Soc. Am.* **94** 334
[8] Zheng T, Zhao L and Chen L 2005 *Phys. Earth Planet. Inter.* **152** 129
[9] Bannister S, Yu J, Leitner B and Kennett B L N 2003 *Geophys. J. Int.* **155** 870
[10] Clithore G, Gudmundsson O and Kennett B L N 2000 *Aust. J. Earth Sci.* **47** 209
[11] Gilbert H J, Sheehan A F, Dueker K G and Molnar P 2003 *J. Geophys. Res.* **108** 2229.
[12] Niu F L, Levander A, Ham S and Obayashi M 2005 *Earth Planet. Sci. Lett.* **239** 9
[13] Herrmann R B 2013 *Seismol. Res. Lett.* **84** 1081
[14] Aki A and Richards P G 2002 *Quantitative Seismology* 2nd Edition (California: University Science Book) p 269
[15] Yu Y, Song J, Liu K H and Gao S S 2015 *J. Geophys. Res.* **120** 3208

Shear wave velocity model beneath CBJI station West Java, Indonesia from joint inversion of teleseismic receiver functions and surface wave dispersion

R H Simanungkalit[1, 2], T Anggono[2], Syuhada[2], A Amran[3] and Supriyanto[1]

[1]Physics Graduate Program, Natural Sciences and Mathematics Faculty, University of Indonesia
[2]Research Center for Physics, Indonesian Institute of Sciences
[3]Indonesian Agency for Meteorology, Climatology and Geophysics

E-mail: kalit.ronald@gmail.com

Abstract. Earthquake signal observations around the world allow seismologists to obtain the information of internal structure of the Earth especially the Earth's crust. In this study, we used joint inversion of receiver functions and surface wave group velocities to investigate crustal structure beneath CBJI station in West Java, Indonesia. Receiver function were calculated from earthquakes with magnitude more than 5 and at distance 30°-90°. Surface wave group velocities were calculated using frequency time analysis from earthquakes at distance of 30°-40°. We inverted shear wave velocity model beneath the station by conducting joint inversion from receiver functions and surface wave dispersions. We suggest that the crustal thickness beneath CBJI station, West Java, Indonesia is about 35 km.

1. Introduction

Earthquake signal observations around the world allow seismologists to know the internal structure of the earth especially the earth's crust. There are various methods of research that used to investigate the subsurface structure of the earth, such as seismic tomography, surface wave dispersion, and receiver function. In this research, the surface wave dispersion method was combined with receiver function method to investigate the shear wave velocity model beneath CBJI Station West Java, Indonesia. Receiver function method has been widely used to investigate the crustal structure [1-3]. Receiver functions are time series, computed from three-component seismograms that shows the response of earth structure beneath a station. Receiver functions primarily are sensitive to shear wave velocity contrasts and vertical travel time. Teleseismic wave contains information about the seismic source structure, the propagation through the mantle, and the local structure beneath the seismic station [4].

Several studies have used surface wave dispersion to model Earth's velocity layer structure [5-7] because Shear wave propagation analysis can explain the difference of medium in the structure of the earth's crust. By conducting joint inversion of receiver function and surface wave dispersion, produces a more tightly constrained crustal structure model than using receiver function or surface wave dispersion separately [8]. Research on the crust and upper mantle with joint inversion of the receiver functions and surface wave dispersion methods has been widely practiced in many countries such as Du et.al in Iceland [9], Chang et.al in Korea [10], and Tkalcic et.al in the Arabian Peninsula [11].

Java Island is located on the southeast edge of Sundaland (figure 1). Java Island has high tectonic activity because lies above subduction zone between Indo-Australian and Eurasian Plate. Although Java Island has high tectonic activity, it is still very few research in crustal and upper mantle in this region. Hidayat [12] investigated velocity structure beneath Tangkuban Parahu West Java from receiver function analysis. Syuhada and Anggono [13] used teleseiscmic receiver function to investigate crustal structure in the Southern part of West Java. In this preliminary study, we carried out joint inversion of teleseismic receiver function and surface wave dispersion to obtain information about the shear wave velocity model beneath one IA-network station in West Java, Indonesia.

2. Methodology

We analyzed surface wave and receiver functions from teleseismic recordings in earthquake recorded at three-component broadband seismometer of CBJI station (figure 1). We choose earthquakes with magnitude more than 5 with distance between 30°-40° and has clear arrival of surface waves. Based on categories above, there are 5 teleseismic earthquakes that used in surface waves analysis. For receiver function analysis, there are 27 events that used in this study. The distance of teleseismic events 30°-90° and magnitude more than 5. Figure 1 shows the location of CBJI station and the location of the telesesimic earthquakes.

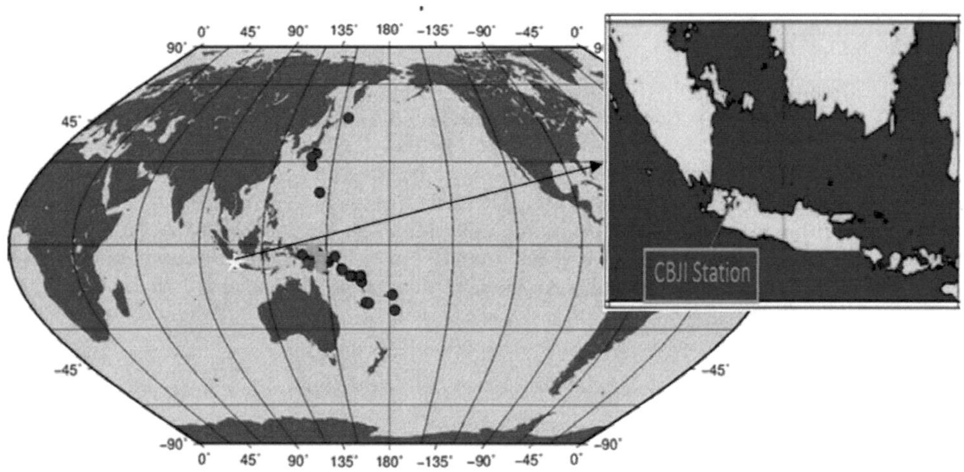

Figure 1. Distribution of earthquakes used in the receiver functions (solid blue) and surface wave (green). Location of CBJI station is also shown in the map (solid yellow triangle).

Rayleigh wave dispersion were analyzed from vertical seismogram's component. Figure 2 shows an example of one earthquake seismograms recorded in CBJI station. In this study, we analyzed group velocity of fundamental mode Rayleigh wave with period from 10-100 s with Gaussian filter parameter (α) 50. We used multiple filter technique to analyze group velocities from recorded seismograms. Multiple filter technique can separates modes of surface waves when arriving at the station. Frequency time analysis (FTAN) diagram is produced from repeating through different periods. A signal is processed by FTAN method to extract the group velocity dispersion curve of the rayleigh fundamental mode. The frequency time analysis (FTAN) method is used to improve in a significant way the multiple filter analysis which Dziewonski et.al developed [14]. This method employs a system of narrow-band Gaussian filters, with varying central frequency, that does not introduce phase distortion and give a good resolution in the time-frequency domain. A FTAN map is the image of a matrix whose columns are the energy values at a certain period and the rows are the energy values at constant group velocity. A sequence of frequency filters and time window is applied to the dispersion curve for an easy extraction of the fundamental mode. The floating filtering technique, combined to a phase equalization, permits to isolate the fundamental mode from the higher modes [15]. Figure 2b shows the frequency time analysis diagram of Rayleigh wave for one of earthquake event recorded at CBJI station.

We calculated the receiver function with iterative deconvolution technique. We used Gaussian Filter parameter alpha 1.5. We used Computer Programs in Seismology 330 software package *(joint96)* in the process of joint inversion. For inversion processing, we used one dimensional (1-D) ak-135 as a starting model.

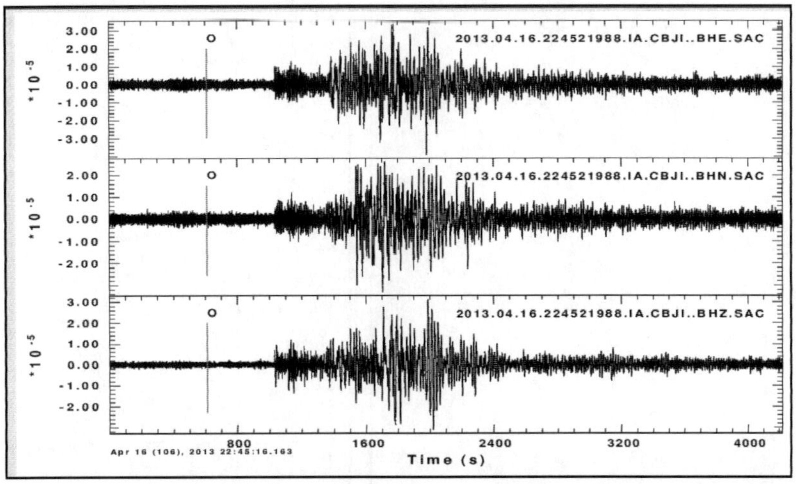

(a)

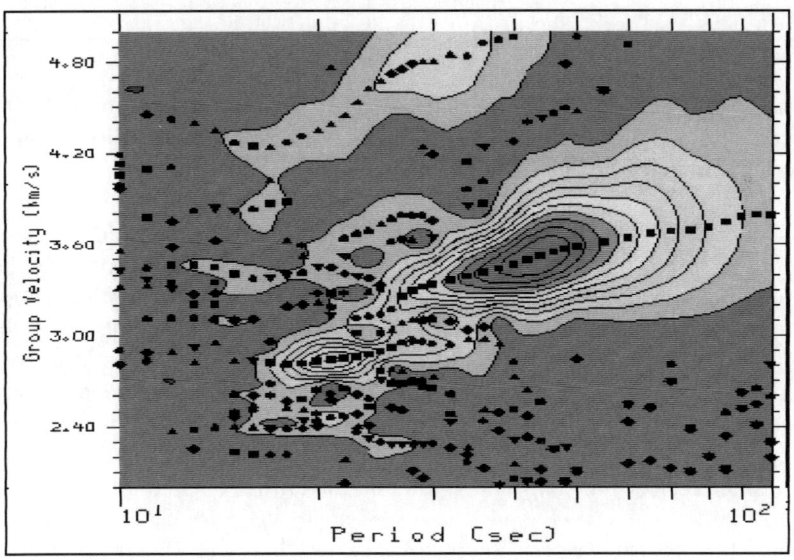

(b)

Figure 2. (a) Seismogram records at CBJI station Seismogram sample that recorded in CBJI Station, 2013/04/16 Earthquake. (b) The frequency time analysis diagram showing the dispersion of rayleigh wave for 2013/04/16 earthquake recorded at CBJI Station.

3. Results and discussions

Measured group velocities of fundamental mode Rayleigh waves and calculated receiver functions are shown in figure 3. Joint inversion of surface wave dispersion and receiver functions was carried out to obtain shear wave velocity model beneath CBJI station. We combined the Rayleigh wave group velocity with receiver functions to determine the subsurface structure. The group velocities of fundamental mode of Rayleigh wave are shown in figure 3a. A total of about twenty seven receiver functions were used for the joint inversion. Figure 3b shows the observed radial components of receiver functions in CBJI station used in this study. Number of iterations is set to be 100.

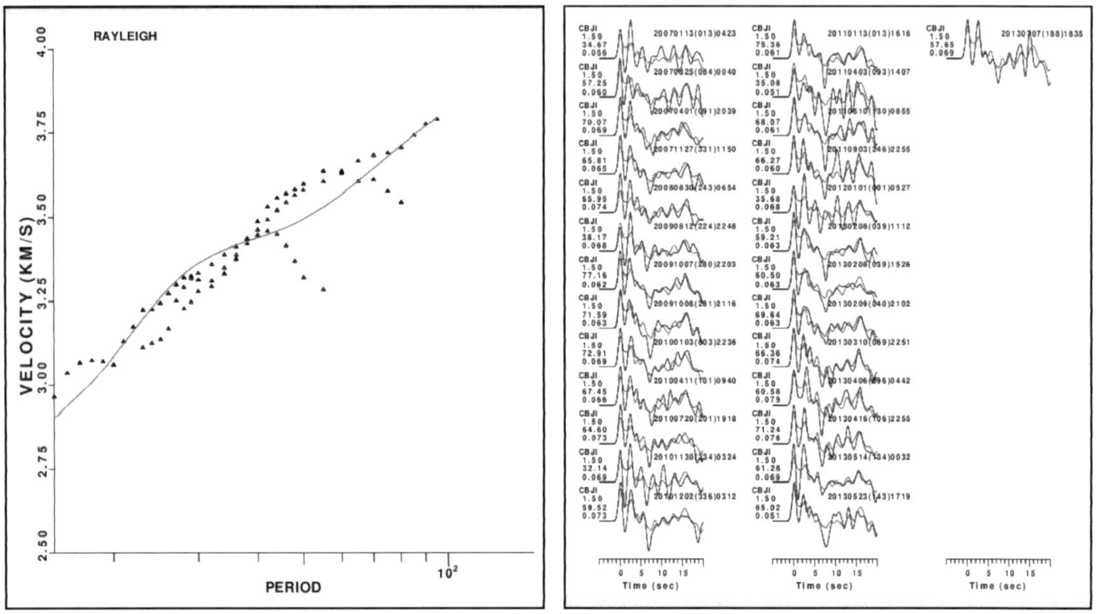

(a) (b)

Figure 3. (a) Measured group velocity of Rayleigh waves at CBJI Station. (b) Calculated radial receiver function at CBJI station. Solid red lines represent theoretical dispersion curve and radial receiver functions from the inverted shear wave velocity model.

Figure 4 shows shear velocity model from joint inversion of teleseismic receiver function and surface wave dispersion. The shear velocity in near surface of CBJI station ~2.5 km/s, which may corresponds to the sediment layer close to the surface. The shear velocity increase up to ~4.1 km/s at the depth ~19 km. We observed shear velocity decrease to about ~ 3.5 km/s at depth of ~26 km and increase to ~4.0 km/s at the depth of ~30 km. The shear velocity increases sharply at depth ~35 km from ~3.8 km/s to 4.25 km/s. From this condition, we suggested that Moho depth beneath CBJI station is about ~35 km. Our result is consistent with previous studies in the region. For example, Hidayat et.al [15] suggested crustal thickness beneath Tangkuban Parahu is about 30-35 km. Amukti and Suryono [16] using receiver functions suggested that the crustal thickness in Central Java is about 32 km.

The velocity structure decreases at lower crustal depth has been investigated in collision zone or subduction zone by Caldwell [17] and Bostock [18]. These studies suggest the low velocity occur in lower crustal depth. Langston in [19, 20] based on receiver function studies showed that low velocity zone present in the forearc structure in Cascadia subduction zone. In the recent studies at Cascadia subduction zone, low velocity zone occur is observed at depth to about 45 km [21]. Other studies also shown the presence of low velocity zone at the subduction zone revealed from the studies of receiver function or seismic tomography, such as Honsu [22] and Chile [23]. The presence of low velocity zone might be attributed to the partial melting [24], presence of over pressured fluid [25], eclogitization [26, 27], serpentinite or graphitic zone [28].

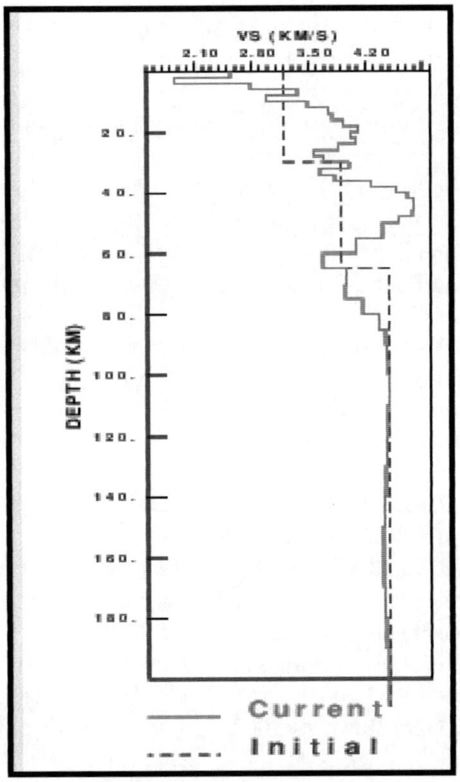

Figure 4. Shear wave velocity model from joint inversion of receiver function and surface wave dispersion.

4. Conclusions

We have carried out the joint inversion of teleseismic receiver function and surface wave dispersion to obtain shear wave velocity model beneath CBJI Station, West Java, Indonesia. The joint inversion of receiver functions and surface wave dispersion data provides constraints on the shear velocity of the propagating medium, which may give better result than inversion of the data separately. The crustal thickness beneath CBJI station, West Java, Indonesia is estimated to be about 35 km.

Acknowledgement

Seismic data were obtained from the GEOFON data centre of the GFZ German Research Centre for Geosciences. I would like to thank Titi Anggono as a supervisor during my stay as research assistant in Indonesian Institute of Sciences.

References

[1] Ammon C J, Randall G E and Zandt G 1990 *J. Geophys. Res.* **95** 15303
[2] Ligorria J and Ammon C 1999 *Bull. Seism. Soc. Am.* **89** 1395
[3] Zhu L and Kanamori H 2000 *J. Geophys. Res.* **105** 2969
[4] Shearer P M 2009 *Introduction to seismology* (Cambridge: Cambridge University Press)
[5] Pasyanos M E 2005 *J. Geophys. Res.* **110** B12301
[6] Mathar J P, Ritter J R R and Friederich W 2006 *Geophys. J. Int.* **164** 377
[7] Kolinsky P 2004 *Acta Geodyn. Geomater.* **1** 165
[8] Julia J, Ammon C J, Herrmann R B and Correig A M 2000 *Geophys. J. Int.* **143** 99
[9] Du Z J and Foulger G R 1999 *Geophys. J. Int.* **139** 419

[10] Chang S-J and Baag C-E 2005 *Bull. Seismol. Soc. Am.* **95** 1516
[11] Tkalcic H, Pasyanos M E, Rodgers J A, Gok R, Walter R W and Al-Amri A A 2006 *J. Geophys. Res.* **111** B11311
[12] Hidayat D, Handayani L, Widiwijayanti C, Suyatno and Sanyoto A 2006 *RISET – Geologi dan Pertambangan* **16** 1
[13] Syuhada and Anggono T 2016 *Proc. Int. Sym. on Frontier of Applied Physics* (Bandung) vol 1711 (New York: AIP Publishing) p 070001
[14] Dziewonski A, Bloch S and Landisman M 1969 *Bull. Seismol. Soc. Am.* **59** 427
[15] Natale M, Nunziata C and Panza G F 2004 *Proc. 13th World Conf. on Earthquake Engineering* (Vancouver) Paper No. 2694
[16] Amukti R and Suryanto W 2013 *Indonesian Journal of Applied Physics* **3** 195
[17] Caldwell W B, Klemperer S L, Rai S S and Lawrence J F 2009 *Tectonophysics* **477** 58
[18] Bostock M G 2013 *Tectonophysics* **609** 547
[19] Langston C A 1977 *Bull. Seismol. Soc. Am.* **67** 713
[20] Langston C A 1981 *J. Geophys. Res.* **86** 3857
[21] Nikulin A, Levin A and Park J 2009 *Geochemistry, Geophysics, Geosystem* **10** Q07004
[22] Kawakatsu H and Watada S 2007 *Science* **316** 1468
[23] Sodoudi F, Yuan X, Asch G and Kind R 2011 *J. Geophys. Res.* **116** B04302
[24] Caldwell W B, Klemperer S L, Rai S S and Lawrence J F 2009 *Tectonophysics* **477** 58
[25] Christensen N I 1984 *Geophys. J. R. Astron. Soc.* **79** 441
[26] Bostock M G, Hyndman R D, Rondenay S and Peacock S M 2002 *Nature* **417** 536
[27] Hyndman R D and Peacock S M 2003 *Earth Planet. Sci. Lett.* **212** 417
[28] Arora B R, Unsworth M J and Rawat G 2007 *Geophys. Res. Lett.* **34** L04307

IOP Conf. Series: Journal of Physics: Conf. Series **985** (2018) 012016 doi:10.1088/1742-6596/985/1/012016

Dimensionality and regional strike analyses of the Cibeber segment, Cimandiri fault zone, West Jawa, Indonesia

F Febriani[1], L Handayani[2], A Setyani[3], T Anggono[1], Syuhada[1] and B Soedjatmiko[1]

[1]Research Center for Physics, Indonesian Institute of Sciences (LIPI), Kawasan Puspiptek Gdg. 440-442, Tangerang Selatan, Banten, Indonesia
[2]Research Center for Geotechnology, Indonesian Institute of Sciences (LIPI), Jl. Sangkuriang Kompleks LIPI Gdg. 70, Bandung, Indonesia
[3]Faculty of Science and Technology, Syarif Hidayatullah State Islamic University, Jl. Ir. H. Juanda No. 95, Ciputat, Tangerang Selatan, Banten, Indonesia

E-mail: febt002@lipi.go.id

Abstract. The dimensionality and regional strike analyses of the Cimandiri Fault, West Java, Indonesia have been investigated. The Cimandiri Fault consists of six segments. They are Loji, Cidadap, Nyalindung, Cibeber, Saguling and Padalarang segments. The magnetotelluric (MT) investigation was done in the Cibeber segment. There were 42 observation points of the magnetotelluric data, which were distributed along 2 lines. The magnetotelluric phase tensor has been applied to determine the dimensionality and regional strike of the Cibeber segment, Cimandiri Fault, West Java. The result of the dimensionality analysis shows that the range values of the skew angle value which indicate the dimensionality of the study area are $-5 \leq \beta \geq 5$. These values indicate if we would like to generate the subsurface model of the Cibeber segment by using the magnetotelluric data, it is safe to assume that the Cibeber segment has the 2-D. While the regional strike analysis presents that the regional strike of the Cibeber segment is about N70-80°E.

1. Introduction

The magnetotellurics (MT) is a passive-surface electromagnetic technique to investigate the electrical resistivity structure from the near surface to the upper mantle [1]. The MT method is useful in fault zone investigations because resistivity in the fault zone is significantly low due to the existence of fluids [2, 3].

There are four distinct morphological and structural units in West Java: (1) the northern coastal plain of West Java, (2) the folded mountains of Bogor, (3) the Bandung zone, which mostly covered by recent volcanic products, and (4) the southern mountain of West Java [4]. The boundary between the Bandung zone and the southern mountain is called the Cimandiri fault zone (from now on, CFZ) the zone is located along the Cimandiri River and is mostly covered by recent volcanic products [4-7]. This area is located eastward of the transitional zone between the frontal subduction of Java and the oblique subduction of Sumatra [7]. The horizontal displacement of the CFZ is about 0.5-1.7 cm/yr on the surface [8].

The Cimandiri fault runs SW-NE, from Pelabuhan Ratu to Padalarang near Bandung and through populated cities such as Pelabuhan Ratu, Sukabumi, and Bandung. The length of the CFZ is about 100 km. The geologic map also shows that the fault zone appears to form a broad deformation zone instead of a narrow zone of fault or faults [9]. The previous study suggested that there are 6 segments of the CFZ [9]. They are Loji, Cidadap, Nyalindung, Cibeber, Saguling, Padalarang segments.

The investigation of the dimensionality and regional strike of the Loji segment has been done [10]. The results presented that the dimensionality of the Loji segment has is 2-D and its regional strike is about N70-80 °E. Since the segments of the CFZ is not continued each other and consist of the numerous micro faults which have the length ranging from a few centimeters to a few meters, therefore it is needed to know the dimensionality and regional strike of the other segments. In this study, we focused to investigate the regional strike and dimensionality of the Cibeber segment based on the magnetotelluric data. This study is the preliminary analysis before we will apply the modeling of the subsurface of the CFZ by using the magnetotelluric data.

2. Data and method

Figure 1b shows the topography map and the location of the 6 segments of the CFZ. The estimated location of the segments based on the recent study [9] is also shown in figure 1b. In this study, the MT investigation was conducted in the Cibeber segment. The MT observation was held in 2015. There were 42 MT observation points which were deployed in the two lines, as presented in figure 1c. The line 1 consisted of the observation points from CM 01- CM 21, whereas, the CM 22 until CM 42 were located along the line 2. The frequency range of all MT data is 0.1 Hz - 10.000 Hz. For further analysis, we used the data which have the coherency ≥ 0.5 as the good data indicator.

To investigate the dimensionality and regional strike of the study area, we applied the phase tensor analysis [11]. According to [11], the magnetotelluric phase tensor, Φ, can be represented as an ellipse which has the principal axes represented by the major and minor axes of the ellipse. The major (Φ max) and minor axes (Φ min) of the ellipse illustrate the principal axes, whereas values of the tensor with the orientation of the major axes are defined by angle $\alpha-\beta$. The α and β values can be formulated as follows:

$$\beta = \frac{1}{2}\arctan\left(\frac{\Phi_{12}-\Phi_{21}}{\Phi_{11}+\Phi_{22}}\right) \quad \alpha = \frac{1}{2}\arctan\left(\frac{\Phi_{12}+\Phi_{21}}{\Phi_{11}+\Phi_{22}}\right) \tag{1}$$

where Φ_{11}, Φ_{12}, Φ_{21}, and Φ_{22} are the components of the phase tensor if it is written as the matrix. To calculate the phase tensor, the analyses does not need an assumption about the dimensionality of the underlying conductivity distribution. Furthermore, it is also applicable if the heterogeneity and the regional conductivity structures of the study area are 3-D [11].

Generally, if the regional conductivity distribution is 3-D, the skew angle (β) value is non-zero and represents the rotation of the major axes of the phase tensor ellipse. Moreover, the phase tensor is symmetric ($\beta = 0$), the principal values of the tensor are equal to its eigenvalues. It occurs if the distribution of the regional conductivity is 1-D or 2-D. In this situation, the orientation of the major axes is defined by α. The 2-D case occurs when Φ max = Φ min and it will be two directions for which a linearly polarized magnetic field will give rise to a linearly polarized electric field. However, when Φ is symmetric and has equal principal values, it will be 1-D. In this case, the electric field will be linearly polarized if the magnetic field is linearly polarized [11].

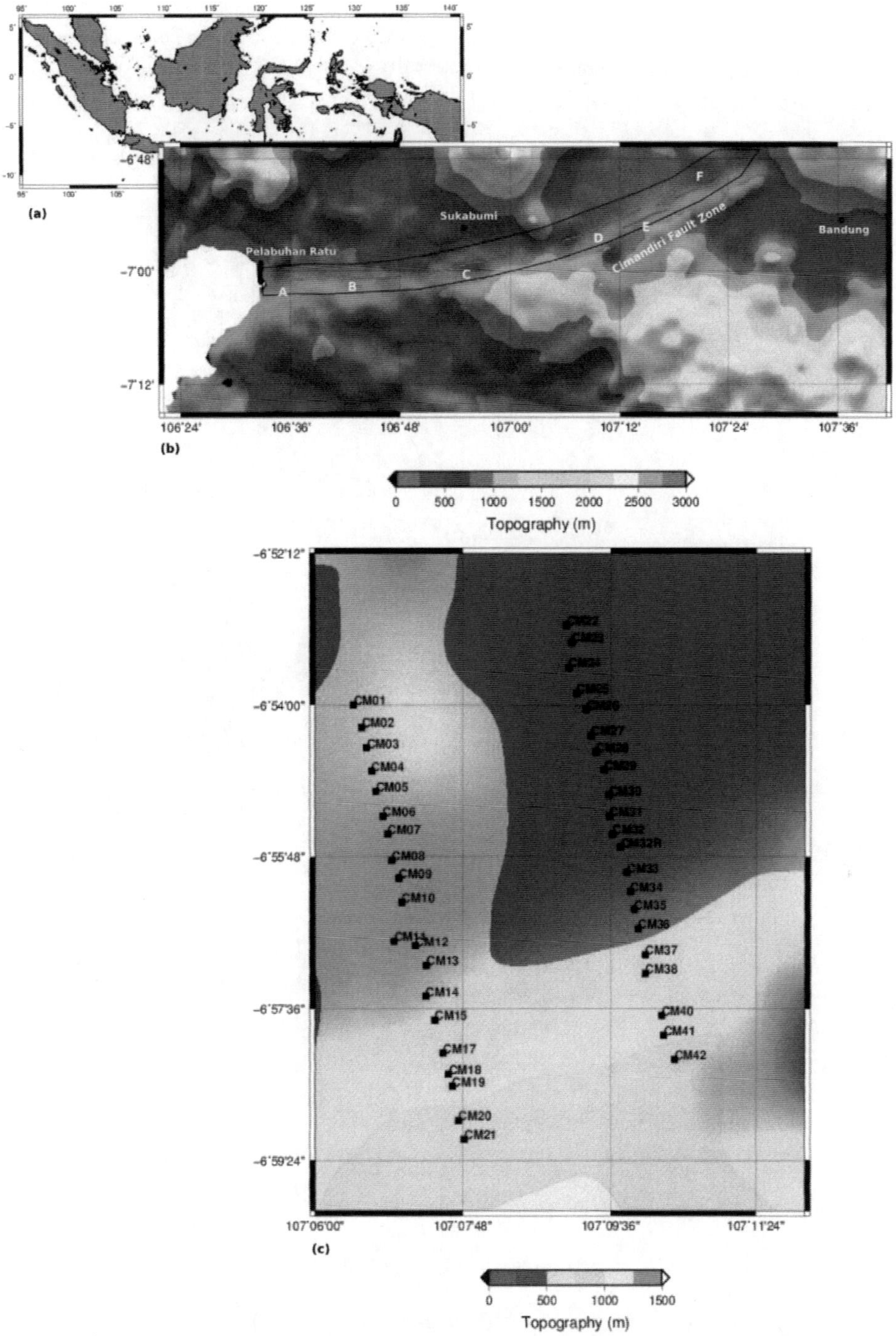

Figure 1. (a) The map of Indonesia. The red color area is the Cimandiri Fault Zone location. (b) The topography map of the Cimandiri Fault Zone; A. Loji segment, B. Cidadap segment, C. Nyalindung segment, D. Cibeber segment, E. Saguling segment, F. Padalarang segment. (c) The MT observation points which were investigated in the Cibeber segment.

IOP Conf. Series: Journal of Physics: Conf. Series **985** (2018) 012016 doi:10.1088/1742-6596/985/1/012016

3. Results and discussion

Figure 2 presents the rose diagram of the major axes orientation which was defined by angle $\alpha-\beta$. Theoretically, the estimated values of the calculated major axes orientation have 90° ambiguity. Figure 2 shows that the dominated general regional strike is generally in the NE-NW direction.

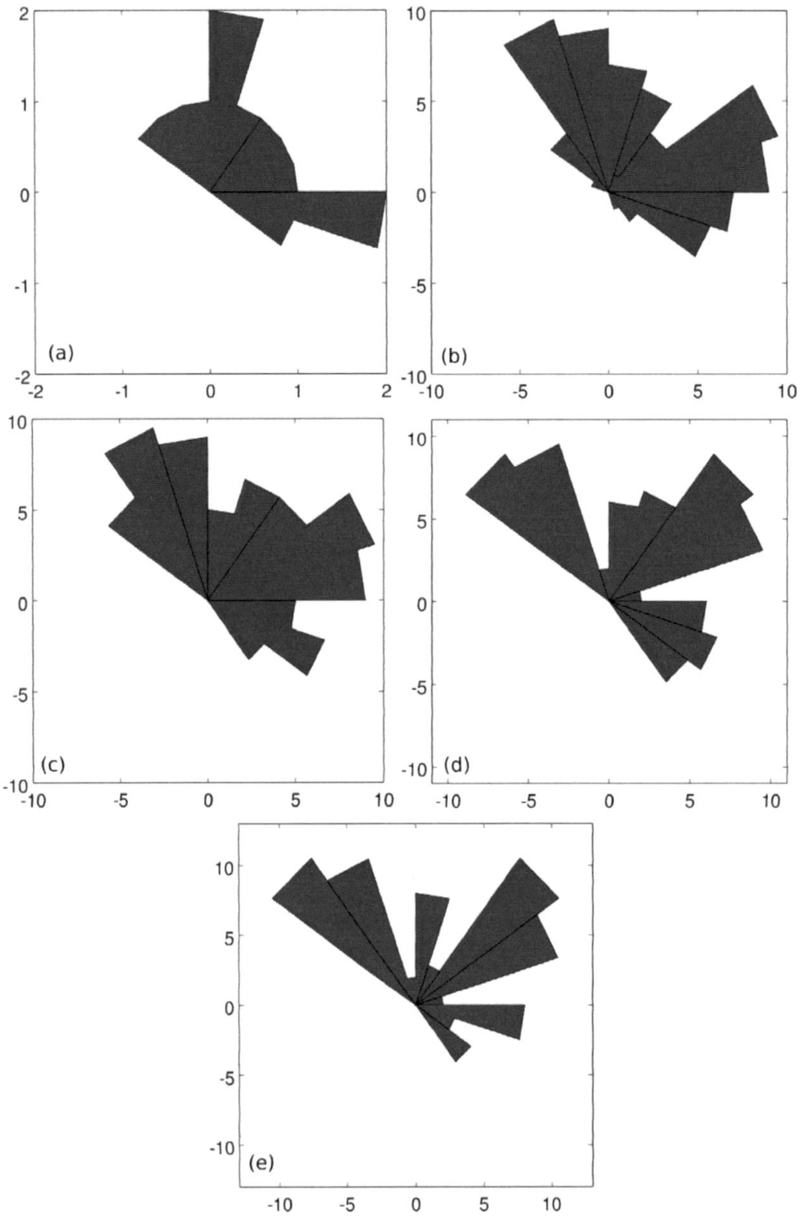

Figure 2. The rose diagrams which represent the orientation of the major axes of the tensor and defined by angle $\alpha-\beta$. (a) for the frequency range 0.1-1 Hz. (b) for the frequency range 1-10 Hz. (c) for the frequency range 10-100 Hz. (d) for the frequency range 100-1000 Hz. (e) for all frequency range

For the frequency range of 1-10 Hz, 10-100 Hz, 100-1000Hz and whole frequency, the results show consistently that the calculated regional strike is about N60-80°E. The results of the regional strike estimation in the Cibeber segment for the lower frequency are different to that of in the Loji segment [10]. However, for the higher frequency, the results are consistent with the previous results [7, 10]. It also supports the previous study that along the CFZ length is expressed in the bedrock by numerous NE, west, and NW trending thrust- and strike-slip faults and folds [9].

Figure 3 presents the skew angle (β) results for all the observation points in this study. The skew angle (β) values can indicate the dimensionality of the study area. Generally, the skew angle values for all observation points are $-5 \leq \beta \geq 5$, as indicated by the color scale in figure 3. These results are also consistent to the dimensionality analysis results in the Loji segment [10].

Figure 3 also shows that the skew angle (β) for the frequency below 1 Hz more complex than that of the frequency more than 1 Hz. For the higher frequency, the phase tensor analysis results reveal the homogeneous skew angle values. The values are about 0. Figure 3 also shows that the observation points of 10, 31, and 32 are the most complex observation points from the lower frequency to higher frequency range based on the magnetotelluric phase tensor analysis. It is possibly associated with the geological feature near the observation points of 10, 31, and 32. The geological map of the study area presents that these observation points are near the transition of the surface geological product in the study area [12].

According to this result, it is safe to assume that the dimensionality of the regional conductivity structures in the study area is 2-D. Therefore, it can be used as the criterion to image the subsurface of the Cibeber segment by using the magnetotelluric data in the next analysis.

The regional strike and dimensionality analyses have been done for the Loji and Cibeber segments. To know more detail about the regional strike and dimensionality of the CFZ, it is important to also investigate other segments.

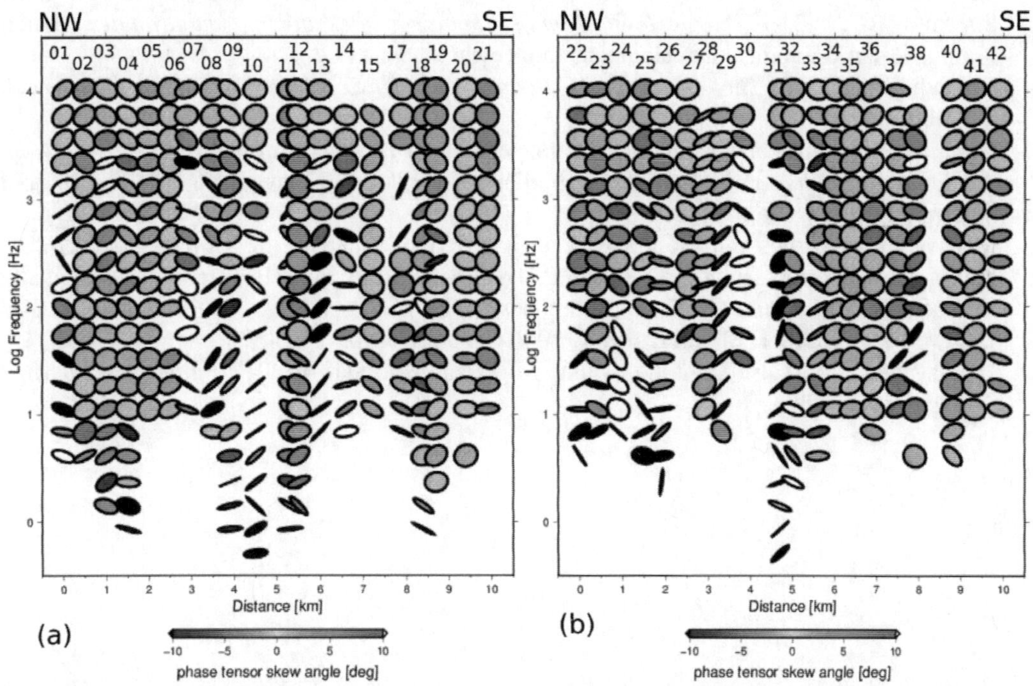

Figure 3. The skew angle (β) of all MT observation points. The number from 01 to 42 related to the CM 01 until CM 42 in the figure 1. The vertical axes indicate the frequency (in log scale). The horizontal axes indicate the observation points distance. The color scale indicates the skew angle β. (a) The skew angle (β) of the CM 01 - CM 21. (β), (b) The skew angle (β) of the CM 22 - CM 42.

4. Conclusion

The investigation of the regional strike and dimensionality in the Cibeber segment, Cimandiri Fault Zone, West Java, by using the magnetotelluric data has been done. All 42 MT observation data which have coherency ≥ 0.5 were analyzed by using the magnetotelluric phase tensor. The results show that the regional strike in the Cibeber segments is NW-NE direction, about N60-80°E. The skew angle values which represent the dimensionality of the study area are $-5 \leq \beta \geq 5$. These results are consistent with the previous results. Based on the results, for the further analysis, it is safe to assume that the dimensionality of the Cibeber segment is 2-D.

Acknowledments

The submission and presentation this manuscript in the ISFAP 2017 are supported by the budget of the Priority Research of the Indonesian Institute of Sciences (LIPI).

References

[1] Vozoff K 1991 The magnetotelluric method *Investigation in geophysics No. 3: Electromagnetic methods in applied geophysics* vol 2, ed M N Nabighian (Oklahoma: Society of Exploration Geophysicists) p 641
[2] Ritter O, Hoffmann-Rothe A, Bedrosian P A, Weckmann U and Haak V 2005 *Geol. Soc. London* **245** 165
[3] Becken M, Ritter O, Bedrosian P A and Weckmann U 2011 *Nature* **480** 87
[4] van Bemmelen R W 1949 *The geology of Indonesia* (The Hague: Government Printing Office)
[5] Sukamto R A B 1975 *Geological map Jampang-Balekambang quadrangle* (Bandung: Indonesian Geological Development and Research Center)
[6] Martodjojo S 2003 *The evolution of Bogor, West Java basin* (Bandung: ITB Bandung Publishing)
[7] Dardji N, Villemni T and Rampnoux J P 1994 *J. Southeast Asian Earth Sci.* **9** 3
[8] Abidin H Z, Andreas H, Kato T, Ito T, Meilano I, Kimata F, Natawidjaya D H and Harjono H 2009 *J. Earthq. Tsunami* **3** 77
[9] Marliyani G I, Arrowsmith J R and Whipple K X 2016 *J. Geophys. Res. Earth. Surf* **121** 2287
[10] Febriani F, Hattori K, Widarto D S, Han P, Yoshino C, Nurdiyanto B, Effendi N, Maulana I and Gaffar E 2013 *Jurnal Geofisika* (in Indonesian) **14** 131
[11] Caldwell T G, Bibby H M and Brown C 2004 *Geophys. J. Int.* **158** 457
[12] Handayani L, Maryati, Kamtono, Mukti M M and Sudrajat Y 2017 *Indonesian Journal on Geoscience* **4** 39

IOP Conf. Series: Journal of Physics: Conf. Series **985** (2018) 012017 doi:10.1088/1742-6596/985/1/012017

Evolution and impact of the 2016 negative Indian Ocean Dipole

I Iskandar[1], D O Lestari[2], P A Utari[2], Supardi[1], Rozirwan[3], M Y N Khakim[1], P Poerwono[1] and D Setiabudidaya[1]

[1]Jurusan Fisika, Fakultas MIPA, Universitas Sriwijaya, Sumatra Selatan, Indonesia
[2]Program Doktor Ilmu Lingkungan, Program Pascasarjana, Universitas Sriwijaya, Sumatra Selatan, Indonesia
[3]Program Studi Ilmu Kelautan, Fakultas MIPA, Universitas Sriwijaya, Sumatra Selatan, Indonesia

iskhaq@mipa.unsri.ac.id

Abstract. Strong negative Indian Ocean Dipole (IOD) event took place in the tropical Indian Ocean during 2016. Based on the Dipole Mode Index (DMI), the event has shown two peaks: in July and September. It is shown that the second peak was stronger than the first peak. Evolution of the event has started in May, reached its first peak in July, weaken in August, but rebounded and came to its second peak in September. The event was terminated in November. Robust sea surface temperature (SST) dipole patterns were observed during both peaks. In July, the SST anomaly in the eastern (western) pole of the IOD reached +1°C (-1.5°C). Meanwhile, during the second peak of the event, the SST anomaly in the eastern (western) pole of the IOD rose (fall) to nearly +2.5°C (-1°C). As a consequence, strong convective activities were observed over the maritime continent causing heavy rainfall during the peak of the event. On the other hand, there was a significant reduce of the rainfall over the eastern Africa during the peak of the event.

1. Introduction

The Indian Ocean Dipole (IOD) is one of dominant coupled ocean-atmosphere mode originated in the tropical Indian Ocean [1,2,3]. A positive IOD has a unique characteristic in which the sea surface temperature (SST) anomaly formed a dipole pattern with negative (positive) SST anomaly covers the southeastern (western and central) tropical Indian Ocean. Associated with these SST patterns, the atmospheric winds anomalously westward, driving anomalous surface westward currents and creating poleward Ekman transport. The divergence of these Ekman transports cause equatorial upwelling and shoal (deepen) the thermocline in the eastern (western) equatorial Indian Ocean leading to cool (warm) SST there. Cool (warm) SST in the eastern (western and central) equatorial Indian Ocean suppressed (enhanced) atmospheric convections. As a consequence, the eastern (western) tropical Indian Ocean and its surrounding continent experience deficit (surplus) rainfall [4,5]. The situation is reverse during the negative IOD event.

The occurrence of the IOD event has a biennial tendency, in which a positive IOD event will be followed by a negative IOD event in the following year [1]. Moreover, the intensity of the event is measured using a Dipole Mode Index (DMI) indicating a gradient in SST anomaly between the western tropical Indian Ocean (10°S - 10°N, 50°E - 70°E) and its eastern counterpart (10°S - 0°S, 90°E - 110°E) [1]. A positive (negative) DMI is referred to a positive (negative) IOD event. During positive IOD event, the SST in the southeastern tropical Indian Ocean experiences cooling associated with anomalously strong southeasterly winds along the southern coast of Java and Sumatra and easterly wind anomalies along the eastern equatorial Indian Ocean. As a consequence, the convective activities shifted westward to the central or even further westward leading to enhanced (decreased) precipitation over the East Africa and India (Indonesia and Australia) [6].

During boreal summer/fall 2016, the DMI indicated a negative value (negative IOD event). In addition, most part of the Indonesian regions have experienced surplus precipitation. This study is designed to describe the evolution of negative IOD event in 2016 and evaluate its possible impact on the Indonesia.

2. Data and Methods

The DMI is defined as a gradient of the SST anomaly between the western tropical Indian Ocean (10°S - 10°N, 50°E - 70°E) and its eastern counterpart (10°S - 0°S, 90°E - 110°E) [1]. In this study, the SST data were obtained from the National Oceanic and Atmospheric Administration (NOAA) Optimum Interpolation Sea Surface Temperature (OISST) Version 2. The data are available from September 1981 until present with a uniform horizontal resolution of 0.25°.

The atmospheric wind data were from the ERA-Interim of the European Centre for Medium-Range Weather Forecasts (ECMWF). This is a low-level wind on the 850 mb pressure level. This level is located above the friction layer where the clouds and precipitation occur. Therefore, this is quite valuable for analyzing all types of synoptic weather patterns. The ECMWF winds have horizontal resolution of 0.5°× 0.5°. The data are available for period of January 1979 until present. Meanwhile, the precipitation data were obtained from the Tropical Rainfall Measuring Mission (TRMM) with horizontal resolution of 0.25° × 0.25° and are available from January 1998 until present. Note that, the anomalies of all variables are defined as the deviation from the monthly climatology over a period of 2001-2015.

3. Results

We first evaluate the evolution of 2016 negative IOD event using the time series of DMI shown in Figure 1. It is shown that the evolution of 2016 negative IOD was started in boreal spring (April/May) as the DMI gradually decreased. By the end of May, the DMI has exceeded its standard deviation and it continued to decrease. The DMI slightly increased from mid-July until mid-August and might be related to the intraseasonal westerly winds during summer monsoon [7]. After a short weakening of the event, the DMI again gradually decreased and reach a minimum value in September. By the end of September, the DMI rapidly increased and the IOD was terminated in November.

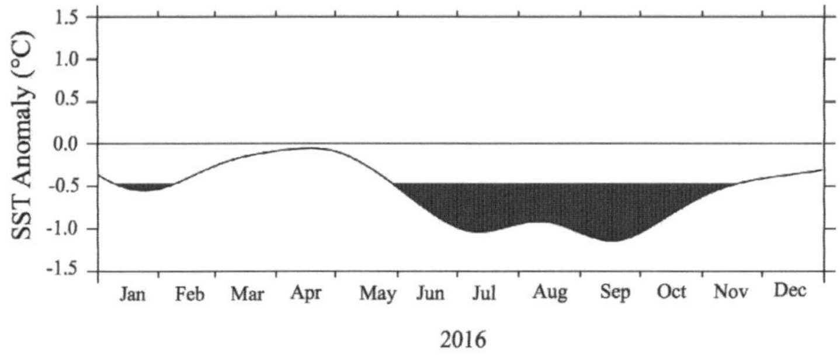

Figure 1. Time series of the Dipole Mode Index (DMI) for a period of January – December 2016. Note that the DMI was defined as a gradient of the SST anomaly (°C) between the western tropical Indian Ocean (10°S - 10°N, 50°E - 70°E) and its eastern counterpart (10°S - 0°S, 90°E - 110°E).

In order to have a spatial description of the evolution of 2016 negative IOD event, monthly SST and winds anomalies are presented in Figure 2. The tropical Indian Ocean usually experiences basin-wide warming following an El Niño event as shown by positive SST anomalies during May 2016 just after the 2015/16 El Niño was terminated (Figure 2a) [8]. Strong westerly wind anomalies were observed, which were associated with the monsoon break [9]. The oceanic response to these westerly wind anomalies is downwelling Kelvin waves propagating eastward across the equatorial Indian Ocean and strong eastward surface currents so-called the Wyrtki jets (*not shown*) [10].

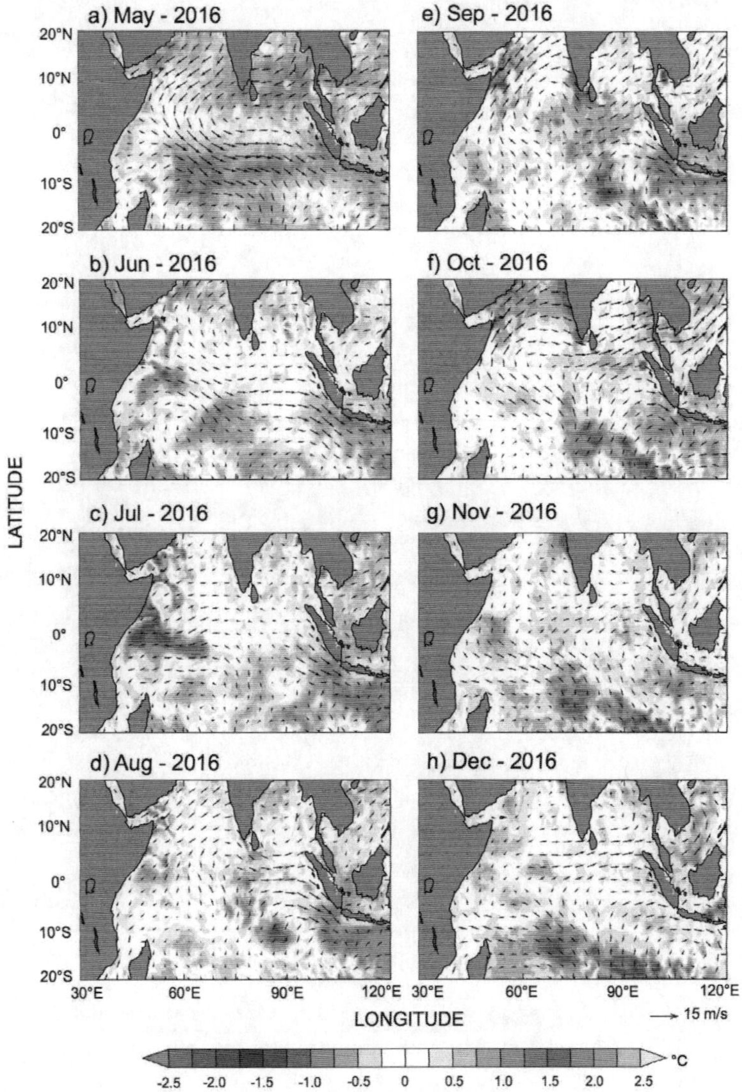

Figure 2. Monthly anomalies of sea surface temperature (*shaded,* °C) and low-level winds (*vector,* m/s) during May – December 2016.

In June, the winds over the equator remained eastward, while along the southern coast of Sumatra and Java strong northwesterly winds were observed (Figure 2b). These northwesterly winds induced downwelling along the coast that warmed the SST and deepened the thermocline. In July, the westerly winds along the equator were significantly weakened while along the coast remained strong (Figure 2c). By August, the westerly winds along the equator and northeasterly winds along the coast were strengthened (Figure 2d). The SST in the eastern tropical Indian Ocean off south Java rose to nearly

+2.5°C, while that in the central Indian Ocean fall to nearly −1°C. The negative IOD event came to its peak in September as the SST anomalies show a dipole pattern with positive (negative) anomalies occupied the eastern (central) tropical Indian Ocean (Figure 2e). The winds anomalies were weakened in the following months and the dipole pattern was completely diminished in December 2016 (Figures 2f-h).

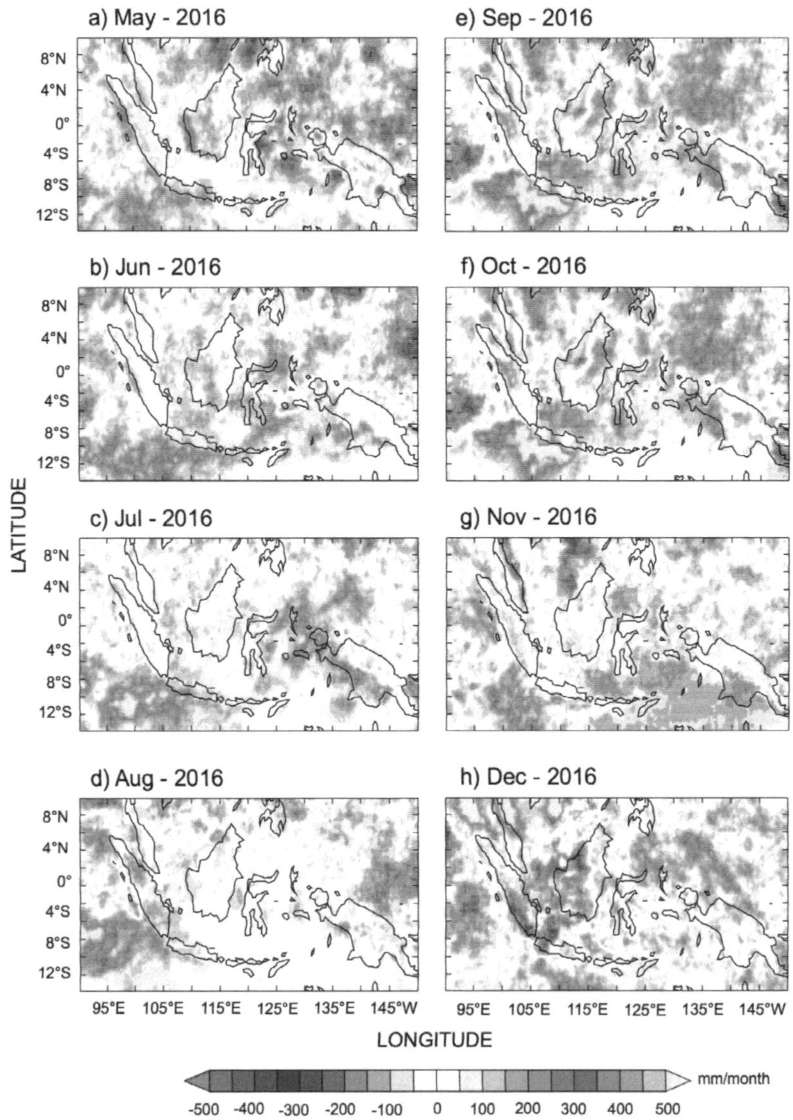

Figure 3. Same as in Figure 2 except for monthly precipitation during May – December 2016 (mm/month).

The 2016 negative IOD event significantly impact the precipitation over the maritime continent. As discussed in previous studies [1-3], the evolution of IOD event is correlated with the convective activities. Warm SST anomalies were associated with enhanced convective activities leading to increased precipitation. During 2016, most part of the Indonesian regions experienced surplus precipitation (Figure 3). Interestingly, it is also shown that anomalously high precipitations were observed over high SST anomalies indicating a strong coupled air-sea interaction in that area (Figures

3d-f). It should be noted that in early spring 2016, the eastern Indonesian regions were still under the influence of 2015/15 El Niño event (figure 3a) [11]. In addition, the local SST could also influence the convective activities over the Indonesian region that could enhanced/decreased the precipitation [12].

4. Conclusion

During 2016, strong negative IOD event took place in the tropical Indian Ocean. The event has shown two peaks: in July and September, in which the second peak was stronger than the first peak. Evolution of the event has started in May, reached its first peak in July, weaken in August, but rebounded and came to its second peak in September. The event was terminated in November, though the SST dipole pattern was completely diminished in December. During the peak of the event in September, the SST anomaly in the eastern (western) pole of the IOD rose (fall) to nearly +2.5°C (-1°C). As a consequence, strong convective activities were observed over the maritime continent causing heavy rainfall during the peak of the event. Note that the eastern Indonesian regions were still under influenced of Pacific mode in early 2016. Moreover, the local SST could also influence the convective activities over the maritime continent.

Acknowledgment
This research is supported by the Ministry of Research, Technology and Higher Education, Indonesia through *Penelitian Berbasis Kompetensi 2017* (No. 454/UN9.3.1/LT/2017) and by the University of Sriwijaya through *Hibah Profesi 2017* (No. 1011/UN9.3.1/PP/2017)

References

[1] Saji N H and Yamagata T 2003*Clim. Res.* **25** 151
[2] Webster P J, Moore A M, Loschnigg J P and Leben R R 1999 *Nature* **401** 356
[3] Murtugudde, R, McCreary J P and Busalacchi A J 2000 *J. Geophys. Res.* **105** 3295
[4] Iskandar I, Mardiansyah W and Setiabudidaya D 2014 *Makara J. Sci.* **18** 106
[5] Endo S and Tozuka T 2015 *Clim. Dyn.* **46** 3371
[6] Yamagata T, Behera S K, Luo J, Masson S, Jury M R and Rao S A 2004 Coupled oceanatmosphere variability in the tropical Indian Ocean *Earth's Climate: The ocean-atmosphere interaction* eds C. Wang, S.P. Xie and J.A. Carton (Washington D C: American Geophysical Union) p 189
[7] Senan R, Sengupta D and Goswami B N 2003 *Geophys. Res. Lett.* **30** 1750,
[8] Yu L and Rienecker M M 2000 *J. Geophy. Res.* **105** 16923
[9] Wyrtki K 1961 *NAGA Report: Physical oceanography of the Southeast Asian waters* (California The University of California)
[10] Wyrtki K 1973 *Science* **181** 262
[11] Iskandar I, Utari P A, Lestari D O, Sari Q W, Setiabudidaya D, Khakim M Y N, Yustian I and Dahlan Z 2017 *Proc. Int. Symposium on earth hazard and disaster mitigation 2016: The 6th annual symposium on earthquake and related geohazard research for disaster risk reduction* (Bandung) vol 1857 (New York: AIP Publishing) 080001
[12] Iskandar I, Mardiansyah W, Setiabudidaya D, Poerwono P, Yusyian I and Dahlan Z 2017 *J. Phys.: Conf. Ser.* **187** 012073

Crustal structure in the southern part of Central Java based on analysis of tele-seismic receiver function using a neighbourhood algorithm

P Ariyanto[1, 3], Syuhada[2], S Rosid[3], T Anggono[2] and Y Januarti[4]

[1]School of Meteorology, Climatology, and Geophysics (STMKG), Perhubungan I no.5 Pondok Betung, Tangerang Selatan 15221, Banten, Indonesia
[2]Research Centre for Physics, Indonesian Institute of Sciences (LIPI), Komplek Puspiptek Serpong, Tangerang Selatan 15314, Banten, Indonesia
[3]Department of Physics, FMIPA Universitas Indonesia, Kampus UI Depok, Depok 16424, Indonesia
[4]Seismic Regional Centre II, Agency of Meteorology, Climatology, and Geophysics (BBMKG Reg. II), Ciputat, Tangerang Selatan 15221, Banten, Indonesia

E-mail: puji.ariyanto@ui.ac.id

Abstract. In this study, we applied receiver functions analysis to determine the crustal thickness, the ratio of Vp/ Vs and the S wave velocity in the southern part of the Central Java. We selected tele-seismic data with magnitude more than 6 (M>6) and epicenter distance 30°-90° recorded from 3 broadband stations: UGM, YOGI, and WOJI station, as part of Indonesia-Geophone Network (IA-GE). Inversions were performed using nonlinear Neighborhood Algorithm (NA). We observed Ps phase conversion on the receiver functions corresponding to Moho depth at around 36-39 km. We also observed strong negative phase arrivals at around 10-12 s which might be associated with Indo-Australian subducting slab underneath the stations. The inversion results show the presence of low velocity zone with high Vp/Vs ratio (>1.78) in the middle crust around the study area which could be related to the Merapi-Lawu Anomaly (MLA).

1. Introduction

The study area is located in the southern part of Central Java, where the Indo-Australian plate is subducted beneath the Eurasian Plate since the Cretaceous at an average speed of 6.7 cm/ year [1, 2]. This subduction activity may produce a great potential natural hazard such as volcanic eruptions, earthquakes and tsunamis. The latest strong subduction activity in the southern part of central Java was a shallow earthquake on 2006 Mei 26 (Mw=6.3) and Mt. Merapi eruption in October - November 2010 causing more than 5000 casualties. In general, the main geologic and tectonic settings in the Central Java are derived from subduction process including Java trenches, magmatic-volcanic arcs, accretionary prisms, back-arc and fore-arc basins forming a complicated geological setting. This tectonic setting suggests that the crust in the study area is composed several segments accumulated in the southern part of eurasian margin in the Cretaceous [2]. Furthermore, two major wrench faults, the Maria-Kebumen Faults and the Pamanukan-Cilacap Faults met in southern part of Central Java and slip each other, caused many geologic changes in this area [2, 3]. Thus, investigating crustal characteristics in Central Java is crucial in order to provide additional information for deformation process in this area.

The previous crustal structure studies in Central Java have been done by several researchers (see e.g. [4-7]). The crustal thickness in Central Java is about 34 km based on receiver functions from tele-seismic earthquakes data recorded at Merapi Amphibious Network (MERAMEX), although the Vp/Vs ratio cannot be reliably inferred due to noisy data [4]. The results also show an exceptionally strong low velocity anomaly (−30%) in the back-arc crust northward of the active volcanoes (MLA) [4, 5]. The significant reduction of seismic P and S wave velocities is mainly observed in the upper 15 km of the crust between Merapi and Lawu volcanoes [5, 6]. Ambient noise tomography further supports the existence of the MLA (Merapi-Lawu Anomaly) suggesting that the low velocity anomaly is concentrated in the upper 5 km of the crust [7].

Receiver function analysis is a tool to characterize the structure of the Earth by utilizing the seismic waveforms derived from tele-seismic events recorded at a three component seismic station [8]. This technique has been widely practiced by some previous researchers to estimate the crustal thickness, Vp/Vs ratio and S wave velocity profile. In this study, we analyze receiver functions to determine the variations of crustal structure: Moho depth, Vp/Vs ratio, and S-wave velocity models beneath YOGI, UGM and WOJI stations using the neighborhood algorithm inversion. The results of this study are expected to improve our knowledge in local crustal models and tectonic development in the southern part of Central Java.

2. Data and method

Our study is based on analysis of tele-seismic waveforms recorded on YOGI, UGM, and WOJI stations as part of IA-Geophone seismic network in southern part of Central Java (figure 1). To compute receiver function, we extracted waveforms of the 3 component broadband sensor (BHZ, BHE, BHN) from tele-seismic events with magnitude 6 or higher and distance between 30° and 90° recorded from 2008 to 2015 (figure 1). All waveforms were windowed between 10 s before and 50 s after the P arrival. We manually inspected the seismograms to select the best waveform with high signal to noise (s/n) ratio and clear direct P arrivals. We then removed the instrument response, rotated the horizontal seismograms into radial and transverse components and deconvolved those seismograms to generate the receiver functions using computer program for seismology [9]. To control the high frequency noise in the receiver functions, we applied the gaussian filter. In this study, we selected radial receiver function with low gaussian width factor of 0.5 to attenuate the frequency bandwidth greater than approximately 0.3 Hz. To assess the quality of receiver functions, we only used the receiver functions with fit criterion between the observed radial component and the convolution of observed vertical component around 90% or above for further analysis. Finally, we analyzed the selected receiver functions using non-linear Neighbourhoods Algorithm inversion (NA) technique of Sambridge [10] to obtain the velocity profiles beneath seismic stations. The NA technique is a fully non-linear inversion technique based on a random walk in the voronoi cells to search the optimum model using the misfit of the previous models.

To estimate the crustal structure beneath the seismic stations, we inverted the receiver function using the following steps. First, we stacked all radial receiver functions at each station to increase the signal to noise ratio. Second, we cut the receiver functions from 5 s before and 20 s after the direct P wave arrival for the inversion process. We chose this time window to assure that all crustal and lithospheric phase arrivals are included in the inversion analysis. For the inversion process, we divided the initial crustal structure into six horizontal layers: sediment, basement, upper crust, middle crust, lower crust and upper mantle. Each layer contains 4 parameters describing the thickness of the layer (km), Vp/Vs, the S-wave velocity at the top and bottom of each layers. We set the initial parameter after trial and error with low misfit. Then, we computed the NA inversion using 5000 iterations to get the best S wave velocity models.

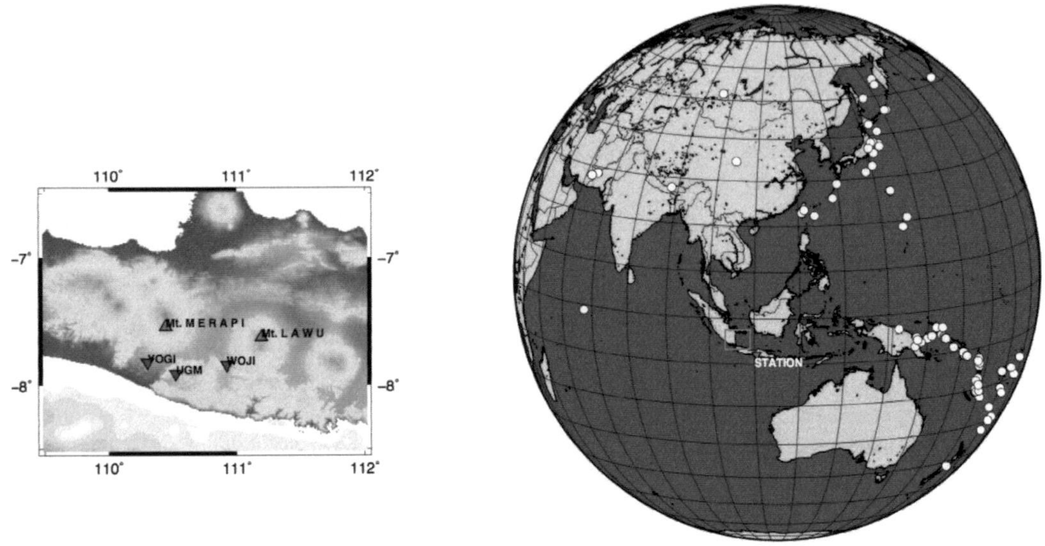

Figure 1. The study area in central java showing the location of YOGI, UGM, and WOJI stations (inverted blue triangles). The red triangles denote the volcano locations. The globe on the right figure depicts the distribution of tele-seismic events used in this study (the white circles). The red rectangle represents the study area enlarged on left figure.

3. Results and discussion

Figure 2 displays the receiver functions obtained for YOGI, UGM, and WOJI stations. In general the receiver functions show similar waveforms in all seismic stations. The phase delay of direct P phase is observed on the radial receiver functions for all seismic stations indicating the presence of low velocity structure in the near surface. This pattern is probably generated from the superposition of the direct P wave and Ps conversion waves at the interface of basement and sediment layers [11]. Ps Moho conversion waves and the multiples (PpPs and may be PpSs + PsPs) are identified at around 6-8 s and 14-16 s for all seismic stations, respectively. We also observed the combination of positive and negative phase polarities at around 4-6 s on the receiver functions beneath YOGI, UGM, and WOJI seismic stations, which may correlate with the presence of a crustal low velocity zone, Merapi Lawu Anomaly (MLA), as revealed by several studies in this area [5-7]. Moreover, we also find interesting feature on the receiver functions of all seismic stations, in which there are strong negative phase arrivals at around 10 to 14 s. We suggest that strong negative phase arrivals observed on these receiver functions might be associated with the subducted Indo-Australian plate underneath Java Island [5, 12].

S-velocity models resulted from the NA inversions of the stacked radial receiver functions in the three seismic stations are shown on figure 3. The white and solid red lines with green background color represent the average and best fitting of S wave velocity model, respectively. The grey shaded area indicates all models (5000 models) searched in the inversion. The red line on the left represents the best fitting Vp/Vs ratio. The inversion result for YOGI station displayed on figure 3a reveals S wave velocity around 0.53 km/s near the surface interpreted as the sedimentary layer. The S wave velocities then fluctuate from 3.09-3.30 km/s at depth 14 km to 2.90-3.12 km/s at depth 18.9. The low velocity zone characterized by negative velocity gradient is identified at 20-26 km depth, with a S-wave velocity as low as ~2.2 km/s and high Vp/Vs ratio (around 2). Beneath this layer, the S wave velocity increases to 2.82-3.86 km/s by 37.3 km depth. Then, the crust mantle discontinuity is identified at 37.3 km depth characterized by a positive gradient velocity contrast from 3.86 to 4.80 km/s.

The S wave velocity profile derived from the NA inversion for UGM station is shown on figure 3b. The near surface is covered by very low velocity material with a velocity as low as 0.51 km/s. The velocities fluctuate down to ~16 km depth, then decrease with a minimum velocity of ~2.2 km/s at ~18-25 km depth. The crust mantle boundary is observed at a depth of 36.7 km marked by a strong positive velocity gradient (3.75 to 4.98 km/s).

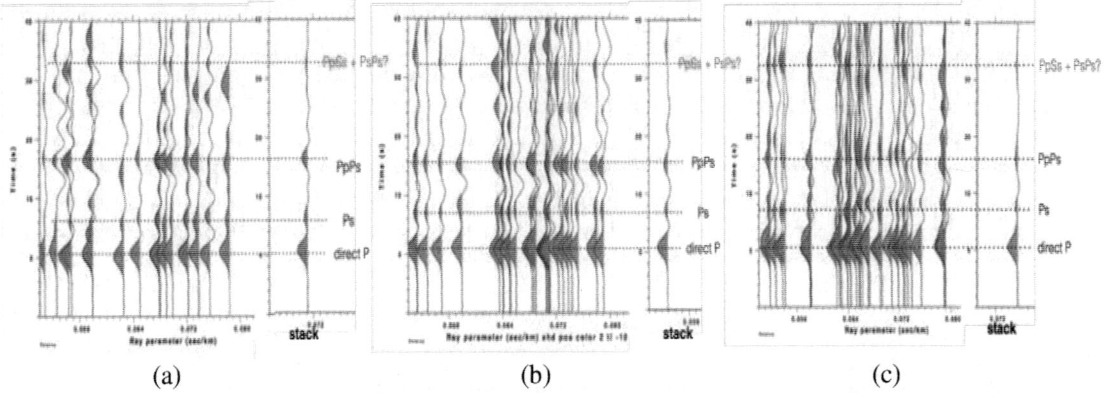

Figure 2. Radial receiver function in (a) YOGI, (b) UGM and (c) WOJI. The right panel is plot the stack of receiver function.

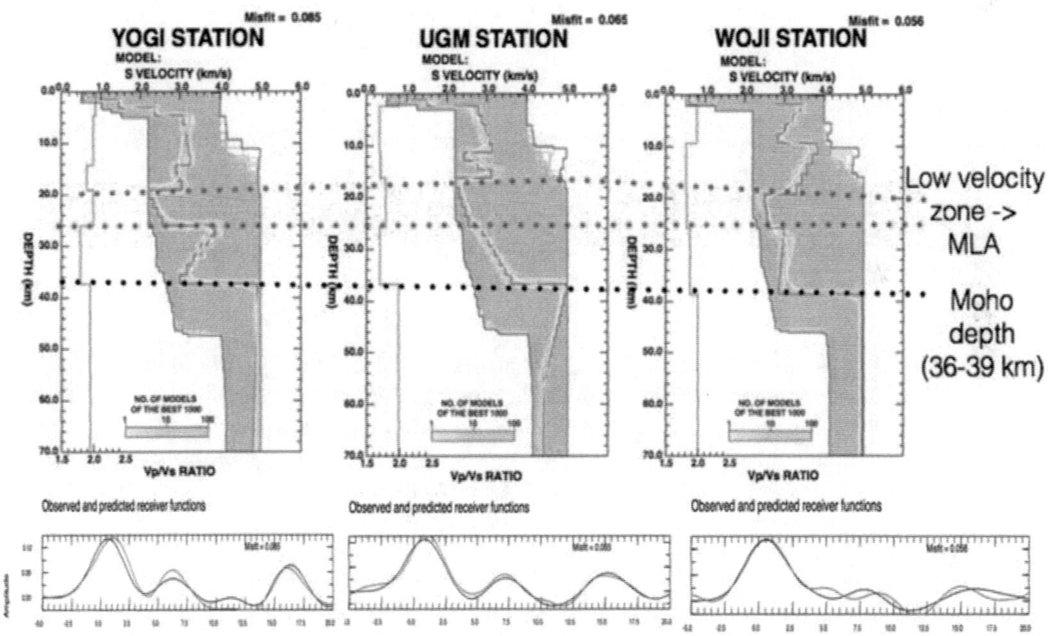

Figure 3. S wave velocity models underneath (a) YOGI, (b) UGM and (c) WOJI stations. The white and solid red lines represent the average and the best fitting model, respectively. The grey shaded area indicates all models (5000 models) seeked in the inversion. The red line on the left represents the best fitting *Vp*/*Vs* ratio. Bottom panels show the observed and predicted receiver function marked by solid black and blue lines, respectively.

The inversion results for WOJI station, depicted in the right panel figure 3, indicate low-velocity sediment in the near-surface layer, with velocities less than 2.00 km/s. Below this layer the velocity increases up to 3.6 km/s at ~10 km depth. Velocity below that depth then decreases from 3.6 to ~2.6 km/s. The low velocity zone is also observed beneath this station at depth of 20-25 km. The Moho discontinuity appears to be reached at 38 km with average Vs ~3.5 km/s.

The low velocity zone observed beneath all seismic stations at 18-28 km depth is probably associated with the presence of MLA (Merapi-Lawu anomaly) as suggested by other previous geophysical studies [5-7]. Seismic tomography study conducted around this area suggests that low velocity zone at depth 20-30 km in MLA can be interpreted as the influence of the fracturing rocks in the mid crust due to stress accumulation from the subducted Indo-Asutralian slab [5]. Furthermore, S velocity models obtained for all seismic stations also show consistence strong velocity contrast at interface between lower crust and the upper mantle identified as Moho depth location at depth ranging from 37 to 39 km [4]. These values are comparable to other geophysical studies conducted in this area [5, 6].

4. Conclusions

In this study, we presented the crustal properties and S-wave velocity structures derived from tele-seismic receiver functions using the non linear NA inversion. We observed that the Moho depth at the southern part in Central Java is ranging from 36 to 39 km. We obtained that the strong negative phase arrivals on the receiver functions at around 10 to 12 s underneath station might be associated with Indo-Australian subducting slab in this area. Low velocity zone can be identified beneath the station in the middle crust (18-28 km) which may be associated with the presence of Merapi-Lawu Anomaly (MLA) with high Vp/Vs ratio (>1.78).

Acknowledgements

This paper is a preliminary result of my master thesis. I am grateful to my supervisor Dr. Syuhada and Syamsu Rosyid Ph.D for their help and guide. This study is supported by LPDP and LIPI. We acknowledge to Agency for Meteorology, Climatology, and Geophysics (BMKG) and GFZ Potsdam for providing the data used in this study. We also acknowledge to R.B. Herrmann, Malcolm Sambridge and K.C. Eagar for providing softwares used in this study.

References

[1] Simon W J F, Socquet A, Vigny C, Ambrosius B A C, Haji Abu S, Promthong C, Subarya C, Sarsito D A, Matheussen S, Morgan P and Spakman W 2007 *J. Geophys. Res.* **112** B06420
[2] Hall R 2009 *Geological Society of London Special Publication* **318** 351
[3] Satyana A H 2007 *Proc. Indonesian Petroleum Association, Thirty-first annual convention and exhibition* (USA: AAPG Datapages, Inc.) **IPA07-G-085**
[4] Wölbern I and Rümpker G 2016 *J. Asian Earth Sci.* **115** 69
[5] Wagner D, Koulakov I, Rabbel W, Luehr B, Wittwer A, Kopp H, Bohm M and Asch G 2007 *Geophys. J. Int.* **170** 923
[6] Koulakov I, Bohm M, Asch G, Luehr B, Manzanares A, Brotopuspito K, Fauzi P, Purbawinata M, Puspito N, Ratdomopurbo A, Kopp H, Rabbel W and Shevkunova E 2007 *J. Geophys. Res.* **112** B08310
[7] Zulfakriza Z, Saygin E, Cummins P R, Widiyantoro S, Nugraha A D, Lu B G and Bodin T 2014 *Geophys. J. Int.* **197** 630
[8] Langston C A 1979 *J. Geophys. Res.* **84** 4749
[9] Herrmann R B 2013 *Seismol. Res. Lett.* **84** 1081
[10] Sambridge M 1999 *Geophys. J. Int.* **138** 479
[11] Bannister S, Bryan C J and Bibby H M 2004 *Geophys. J. Int.* **159** 291
[12] Syuhada, Hananto N D, Abdullah C I, Puspito N T, Anggono T and Yudistira T 2016 *Acta Geophys.* **64** 2020

IOP Conf. Series: Journal of Physics: Conf. Series **985** (2018) 012019 doi:10.1088/1742-6596/985/1/012019

Analysis and application of impedance polar diagram and zstrike rose diagram of magnetotellurics data in southern part of the Wayang Windu geothermal field

O R Rohayat[1], R A Wicaksono[2] and Y Daud[1, 3]

[1] Physics Department, University of Indonesia, Depok, 16424, Indonesia
[2] Star Energy Geothermal Wayang Windu Ltd., Indonesia
[3] PT. NewQuest Geotechnology, Indonesia

Email: okkyrizki27@outlook.com

Abstract. In this study, we determined the main direction of geoelectric strike in the southern part of the Wayang Windu geothermal field using magnetotellurics (MT) data. The strike direction was obtained by analyzing data using impedance polar and Zstrike rose diagram. We investigated 51 MT data at different sites of the southern part of the Wayang Windu geothermal field. Determination of geoelectric strike direction is important since the strike is the rotation references in MT data processing. Our findings had pointed out that the geoelectric strike direction in this study area is in accordance with the direction of geological structure and has a good correlation with structures delineated from 3D MT inversion model.

1. Introduction

The magnetotelluric (MT) technique is a passive electromagnetic (EM) technique that involves measuring fluctuations in the natural electric, E, and magnetic, B, fields at the surface of the Earth as a means of determining the subsurface conductivity structure [1]. This is a passive EM method because the fluctuations of EM fields on the Earth's surface are generated by the fluctuations of EM field in the atmosphere due to the solar wind and lightning activity, and the fluctuations of secondary inductive EM field in the Earth's layer due to the resistivity variations of rocks. The EM that included electric (E) and magnetic (H) fields is in orthogonal directions at the Earth's surface [1]. Therefore, the MT data included two electric (E_x, E_y) and two magnetic (H_x, H_y) components, both in horizontal layouts with the north-south and east-west directions, and one vertical magnetic (H_z component [2].

The MT method is the most common geophysical method used in geothermal exploration for mapping the resistivity structure below the surface. It is a powerful method to delineate a conductive layer (clay cap) that indicates the existence of geothermal system in the survey area. Moreover, MT method also can delineate geoelectric strike, the difference in lateral resistivity structure. The existence and directions of the geoelectric strike can be delineated by analyzing the already existing information in MT data: impedance polar diagram and impedance strike (Zstrike).

Determination of geoelectric strike direction in MT method is valuable to be rotation references in MT data processing. It is supposed to make the 1D inversion model more reliable after rotation applied at a certain angle depending on the direction of the geoelectric structure near the MT site. In the other hand, the interpreted geoelectric strike should correlate with 3D inversion model layer to increase our confidence in the interpretation of reservoir zone and reservoir boundary.

In this study, we analyzed the impedance polar diagram and Zstrike rose diagram of MT data in southern part of the Wayang Windu geothermal field. We investigated the 51 MT data at different sites in order to determine the main direction of geoelectric strike in this area. We also made a MT forward modeling as a reference to understand the pattern of impedance polar diagram from the synthetic data to be compared with the real data.

2. Overview of Wayang Windu geothermal field

The Wayang Windu geothermal field is located approximately 40 km south of Bandung, Jawa Barat, Indonesia. The field is surrounded by high topography geothermal systems such as the Kamojang, Darajat, Patuha, Salak and Karaha Bodas fields. It is interpreted to be in transitional between vapour-dominated and liquid-dominated system [3]. It is high-enthalpy geothermal system, with reservoir temperatures above 225 °C. It has operated since 2000 with total installed capacity currently at 227 MWe from unit I and unit II [4]. Currently, it is included in the work area of Star Energy Geothermal Wayang Windu Ltd.

In this study, we define Wayang Windu geothermal field in two zones: the northern and southern field. The northern field is interpreted to be more vapor -dominated system while the southern field is more liquid-dominated system [3]. In this study, we focus only on the southern field to analyze analyse the impedance polar diagram and Zstrike rose diagram to understand the main direction of geoelectric strike.

2.1. Geology of the southern field

The Southern Wayang Windu geothermal field is located in an active tectonic zone where major structural faults and fractures are dominated by WNW, NW, and NE trends. The trends are the best result of combination of many methods based on the integration of surface and subsurface data. Figure 1 shows the geological structure at the southern field. It also shows the location of 51 MT sites in the southern part of the Wayang Windu geothermal field which are used in this study.

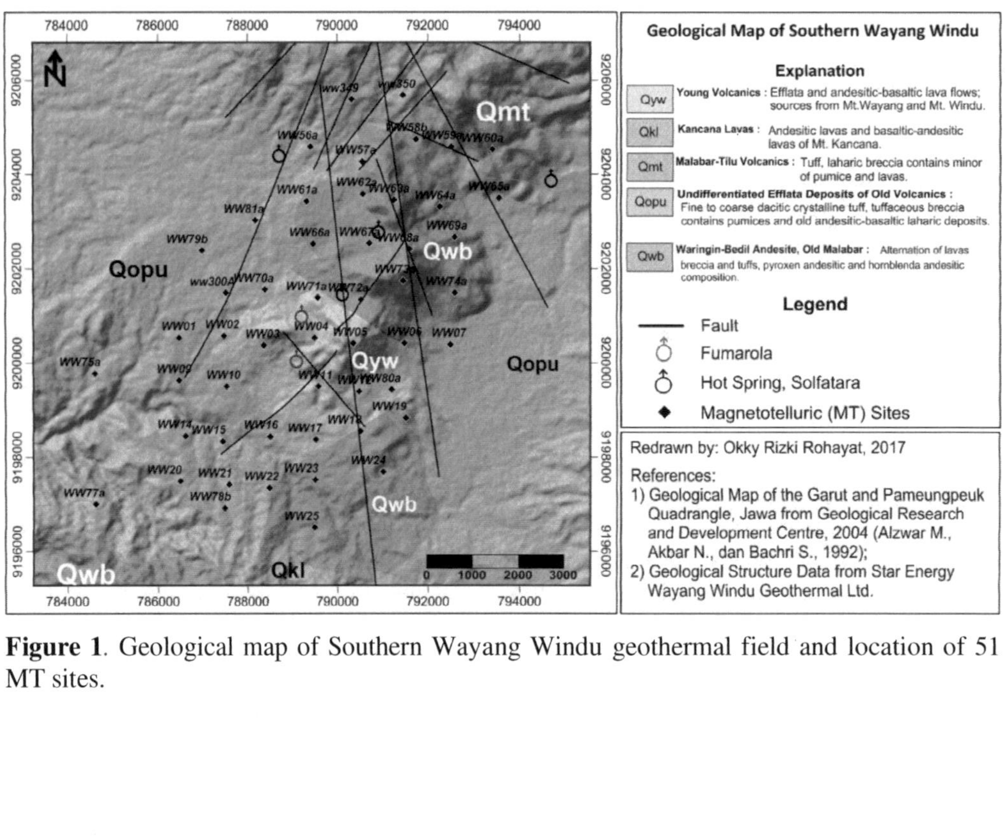

Figure 1. Geological map of Southern Wayang Windu geothermal field and location of 51 MT sites.

IOP Conf. Series: Journal of Physics: Conf. Series **985** (2018) 012019 doi:10.1088/1742-6596/985/1/012019

3. Impedance tensor of MT data

The impedance tensor describes the relation between the electric and magnetic fields, where both are frequency dependent equations. It can be given by

$$E_i(\omega) = Z_{ij}(\omega) * H_j(\omega) \tag{1}$$

where ω is the frequency and Z is the impedance tensor related to the electrical properties of subsurface materials. Equation 1 may also be formatted in matrix form as

$$\begin{pmatrix} E_x \\ E_y \end{pmatrix} = \begin{pmatrix} Z_{xx} & Z_{xy} \\ Z_{yx} & Z_{yy} \end{pmatrix} \begin{pmatrix} H_x \\ H_y \end{pmatrix} \tag{2}$$

where Z_{xy} and Z_{yx} are the principal or off-diagonal impedances, Z_{xx} and Z_{yy} are the diagonal impedances [5, 6].

The elements of the impedance tensor are in general dependent on the measurement directions of x and y. The impedance can be rotated through an angle θ from the positive x-axis (north) to a new coordinate system according to:

$$Z'(\omega) = R(\theta)Z(\omega)R(\theta)^T = RZR^T \tag{3}$$

where $R(\theta) = \begin{pmatrix} cos\theta & sin\theta \\ -sin\theta & cos\theta \end{pmatrix}$ is the rotation operator and R^T is the transpose of R.

$$E' = Z'H' \tag{4}$$

and the elements are

$$Z'_{xx} = Z_{xx}cos^2\theta + (Z_{xy} + Z_{yx})sin\theta cos\theta + Z_{yy}sin^2\theta \tag{4a}$$

$$Z'_{xy} = Z_{xy}cos^2\theta + (Z_{yy} - Z_{xx})sin\theta cos\theta - Z_{yx}sin^2\theta \tag{4b}$$

$$Z'_{yx} = Z_{yx}cos^2\theta + (Z_{yy} - Z_{xx})sin\theta cos\theta - Z_{xy}sin^2\theta \tag{4c}$$

$$Z'_{yy} = Z_{yy}cos^2\theta - (Z_{xy} + Z_{yx})sin\theta cos\theta + Z_{xx}sin^2\theta \tag{4d}$$

The dependence of the impedance tensor upon the direction of the coordinate axes x, y can be displayed by polar diagrams [7]. Impedance polar plots provide a measure for the MT data dimensionality.

The most common approximation is based on the maximization of the off-diagonal components of the MT tensor or minimization of the diagonal ones [5]. Then, the strike angle equation can be written as

$$\propto = \frac{1}{4} arctan\left(\frac{(Z_{xx}-Z_{yy})(Z_{xy}+Z_{yx})^* + (Z_{xx}+Z_{yy})^*(Z_{xy}-Z_{yx})}{|Z_{xx}-Z_{yy}|^2 - |Z_{xy}+Z_{yx}|^2} \right) \tag{5}$$

where $*$ means the complex conjugate. Equation 5 is also known as Swift angle [8] or Zstrike. Zstrike or the Swift angle gives the electrical strike [9] or we called it the geoelectric strike. Note that the solved strike direction from equation 5 contains a 90° ambiguity, because rotation by 90° only exchanges the location of the two principal impedance tensor elements within the tensor [1].

4. Methodology

The study was divided into several stages: forward modeling, analysis and interpretation of impedance polar diagram of Wayang Windu MT data refer to forward modeling result, making Z strike rose diagram to compare with the result of impedance polar diagram analysis and interpretation, and comparing the geoelectric strikes that obtained from the analysis and interpretation with the geological map and 3D inversion model layer.

4.1. Forward modeling

This forward modeling aims to understand the pattern of impedance polar diagrams from a given resistivity structure model. This pattern is used as a reference to be applied to real MT data analysis. The EDI (Electrical Data Interchange) files of synthetic MT data was generated by forward modeling run using WinGLink for a given 3D resistivity structure model. The EDI files were extracted and used to run the impedance polar diagram map. The pattern of impedance polar diagrams was obtained from these maps.

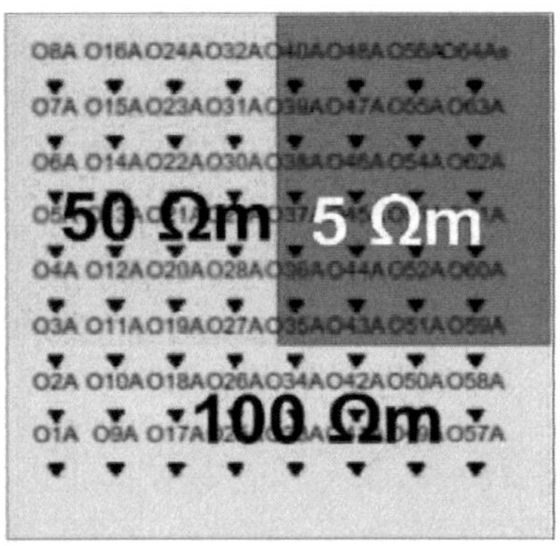

Figure 2. A given 3D resistivity structure model with three different resistivity values.

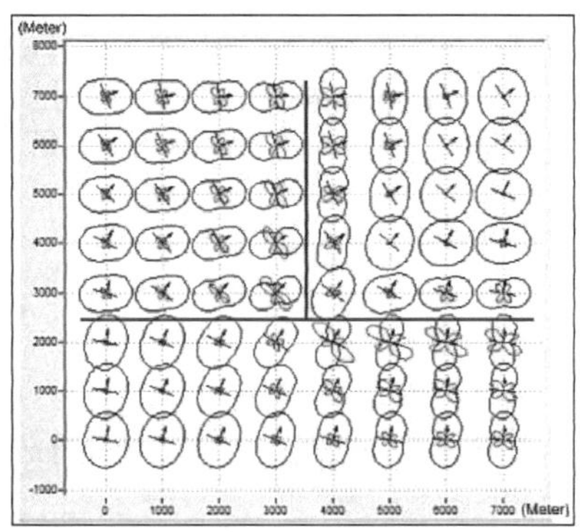

Figure 3. Map of impedance polar diagrams from forward modelling and position of geoelectric strike (blue line).

Figure 2 shows a given 3D resistivity structure model in forward modelling with three different colors representing the three different resistivity values. This model's length, width, and depth are 7000 m, 7000 m, and 3000 m, respectively. Figure 3 shows the impedance polar diagrams as a result of a given model. The geoelectric strike (blue line) distinct two different zones of resistivity. The pattern of the impedance polar diagram elongation is parallel to the strike for the MT stations located in more conductive zone, and perpendicular to the strike for the MT stations located in more resistive zone. Therefore, by using impedance polar diagram, not only the direction of the geoelectric strike can be found, but also the location where the impedance strike shows particular direction.

5. Result and discussion

In this study, we used MT data in EDI file format. We analyzed the 51 MT data in southern Wayang Windu geothermal field to determine the geoelectric strike from impedance polar diagrams map (figure 4) and Zstrike rose diagrams map (figure 5).

Figure 4 shows the impedance polar diagrams in the frequency 0.1 Hz of 51 MT data of southern part of the Wayang Windu geothermal field. The blue lines are the geoelectric strike locations as a result of analysis and interpretation. The blue lines drawn on the map between polar diagrams that relatively perpendicular to each other, which referred to the result of forward modeling. Some of the blue lines have North-West (NW) and the others have North-East (NE) direction trends.

Figure 5 shows Zstrike rose diagrams of southern Wayang Windu geothermal field in the frequency greater than 0.1 Hz. Majority of Zstrike rose diagrams display the direction of geoelectric strike in NW and NE which is correlated with the interpretation of the impedance polar diagram. Briefly, by comparing figure 4 and figure 5, the strike position of the interpretation of the impedance polar diagram is confirmed by the Zstrike rose diagram.

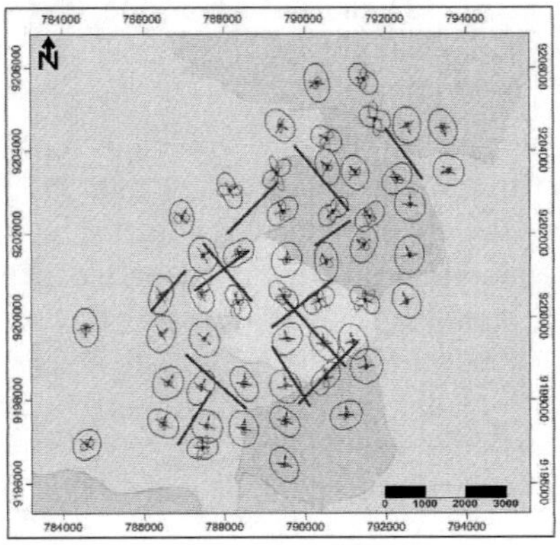

Figure 4. Impedance polar diagram map in the frequency 0.1 Hz and the geoelectric strikes (blue line).

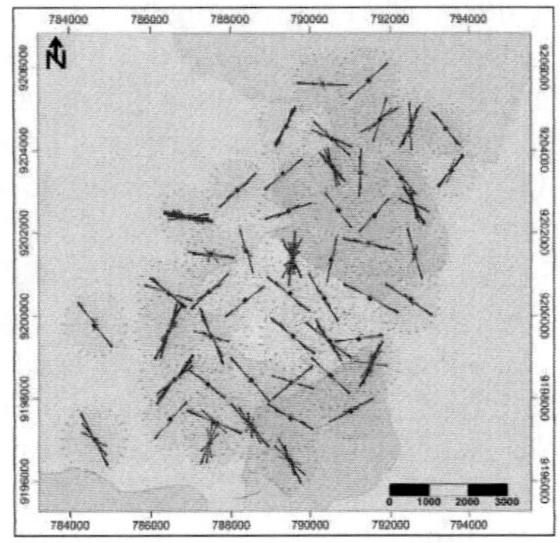

Figure 5. Zstrike rose diagram map in the frequency greater than 0.1 Hz.

The direction of the geoelectric strike obtained from the polar impedance diagram and the Zstrike rose diagram correlate to the direction of the geological structure as shown in figure 1, i.e. the direction of the geological structure to the NW and NE trends. We interpreted that this is may be caused by the fluid flow mechanism and alteration process within geological structures which make the resistivity differences laterally.

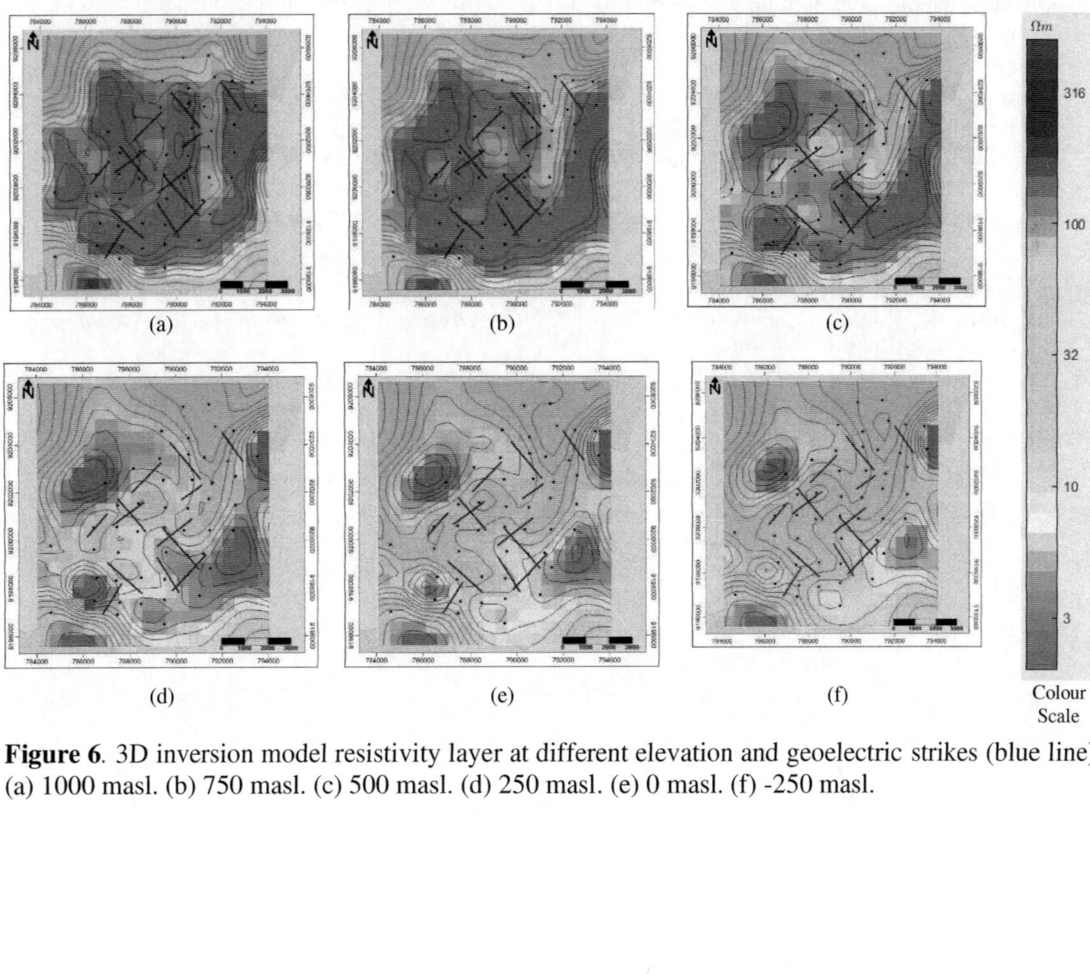

Figure 6. 3D inversion model resistivity layer at different elevation and geoelectric strikes (blue line). (a) 1000 masl. (b) 750 masl. (c) 500 masl. (d) 250 masl. (e) 0 masl. (f) -250 masl.

To compare with the 3D inversion model, the results of the geoelectric strike from the impedance polar diagram confirmed by Zstrike rose diagram were overlaid with 3D inversion model resistivity layer at different elevation (figure 6). Figure 6 (a), (b), (c), (d), (e), and (f) show the correlation between the geoelectric strike with the resistivity layer at elevation 1000 masl, 750 masl, 500 masl, 250 masl, 0 masl, and -250 masl, respectively. At shallow depth, the clay cap obviously covers the geothermal system and therefore, the correlation between the geoelectric strike and the resistivity structure in the reservoir still cannot be observed. From the figures, the reservoir section begins to reveal at a 500 masl elevation (figure 6c). It can be seen that the strikes represent a lateral resistivity difference in the resistivity layer. At some point, several interpretation of geoelectric strikes have no correlation with the model at a deeper depth, indicating that lateral resistivity variations occur only until a certain depth. However, there are several geoelectric strikes that always correlate consistently with the model in the deep elevation, indicating the possibility that these geoelectric strikes are the reservoir boundary that separates the reservoir zone from its surroundings.

6. Conclusions

This study has yielded results that the direction of geoelectric strike in the southern part of the Wayang Windu geothermal field is in the NW and NE trends. These trends correlate to the direction of the major geological structure. The position of the geoelectric strikes also has a good correlation with 3D inversion model resistivity layer. This is expected to assist a further interpretation of the geothermal area delineation.

Acknowledgement

We would like to thank to the Management of Star Energy Geothermal Wayang Windu Ltd., Indonesia for their support of the data and permission to present this paper. Special thanks go to PT NewQuest Geotechnology that helped us by providing 3D inversion software for MT data.

References

[1] Simpson F, and Bahr K 2005 Practical magnetotelluric *Surveys in Geophysics* (Cambridge: Cambridge University Press) **28** 325
[2] Telesca L, Lovallo M, Hsu H L and Chen C C 2012 *J. Asian Earth Sci.* **54** 72
[3] Bogie I, Kusumah Y I and Wisnandary M C 2008 *Geothermics* **37** 347
[4] Masri A, Barton C, Hartley L and Ramadhan Y 2015 *Proc. Fourtieth Workshop on Geothermal Reservoir Engineering* (Stanford: Stanford University)
[5] Vozoff K 1972 *Geophysics* **37** 98
[6] Reddy I K, Rankin D, and Phillips R J 1977 *Geophysics Journal of the Royal Astronomical Society* **51** 313
[7] Berdichevsky M N 1968 *Electrical prospecting by the method of magnetotelluric profiling* (Moscow: Nedra Publ. House)
[8] Swift C M 1967 *A Magnetotelluric investigation of an electrical conductivity anomaly in the South Western United States* Doctoral Thesis (Massachusetts: Massachusetts Institute of Technology)
[9] Teklesenbet A 2012 *Multidimensional inversion of MT data from alid geothermal area, eritrea; comparison with geological structures and identification af a geothermal reservoir* Master Thesis (Reykjavík: University of Iceland)

Structural and crystal orientation analysis of Al-Si coating on Ni-based superalloy by means of EBSD technique

A N Muslimin[1], E Sugiarti[1,3], T Aritonang[2], R. I Purawiardi[1] and R D Desiati[3]

[1]Laboratory of Focused Ion Beam, Research Centre for Physics, Indonesian Institute of Sciences (LIPI), Serpong 15314, Indonesia
[2]Department of Metallurgical Engineering, Faculty of Engineering, University of Sultan Ageng Tirtayasa, Cilogen 42124, Indonesia
[3]Laboratory of High Temperature Materials and Coatings, Research Centre for Physics, Indonesian Institute of Sciences (LIPI), Serpong 15314, Indonesia

E-mail: enis002@lipi.go.id

Abstract. Ni-based superalloy is widely used for high performance components in power generation turbine due to its excellent mechanical properties. However, Ni-based superalloy has low oxidation resistantance. Therefore, surface coating is required to improve oxidation resistance at high temperatures. Al-Si as a coting material was successfully co-deposited on Ni-based substrate by pack cementation method at 900 °C for about 4 hours. The oxidation test was carried out at high temperature of 1000 °C for 100 hours. Micro structural characterization and analysis on crystal orientation were perfomed by using Field Emission Scanning Electron Microscope (FE-SEM) and Electron Back Scatter Diffraction (EBSD) technique, respectively. The results showed that the coating layer with a homogenous layer and had a thickness of 53 μm consisting of β-NiAl with cubic structure and Ni_2Al_3 with hexagonal structure. TGO layer was developed after oxidation and had a thickness of about 5 μm consisting of α-Al_2O_3 and spinel $NiCr_2O_4$. The phase composition map and crystal orientation acquired by EBSD technique was also discussed both in TGO and coating layers.

1. Introduction

Ni-based superalloy is used in gas turbine engine component that require high creep resistance and fatigue strength at high operating temperatures. The applications of Ni-based superalloy in aggressive environtment are subjected to oxidation and hot corrosion attacks. Surface coating has been one of solution to solve these problems for superalloys in harsh environtment [1]. Aluminium (Al) is the most commonly used element for coating material due to the ability to form highly stable protective layer of Al_2O_3 in high temperature [2-4]. However, simple aluminide coating was not sufficient to withstand against high temperature oxidation and hot corrosion. Therefore, a combination of aluminide coating with silicon (Si) element was used to increase oxidation and hot corrosion resistance at high temperatures [5-11]. It has been known that Al-Si coating applied by a pack cementation method would formed β-NiAl and δ-Ni_2Al_3 phases in coating layer [12-13]. A thermally grown oxide (TGO) layer is formed at the surface of coating layer during high temperature oxidation. The TGO was mainly composed of mixed oxide containing α-Al_2O_3 and spinels $Ni(Cr,Al)_2O_4$ [1].

Recently, many researchers have studied the phase distribution and correlation towards microstructure of coating and TGO layer by FE-SEM and EBSD techniques [1, 14]. Wollmer et al.

[14] investigated the micro structural and crystal orientations of β-NiAl and γ-Ni_3Al formed in the NiAl coating on superalloy substrate. In addition, Chmiela et al [1] studied microstructural and crystal orientation of thermal barrier coating with addition of $Gd_2Zr_2O_7$ as reactive element. It showed that EBSD is one of powerfull technique that can provide detailed information about the crystalline structure and crystallographic orientations of the grains.

Therefore, the present research will investigate the microstructure, phase distribution, crystal structure, and crystal orientation of Al-Si coating as well as TGO layer before and after oxidation by using FE-SEM and EBSD technique.

2. Experimental

The Al-Si powder was coated on Ni-based superalloy by pack cementation methode. The coating material has a composition of 2.6 wt.% Al, 10 wt.% Si, 4 wt.% NH_4Cl and 83.4 wt.% Al_2O_3. The pack cementation process was carried out at temperature of 900 °C for approximately 4 hours. The Al-Si coating samples were subjected to oxidation test at 1000 °C for 100 hours. A microstructural characterization was performed on cross sectional of Al-Si coating sample before and after oxidation by FE-SEM (JEOL JFIB-4610F) equipped with Energy Dispersive X-ray Spectroscopy (EDS, Oxford X-Max 50 detector). Crystal structure, crystal orientation and phase composition were evaluated by EBSD technique (Oxford Nordlys Max^3 detector). Prior to EBSD analysis, the sample was cut in cross section with a size of 3×5×0,5 mm followed by mechanical polishing by SiC paper with a grid size from #100 to #3000 and final polished using Al_2O_3 suspension with a grid size 0.10 μm and 0.05 μm. A quantitative analysis was acquired by EDS technique with primary electron beam 15 kV and probe current number 13. Moreover, EBSD analysis was collected with condition of 20 kV electron beam and 14 probe current in order to optimize the data acqusition.

3. Results and discussions

3.1 Microstructure of Al-Si coating layer

Figure 1 shows cross sectional morphologies of Al-Si coating layer before and after oxidation. The result showed that the coating comprised of two layers: interdiffusion layer and Al-Si coating layer. Complex structure with irregular shape and different contrast in the interdiffusion zone (figure 1a) might indicated the inward and outward diffusion of nickel from the substrate and aluminium-silikon from the coating element.

A homogeneous layer with the thickness of 35 μm was formed on the Al-Si coating layer as shown in figure 1b. A TGO layer was entirely formed on the top surface of Al-Si coating layer as shown in figure 1c. A lot of voids were formed at the interface between Al-Si coating and TGO layers as clearly shown in SEI image of figure 1(c). The thickness of TGO layer was approximate 5 μm.

3.2 Quantitative analysis of Al-Si coating layer

EDS analysis were performed on the cross sectional of Al-Si coating layer before and after oxidation as shown in figures 2 and figure 3, respectively. According to the previous result, β-NiAl and δ-Ni_2Al_3 phases have been identified in the Al-Si coating layer on Ni-based superalloy [12-13]. The results obtained in figure 2 revealed that Al-Si coating layer contained high intensity of nickel and aluminium (figures 2b - 2d). The compositions of each elements was notified in about 45 at.% Ni and 55 at.% Al. It is known that Ni with composition range of 45-60 at% formed at temperature of 900 °C could identified as cubic structure of β-NiAl phase based on the binary phase diagram of Ni-Al system [15]. Therefore, the present results might indicated the formation of homogenous and stable β-NiAl in the whole part of Al-Si coating layer as shown in figure 2a. It is noticed that the composition of initial coating powder and temperature of pack cementation used in the present study would promote the formation of silicon-modified NiAl through inward diffusion of Al and Si as well as the outward diffusion of Ni. These results have a good agreement with the statement that Al is main element diffusion in NiAl phase when the content of Al in this phase is more than 41 at.% [15].

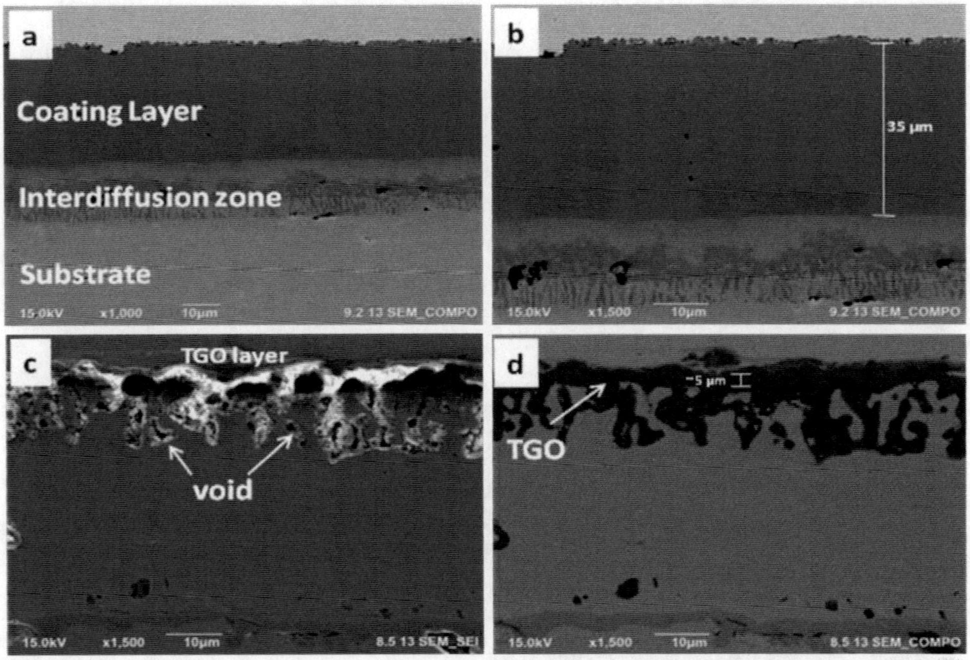

Figure 1. FE-SEM cross-section morphology of Al-Si coating layer: (a) and (b) BSE images of Al-Si coating sample before oxidation with magnification of 1000x and 1500x, respectively. (c) and (d) SEI and BSE images of Al-Si coating sample after oxidation.

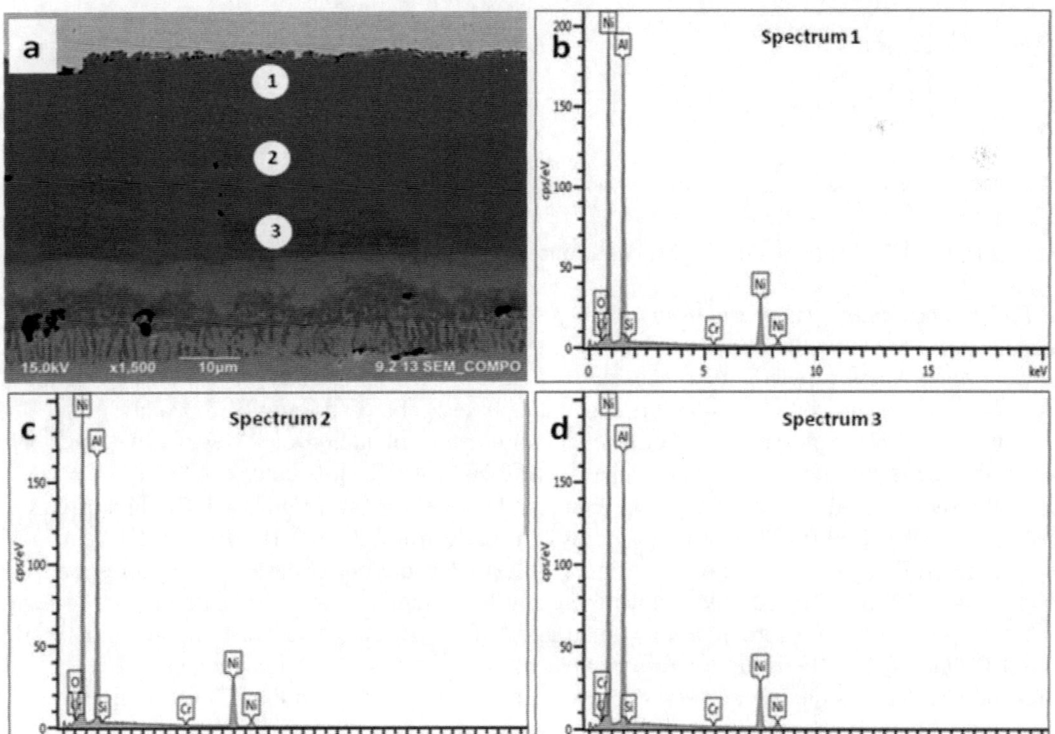

Figure 2. Three points of EDS spectra obtained from Al-Si coating layer before oxidation.

Figure 3 shows the EDS analysis of Al-Si coating layer after oxidation. The peak intensity in point (1) showed high signal for aluminium and oxygen content which indicated by dark area as shown in figure 3a. In contrast, high intensity of nickel, aluminium, and small amount of chromium and silicon were obtained in the grey area of coating layer as shown in figure 3a. The high composition of nickel in the gray area might indicated the existance of stable β-NiAl phase and the formation of spinel $NiCr_2O_4$ [1]. The results revealed that the outer layer was comprised of α-Al_2O_3 protective oxide. By the formation of a dense and continuous layer, a protective oxide could play an important part as diffusion barrier of oxygen to diffuse inward into the coating layer and substrate. The reason is because alpha alumina (α-Al_2O_3) is very stable because of its crystalline structure, thus it can reduce the diffusion rate of anions and cations [8].

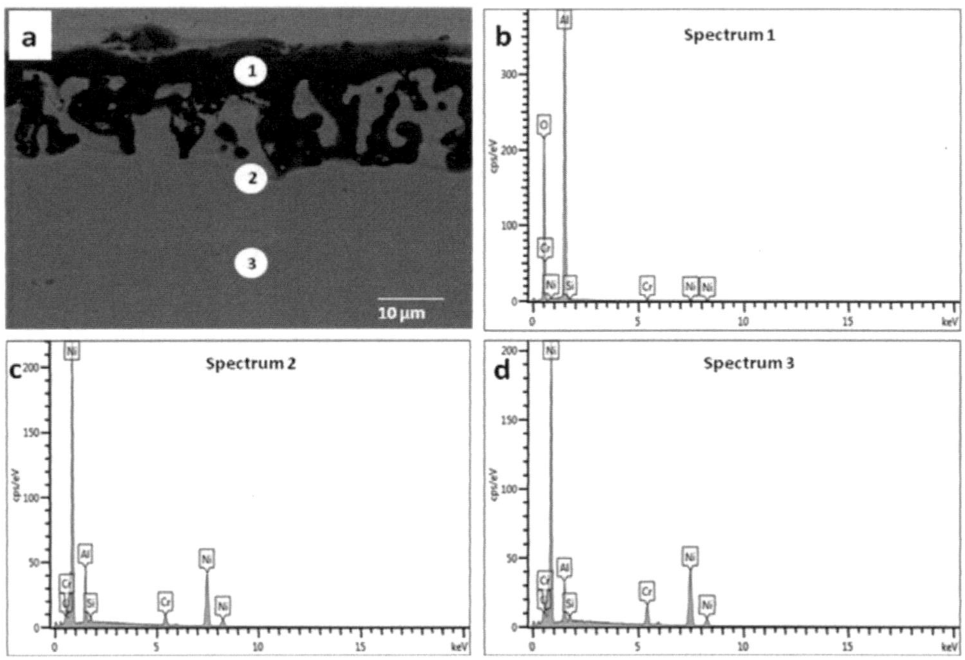

Figure 3. Three points of EDS spectra obtained from Al-Si coating layer after oxidation.

3.3 Phase composition and crystal orientation of Al-Si coating layer before and after oxidation

EBSD analysis was initially obtained by acquiring the kikuchi pattern of selected area and by using a software, the kikuchi pattern is then considered indexed when its orientation and phases are known. Therefore, it is very important to store the candidate phases (table 1) prior to EBSD analysis.

Figure 4 shows the phase composition map and crystal orientation of Al-Si coating layer before oxidation. The result revealed the major phase was β-NiAl which distributed 99.85 % in the selected area (20×20 µm) of Al-Si coating layer as indicated by blue region in figure 4a. The minor phases of Ni_2Al_3 was notified by 0.15% and indicated by dot marked in red color (figure 4a). Further, inverse pole figure (IPF) in z axis as shown in figure 4b revealed the crystal orientation correspond to its grain. A large grain with red color showed the crystal orientation of β-NiAl in <0 0 1> direction. Whereas, inside the β-NiAl grain, a very small grain of Ni_2Al_3 co-existed with the direction between <0 0 0 1> and <0 1 -1 0> as indicated by yellow color in figure 4b. A homogeneous β-NiAl with big grain showed that this phase was very stable and uniformly observed in the Al-Si coating layer. This result indicated that the existance of β-NiAl with cubic structure would increase the oxidation resistantance and the life time of the sample for the application [5].

Table 1. Crystal structure of identified phases.

Phase name	Crystallographic system	Space group	Unit cell parameter (nm)		
			a_0	b_0	c_0
β-NiAl	Cubic	Pm-3m	2.88	2.88	2.88
Ni_2Al_3	Hexagonal	P-3m1	4.065	4.065	4.906
α-Al_2O_3	Rhombohedral	R-3c	4.759	4.759	12.99
$NiCr_2O_4$	Tetragonal	I 41/amd	5.760	5.760	8.50

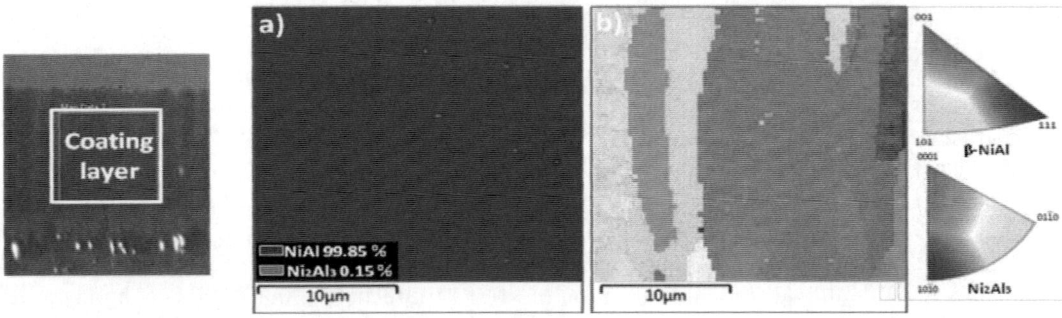

Figure 4. EBSD investigation of Al-Si coating sample before oxidation: (a) phase composition map and (b) IPF Z crystal orientation.

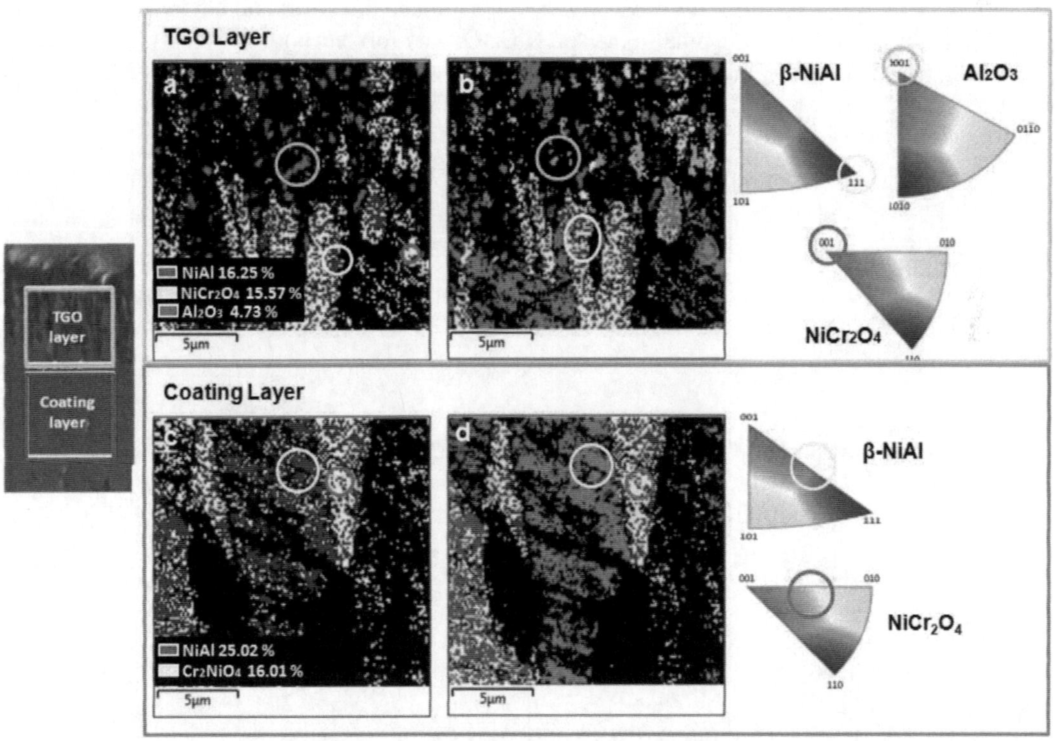

Figure 5. EBSD investigation of Al-Si coating sample after oxidation: (a) and (c) phase composition map of TGO and coating layer, respectively. (b) and (d) IPF Z crystal orientation of TGO and coating layer, respectively.

EBSD analysis in Al-Si coating after oxidation is shown in figure 5. EBSD analysis was captured in TGO and coating layer with observing area of 15×15 μm. It has been understood from SEM result that the thickness of TGO was about 5 μm. Therefore, EBSD analysis in TGO layer was captured in the interface between TGO and coating layer as shown in figures 5a and 5b. The results showed that NiAl, Cr_2NiO_4 and Al_2O_3 had a percentage of 16.25%, 15.57% and 4.73% in the observing area, respectively. Al_2O_3 with red region in figure 5a had a smaller grain structure than NiAl and Cr_2NiO_4. Moreover, Cr_2NiO_4 coexisted in NiAl grain structure which indicated by yellow and blue region in figure 5a. Black region in phase composition map and inverse pole figure (IPF) image indicated non-indexed area. By IPF in z axis, it showed that the crystal orientation of Al_2O_3 was on <0 0 0 1> plane direction, β-NiAl was on <1 1 1> plane direction, and $NiCr_2O_4$ was on <0 0 1> plane direction in TGO layer. Moreover, the crystal orientation of β-NiAl was between <1 1 1> and <0 0 1> plane direction and $NiCr_2O_4$ was between <0 0 1> and <1 1 1> plane direction.

4. Conclusion

Microstructure, phase composition, and crystal orientation of Al-Si coating sampels were succesfully investigated by using FE-SEM and EBSD techniques. Al-Si coating layer developed by pack cementation process at temperature of 900 °C had a homogeneous layer with the thickness of 35 μm. The quantitative of EDS analysis showed that the composition of Ni-Al in the coating layer indicated the existance of β-NiAl with cubic structure having 99.85% of phase map from EBSD analysis. β-NiAl had a large grain with crystal orientation of <0 0 1> direction in the Al-Si coating layer before oxidation. Ni_2Al_3 had also been identified in the coating layer before oxidation with the phase composition of 0.15%. After oxidation, TGO layer was growth on the top surface of Al-Si coating layer with a thickness of about 5 μm which comprised of protective oxide of Al_2O_3 and spinel oxide of $NiCr_2O_4$. Crystal orientation of Al_2O_3 was on <0 0 0 1> plane direction and $NiCr_2O_4$ was on <0 0 1> plane direction in TGO layer. Spinel oxide of $NiCr_2O_4$ was formed both in TGO and coating layers after oxidation. This phase is formed from reaction between Cr_2O_3 and NiO before the continuous oxida scale of Cr_2O_3 formed. Spinel Cr_2NiO_4 can act as a diffusion barrier of Ni ions thus preventing the formation of excessive unprotective Ni oxide. In addition, a protective oxide of Al_2O_3 formed in TGO layer could improved the oxidation resistance of Ni-based superalloy for high temperature applications.

Acknowledgments

The authors would like to thank to Kegiatan Unggulan LIPI, INSINAS IRPI grant No. 04/INS-2/PPK/E/E4/2017 and 2016 Toray Science Technology Research Grant for financial support of this work.

References

[1] Chmiela B, Sozanska M and Moskal G 2012 *IOP: Conf. Ser.: Mater. Sci. Eng.* **32** 012006
[2] Xiang Z D, Datta P K 2003 *Mater. Sci. Technol.* **19** 935
[3] Muktinutalapati N R 2011 *Advances in Gas Turbine Technology* 22905
[4] Tong L, Dengzun Y and Chungen Z 2010 *Chinese. J. Aeronaut.* **23** 381
[5] Tu X, Peng H, Zheng L, Qi W, He J, Guo H and Gong S 2015 *Appl. Surf. Sci.* **325** 20
[6] Fu C, Kang W K and Cao G H 2014 *Surf. Coat. Technol.* **258** 347
[7] Wang K L, Chen F S and Leu G S 2003 *Mater. Sci. Eng. A* **357** 27
[8] Fan Q X, Jiang S M and Wu D L, Gong J 2013 *Corros. Sci.* **76** 373
[9] Liu C T, Ma J, Sun X F and Zhao P C 2010 *Surf. Coat. Technol.* **204** 3641
[10] Xiang Z D, Burnell-Gray J S and Datta P K 2001 *J. Mater. Sci.* **36** 5673
[11] Mansuri M, Hadavi S M M and Zare E 2016 *Metall. Mater. Trans. A* **47** 293
[12] Bozza F, Bolelli G, Giolli C, Giorgetti A, Lusvarghi L, Sassatelli P, Scrivani A, Candeli A and Thoma M 2014 *Surf. Coat. Technol.* **239** 147
[13] Arabi H, Rastgari S, Salehpour Z and Bakhshi A 2008 *Inter. J. Eng. Sci.* **19** 39
[14] Wöllmer S, Zaefferer S, Göken M, Mack T and Glatzel U 2003 *Surf. Coat. Technol.* **167** 83
[15] Kitahara H, Ueji R, Ueda M, Tsuji N and Minamino Y 2005 *Mater. Charact.* **54** 378

IOP Conf. Series: Journal of Physics: Conf. Series **985** (2018) 012021 doi:10.1088/1742-6596/985/1/012021

Fe K-Edge X-ray absorption near-edge spectroscopy (XANES) and X-ray diffraction (XRD) analyses of $LiFePO_4$ and its base materials

C Latif[1], V S I Negara[1], W Wongtepa[2], P Thamatkeng[2], M Zainuri[1] and S Pratapa[1]

[1] Department of Physics, Institute of Technology Sepuluh Nopember (ITS), Surabaya, Indonesia
[2] Synchrotron Light Research Institute (SLRI), Nakhon Ratchasima, Thailand

E-mail: suminar_pratapa@physics.its.ac.id

Abstract. XANES analysis has been performed with the aim of knowing the Fe oxidation state in a synthesized $LiFePO_4$ and its base materials. XANES measurements were performed at SLRI on energy around Fe K-edge. An XRD analysis has also been performed with the aim of knowing the phase composition, lattice parameters and crystallite size of the $LiFePO_4$ as well as the base materials. From the XRD analysis, it was found that the dominating phase in the iron sand sample was Fe_3O_4 and the only phase found after calcination was $LiFePO_4$. The latter phase exhibited crystallite size of 100 nm and lattice parameters a = 10.169916 Å, b = 5.919674 Å, c = 4.627893 Å. Qualitative analysis of XANES data revealed that the oxidation number of Fe in the sample before calcination was greater than that after calcination and Fe in the natural iron sand, indicated by the E_0 values of 7129.2 eV, 7120.6 eV and 7124.4 eV respectively.

1. Introduction

$LiFePO_4$ (LFP) is one of the cheapest, safe, and highly promising lithium battery cathode materials with a high theoretical capacity of approximately 170 mAh/g [1]. Previously, LFP with commercial raw materials have been synthesized by sol gel [2], solid state reaction [3], hydrothermal [4], and other methods. We found hardly any literature discussing a synthesis of LFP using natural mineral as a source of Fe. Nevertheless, iron sand as the source of Fe can easily be mined in Indonesia and its purification or uses as raw materials for the synthesis of functional materials are possible [5]. Meanwhile, it is known that structural studies, namely microstructure using electron microscopy technique and crystal structure using diffraction method, favors on explaining some physical characteristics of materials. On another hand, there are other measurements namely local structure using XANES that can provide local chemical information on complex or regular materials, such as oxidation state, bond angle, and site symmetries [6]. Therefore, the combination of XRD and XANES for LFP synthesis using iron sand as Fe source will be an interesting study to find out the structure of the material.

In this work, we report the XANES and XRD analysis of LFP with iron stone as the base material. The iron stone was taken from Tanah Laut, South Borneo, Indonesia. XANES measurements were

IOP Conf. Series: Journal of Physics: Conf. Series **985** (2018) 012021 doi:10.1088/1742-6596/985/1/012021

performed at SLRI, Thailand. The iron stone contained approximately 89.8% Fe according to XRF measurement. The minor elements included Si (7.1%), Ca (1.4%), and Mn (1.4%).

2. Experimental

The LFP synthesis began with obtaining the Fe precursor by dissolving the natural iron stone in the HCl solution while stirring at a temperature of 100 °C for 1.5 h. After the solution was formed, filtration was made using filter paper to separate the solution with impurity elements. This solution was dissolved overnight and called solution I. Furthermore, Li_2CO_3 was dissolved with HCl and called solution II. Solutions I and II were mixed and further added with $NH_4H_2PO_4$ powder which has been dissolved in distilled water and called solution III. Solution III was stirred for 1 h, then NH_4OH was drop-wisely introduced to control the pH until a gel formed. The gel is then dried and wet milled for 9 hours with alcohol as the mixing medium. After that, the sample was dried and calcined at 600 °C for 10 h in a nitrogen gas atmosphere. XRD was used for characterizing crystalline structures whereas XANES was used to characterize oxidation states. XRD measurements were carried out using PANalytical X'pert PRO diffractometer. This measurement used the target Cu (λ = 1.54060Å) at scanning range of 10-70 °2θ. Meanwhile, the XANES measurements were made at the Beam-line 8 of the Synchrotron Light Research Institute (SLRI), Nakhon Ratchasima, Thailand. This measurement used energy from 7012-7312 eV.

3. Results and discussions

Figure 1 shows the XRD patterns of the samples. Analysis of XRD data of the base material showed that the phases present in iron stone were Fe_3O_4 magnetite (Reference Code 96-900-5842) and Fe_2O_3 maghemite (96-900-0140). A strong indication of the presence of the amorphous substance is evident. Furthermore, the XRD data of the sample before calcination showed that the LFP phase has not yet been formed, but the identified phase is NH_4Cl (96-101-1130). This is due to the acid-base mixing. The mixing caused the dissolution of the iron stone and amorphization of the crystals. The thermal energy during drying was insufficient to favor reaction of the composing substances and only able to crystallizes NH_4Cl. Further analyzing the XRD data, we found that the $LiFePO_4$ (96-110-1112) phase was formed after calcination, where the NH_4Cl phase was no longer found as a result of evaporation during calcination. We note here that a single LFP phase has been achieved.

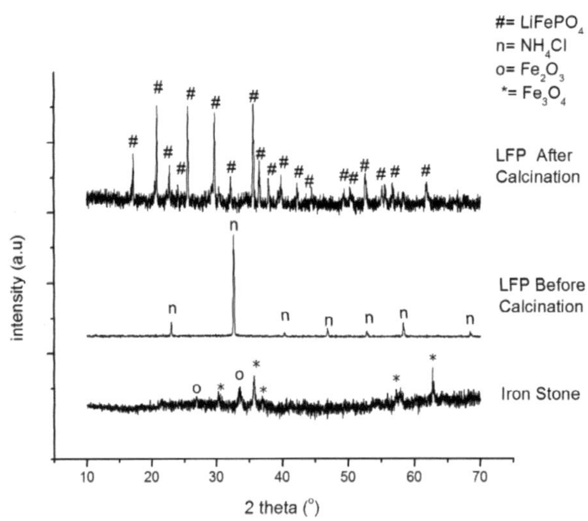

Figure 1. XRD patterns for the samples.

IOP Conf. Series: Journal of Physics: Conf. Series **985** (2018) 012021 doi:10.1088/1742-6596/985/1/012021

Table 1 shows the output of the Rietveld analysis on the XRD data for the after calcination LFP sample. Olivine $LiFePO_4$ crystal data (Code 96-110-1112) was used in the analysis. The goodness-of-fit values are GoF = 1.5% (Rietica) and Sig = 1.7% (MAUD) indicating that the Rietveld refinement using both softwares were successful. The result shows that the orthorhombic $LiFePO_4$ phase exhibits lattice parameters of a = 10.169916 Å, b = 5.919674 Å, c = 4.627893 Å, which are comparable to those in the literature [7]. From the analysis using MAUD, we obtained the crystallite size of $LiFePO_4$ of approximately 100 nm. The crystals also exhibited relatively low residual micro-strain, i.e. 4.034×10^{-5}. Therefore, the sol-gel method, making use of local iron stone as the source of Fe, can be used to produce a nano-metric, single-phase LFP powder.

Table 1. Rietveld analysis output of the XRD data of the calcined sample using *Rietica* and *MAUD*.

	Rietica				MAUD		
Sample	*a* (Å)	*b* (Å)	*c* (Å)	GoF (%)	Size (nm)	Microstrain	Sig. (%)
LFP After Calcination	10.169916	5.919674	4.627893	1.5	100	4.0×10^{-5}	1.7

Figure 2 shows the XANES spectra of the LFP material and its base samples. The pre-edge and edge regions can be used to explain the oxidation state of Fe in the samples. The pre-edge region is formed by the transition of electrons from the 1s orbital to the outer shell depending on the absorber used [8]. Until now there has not been an equation for modeling the XANES spectrum thereforeusually the analysis is only done through matching measurable patterns of components to the combined pattern [9].

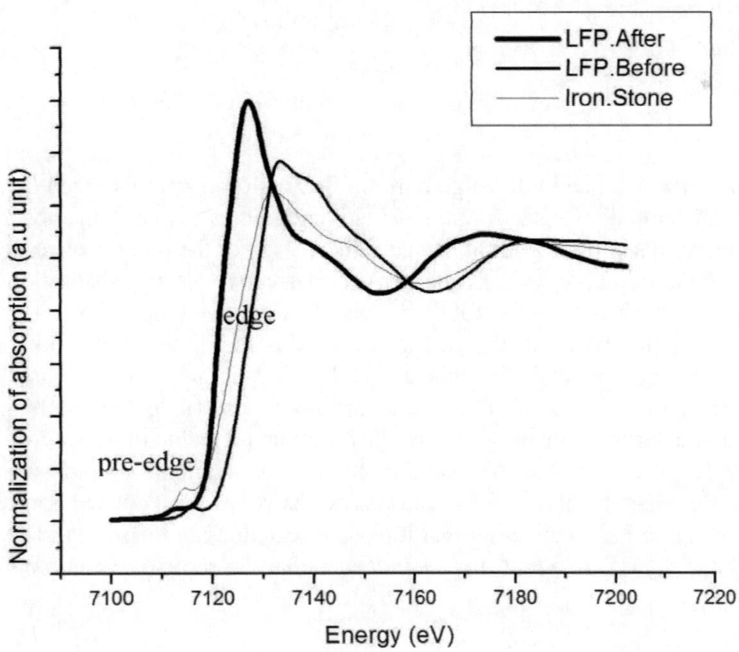

Figure 2. XANES spectra at the K-edge of iron for the samples.

Figure 3 shows the pre-edge region of the samples. The pre-edge peak of the iron stone is about 7114 eV. This result corresponds to the phase identification with XRD, e.g. that the iron stone sample contains dominant Fe^{3+} ions from Fe_3O_4 and Fe_2O_3. Meanwhile, the pre-edge peak for the calcined samples is about 7112 eV, indicating that the Fe atoms in the samples are Fe^{2+} ionic status indicating the formation of LFP containing Fe^{2+} ions. This is consistent with the results obtained by Berlier et al. where $LiFePO_4$ (Fe^{2+} oxidation) has a pre-edge peak at 7111.8 eV. This pre-edge peak position is smaller than that of $FePO_4$ (Fe^{3+} oxidation), ca. 7114.2 eV [10]. For the uncalcined samples, the pre-edge peak position is about 7115. This indicates that the oxidation state of the uncalcined sample is larger than the other two samples.

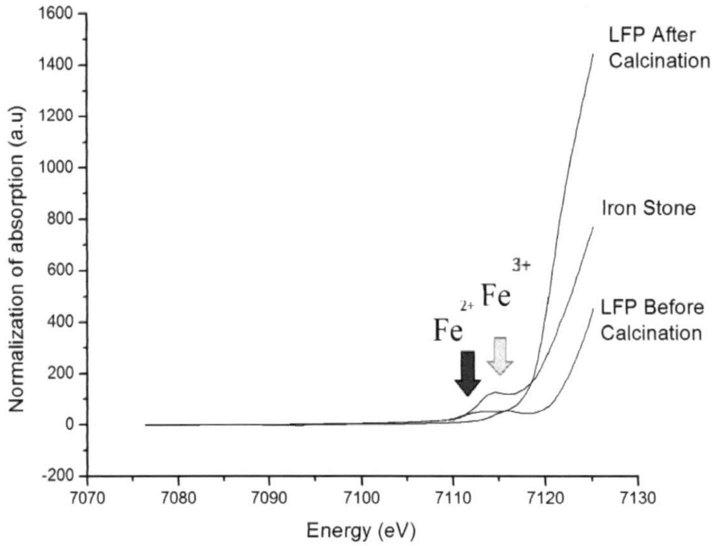

Figure 3. Pre-edge Fe absorption of the samples.

The second analysis is done in the edge region. This region is characterized by a sharp increase in absorption with discontinuities – see figure 2. This increase occurs due to the absorption of X-rays by the electrons of Fe atoms in the K orbital, because the energy of the photon corresponds to the electron binding energy of the orbital ($E = E_0$). The peaking derivative for the absorption plot in figure 2 is associated to E_0. The absorption is followed by electron excitation from 1s (orbital K) to the unoccupied state. Another term often used for any edge is the dipole transition from this level (core levels) to the empty valence band. For this study, the XANES plots of the observed materials were compared with those of the FeO and Fe_2O_3 standard materials. Table 2 presents the E_0 values for all samples including the standard materials. It can be seen that the value of E_0 for iron stone lies between FeO and Fe_2O_3. It means that the iron stone contains dominating Fe atoms which have oxidation numbers which are a combination of 2+ and 3+. Meanwhile, the calcined sample shows E_0 value which is almost equal to FeO, indicating that it has an oxidation number of 2+. The uncalcined sample exhibits an E_0 value greater than Fe_2O_3. This indicates that the sample has an oxidation number of 3+ or more.

Table 2. E_0 values of the samples.

Sample	E_0 (eV)
Iron Stone	7124.4
LFP Before Calcination	7129.2
LFP After Calcination	7120.6
FeO (standard)	7119.0
Fe_2O_3 (standard)	7125.2

4. Conclusion

It can be concluded from this study that XRD and XANES analyses can show the crystal structure and oxidation state of $LiFePO_4$ and its base materials. The iron stone contained Fe_3O_4 and Fe_2O_3 phases which have 2+ and 3+ oxidation numbers. The uncalcined sample had NH_4Cl and amorphous "phases" of LFP with Fe oxidation numbers of 3+ or more. The calcined sample, on the other hand, exhibited LFP phase with a crystallite size of approximately 100 nm and lattice parameters of a = 10.169916 Å, b = 5.919674 Å, c = 4.627893 Å, as well as oxidation number of 2+.

Acknowledgments

The authors would like to thank the directorate of higher education of Indonesia and LPPM ITS to financially support the research through Program PMDSU Number 490/PKS/ITS/2017 given CL through SP. The authors also thanked the Synchroton Light Research Institute (SLRI) of Thailand who provided the XAS facility under Beam Time Project Number 2618 (2017).

References

[1] Padhi A K, Nanjundaswamy K S and Goodenough J B 1997 *J. Electrochem. Soc.* **144** 1188
[2] Xie G, Zhu H J, Liu X M and Yang H 2013 *J. Alloys Compd.* **574** 155
[3] Ding K, Li W, Wang Q, Wei S and Guo Z 2012 *J. Nanosci. Nanotechnol.* **12** 3812
[4] Bolloju S, Rohan R, Wu S T, Yen H X, Dwivedi G D, Lin Y A and Lee J T 2016 *Electrochim. Acta* **220** 164
[5] Angela R, Islam H, Sari V, Latif C, Zainuri M and Pratapa S 2017 *AIP Conf. Proc.* **1788** 030102
[6] Ankudinov A L, Ravel B, Rehr J J and Conradson S D 1998 *Phys. Rev. B* **58** 7565
[7] Dong Y Z, Zhao Y M and Duan H 2011 *Mater. Chem. Phys.* **129** 756
[8] Trofimova N N, Veligzhanin A A, Murzin V Y, Chernyshov A A, Khramov E V, Zabluda V N, Edel'man I S, Slovokhotov Y L and Zubavichus Y V 2013 *Nanotechnol. Russ.* **8** 396
[9] Rehr J J, Kas J J, Vila F D and Newville M 2017 Theory and Analysis of XAFS (*XAFS Techniques for Catalysts, Nanomaterials, and Surfaces*) ed Y Iwasawa, K Asakura and M Tada (USA: Springer International Publishing Switzerland)
[10] Berlier G, Spoto G, Fisicaro P, Bordiga S, Zecchina A, Giamello E and Lamberti C 2002 *Microchem. J.* **71** 101

Characterization of sonicated natural zeolite/ferric chloride hexahydrate by infrared spectroscopy

T A B Prasetyo and B Soegijono

Department of Physics, Materials Science Program, Universitas Indonesia, Depok, Indonesia 16424

E-mail: naufal@ui.ac.id

Abstract. The characteristics of sonicated Bayah natural zeolite with and without ferric chloride hexahydrate solution using infrared method has been studied. High intensity ultrasonic waves were exposed to the samples for 40 min, 80 min and 120 min. Infra red spectra analysis was conducted to evaluate zeolite vibrational spectrum contributions, namely, the vibrations from the framework of the zeolite, from the charge-balancing cations, and from the relatively isolated groups, such as the surface OH groups and their behavior after sonication process. An addition of $FeCl_3.6H_2O$ and sonication process on natural zeolite improved secondary building units link by forming oxygen bridges and also close relationship with duration of applied high intensity ultrasonic process. Longer ultrasonic process resulted in more increment of O-H absorbance.

1. Introduction

In an attempt to improve the properties of Indonesian natural zeolite, we have undertaken the characterization and modification of natural zeolite produced in Bayah, Indonesia. This Zeolite consists mainly of crystalline mordenite type [1]. Bayah natural zeolite has been used as adsorbates for various purposes. It has been studied previously and standardized by Indonesian standardization body (SNI). As an adsorbent material, it is categorized as a material with the highest cation exchange capacity compared with other Indonesian natural zeolites from Sukabumi, Tasikmalaya and Cikalong [1, 2].

High intensity ultrasonic wave processing has been used to produce nanomaterials and chemical reaction. High intensity ultrasonic liquid processor produces microbubbles in the solution between of the primary and secondary flow. This sonication process may increase or change the physical and chemical properties of a material [3]. It has been studied and understood that high power ultrasonic irradiation is very effective in homogenizing the reactants in the suspension, leading to the improvement of the reactivity of both solid and liquid by simulating their active surface [4]. Purification of Natural Zeolite by using ultrasonic technique is also applied by researchers to remove and separate impurities materials; this technique is believed to deliver faster process and result [5].

Infra red spectra analysis has been carried out to find out specific improvement of existing natural zeolite properties after introducing ultrasonic wave and its modifications by adding $FeCl_3.6H_2O$ and performing sonication process for 40 min, 80 min and 120 min. Spectra were recorded in the range of 4,000-600 cm^{-1} and evaluated on most of three spectra region; O–H spectra region is very interesting to be observed as gas-absorbent candidate material [6].

2. Experimental procedure

Natural zeolite produced in Bayah, Indonesia was ground, homogenized and sieved below 150 μm (100 Mesh Tyler ™) as part of sample preparation and characterization [7]. Addition of Ferric chloride hexahydrate ($FeCl_3.6H_2O$) produced by Merck Millipore, Chemical Abstracts Service (CAS) #10025-77-1 with specified weight was diluted in 100 ml of distilled water.

To remove impurities, sieved natural zeolite were washed and stirred with distilled water. Washing process was conducted four times with two hours long for each sequence [8]. Air drying was performed prior to initial characterization to find out the baseline of zeolite infrared spectroscopy. The natural zeolite sample is denoted as ZA (natural zeolite). Natural zeolite samples were exposed with high ultrasonic wave for 40 min, 80 min and 120 min with the sample code of ZA40, ZA80, and ZA120 respectively. Another zeolite samples were added with 10 gram $FeCl_3.6H_2O$ and diluted in 100 ml distilled water. The solution was exposed to high intensity ultrasonic liquid processor VCX750 at 20 kHz, 40 Amp [9] for 40 min (sample code ZAM1), 80 min (sample code ZAM2) and 120 min (sample code ZAM3). Slurry product of the sonication process were precipitated and subsequently heated by microwave furnace at temperature of 90 °C for 30 min [8].

Fourier transform infrared (FT-IR) spectroscopy was used to study zeolite and zeolite/$FeCl_3.6H_2O$ in various sonication processes. Pure KBr powder was added to the slightly solid sample and then blended. This mixture was placed in a mold and pressurized by a mechanical pressure apparatus for a few minutes. The FT-IR spectra were recorded in the range of 4,000-600 cm^{-1} using a Perkin Elmer GX-FT-IR spectrophotometer. Standard practice of general techniques for obtaining Infrared Spectra for Qualitative Analysis was referred to ASTM E1257-13. Figure 1 shows the flow diagram of the experiment.

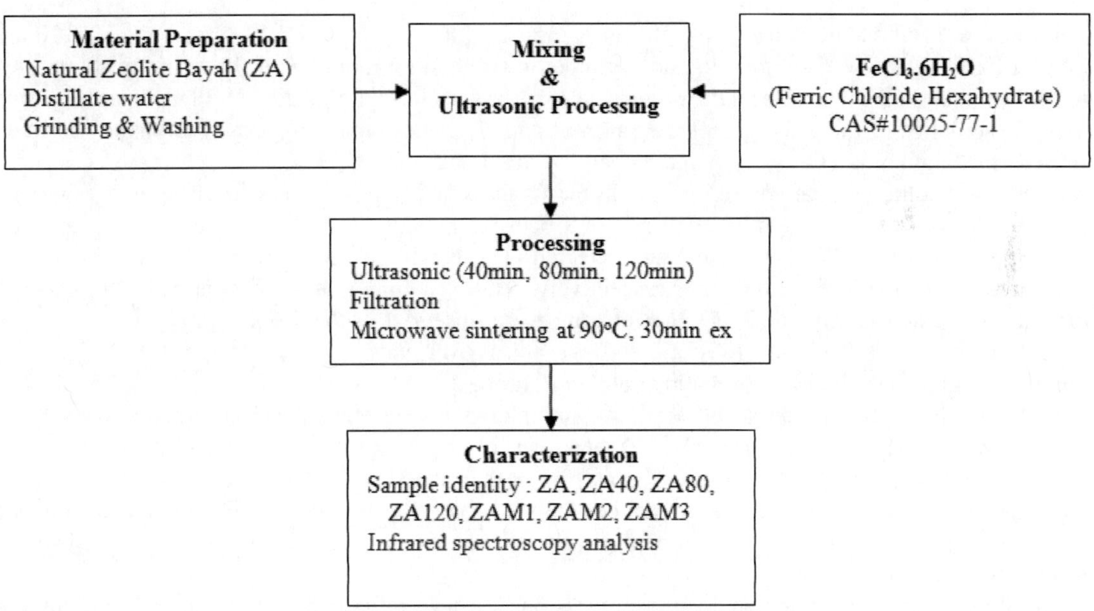

Figure 1. Research flow diagram.

3. Results and discussion

Bayah natural zeolite sample (ZA) was characterized using FT-IR spectroscopy and set as a baseline to be compared with another sample as shown on figure 2. It can be observed that the band associated with the symmetric and asymmetric stretching vibrations of the hydroxyl functional group is situated at 3,406 cm^{-1}. Additionally, the band located at 1,630 cm^{-1} is due to the vibration of the Si-O bond. The generated vibrations by Al-O bonds are shown at 1,033 cm^{-1}. The allotropic phase of SiO_2 is identified by the wave number and observed in 795 cm^{-1}. It is important to mention that the wave number associated with the Si-O and O-Al bonds, which is typical of tectosilicates and has high intensity suggesting that the zeolite has large surface area [10].

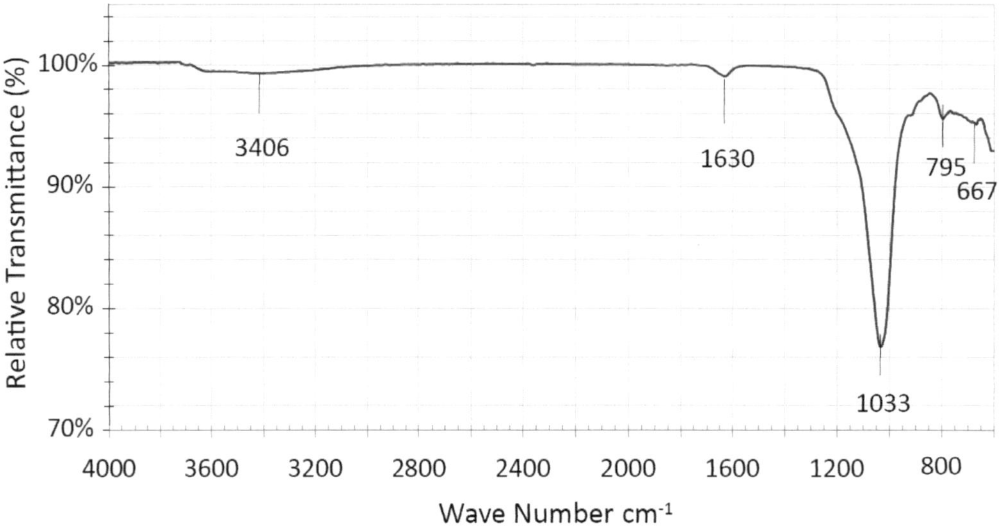

Figure 2. FT-IR spectroscopy obtained from Bayah natural zeolite (ZA).

Exposure of ultrasonic wave to the Bayah natural zeolite on sample ZA40, ZA80 and ZA120 are seen on figure 3. Second symmetric and asymmetric stretching vibrations of the hydroxyl functional group found at 3,630 cm^{-1}. Second bridged hydroxyl associated with acidity and responsible for the catalytic activities of the acid sites of the samples and Si/Al ratio may suggest a correlation between the acidic strength of the OH groups and its band wavenumber [6].

Modified zeolite with an addition of $FeCl_3.6H_2O$ and sonification process for 40 min, 80 min and 120 min shown in figure 4. Infrared bands of silicon compounds O-Si-O asymmetric stretching on wavenumber ranges 1,130–1,000 cm^{-1} was detected [11]. Modified zeolite wavenumber shifted from 1,033 into 1,030, 1,031 and 1,032 cm^{-1} respectively. Halogen compound C–F stretching shifted from 795 into 790 and 795 cm^{-1} [4,5], C–H stretching were observed in the 1,600–1,800 cm^{-1} range, the second overtones at 1,150–1,210 cm^{-1} and the third overtones at 880–915 cm^{-1} [10,11], O–H stretching region was observed on 3,400–3,800 cm^{-1} range[12,13].

C–F stretching observation at 795 cm^{-1} wavenumber showing that ZAM3 absorbance was higher than ZAM2, ZAM1 and ZA. The O-Si-O asymmetric stretching observation at 1,033-1,030 cm^{-1} wavenumber showing ZAM1 absorbance was higher than ZAM3, ZA and ZAM2. O–H stretching observation at 3,800–3,400 cm^{-1} wavenumber showed that modified zeolite has higher absorbance compared to natural zeolites as well as multiple absorbance found at ZAM3, and it is often utilized as absorbent raw material [14].

Additional $FeCl_3.6H_2O$ and sonication process on ZA have influence on shifting of wave number for stretching of C–F, O-Si-O and O–H. The O–H absorbance increment on ZAM1, ZAM2 and ZAM3 may be due to high intensity ultrasonic expossure. Ultrasonic produces microbubblesin between the primary and secondary flow cracking of $FeCl_3.6H_2O$ [3]. Table 1 shows that ultrasonic process contributes to create second and third peaks overtones .

IOP Conf. Series: Journal of Physics: Conf. Series **985** (2018) 012022 doi:10.1088/1742-6596/985/1/012022

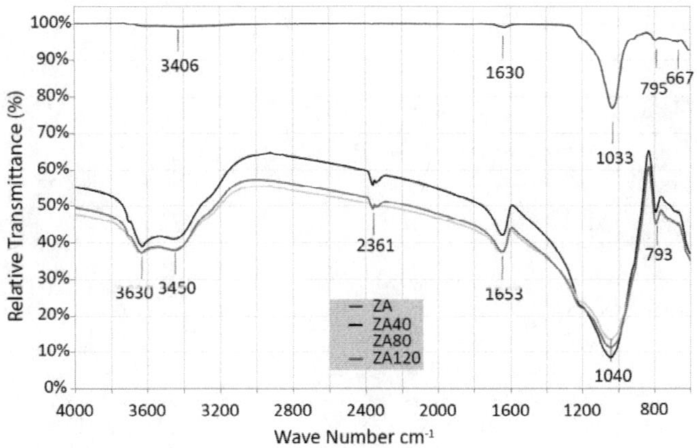

(a)

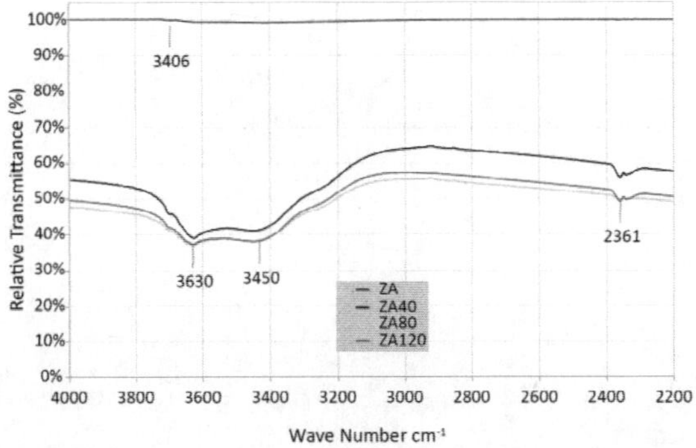

(b)

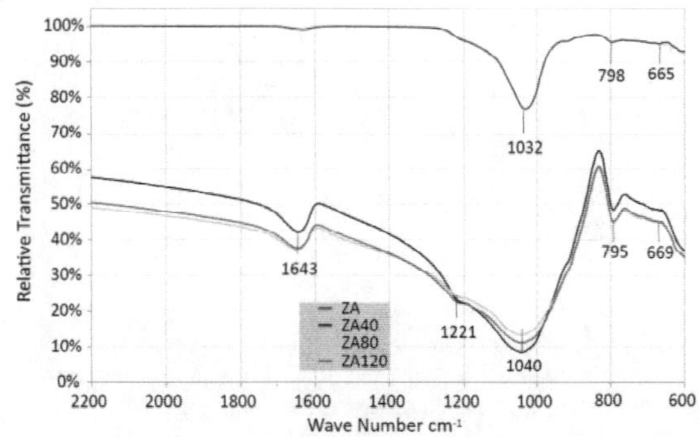

(c)

Figure 3. FT-IR spectroscopy obtained from (a) ZA, ZA40, ZA80 and ZA120. (b) Second symmetric and asymmetric stretching at 3,630 cm^{-1}, and (c) O-Si-O asymmetric stretching.

Table 1. IR intensities adsorbent of ZA and its derivatives. The notations are as follows: w is weak, m is medium, s is strong.

Sample	Ultrasonic (min)	Symmetric stretching O-Si-O and O-Al-O	Asymmetric stretching O-Si-O and O-Al-O	H-O-H Bending	OH Stretching Symmetric & Asymmetric
ZA	0	795 s	1,033 s	1,630 s	3,406 s
ZA40	40	793 m	1,042 w	1,645 m	3,627 m; 3,455 m
ZA80	80	792 m	1,040 w	1,653 m	3,630 m; 3,449 m
ZA120	120	793 m	1,045 w	1,647 m	3,630 m; 3,451 m
ZAM1	40	795 s	1,030 s	1,629 s; 1,741 s	2,970 s; 3,022 s; 3,408 s
ZAM2	80	793 s	1,032 s	1,626 s; 1,740 s	2,971 s; 3,023 s; 3,381 s
ZAM3	120	790 s	1,031 s	1,630 s; 1,741 s	2,971 s; 3,022 s; 3,434 s

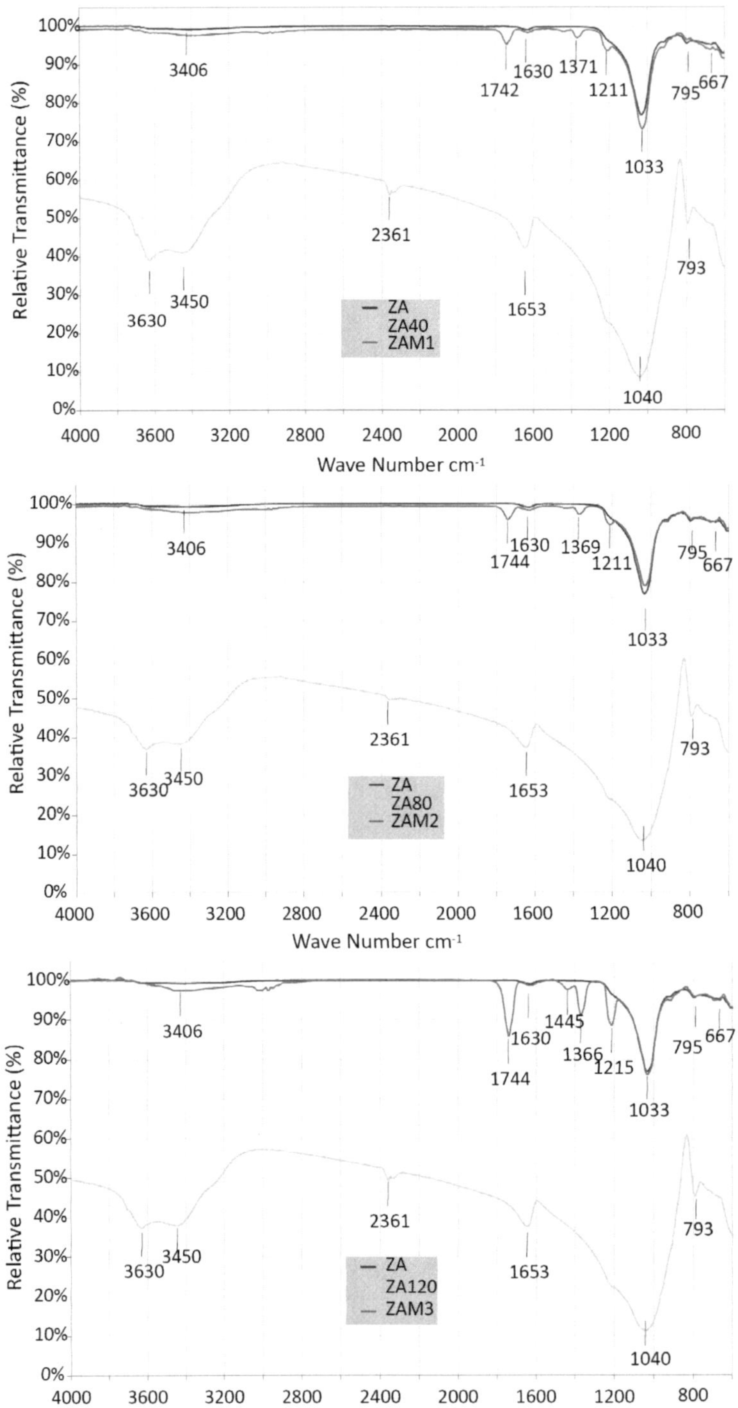

Figure 4. FT-IR spectroscopy transformation of ZA, ZA40, ZA80, ZA120 and modified Zeolites with $FeCl_3.6H_2O$ (ZAM1, ZAM2, ZAM3).

IOP Conf. Series: Journal of Physics: Conf. Series **985** (2018) 012022 doi:10.1088/1742-6596/985/1/012022

4. Conclusion

Natural and modified Zeolite from Bayah have been well characterized and the difference was distinctive. Ultrasonic exposure on Zeolite affect the absorbance properties. Modified ZA with ultrasonic process and Ferric Chloride Hexahydrate addition are improving secondary building units linked by forming oxygen bridges and strong intensities of infrared adsorbtion band. Ultrasonic process for natural zeolite develops second symmetric and asymmetric stretching vibrations of the hydroxyl functional group. Second bridged hydroxyl are associated with acidity and responsible for the catalytic activities of the acid sites of the materials and Si/Al ratio may suggest a correlation between the acidic strength of the OH groups and its band wavenumber. High ultrasonic exposure and addition of $FeCl_3.6H_2O$ may enhance the properties of zeolite as absorbent material.

Acknowledgments

This research is supported financially by "Hibah Publikasi Internasional Terindeks untuk Tugas Akhir Mahasiswa University of Indonesia" through letter of decree no. 630/UN2.R3.1/HKP.05.00/2017.

References

[1] Trisunaryanti W, Syoufian A and Purwono S 2013 *J. Chem. Eng.* **7** 175
[2] Al-Jabri M 2008 *Journal Standarisasi* **10** 59
[3] Bang J H and Suslick K S 2010 *Adv. Mater.* **22** 1039
[4] Kim W, Choi D and Kim S 2010 *Mater. Trans.* **51** 1694
[5] Supardi S and Parikin 2010 *Proc. Int. Conf. on Material Science and Technology* Jakarta p 229
[6] Li C and Wu Z 2003 Microporous Materials Characterized by Vibrational Spectroscopy *Handbook of Zeolite Science and Technology* ed SM Auerbach, K A Carrado and P K Dutta (New York: Marcel Dekker)
[7] Tomašević-Čanović M 2005 *J. Serb. Chem. Soc.* **70** 1335
[8] Andac O, Tather M, Sirkecioglu A, Ece I and Erdem-Senatalar A 2005 *Micropor. Mesopor. Mater.* **79** 225
[9] Peters D 1996 *J. Mater. Chem.* **6** 1605
[10] Ruíz-Baltazar A, Esparza R, Gonzalez M, Rosas G and Pérez R 2015 *J. Nanomater.* 364763
[11] Jahangirian H et al. 2013 Dig. *J. Nanomater. Biostruct.* **8** 1405
[12] Miller F A and Wilkins C H 1952 *Anal. Chem.* **24** 1253
[13] Byrappa K and Kumar B V S 2007 *Asian J. Chem.* **19** 4933
[14] Prasetyo T A B and Soegijono J 2017 *J. Phys.: Conf. Ser.* **817** 012032

IOP Conf. Series: Journal of Physics: Conf. Series **985** (2018) 012023 doi:10.1088/1742-6596/985/1/012023

Effect of high current density to defect generation of blue LED and its characterization with transmission electron microscope

R Gunawan[1,2], E Sugiarti[2], Isnaeni[2], R I Purawiardi[2], H Widodo[2], A N Muslimin[2], Yuliasari[2], C E Ronaldus [2], N Prastomo[1] and S Hastuty[3]

[1]Physics Energy Engineering, Surya University, Jl. Boulevard Gading Serpong kav. M5 no. 21, Summarecon Serpong, Tangerang, Indonesia 15810
[2]Research Center for Physics, Indonesian Institute of Sciences (LIPI), Kompleks Puspitek, Serpong, Indonesia 15314
[3]Mechanical Engineering, Pertamina University, Jl. Teuku Nyak Arief, Simprung, Jakarta, Indonesia 12220

E-mail: robertogunawan95@gmail.com

Abstract. The optical, electrical and structural characteristics of InGaN-based blue light-emitting diodes (LEDs) were investigated to identify the degradation of LED before and after current injection. The sample was injected by high current of 200 A/cm^2 for 5 and 20 minutes. It was observed that injection of current shifts light intensity and wavelength characteristics that indicated defect generation. Transmission Electron Microscopy (TEM) characterization was carried out in order to clarify the structure degradation caused by defect in active layer which consisted of 14 quantum well with thickness of about 5 nm and confined with barrier layer with thickness of about 12 nm. TEM results showed pre-existing defect in LED before injection with high current. Furthermore, discontinue and edge defect was found in dark spot region of LED after injection with high current.

1. Introduction

Our daily lives cannot be separated from energy consumption. One of the biggest energy consumption is lighting. According to International Energy Agency (IEA), in 2013, lighting consumed 20% of world's total building electric consumption [1]. Therefore, more efficient lighting technology becomes essential in our society. Advancement in research and development of high efficient lighting technology has make LED as emerging technology that can work in low operational voltage and current. LED applications is not limited for general lighting, but also applied as back light unit (BLU) of LCD display, for transportation [2], and also for communication [3].

Light Emitting Diode (LED) is lighting technology based on solid state. LED die is comprised of three important parts which is p-junction, quantum well/multi quantum well, and n-junction. Light from LED is produced as photon emission of electron and hole recombination that each of them has obtained enough energy to flow through p-n junction to active layer. Light spectrums of LED are influenced by epiwafer type. InGaN/AlGaN produces blue, green and UV spectrum, InAlGaP produces red and yellow spectrum, and AlGaAs produces red and infrared spectrum. These different light spectrums emitted by LED are determined by energy gap between excitation state and ground state in LED material [2].

White spectrum is combination of all color spectrums. White color in LED can be produced by merging blue, red, and green color of LED in one package. In recent trend, white LED uses blue LED die mixed with phosphor in its package where not only the LED package will emit blue color but also other colors such as yellow, green, and red [2]. Therefore, blue LED has great importance in our daily life for lighting. However, defects in active layer affect the performance of LED by causing degradation of light intensity, color change, shifting of electrical parameter, to total blackout of LED. In addition, one of the problems that leads to LED failure is high current. Therefore, the aim of the present research is to examine failure analysis of commercial LED that was injected with high current. Optoelectronic characterization was investigated to clarify the performance and degradation of LED. Accordingly, the identification of defect generation in its active layer will be examined using Transmission Electron Microscope (TEM).

2. Research method

Commercial blue LED die sample with dimension of 200×500×400 μm was provided by Research Center for Physics – Indonesian Institutes of Sciences (LIPI) which has optimum injected current at 90 A/cm^2. The sample was treated by high current injection of 200 A/cm^2 for 5 and 20 minutes. It was found from pre-eliminary test that 200 A/cm^2 was the highest injected current that can be given before LED have total blackout/will not produce light emission. 5 minutes was chosen as minimum aging time when dark spot was first detected. Increased aging time with 5 minutes interval, it was found that total blackout happened at 25 minutes treatment time. Therefore, 20 minutes was chosen as maximum treatment time to analyze the optoelectrical degradation of LED. Optical degradation was examined with electroluminescence test (EL) and optical microscope. Electrical degradation was also observed with measurement of current-voltage (I-V). Lamellae sample of LED die was prepared to have a thickness below 50 nm by using Focused Ion Beam (FIB) JEOL, JIB 4610F. Furthermore, defect structure of LED die was observed by Transmission Electron Microscope (TEM) FEI, Tecnai G2 200 kV.

3. Results and discussion

Figure 1 shows light intensity characteristic of blue LED before aging or deteriorating by an application of high current density. The result showed that in area I, light intensity increases linearly with increasing current until it reached 50 mA. In area II, the increment rate of light intensity slightly decreased compared to area I. The end of area II is marked by current at 90 mA. This change is caused by efficiency drop due to the lack of auger recombination that is responsible for photon production [4]. After reaching the highest current at 90 mA, the light intensity decreased significantly. This phenomenon is caused by the dominance of non-radiative recombination rather than radiative recombination [5]. The non-radiative recombination in the form of the increment of temperature is caused by current localization of Au electrode in n junction (n-pad) [6]. This current localization caused uneven proportion between average current density and operational current [7].

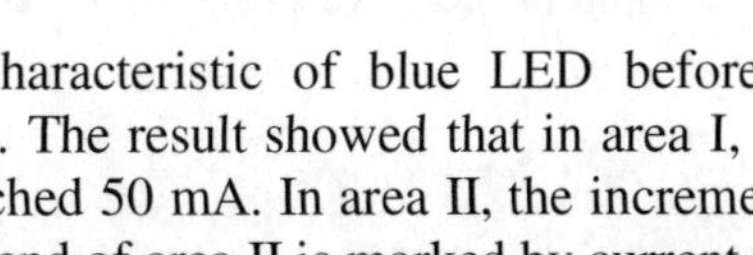

Figure 1. Light intensity characteristic of untreated blue LED.

Figure 2 shows light intensity characteristic after the LED is being injected with high current density 200 A/cm^2 for 5 and 20 minutes. The graph showed shifting of optimum point before the light intensity decreased along with increasing injected current from 90 mA to 70 mA. This shifting indicated the efficiency drop occurred earlier before aging. This performance degradation of LED is probably occurred with the reduction of electron transport rate due to the disappearance of electron path and eventually it might cause electron overflow [8].

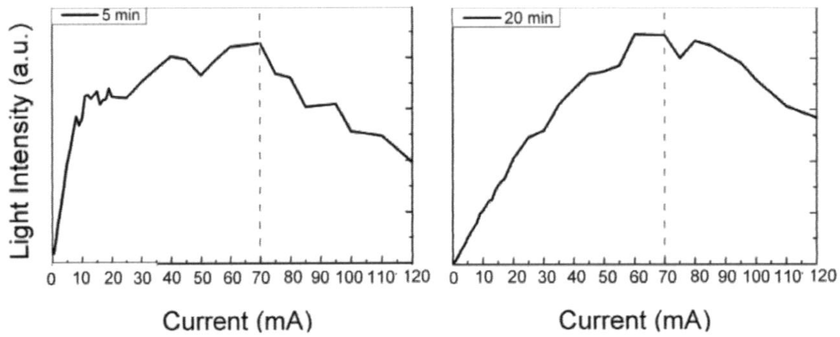

Figure 2. Light intensity of treated blue LED with high current density for 5 and 20 minutes.

Colour emission of LED is also affected by injection of high current density. Figure 3 shows peak colour spectrum emission characteristic correspond to current. It could be observed that the wavelength dropped from about 450 nm at the beginning. Furthermore, the wavelength was steady for the non-aging treated sample, while the wavelength increased slightly after it reached the lowest point for the aging treated sample. The decrease of the wavelength is caused by band gap filling effect [9], where the injected electrons are forced to be located at a higher energy band as the lower energy band has been congested. This phenomenon is called blueshifting, caused by electrons that release photon emission from higher energy state at shorter wavelength. Furthermore, the increase of peak wavelength could be caused by the rise of temperature along with the increasing of current density that shortened the energy gap between conduction and valence band [10]. The change of wavelength at this stage is referred as redshifting.

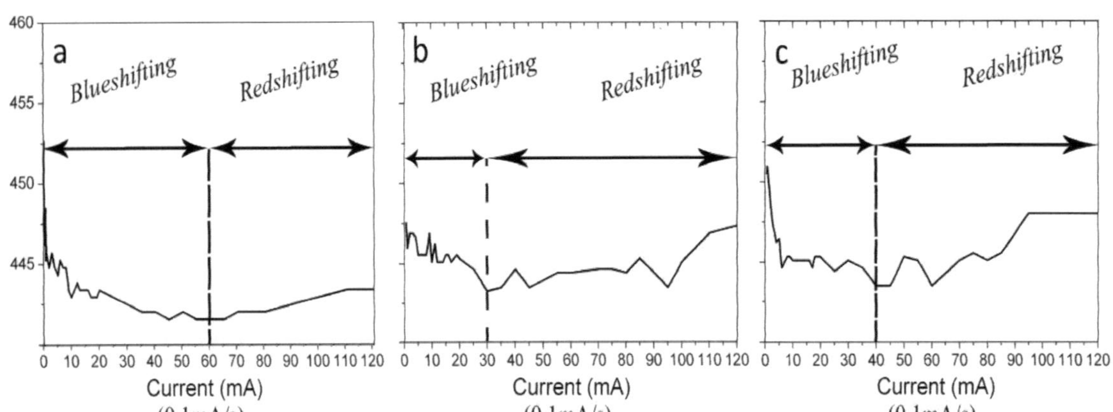

Figure 3. Peak color spectrum emission of blue LED: (a) untreated LED, (b) treated with high current density for 5 minutes, and (c) 20 minutes.

As shown in graph, the lowest point of aging treated sample moved to smaller current value. Shifting of the current indicates significant increase in LED junction temperature with current that affects the occurrence of redshifting. This suggests phonon assisted recombination due to performance degradation of aging treated LED.

Figure 4 shows illumination mapping of non-aging dan aging treated LED injected with operational current of 0.1 mA, observed using optical microscope to examine bright and dark region of LED die. Normally, the overall light emission of LED surface had the same average level of radiance (figure 4(a)). However, faded blue color region (leftside of the die) which is also known as faint darkspot (FDS) was spotted, as indicated by circled area in figure 4a. The FDS area implies the existence of pre-existing defect that acts as non-radiative recombination center [11]. Figures 4b and 4c reveals the growth of darkspot (DS) region of aging treated LED. The presence of DS region reveals defect generation that acts as non-radiative recombination center.

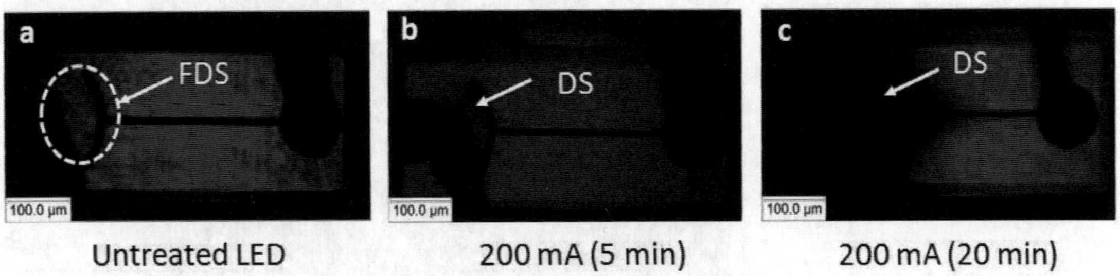

Figure 4. Illumination mapping of blue LED operate with current 0.1 mA: (a) untreated LED, (b) treated with high current density for 5 minutes, and (c) for 20 minutes.

Degradation of electrical performance is also observed in figure 5 and can be classified into low bias and high bias regions. The appearing current begin at lower voltage after LED was aged with high current density in low bias region (<2.5 V) as shown in figure 5 (inset). Non-aging treated sample has high bias region with current starting at 2.3 V. On the contrary, aging treated sample begin at 2 V and 0.1 V after treated for 5 and 20 minutes, respectively. The change of current appearance suggests the existence of leakage current in low bias region of treated LED. It implies defect generation in active layer that acts as non-radiative recombination center [5]. Furthermore, in high bias region (>2.5 V), the reduction of operational current indicated interface degradation between metal and die that increased the resistance thus increase the operational voltage of LED [12, 13].

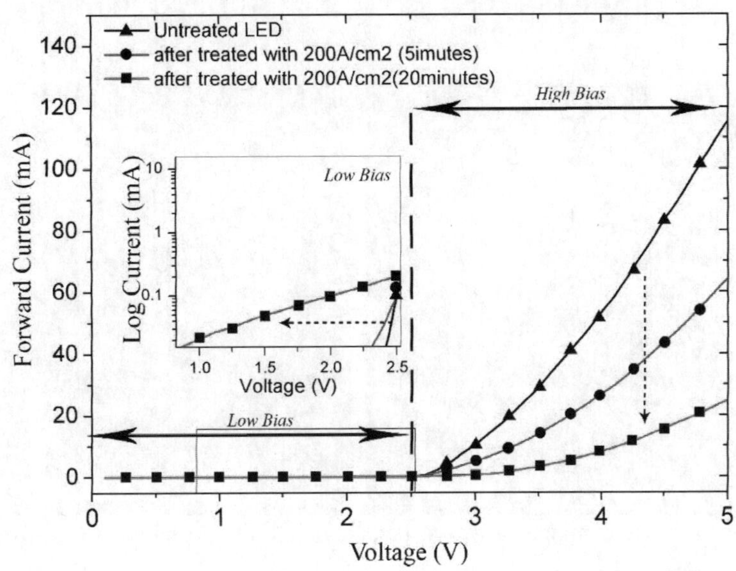

Figure 5. I-V curve characteristic of untreated and treated blue LED.

Figure 6 shows TEM analysis in active layer (darkspot) of untreated LED. The result shows that the fringe pattern is horizontally constructed without crystal defect. Dark contrast indicated the element with higher atomic number compared to brighter area. In the LED die, indium has higher atomic number (49) than gallium (31) and nitrogen (7) that caused Coulomb force between electron and indium higher than Ga and N. Furthermore, the results confirmed the active layer consists of 14 quantum well with thickness of about 5 nm and confined with barrier layer with thickness of about 12 nm.

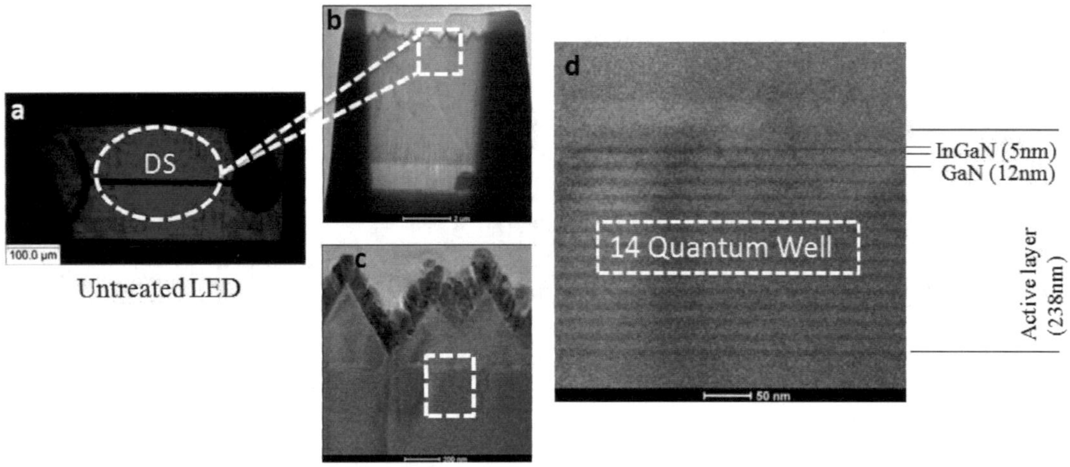

Figure 6. Bright Field (BF) image of blue LED active layer nanostructure without crystal defect.

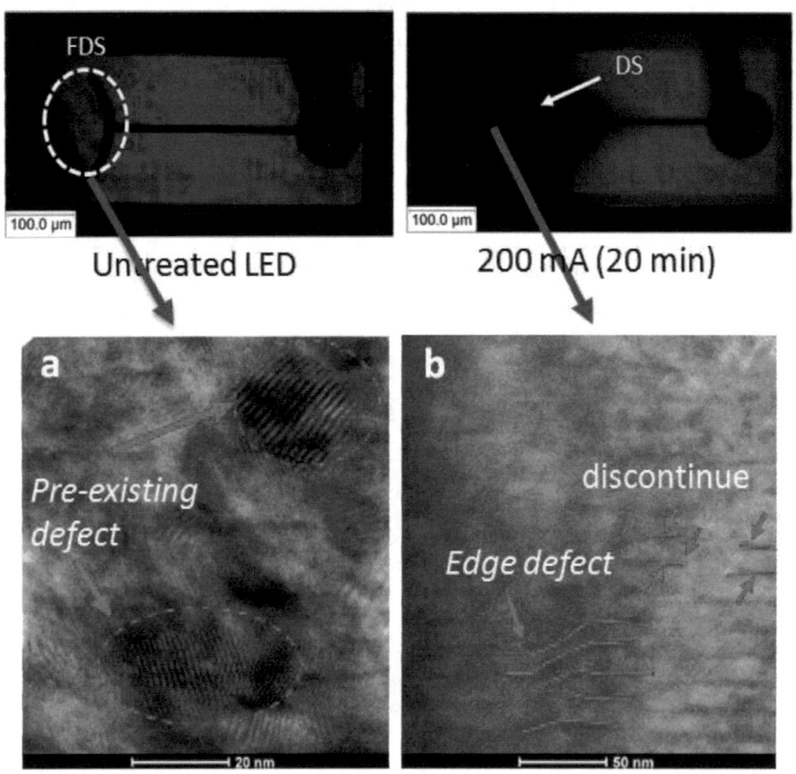

Figure 7. Bright Field (BF) image in active layer nanostructure of blue LED (a) FDS of untreated LED and (b) DS of treated LED.

Figure 7a reveals the existence of pre-existing defect in faint darkspot (FDS) region of untreated LED. The pre-existing defect forms as cluster in active layer. This defect performed as non-radiactive recombination center that affects the light emission in this region. Figure 7b represents an active layer structure in darkspot (DS) region after of treated LED with high current density for 20 minutes. Defect generation is observed and could be seen in multi quantum well layer. These defects are formed as the result of high current injection which generated heat in LED die. It gives enough energy to move the lattice crystal structure in active layer and separate In-N bond because of its low bonding energy [14].

4. Conclusion

The changes in optoelectric characteristic of blue LED occurred after injection of high current density with value of 200 A/cm^2. In case of optical characteristic, shifting of optimum intensity occurred to lower current from 90 mA to 70 mA. Transition point between blueshifting and redshifting of peak blue color of LED also shifted to lower current which is from 60 mA for untreated LED and 30 mA after 5 minutes injection and 40 mA after 20 minutes injection. Furthermore, electrical characteristic of LED was also changed. In lower bias region (<2.5V), the existence of leakage current for LED after high current injection was observed. Resistance was increased in high bias region (>2.5V) of LED after high current injection. Illumination map of light emission of LED die showed faint darkspot region on untreated LED and generation of dark spot region after LED was injected with high current. TEM results showed pre-existing defect formed as cluster in multi quantum well layer in faint darkspot region. Discontinue and edge defect were found in darkspot region. The mechanism of defect generation could be caused by heat generation that was produced as the result of high current injection.

Acknowledgments

The authors would like to thank to Unggulan Program of Research Center for Physics-Indonesian Institute of Sciences (LIPI) and INSINAS IRPI of Ministry of Research, Technology and Higher Education for financial support of this work.

References

[1] International Energy Agency, https://www.iea.org/Sankey/#?c=World Final consumption (last accessed on 21 March 2017)
[2] Chang M H, Das D, Varde P V and Pecht M 2012 *Microelectron. Reliab.* **52** 762
[3] Medina C, Zambrano M and Navarro K 2015 International Journal of Advances in Engineering and Technology **8** 482
[4] Iveland J, Martinelli L, Peretti J, Speck J S and Weisbuch C 2013 *Phys. Rev. Lett.* **110** 177406
[5] Uddin A, Wei A C and Andersson G 2005 *Thin Solid Films* **483** 378
[6] Bogdanov M V, Bulashevich K A, Khokhlev O V, Evstratov I Y, Ramm M S and Karpov S Y 2010 *Phys. Status Solidi C* **7** 2124
[7] Karpov S Y 2015 *Opt. Quant. Electron.* **47** 1293
[8] Vampola K J, Iza M, Keller S, DenBaars S P and Nakamura S 2009 *Appl. Phys. Lett.* **94** 061116
[9] Wang T, Nakagawa D, Lachab M, Sugahara T and Sakai S 1999 *Appl. Phys. Lett.* **74** 3128
[10] Pong B J, Chen C H, Yen S H, Hsu J F, Tun C J, Kuo Y K, Kuo C H and Chi G Ch 2006 *Solid State Electronics* **50** 1588
[11] Egawa T, Jimbo T and Umeno M 1997 *J. Appl. Phys.* **82** 5816
[12] Moon S M and Kwak J S 2009 *J. Korean Phys. Soc.* **55** 1128
[13] Meneghini M, Podda S, Morelli A, Pintus R, Travisanello L, Meneghesso G, Vanzi M and Zanoni E 2006 *Microelectron. Reliab.* **46** 1720
[14] Cao X A, Sandvik P M, LeBoeuf S F and Arthur S D 2003 *Microelectron. Reliab.* **43** 1987

IOP Conf. Series: Journal of Physics: Conf. Series **985** (2018) 012024 doi:10.1088/1742-6596/985/1/012024

Fabrication and characterization of fine ceramic based on alumina, bentonite, and glass bead

P Sebayang[1], Nurdina[2], S Simbolon[1], C Kurniawan[1], M Yunus[1], E A Setiadi[1] and Z Sitorus[2]

[1]Research Center for Physics, Indonesian Institute of Sciences, Kawasan PUSPIPTEK, Tangerang Selatan, Indonesia
[2]Physics Department, Faculty of Mathematics and Natural Science, University of North Sumatera, Medan, Indonesia

E-mail: pard003@lipi.go.id

Abstract. Fabrication of fine ceramics based on alumina, bentonite and glass bead has been carried out by powder metallurgy. The preparation of powder has been performed using High Energy Milling (HEM) with wet milling process and using toluene as medium for 2 hours. The powder milling result was dried in oven at 100 °C for 24 hours. After that, the powder was compacted into pellet by using hydraulic press with 80 kgf/cm^2 pressure at room temperature. Then, the pellet was sintered at 900 °C for 4 hours. Materials characterization such as physical properties (true density, bulk density, porosity, and water absorption), average particle diameter, hardness, microstructure and phase were measured by Archimedes method, Particle Size Analyzer (PSA), Hardness Vickers (HV), Scanning Electron Microscope (SEM-EDX) and X-Ray Diffraction (XRD). From the result, the optimum condition is sample D (with addition of 30 wt.% γ-Al_2O_3) with sintering temperature of 900 °C for 4 hours. At this condition, these properties were measured: average particle diameter of 4.27 μm, true density of 2.32 g/cm^3, porosity of 5.57%, water absorption of 2.46%, bulk density of 2.39 g/cm^3, and hardness of 632 HV. The fine ceramic has four phases with albite ($Al_2NaO_8Si_3$) and quartz (SiO_2) as dominant phases and corundum (Al_2O_3) and nepheline ($AlNaO_4Si$) as minor phases.

1. Introduction

Ceramics can be classified into traditional and fine ceramics. Traditional ceramics are generally made from naturally occurring raw materials such as clay, mica, quartz, porcelain, and pottery. Most of the traditional ceramics are based on porous silicates whose microstructures are not homogeneous and consist of many phases. Meanwhile, the fine ceramics are mostly developed through synthesis process or powder metallurgy. The fine ceramics can be produced predominantly from alumina, titanium, zirconia, magnesia, and other oxides [1, 2]. Alumina (Al_2O_3) is an oxide or fine ceramics that can be applied in various fields such as electronic and mechanical component [3].

The Al_2O_3 ceramics has melting point of about 2000 °C, density of 3.96 g/cm^3, and hardness of 14 kg/mm^2. The Al_2O_3 ceramics is electrically insulating, strong, hard, and resistant to high temperatures [4]. However, the growth of alumina ceramic grains becomes abnormal at high temperatures. This causes inhomogeneous microstructures of formed grains which can degrade the quality of ceramics [5]. One way to lower the sintering temperature is to reduce grain sizes into nanoparticle or to incorporate certain additives which have lower melting point than alumina, such as MgO and SiO_2 [6].

An addition of silica (SiO_2) materials to alumina ceramics can form a liquid phase and enable to repair densification of the materials. Excessive addition of SiO_2 will impact to the decreasing mechanical strength on alumina ceramics [7]. Besides an addition of silica, bentonite can also be used as a mixture to enhance the flexibility of ceramic materials.

The bentonite is a type of clay with 85% montmorillonite minerals and 15% others such as cristobalite, feldspar, calcite, gypsum, kaolinite, plagioclase, and illite [8]. The chemical formula of bentonite is (Na, Ca) $O.Al_2O_3.4SiO_2.H_2O$ [9]. The special properties of bentonite are the ability to form thixotrophic gel with water, high capacity of exchange cation, and to absorb water in large quantity until it increases its volume around 12-15 times. The ability of montrmorillonite to absorb water will decrease after heating at critical temperature at 105 – 390 °C [8].

The manufacture of fine ceramics based on bentonite, SiO_2, and γ-Al_2O_3 powders is widely used as ceramic filter and furnace equipment material. Besides, glass bead is also used as additive of ceramics. The addition of glass bead can improve the thermal and mechanical properties of ceramics [10].

In this research, we studied the fabrication of fine ceramics based on bentonite, glass bead, and γ-Al_2O_3 powders. The ceramic was prepared by metallurgy method with wet milling process. It was compacted into pellet by die pressing method and sintered at 900 °C for 4 hours [11]. The manufacture of fine ceramics based on bentonite, glass bead, and alumina is expected to improve the physical and mechanical properties of this material.

2. Experimental method

Bentonite, glass bead, and γ-Al_2O_3 powders were used as raw materials and the compositions are shown in table 1.

Table 1. Composition of raw material for fine ceramics process.

Sample code	Composition (wt.%)		
	Alumina	Bentonite	Glass bead
A	0	50	50
B	10	45	45
C	20	30	30
D	30	35	35

The fabrication of fine ceramics was carried out by powder metallurgy method. The raw materials were mixed and milled by using High Energy Milling (HEM). This process was done by wet milling for 2 hours and using toluene as liquid medium. Comparison of ball mill to powder was fixed at 10:1. After the milling process, the powder was dried in the oven at 100 °C for 24 hours. Furthermore, the powder was compacted into pellet with pressure of 80 kgf/cm^2 at room temperature. The pellet was then sintered at 900 °C temperature for 4 hours.

Characterizations of the materials were performed including true density, bulk density, porosity, and water absorption using Archimedes method [12, 13]. The true density measurement of the powder mixture was carried out using Pycnometer. While the average particle diameter was measured using Particle Size Analyzer (PSA) Cilas-1190, hardness (HV) was performed using Micro Hardness Tester (LECO LM-100AT), morphological using Hitachi SU-3500 Scanning Electron Microscopy (SEM), and phase was analyzed by using X -Ray Diffraction (XRD) Rigaku Smart Lab type, λ = 1.5418 Å.

3. Results and discussion

Diameter distribution results of the powder mixture are shown in figure 1. The diameter powders of the samples are 5.86, 5.17, 5.13, and 4.27 μm with various alumina additions of 0, 10, 20, and 30 wt.%, respectively. It shows that the variation of alumina addition (Al_2O_3) does not give any effect to the particle size of resulted sample. The same particle size of sample shows that the milling time of the fine ceramic fabrication is optimum.

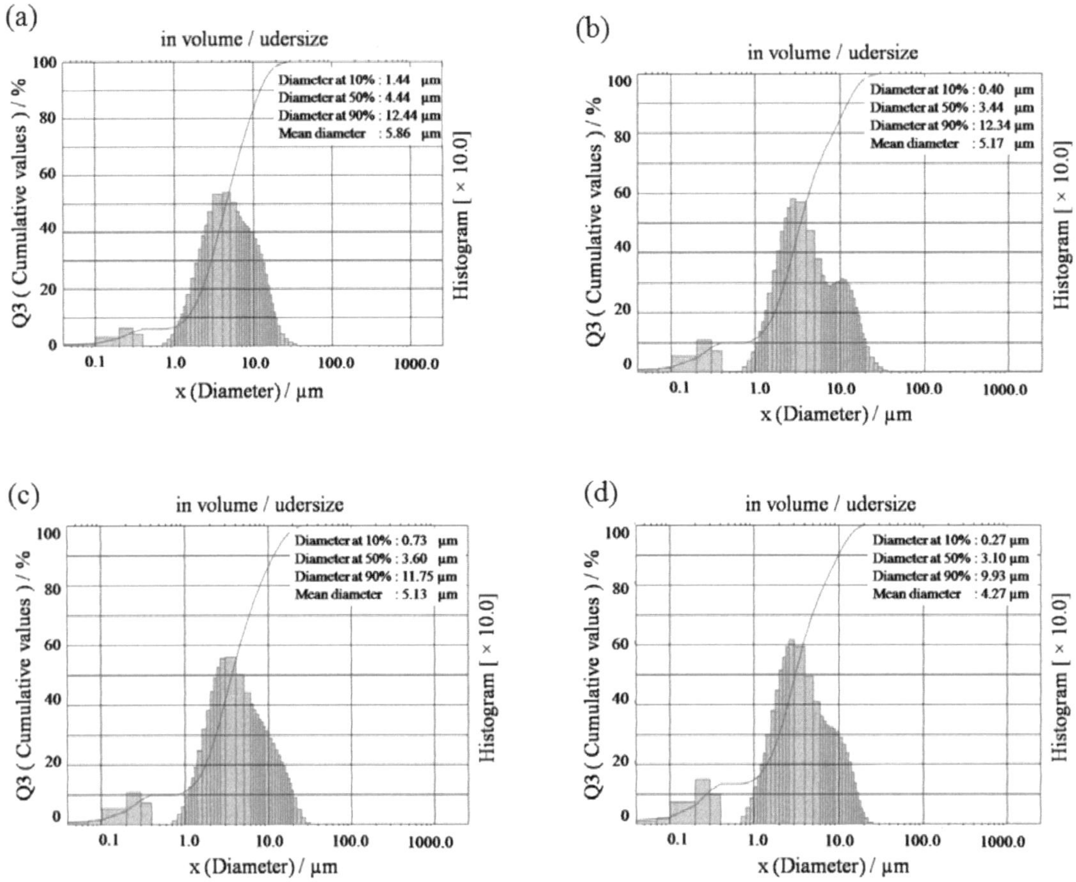

Figure 1. Particle size distribution after wet milling with various addition of alumina (Al_2O_3): (a) 0, b) 10, c) 20, and d) 30 wt.%.

Figure 2a shows the percentage of alumina addition (Al_2O_3) versus bulk and true density, while figure 2b shows porosity and water absorption. The bulk density value is around 2.16 - 2.39 g/cm^3, while true density value is around 2.28 - 2.32 g/cm^3. The porosity value is obtained in range of 5.57 - 9.82% and water absorption is around 2.46 - 5.01%. The effect of alumina addition (Al_2O_3) tends to increase true and bulk density values. Alumina has higher true density than bentonite (true density of alumina = 3960 kg/m^3, true density of glass bead = 120 kg/m^3, and true density of bentonite = 25 – 40 kg/m^3), so the variation of alumina addition contributes to the change of density. In the other hand, from figure 2b it can be seen that the addition of alumina decreases porosity and water absorption. The porosity is inversely proportional with the density value. As the density value increases, the porosity of the sample decreases.

SEM-EDX results are shown in figure 3. Figures 3a and 3b are for Sample A, and 3c and 3d for Sample D. The aluminum element Al increases with the addition of γ-Al_2O_3. In Sample A, it has flake-like morphology with porous structure. The interlayer spaces of bentonite, which was thermally activated, can be collapsed and result in more tightly bound structure with reduction in porous structure [14]. However, in this case the addition of glass beads decreases the porosity of sample, and the surface becomes rather flat caused by melting of glass bead. The dominant elements composition of Sample A (without addition of γ-Al_2O_3) are O = 79.4 at.%, Si = 13.18 at.%, Na = 4.15 at.%, and Al

= 2.21 at.%. Meanwhile, the dominant elements composition of Sample D (with addition of γ-Al_2O_3) are O = 67.62 at.%, Si = 14.40 at.%, Na = 3.07 at.%, and Al = 12.73 at.%. The others element of fine ceramics consist of Mg, K, and Ca are under 2 at.%.

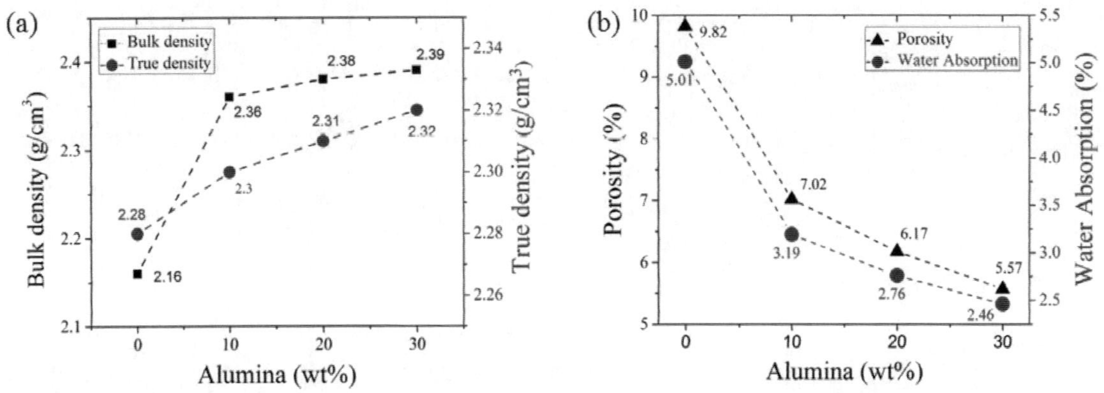

Figure 2. The effect of percentage of alumina addition (Al_2O_3) to (a) average particle diameter and true density and (b) porosity and water absorption.

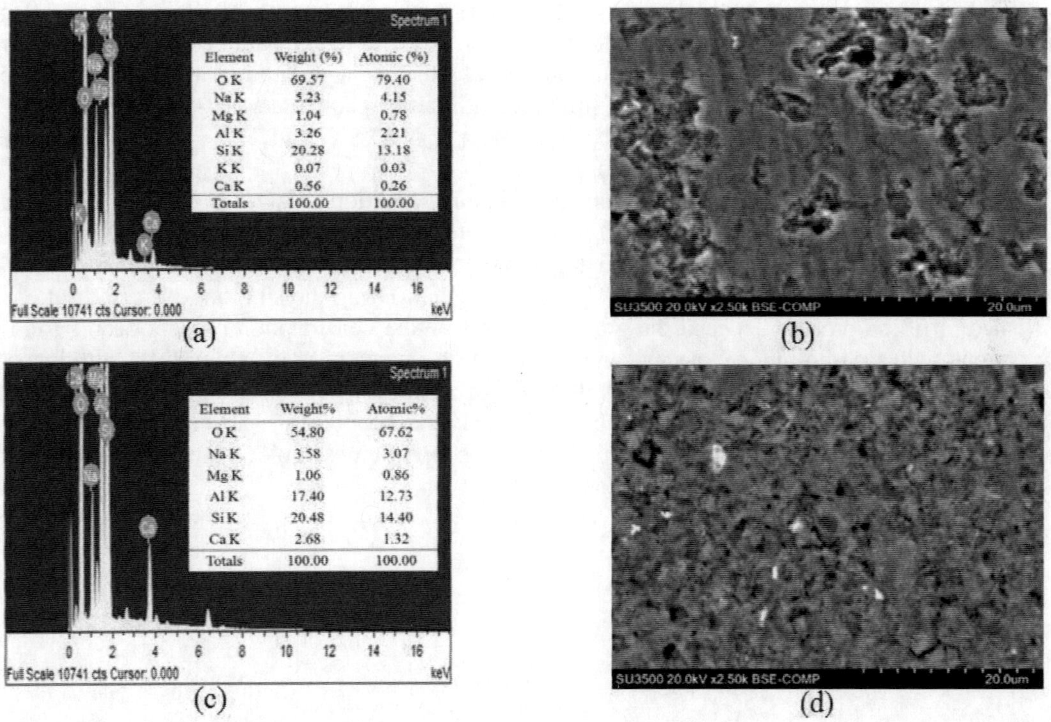

Figure 3. SEM-EDX results of (a) element composition and (b) morphology from Sample A. (c) Element composition and (d) morphology from Sample D.

The result of fine ceramics microstructure analysis using X-Ray Diffraction (XRD) is shown in figure 4. The performed analysis was limited before and after the addition of 30 wt.% Al_2O_3 with sintering temperature of 900 °C for 4 hours. Figure 4 shows that with or without addition of 30 wt.% Al_2O_3 the ceramic has four phases, which are albite ($Al_2NaO_8Si_3$), corundum (Al_2O_3), nepheline

($AlNaO_4Si$), and quartz (SiO_2). This result is consistent with EDX analysis. Before sintering temperature, no nepheline phase was observed. As the alumina was being added, the new phase of nepheline was formed. We suggest that the excess of additive alumina and the existence of silica in bentonite and glass bead tend to form nepheline than albite.

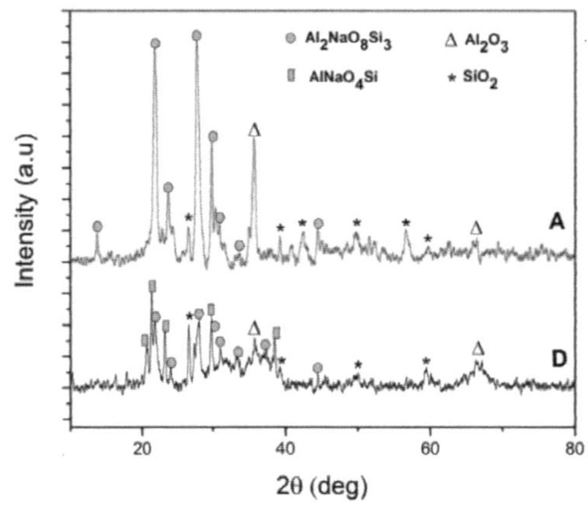

Figure 4. X-ray diffraction pattern of Samples A and D.

The $Al_2NaO_8Si_3$ phase has a triclinic crystal structure with parameters a = 8.259 Å, b = 12.975 Å, and c = 7151 Å. Al_2O_3 phase has a trigonal crystal structure with lattice parameters a = 4.766 Å and c = 12.980 Å. $AlNaO_4Si$ phase, it has a hexagonal crystal structure with lattice parameters a = 9.995 Å, c = 24.797 Å. SiO_2 phase has a hexagonal structure with lattice parameters a = 4.918 Å, c = 5.470 Å. The dominant phases of the fine ceramic are albite ($Al_2NaO_8Si_3$) and SiO_2, while the corundum (Al_2O_3) and nepheline ($AlNaO_4Si$) are minor phase.

Figure 5 shows the effect of the percentage of alumina addition (Al_2O_3) to hardness (HV) on the manufacture of fine ceramics. The obtained hardness value increases from 495 to 632 HV. This is due to alumina, which has relatively higher hardness [4]. Therefore, an addition of Al_2O_3 results in increasing of fine ceramics densification. The reason is that during the sintering process heat energy activated diffusion process between the granules, therefore the densification process occurs followed by volume reduction but not by mass change. The optimum condition is achieved by addition of 30 wt.% Al_2O_3 and sintering temperature of 900 °C for 4 hours [11], resulting in average particles diameter of 4.23 μm, true density of 2.32 g/cm^3, porosity of 5.57%, water absorption of 2.46%, bulk density of 2.39 g/cm^3, and hardness of 632 HV. The optimum sample is obtained at the higher hardness condition.

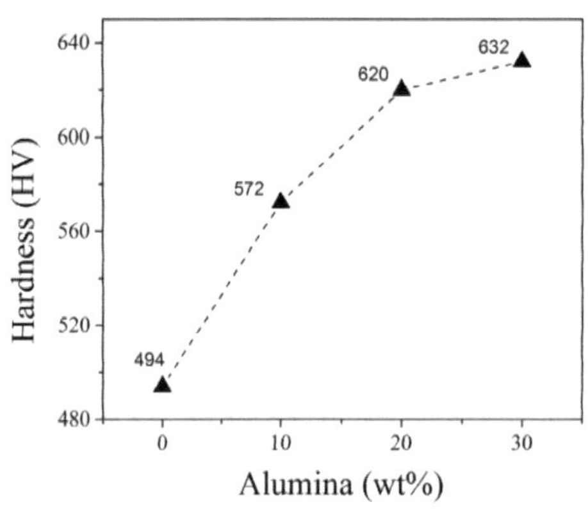

Figure 5. The effect of alumina (Al_2O_3) addition to bulk density and hardness (HV).

4. Conclusions

Fine ceramics based on alumina (γ-Al_2O_3), bentonite, and glass bead have been successfully fabricated using HEM with wet milling process. The resulting powder from 2 hours milling has average diameter of 4.23 μm and true density of 2.32 g/cm^3. The optimum condition is obtained in Sample D with the addition of 30 wt.% Al_2O_3 and sintering temperature of 900 °C for 4 hours because it has the higher hardness. The properties of Sample D are: porosity of 5.57%, water absorption of 2.46%, bulk density of 2.39 g/cm^3, and hardness of 632 HV. Fine ceramics produced from alumina material (γ-Al_2O_3), bentonite and glass bead have dominant phase albite ($Al_2NaO_8Si_3$) and quartz (SiO_2), and minor phases corundum (Al_2O_3) and nepheline ($AlNaO_4Si$).

Acknowledgments

The authors acknowledge to P2F - LIPI 2017 for fund and characterization facility.

References

[1] Rasin F A and Hamad E A 2012 *British Journal of Science* **7** 56
[2] Heimann R B 2010 *Classic and Advanced Ceramics: From Fundamentals to Applications* (Weinheim: WILEY-VCH Verlag GmbH & Co. KGaA)
[3] Buchanan R C 1986 *Ceramic Materials for Electronics* (New York: Marcel Dekker, inc)
[4] Davis K 2010 *School of Doctoral Studies (European Union) Journal* 109
[5] Montanaro L, Tulliani J M Perrot C and Negro A 1997 *J. Eur. Ceram. Soc.* **17** 1715
[6] Rhamdhani M A, Soepriyanto S, Ramelan A and Barliansya A 2005 *Jurnal Teknologi Mineral* **12** 148
[7] Louet N, Reveron H and Fantozzi G 2008 *J. Eur. Ceram. Soc.* **28** 205
[8] Adamis Z and Williams R B 2005 *Bentonite, Kaolin, and Selected Clay Minerals* Environmental Health Criteria vol 231 (Geneva: World Health Organization)
[9] Alemdar A, Öztekin N, Erim F B, Ece Ö I and Güngör N 2005 *Bull. Mater. Sci.* **28** 287
[10] Huang L, Yuan Q, Jiang W, An L, Jiang S and Li R K Y 2004 *J. Appl. Polym. Sci.* **94** 1885
[11] Chen Y F, Wang M C and Hon M H 2004 *J. Eur. Ceram. Soc.* **24** 2389
[12] Kurniawan C, Nainggolan M M, Sebayang K, Ginting M and Sebayang P 2017 *J. Phys.: Conf. Ser.* **817** 012057
[13] Setiadi E A, Kurniawan C, Sebayang P and Ginting M 2017 *J. Phys.: Conf. Ser.* **817** 012054
[14] Toor M and Jin B 2012 *Chem. Eng. J.* **187** 79

IOP Conf. Series: Journal of Physics: Conf. Series **985** (2018) 012025 doi:10.1088/1742-6596/985/1/012025

Synthesis, characterization, and photocatalytic properties of nanocrystalline NZO thin films

D Aryanto[1], E Hastuti[2], N Husniya[2], T Sudiro[1] and B W Nuryadin[3]

[1] Research Center for Physics, Indonesian Institute of Sciences, Tangerang Selatan, Indonesia
[2] Physics Department, Faculty of Science and Technology, UIN Maulana Malik Ibrahim, Malang, Indonesia
[3] Physics Department, UIN Sunan Gunung Jati, Bandung, Indonesia

E-mail: didik_phys@yahoo.co.id

Abstract. Nanocrystalline Ni-doped ZnO (NZO) thin films were synthesized on glass substrate using sol-gel spin coating methods. The effect of annealing on the structural and optical properties of nanocrystalline thin film was studied using X-ray diffractometer (XRD), field emission scanning electron microscopy (FESEM), UV-VIS spectrophotometry, and photoluminescence (PL). The results showed that the annealing temperature strongly influenced the physical properties of nanocrystalline NZO thin films. The photocatalytic properties of nanocrystalline NZO thin films were evaluated using an aqueous solution of Rhodamine-B. The photocatalytic activity of nanocrystalline NZO thin films increased with the increase of annealing temperature. The results indicated that the structure, morphology, and band gap energy of nanocrystalline NZO thin films played an important role in photocatalytic activity.

1. Introduction

Photocatalysis based on metal oxide semiconductor has been considered as alternative materials for degrading recalcitrant organic compound in wastewaters [1]. Many metal oxides including CdS, TiO_2, Fe_2O_3, WO_3, and ZnO have been studied as fine powder for photocatalysis application [2 - 5]. The problem of using the powder photocatalysts is that they are difficult to separate, recovery, and recycle [1]. It generates potential health and environmental risks. These problems can be overcome by using thin films as photocatalysts. However, their photocatalytic activity is lower than the powder photocatalysts [1], caused by the reduction of the surface area. Accordingly, the surface morphology modification of thin films such as porosity, film homogeneity, and pore size distribution is indispensable to improve the photocatalytic activity [3, 6]. Besides, the doping of metallic and rare earth element can be used to modify films surface and enhance the photocatalytic performance [2, 7 - 10].

Among the metal oxide semiconductor, ZnO has several advantages such as relatively cheap, non-toxic and chemically stable [9]. ZnO is II-VI semiconductor compound with direct wide-band-gap of 3.37 eV and large exciton binding energy of 60 meV at room temperature [3, 11, 12]. It makes ZnO applicable as solar cell [13], laser [14], gas sensors [15], piezoelectric materials [16], and light emitting diodes (LEDs) [17]. In recent years, the function of ZnO as photocatalyst [2, 3, 7-9, 18-20] and antibacterial agent [8, 11] has been investigated. This is owing to its excellent photocatalytic efficiency and good stability. Several studies show that the ZnO exhibits greater photocatalytic efficiency than TiO_2 to degrade organic pollutants [4]. However, due to its wide band gap, ZnO exhibits photocatalytic activity only under UV light [4, 8, 9]. It greatly inhibited their applications. Some studies reported that the optical response of ZnO into the visible light can be improved by doping it with transition metal elements such as Ag [9], Cu [2], Mn [21], Co [8], Mg [22], Cd [9] and Ni [19]. In particular, Ni is one of the most efficient doping elements to modify the band-gap of ZnO. The advantage of Ni metal over other metals like Ag, Mn, Pd, and Cd is due to its low cost. Furthermore, Ni ion has the same radius and valence-band with Zn ion, which results in large solubility in the ZnO matrix [23]. Some works on Ni-doped ZnO reported that the optical, electrical, and magnetic properties of ZnO were changed after Ni doping [24, 25]. However, little reports have been found to investigate the photocalaytic activity of Ni-doped ZnO thin films [12, 19]. In this work, Ni-doped ZnO (Ni = 3 at.%) thin films was fabricated using sol-gel methods. The effect of annealing on the structural, optical, and photocatalytic properties of ZnO thin films was discussed.

2. Experimental details

2.1. Synthesis of NZO thin films

Nanocrystalline NZO thin films were prepared on corning glass by sol-gel spin coating methods. First, corning glass substrate was cleaned sequentially with acetone, methanol and DI water in an ultrasonic bath and then dried. The NZO solution was prepared by mixing zinc acetate dihydrate ($Zn(CH_3COO)_2.2H_2O$), nickel acetate tethrahydrate ($Ni(OCOCH_3)_2.4H_2O$), isopropanol and ethanolamine as precursor, dopant, solvent, and stabilizer, respectively. The molar ratio of isopropanol and zinc acetate dihydrate was maintained at 1.0 and the concentration of nickel was 3% from zinc. The resulted solution was stirred at 60 °C for 90 min and then aged for 24 hours at room temperature before usage. The NZO solution was then dropped on the corning glass substrate and rotated at 2500 rpm for 15 seconds using VTC-100 spin coater system. After deposition, the NZO films were preheated at 300 °C for 30 minutes to remove the solvent and then heated in air at 400, 500 and 600 °C for 1 hour.

2.2. Characterizations

The structure of nanocrystalline NZO thin films was studied with XRD-Smartlab Rigaku system. The surface morphology of the thin films was investigated by FESEM JIB-4610F system. The optical properties were determined using UV-Vis spectrophotometer UV-2600 Shimadzu and Photoluminescence Cary Eclipse Spectrophotometer Agilent Technology. All characterization was performed at room temperature.

2.3. Photocatalytic activity

The photocatalytic activity of NZO thin films was investigated by photodecolorization of the Rhodamine-B (RhB) solution under UV light. The experiment was carried out in home-made reactor at room temperature with Royalux T5 6 W lamps as UV source. The obtained films were placed inside 100 mL of a 10 ppm RhB solution with eight UV lamps. The photocatalytic degradation was evaluated by measuring the absorbance of RhB solution for every 30 min using visible spectrophotometer AMTAST AMV01. The degradation efficiency of RhB was calculated using equation in reference [8, 12].

3. Results and discussion

3.1. Structural properties

Figure 1 shows XRD patterns of nanocrystalline NZO thin films as a function of heat treatment temperature of 400, 500, and 600 °C. All samples show seven diffraction peaks corresponding to

(100), (002), (101), (102), (110), (103), and (112) planes of the ZnO (ICCD 01-078-3315). It indicates that all films have polycrystalline nature and hexagonal wurtzite structure [20]. The nickel metal and other oxide phases are not detected in the XRD pattern, suggesting that the Ni dopant has entered into the ZnO lattice. The atomic radii of Ni and Zn are almost equal which lead to no effect on the crystal orientation of nanocrystalline NZO films. Further observations show that there is no peak shifted from nanocrystalline NZO thin film after heat treatments, but slight changed in FWHM and intensity. The result reveals that the heat treatment has an effect on the crystal parameters, as shown in table 1.

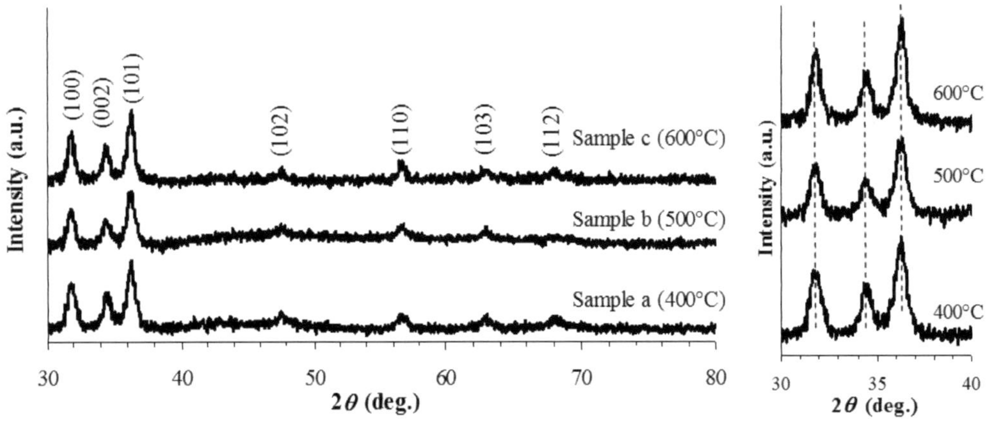

Figure 1. XRD patterns of NZO thin films.

The crystalline size (*D*), lattice strain (ε), residual stress (σ), and dislocation density (ρ) were estimated from equation 3 of reference [8], equation 4 of reference [26], equation 6 of reference [27], and equation 4 of reference [28], respectively. It can be seen in the table 1 that the crystalline size of film increases with increasing temperature. The mobility of the species on the surface occurs as the annealing temperature increases [26, 29, 30] which causes the diffusion at the grain boundaries. This phenomenon causes an agglomeration of small grains into larger grain and reduces the crystal defect. It was indicated by the decreasing in lattice strain and dislocation density as shown in the table 1. The calculation result of residual stress reveals that the Ni^{2+} on the ZnO lattice affects the compressive stress in the films [26]. The difference in the ionic radii between host material and the dopant influences the stresses arising. The crystal parameters of the film fabricated by sol-gel method are strongly influenced by heat treatment.

Table 1. Crystal parameter of nanocrystalline NZO thin films as effect of heating temperature.

Crystal parameters	Orientation								
	100			*002*			*101*		
	400 °C	500 °C	600 °C	400 °C	500 °C	600 °C	400 °C	500 °C	600 °C
FWHM	0.728	0.614	0.542	0.605	0.581	0.569	0.724	0.667	0.577
D (nm)	11.35	13.45	15.25	13.73	14.33	14.63	11.55	12.53	14.50
ε (%)	0.0112	0.0094	0.0083	0.0085	0.0082	0.0080	0.0096	0.0089	0.0077
σ (Gpa)	-2.608	-2.192	-1.931	-1.993	-1.906	-1.871	-2.245	-2.066	-1.788
ρ (line/nm^2)	0.0121	0.0086	0.0067	0.0563	0.0515	0.0496	0.0165	0.0140	0.0105

3.2. Morphological studies

The plane view of FESEM micrograph of nanocrystalline NZO thin films at different heating temperature is shown in figure 2. All films morphology consists of inhomogeneous grains which suggest that the nucleation of nanocrystalline NZO thin film on the substrate surface is non-uniform during crystal growth. Some pores can be seen in the morphology of nanocrystalline NZO thin film at heating temperature of 500 °C. The film becomes denser and the mean grain size of the films increases with increasing heating temperature. The increase of grain size was caused by coalescence of small grain [29]. This result corresponds to the XRD results as presented above.

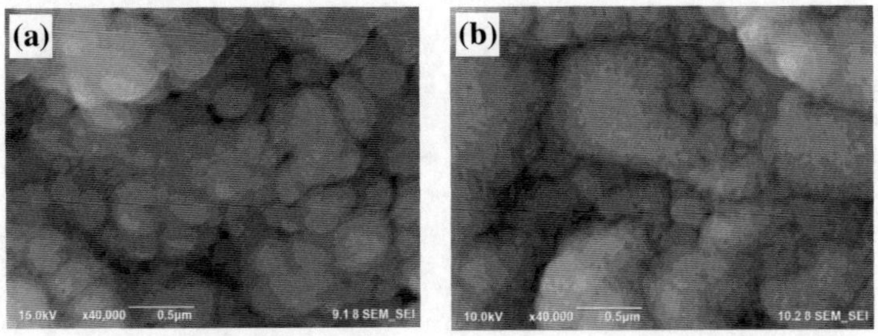

Figure 2. FESEM images of nanocrystalline NZO thin film heated at (a) 500 °C and (b) 600 °C.

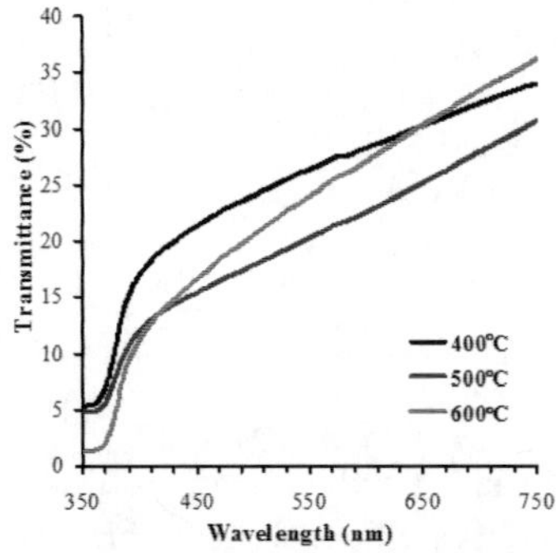

Figure 3. Transmission spectra of nanocrystalline NZO thin films heated at different temperatures.

3.3. Optical properties

Figure 3 shows the optical transmittance spectra of nanocrystalline ZnO thin films in the wavelength range of 350-750 nm. The transmittance spectra of all sample show a quite sharp absorption edge below 400 nm which is attributed to the fundamental absorption of zinc oxide [31, 32]. In figure 3, it can also be seen that all films exhibit low transmittance (less than 40%). The presence of Ni in the ZnO lattice promotes crystal defect as shown in XRD characterization, thus reducing the transmittance [33]. Besides that, inhomogeneous grains of NZO (as shown in figure 2) affect the optical scattering and contribute to the low transmittance [34].

The room temperature PL spectra of nanocrystalline NZO thin film as function of heating temperature are shown in figure 4. The PL spectra of all sample show several UV and visible emission, such as 361 nm (UV), 377 nm (UV), 421 nm (violet), 485 nm (blue), 495 nm (green), and 596 nm (orange). The UV emission originates from the recombination of free excitons and attributed to the near-band edge emission of ZnO [13, 27, 34]. The violet emission is due to the exciton transition from the localized level below the conduction band to the valance band [10]. Zhang et al. [35] reported that the blue emission originated from the electron transition from the shallow donor level of oxygen vacancies to the valence band, and electron transition from the shallow donor level of zinc interstitials to the valence band. The green emission may originate from the electron transition from the level of ionized oxygen vacancies to the valence band [18, 35]. The orange emission is attributed to oxygen interstitial in the ZnO system [30]. The PL intensity of nanocrystalline NZO thin film decreases with increasing heating temperature. It indicates that the defect concentration in the nanocrystalline NZO film decreases with the increasing heating temperature [13].

3.4. Photocatalytic activity

The photocatalytic activity of nanocrystalline NZO thin films on RhB solution under UV light irradiation is shown in figure 5a. No degradation of RhB solution in the absence of NZO films can be observed in figure 5a. The degradation also did not occur when UV lamp was not used (dark condition). This result indicates that the NZO thin films [36] and UV light [5] are essential for the degradation of RhB. The significant enhancement in photocatalytic activity happened in nanocrystalline NZO thin film with heating temperature of 500 °C and 600 °C. Similar result has been reported by Lv et al. [10] where the photocatalytic activity of nanocrystalline ZnO thin films enhances with the increase of annealing temperature. The increase of structural, surface morphology, and optical properties are believed to have contributed in the acceleration of the photocatalytic activity.

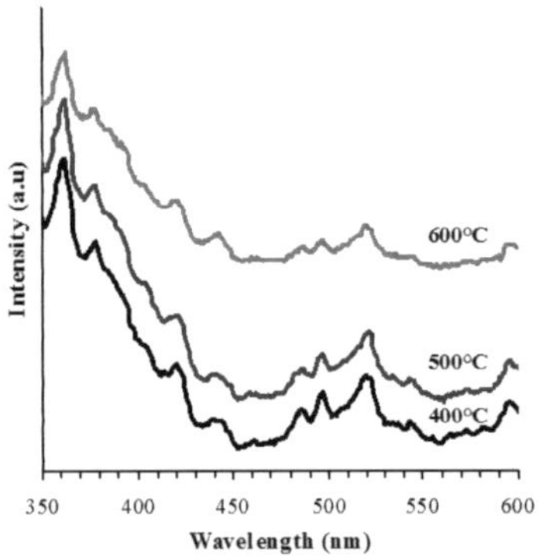

Figure 4. PL spectra of nanocrystalline NZO thin films heated at different temperatures.

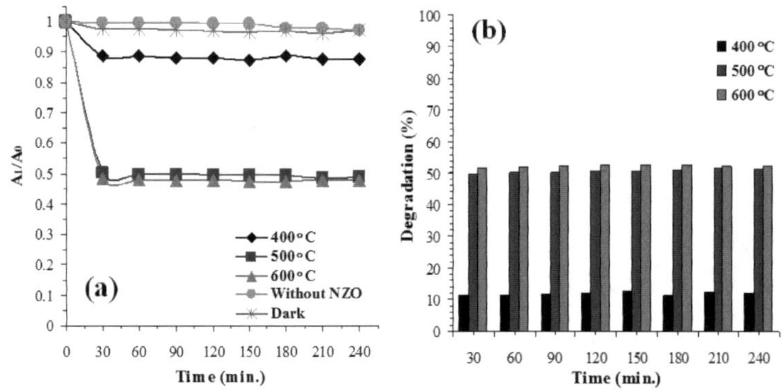

Figure 5. Photocatalytic (a) degradation and (b) degradation efficiency of RhB for nanocrystalline NZO thin films.

Figure 5b shows the photocatalytic degradation of RhB solution with nanocrystalline NZO thin films for an irradiation time of 240 min. The degradation efficiency of nanocrystalline NZO thin films increases to 12.15%, 51.23%, and 52.50% at elevated heat treatment temperature of 400 °C, 500 °C, and 600 °C, respectively. Increasing the crystallinity and mean grain size at heat treatment temperature of 500 °C and 600 °C affected the acceleration of photocatalytic activity. Besides, the Ni dopant in ZnO system generates the crystal defect such as oxygen vacancies, zinc interstitials, and oxygen interstitial. This result is shown in PL characterization (figure 4). The defect is considered to be the active sites of the ZnO photocatalyst because the redox reaction was probably occured on the surface defects [10]. In general, the photocatalytic activity of thin films was influenced by the crystallinity and surface morphology of the film [3, 10].

4. Conclusion

Nanocrystalline NZO thin film has been synthesized by sol-gel spin coating method as a function of heat treatment temperature. All films have polycrystalline hexagonal wurtzite structure with inhomogeneous grains. The transmittance spectra of all film are below than 40% at the visible light. All films show several UV and visible emission in the PL spectra. The structural, surface morphology, and optical properties of nanocrystalline NZO thin film increased with increasing heating temperature. The photocatalytic activity of nanocrystalline NZO thin films enhances with increasing annealing temperature. It can probably be ascribed to the increase of crystallinity and the change in surface morphology. The presence of defect in the nanocrystalline NZO thin films also contributes to the photocatalytic activity.

Acknowledgments
The authors are very much thankful to Research Center for Physics, Indonesian Institute of Sciences for the financial support and thank also for Materials Research Laboratory- Physics Department, Faculty of Science and Technology, UIN Maulana Malik Ibrahim for the photocatalys test facilities.

References

[1] Ahumada-Lazo R, Torres-Martínez L M, Ruíz-Gómez M A, Vega-Becerra O E and Figueroa-Torres M Z 2014 *Appl. Surf. Sci.* **322** 35
[2] Saidani T, Zaabat M, Aida M S and Boudine B 2015 *Superlattices Microstruct.* **88** 315
[3] Mclaren A, Valdes-Solis T, Li G and Tsang S C 2009 *J. Am. Chem. Soc.* **131** 12540
[4] Hunge Y M, Mahadik M A, Moholkar A V and Bhosale C H, 2017 *Appl. Surf. Sci.* **420** 764
[5] Wang C and Cao L 2011 *J. Rare Earth* **29** 727
[6] Choi H, Stathatos E and Dionysiou D D 2007 *Desalination* **202** 199
[7] Alam U, Khan A, Raza W, Khan A, Bahnemann D and Muneer M 2017 *Catal. Today* **284** 169
[8] Poongodi G, Anandan P, Kumar R M and Jayavel R 2015 *Spectrochim. Acta A* **148** 237
[9] Weng Y C and Hsiao K T 2015 *Int. J. Hydrog. Energy* **40** 3238
[10] Lv J, Gong W, Huang K, Zhu J, Meng F, Song X and Sun Z 2011 *Superlattices Microstruct.* **50** 98
[11] Podporska-Carroll J, Myles A, Quilty B, McCormack D E, Fagan R, Hinder S J, Dionysiou D D and Pillai S C 2017 *J. Hazardous Mater.* **324** 39
[12] Kaneva N V, Dimitrov D T and Dushkin C D 2011 *Appl. Surf. Sci.* **257** 8113
[13] Bedia A, Bedia F Z, Aillerie M, Maloufi N and Benyoucef B 2015 *Energy Procedia* **74** 529
[14] Chu S, Olmedo M, Yang Z, Kong J and Liu J 2008 *Appl. Phys. Lett.* **93** 181106
[15] Cittadini M, Sturaro M, Guglielmi M, Resmini A, Tredici I G, Anselmi-Tamburini U, Koshy P, Sorrell C C and Martucci A 2015 *Sens. Actuator B-Chem.* **213** 493
[16] Fang T H and Kang S H 2010 *Curr. Nanosci.* **6** 505
[17] Cole J J, Wang X, Knuesel R J and Jacobs H O 2008 *Nano Lett.* **8** 1477
[18] Wu S, Chen Z, Wang T and Ji X 2017 *Appl. Surf. Sci.* **412** 69
[19] Abdel-wahab M Sh, Jilani A, Yahia I S and Al-Ghamdi A A 2016 *Superlattices and Microstruct.* **94** 108
[20] Salah N, Hameed A, Aslam M, Abdel-wahab M Sh, Babkair S S and Bahabri F S, 2016 *Chem. Eng. J.* **291** 115
[21] Türkyılmaz Ş Ş, Güy N and Özacar M 2017 *J. Photochem. Photobiol. A* **341** 39
[22] Abed C, Bouzidi C, Elhouichet H, Gelloz B and Ferid M 2015 *Appl. Surf. Sci.* **349** 855
[23] Farag A A M, Cavaş M, Yakuphanoglu F and Amanullah F M 2011 *J. Alloys Compd.* **509** 7900
[24] Rajeh S, Mhamdi A, Khirouni K, Amlouk M and Guermazi S 2015 *Opt. Laser Technol.* **69** 113
[25] Yu M, Qiu H, Chen X, Li H and Gong W 2011 *Mater. Chem. Phys.* **126** 797
[26] Chaitra U, Kekuda D and Rao K M 2017 *Ceram. Int.* **43** 7115
[27] Fang D, Lin K, Xue T, Cui C, Chen X, Yao P and Li H 2014 *J. Alloys Compd.* **589** 346
[28] Rao T P and Kumar M C S 2010 *J. Alloys Compd.* **506** 788
[29] Belkhalfa H, Ayed H, Hafdallah A, Aida M S and Ighil R T 2016 *Optik* **127** 2336
[30] Hasabeldaim E, Ntwaeaborwa O M, Kroon R E, Coetsee E and Swart H C 2017 *Opt. Mater.* **74** 139
[31] Singh S, Srinivasa R S and Major S S 2007 *Thin Solid Films* **515** 8718
[32] Hadri A, Taibi M, Ioghmarti M, Nassiri C, Tlemçani T S and Mzerd A 2016 *Thin Solid Films* **601** 7
[33] Ying M, Wang S, Duan T, Liao B, Zhang X, Mei Z, Du X, Gerriu F M, Fox A M and Gehring G A 2016 *Mater. Lett.* **171** 121
[34] Pati S 2017 *J. Alloys Compd.* **695** 3552
[35] Zhang D H, Wang Q P and Xue Z Y 2003 *Appl. Surf. Sci.* **207** 20
[36] Ali M B, Barka-Bouaifel F, Sieber B, Elhouichet H, Addad A, Boussekey L, Férid M and Boukherroub R 2016 *Superlattices and Microstruct.* **91** 225

IOP Conf. Series: Journal of Physics: Conf. Series **985** (2018) 012026 doi:10.1088/1742-6596/985/1/012026

Effect of surface density silver nanoplate films toward surface-enhanced Raman scattering enhancement for bisphenol A detection

N A Bakar[1], M M Salleh[1], A A Umar[1] and J G Shapter[2]

[1] Institute of Microengineering and Nanoelectronics (IMEN), Universiti Kebangsaan Malaysia, 43650 Bangi, Selangor, Malaysia
[2] School of Chemical and Physical Sciences and Flinders Centre for Nanoscale Science and Technology, Flinders University, Bedford Park, Adelaide South Australia 5001, Australia

E-mail: norhayati.ab@ukm.edu.my

Abstract. This paper reports a study on surface-enhanced Raman scattering (SERS) phenomenon of triangular silver nanoplate (NP) films towards bisphenol A (BPA) detection. The NP films were prepared using self-assembly technique with four different immersion times; 1 hour, 2 hours, 5 hours, and 8 hours. The SERS measurement was studied by observing the changes in Raman spectra of BPA after BPA absorbed on the NP films. It was found that the Raman intensity of BPA peaks was enhanced by using the prepared SERS substrates. This is clearly indicated that these SERS silver substrates are suitable to sense industrial chemical and potentially used as SERS detector. However, the rate of SERS enhancement is depended on the distribution of NP on the substrate surface.

1. Introduction

Bisphenol A (BPA) is an organic chemical that is used domestically by the manufacturing plant to produce polycarbonate plastic [1], stabilizer in polyvinyl chloride, epoxy coating, and antioxidant [2]. The use of BPA could negatively impact human health, harm to the environment, flora, and fauna. However, prevention in the production of BPA for human consumption is still very difficult to do because of its importance in everyday life. Therefore, monitoring techniques on BPA chemical need to be done so that our future generations will live in a clean environment and not suffer from chronic diseases such as cancer.

Recently, surface-enhanced Raman scattering measurement was rapidly developed in sensor technology especially for monitoring contamination of environment and food, in medical and crime. SERS sensor is an important measurement technique in chemistry and biology fields due to it can directly detect a specific molecule precisely. For instance, SERS was studied for detection of melamine in dairy products [3], pesticides [4], kidney disease diagnosis [5], cancer and diabetes [6], forensic analysis for blood classes [7], urea [8], and human semen [9].

SERS phenomenon occurs when the signal from the Raman spectrum of a specific molecule near the defined metal surface was amplified by a factor of 10^3 and above [10]. This phenomenon is due to a result from plasmonic effect of the metal nanoparticles [11]. Before doing SERS measurement, analyte was mixed with nanometal colloidal or absorbed on a deposited nanometal substrates which are called SERS substrates. Nanometal elements that are commonly used to get a good plasmonic effect are gold and silver. However, nanosilver has attracted attention in SERS due to low synthesis preparation cost, sharp and high plasmon resonance curve, and strong SERS effect [12].

In this paper, we attempt to study the detection of BPA molecule on four SERS substrate using self-build SERS sensor system. These SERS substrates were prepared with various immersion times to obtain different morphologies on substrate surface. It was observed that the distribution of NP on the surface will be affected by the sensitivity of the sensor towards BPA detection. The result of this work suggested that NP films can be used as a good indicator for an industrial chemical detection.

2. Experiment

Silver triangular nanoplate (NP) substrates were prepared by depositing silver nanoplate on the quartz surface from a colloidal using self-assembly technique. This preparation was described in a previous report [13]. Normal grade quartz substrates without special specification have been purchased from Latech Scientific Supply (Kuala Lumpur, Malaysia) and cut to size 1.2×1.2 cm. Quartz substrate has been chosen because it does not have a Raman spectrum and suitable for implementing in the use of surface-enhanced Raman scattering (SERS). In the study of NP film formation, the experiment was conducted by varying the immersion time of the quartz surface in the colloidal NP for 1 hour, 2 hours, 5 hours, and 8 hours to obtain different film morphologies. The films were labeled as S1, S2, S5, and S8 for 1 hour, 2 hours, 5 hours, and 8 hours of immersion time process respectively.

Bisphenol A (BPA) was supplied by Sigma Aldrich (Kuala Lumpur, Malaysia) with purity of 99.9% and was used without further purification. BPA solution was prepared by dissolving it in ethanol with eight different concentrations, namely 0.001 M, 0.005 M, 0.01 M, 0.025 M, 0.05 M, 0.1 M, 0.5 M, and 1.0 M. In this study, SERS substrate was prepared by depositing BPA solution onto S1, S2, S5, and S8 surface. These substrates were prepared by dropping 0.02 ml of BPA solution and allowed to dry at room temperature for 5 minutes. After that, the SERS substrates were measured to study the presence of BPA molecule on the surface using home-made SERS sensor. Raman spectra of a bare quartz, BPA on quartz surface, NP films were recorded as reference spectra.

Basic characterizations were performed on the NP films to study the change in morphologies of the film's surface. The S1, S2, S5, and S8 were characterized using field emission scanning electron microscopy (FESEM) (Zeiss Supra 55-VP, Selangor, Malaysia) and optical absorbance (Shimadzu UV-160A, Selangor, Malaysia). The SERS measurement was recorded using a self-build SERS sensor system with a 785 nm excitation laser wavelength with 10 seconds integration time.

3. Results and discussion

Triangular silver NP films were simply prepared on a quartz surface using self-assembly technique with four different immersion times which were varied from 1 hour to 8 hours. FESEM was employed to study the successful attachment of the NP on the surface as presented in figure 1. Silver NP particles were starting to attach on the surface toward increasing immersion time after exposing the surface into the colloidal silver NP for 1 hour. The distribution of the NP attachment on the quartz surface was subsequently increasing towards the longer immersion time of the surface into the colloidal. Hence, the calculated of NP distribution on the surface for S1, S2, S5, and S8 are 60, 113, 209, and 430 counts/μm respectively. Besides, the FESEM images showed that the triangular-shape NP on the surface was maintained with the increasing of the immersion times. The average size of the triangular NP on the surface is from 18 nm to 25 nm. These images also revealed the deposited nanoplates particles were well organized on the surface without overlapping each other and formed monolayer films on the surface. The high coverage of the nanoplates with slightly overlapping each plate was obtained after exposing the surface for eight hours into the colloidal.

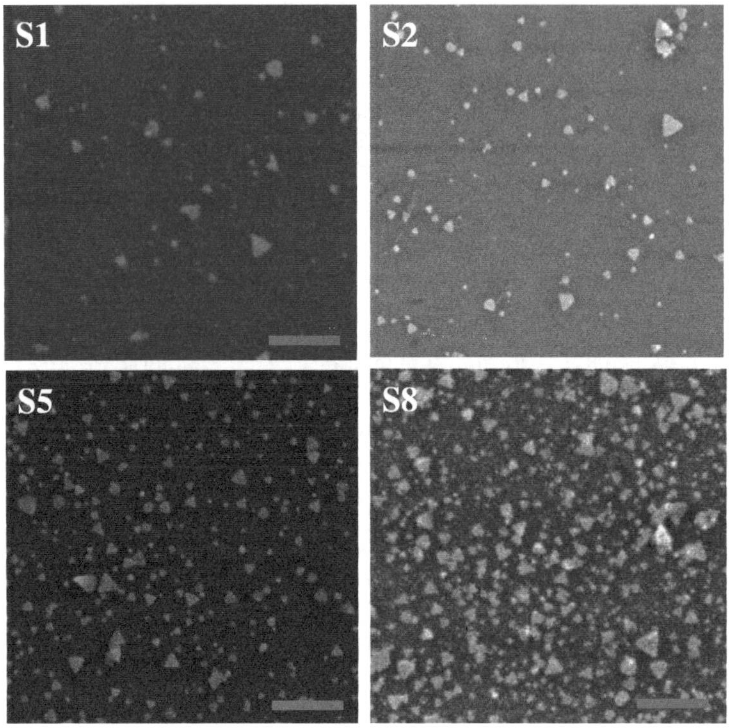

Figure 1. FESEM images of the triangular silver NP on the quartz surface. Scale bars are 100 nm.

Optical spectroscopy has been used to study surface plasmon resonance properties of the triangular silver NP. Figure 2 shows typical absorbance spectra of the NP for S1, S2, S5, and S8. It was found that the triangular silver NP has two absorbance peaks at 330 nm and 640 nm with an absence of a shoulder at 440 nm. A weak intensity of 330 nm peaks was existed due to the small percentage of deposited quasi-sphere on the surface. Meanwhile, an intense broad surface plasmon at 640 nm peak and the shoulder are attributed to Localized Surface Plasmon Resonance (LSPR) and Transverse Surface Plasmon Resonance (TSPR) respectively. The shoulder situated at 440 nm also refers to the successful formation of anisotropic nanoparticles. The intensity of LSPR and TSPR were generally increased with the increase of the immersion times, which is corresponding to the increase of the NP distribution on the surface. However, the position peaks of the absorbance did not have significant changes for S1, S2, S5, and S8.

These four NP films were carried out to sense BPA molecule using SERS sensor. BPA solution at 0.5 M of concentration was dropped on a bare quartz surface as a reference spectrum and dropped on the each NP films. BPA was then let dry under room temperature for 5 minutes. SERS sensor was measured by capturing the Raman spectra of 0.5 M BPA concentration on the bare quartz substrate and on the S1, S2, S5, and S8 as shown in figure 3. Raman spectra were also recorded for bare quartz and the NP substrates without BPA molecule but no Raman peaks existed. Based on the Raman spectra in figure 3, BPA has exhibited seven major Raman peaks at 642 cm^{-1}, 819 cm^{-1}, 832 cm^{-1}, 1114 cm^{-1}, 1183 cm^{-1}, 1600 cm^{-1}, 1614 cm^{-1} [14]. These Raman peaks were enhanced when the BPA molecule absorbed on S1, S2, S5, and S8, clearly demonstrated as a result from plasmonic effect of the triangular silver NP which referred to SERS phenomenon. The intensity of SERS signal was increasing towards increasing of thin films NP distribution on the surface except for S5 and S8 but the Raman peaks position of BPA molecule remain.

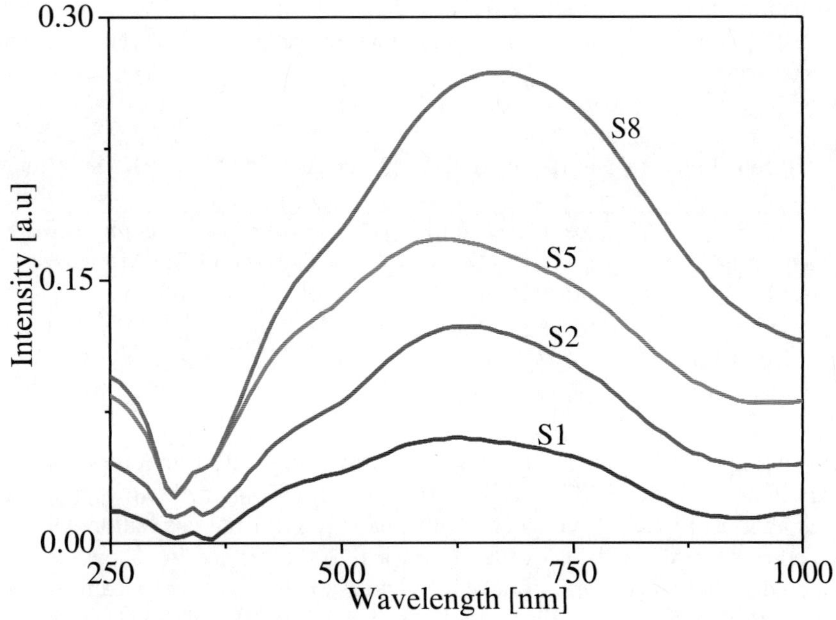

Figure 2. Optical absorbance of the triangular silver NP films.

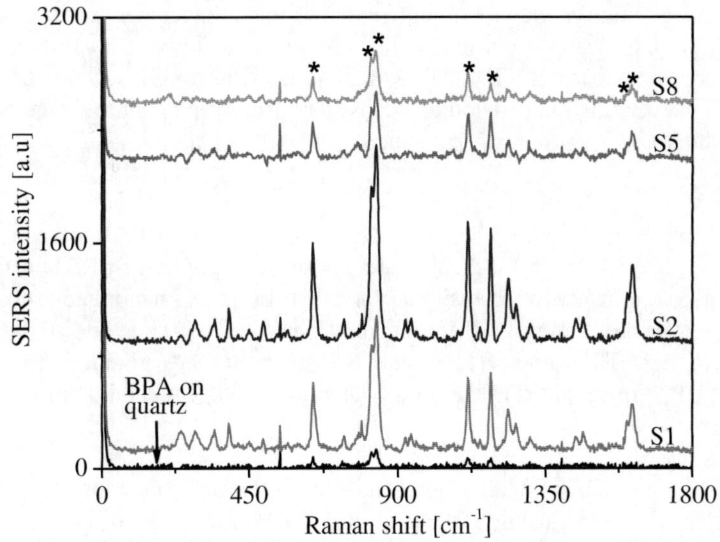

Figure 3. SERS spectra of BPA molecule on the bare quartz and on the NP films surface. BPA peaks are represented as stars.

Referring to SERS analysis in figure 3, it was found that the different morphologies of NP films have given different performance towards bisphenol A detection. Here, we study the sensitivity of the NP films as a SERS substrates towards detection of BPA molecule by calculating the Enhancement Factor (EF). This EF calculation was reported previously in [15]. The analysis of EF values for 642 cm^{-1} peaks and distribution of surface density of each film were arranged in table 1.

The EF value for BPA peaks at 642 cm^{-1} for S1 is 1.04×10^{3} which demonstrated that S1 has been giving a good sensitivity to sense BPA molecule. S2 has shown more sensitivity by increasing EF value and distribution of NP on the surface. However, the EF value of S5 and S8 was decreasing with increasing NP distribution on the surface, indicating that S5 and S8 are less sensitive than S1 and S2 towards BPA detection.

Table 1. Enhancement factors of 642 cm^{-1} peaks using four SERS substrates.

SERS substrate	Enhancement factor [EF]	Distribution [count/μm] (based on FESEM images)
S1	1.04×10^3	60
S2	1.52×10^3	113
S5	5.53×10^2	209
S8	3.87×10^2	430

To explain why high distribution surface gave low EF value, we need to refer to absorption spectra and morphology images of S1, S2, S5, and S8. NP structure has obtained two oscillation modes after the NP particles irradiated by electromagnetic wave, namely localized oscillation which is horizontal to the surface and transverse oscillation which is vertical to the surface. As shown in figure 2, SERS substrate has high distribution and will produce high-intensity absorbance of localized mode.

In SERS mechanism, electromagnetic wave that irradiated on NP will be absorbed and transferred to analyte molecule and then Raman intensity signal will be enhanced. However, induce oscillation mode on NP that contributes to the energy transfer is transverse mode. At the same time, localized oscillation mode interacted with same oscillation mode from the nearest neighbouring nanoplate. As the result, localized oscillation mode is not contributed to the enhancement of intensity Raman signal. Therefore, closer nanoplate arrangement in the film surface gives lower EF value.

Generally, in previous research, EF value increased with rough surface or dense nanometal distribution [16]. However, EF and distribution density relationship is contradicted for metal materials which have a thin thickness or monolayer film such as NP film [17].

4. Conclusion

Study on using NP substrates in SERS sensor successfully detected BPA molecule using four SERS substrates namely S1, S2, S5, and S8. S2 substrate has shown optimum SERS substrates with highest sensitivity of EF value, i.e. 1.52×10^3. Density surface distribution of nanoplates substrates determined EF value and lowest detection limit of BPA concentrations. The low-density distribution of nanoplates on the surface gave high EF value. Hence, the NP substrates are potential to be used as SERS substrates to detect BPA for monitoring the food, water, and environmental quality.

Acknowledgments

This work has been supported by Malaysian Ministry of Higher Education, Universiti Kebangsaan Malaysia and Flinders University using a financial grant of MRUN-2015-002.

References

[1] Brock J W, Yoshimura Y, Barr J R, Maggio V L, Graiser S R, Nakazawa H and Needham L L 2001 *J. Expo. Sci. Environ. Epidemiol.* **11** 323
[2] Staples C A, Dorn P B, Klecka G M, O'Block S T, Branson D R and Harris L R 2000 *Chemosphere* **40** 521
[3] Cheng Y, Dong Y, Wu J, Yang X, Bai H, Zheng H, Ren D, Zou Y and Li M 2010 *J. Food Compos. Anal.* **23** 199
[4] Vongsvivut J, Robertson E G and McNaughton D 2010 *J. Raman Spectrosc.* **41** 1137
[5] Abdullah N A, Bakar N A, Shapter J G, Salleh M M and Umar A A 2017 *Adv. Nat. Sci.: Nanosci. Nanotechnol.* **8** 025015

[6] Chi J, Zaw T, Cardona I, Hosnain M, Garg N, Lefkowitz H R, Tolias P and Du H 2015 *Biomed. Opt. Express* **6** 761
[7] Boyd S, Bertino M F, Ye D, White L S and Seashols S J 2013 *J. Forensic Sci.* **58** 753
[8] Abdullah N A, Bakar N A, Salleh M M and Umar A A 2016 *KnE Engineering 6 pages DOI 10.18502/keg.v1i1.484*
[9] Irvine J 2015 *Surface-enhanced Raman spectroscopy for forensic analysis of human semen* Master Thesis *(* Boston: Boston University)
[10] Le Ru E C, Blackie E, Meyer M and Etchegoin P G 2007 *J. Phys. Chem. C* **111** 13794
[11] Halas N J and Moskovits M 2013 *MRS Bull.* **38** 607
[12] Xu L, Yan W, Ma W, Kuang H, Wu X, Liu L, Zhao Y, Wang L and Xu C 2015 *Adv. Mater.* **27** 1799
[13] Bakar N A, Shapter J G, Salleh M M and Umar A A 2015 *Appl. Sci.* **5** 209
[14] Yao W, Wang S and Wang H 2012 *Method for detecting BPA by surface enhanced Raman spectroscopy* US Patent US20120236300A1
[15] Bakar N A, Salleh M M, Umar A A and Shapter J G 2017 *Adv. Nat. Sci.: Nanosci. Nanotechnol.* **8** 025008
[16] Brolo A G, Irish D E, Szymanski G and Lipkowski J 1998 *Langmuir* **14** 517
[17] Stolberg L, Lipkowski J and Irish D E 1991 *J. Electroanal. Chem.* **300** 563

IOP Conf. Series: Journal of Physics: Conf. Series **985** (2018) 012027 doi:10.1088/1742-6596/985/1/012027

In-situ deposition of hematite (α-Fe_2O_3) microcubes on cotton cellulose via hydrothermal method

M Gili[1], G Latag[2] and M Balela[2]

[1] Material Science and Engineering Program, College of Science, University of the Philippines Diliman, Quezon City 1101, Philippines
[2] Department of Mining, Metallurgical and Materials Engineering, College of Engineering, University of the Philippines Diliman, Quezon City 1101, Philippines

E-mail: gilimonbryan@yahoo.com

Abstract. Hematite microcubes with truncated edges have been successfully deposited on cotton cellulose via one-step hydrothermal process using anhydrous $FeCl_3$ and glycine as Fe(III) precursor and chelating agent, respectively. The amount of glycine significantly affects the morphology and yield of hematite. The addition of 0.495 g of glycine to 50 ml of 0.1 M $FeCl_3$ solution with 0.400 g of cotton resulted to hematite-deposited cellulose having ~15% hematite content. The reduction of glycine to 0.247 g increased the amount of hematite on the surface of the cotton cellulose to ~20% by weight. However, the hematite microcubes have a wide size distribution, with particle size in the range of 0.684 μm to 1.520 μm. Without glycine, hematite cannot be formed in the solution.

1. Introduction

Hematite (α-Fe2O3) is the most stable oxide of iron under ambient conditions [1]. It has acquired considerable attention due to its wide range of applications being a cheap, n-type semiconductor with high corrosion resistance, high stability, substantial absorption of visible light, low bandgap of 2.1 eV, and low toxicity [2]. Numerous studies have shown its potential, such as in photocatalysis [3], gas sensing applications [4], lithium batteries [5 - 6], and water treatment [7 - 8].

The size and shape of the synthesized hematite nanoparticles play important role in the properties and performance of the material [1]. Extensive studies regarding hematite synthesis have produced different morphologies ranging from microflowers [8], nanocubes [9], nanotubes [10] and nanospheres [6] among others. Numerous techniques have been developed to produce these structures and achieved the desired properties [1]. Some of these methods are direct thermal decomposition [11], electrochemical synthesis [10], sonochemical synthesis [12] and chemical co-precipitation [13]. However, most of these methods require multiple steps, massive amount of energy and expensive chemicals while others are environmentally unsafe. Moreover, wide size distribution of particles is obtained for some processes, which is why hydrothermal method is preferred. In hydrothermal synthesis, glycine is commonly used as chelating agent in preparing hematite [1, 14]. However, there are only few comprehensive studies on its role in the formation of hematite.

Deposition of hematite on cellulosic scaffolds has been performed to cater specific applications ranging from drug delivery [15] to pigment science [16]. It is because the use of cellulose as a substrate can promote nucleation and growth of the metallic oxide in aqueous medium [16]. Hematite with microcube morphology has been known to create an impact in drug delivery, wherein the α-Fe_2O_3 microcubes are embedded in drug fillers commonly made of cellulosic materials [15, 17]. The deposition of hematite micro and nanoparticles in various substrates makes product recovery easier due to the apparent bigger structure.

In this study, a simple, cheap and environment friendly way to synthesize and deposit α-Fe_2O_3 microparticles in cotton cellulose is presented. A single-step hydrothermal method has been performed with the use of anhydrous $FeCl_3$ as precursor and glycine as chelating agent. The effects of varying the amount of glycine in the formation of hematite have been investigated. The crystal structure and morphology of the products were analyzed and the amount of deposited hematite on the cellulose fibers was determined from the thermogram of each sample.

2. Methodology

In a typical preparation, 0.811 g of anhydrous ferric chloride ($FeCl_3$, HIMEDIA) was dissolved in 50 ml deionized water to produce 0.1 M $FeCl_3$ yellow-brown transparent solution. Then 0.495 g of glycine (NH_2CH_2COOH, UNIVAR, Ajax Finechem) was added slowly upon continuous stirring, producing a black solution. It was mixed thoroughly for 30 min using a magnetic stirrer at 500 rpm. Then 0.400 g of commercial cotton (Cleene, Selair Manufacturing Corp.) was added and the mixture was heated in a hot plate. It was allowed to boil for 15 min where it turned brown-orange. The solution was transferred to a 100 ml Teflon-lined stainless steel autoclave and heated to 160 °C in oven (Yamato, ADP 210C) for 5 h. After the hydrothermal process, the cotton fibers fragmented into small pieces fine enough for the final product to be a suspension. The solid products were collected via decantation and washed with DI water, shaken vigorously to remove loose hematite. After three times of washing, the solid products were collected and dried in oven at 85 °C for 4 h. To determine the effect of glycine, the amount was reduced in other experiments. Another experiment was also carried out without the cotton to determine its effect in the morphology of the hematite.

The structure and crystallinity of the formed hematite was analyzed using XRD (SHIMADZU, XRD-7000 Maxima). The morphology was observed using FESEM (JSM-7800fPrime Japan) and the thermal behavior and stability of the samples were investigated via TG-DTA (SHIMADZU TA-60WS) run at 10 °C/min ramp rate from room temperature to 1000 °C in static air.

3. Results and discussion

When 0.495 g of glycine was added, the as-prepared powder has a deep red color. The red color indicates the presence of hematite particles as commonly reported [7]. In the absence of glycine, black solid products was produced which suggest nonexistence of hematite.

Figure 1 shows the corresponding XRD patterns of the final products with 0.495 g glycine and without the added glycine. In figure 1a, sharp peaks attributed to hematite are indexed. Some notable peaks are at 2θ = 24.12°, 33.3°, 35.74°, 40.94°, 49.54° and 54.14° which correspond to planes (0 1 2), (1 0 4), (1 1 0), (1 1 3), (0 2 4) and (1 1 6) [18]. This confirms the formation of hematite as indicated by the reddish appearance of the powder. Other peaks at 2θ = 15.0°, 16.5°, 22.86° and 34.5° belong to cellulose I of the cotton due to planes (1 0 1), (1 0 $\bar{1}$), (0 0 2) and (0 0 4), respectively [19].

No peaks corresponding to hematite are observed in the XRD pattern of the sample without glycine as seen in figure 1b. Only broad peaks corresponding to cellulose I are clearly present. Notice that there is a significant increase in the intensity of planes (1 0 1) and (1 0 $\bar{1}$).

Figure 2a shows the SEM micrographs of the products formed upon addition of 0.495 g of glycine without the cotton substrate. The particles have cubic shape with truncated edges. Additionally, the cubic hematite particles have sponge-like structures. The size of the particles ranges from 0.851 μm to 1.269 μm. Agglomeration of these particles have been observed which resulted to bulk structures that reaches several micrometers in diameter.

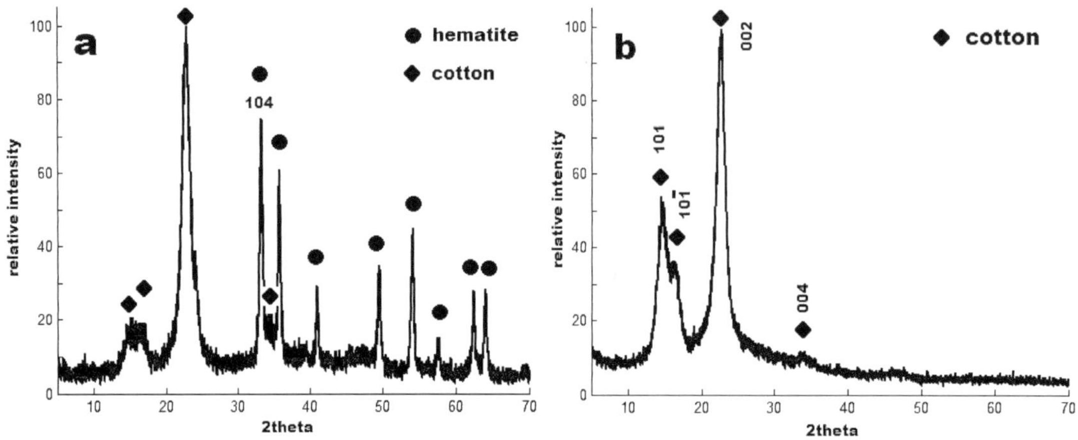

Figure 1. X-ray diffraction pattern of the samples with a) 0.495 g glycine and b) without glycine.

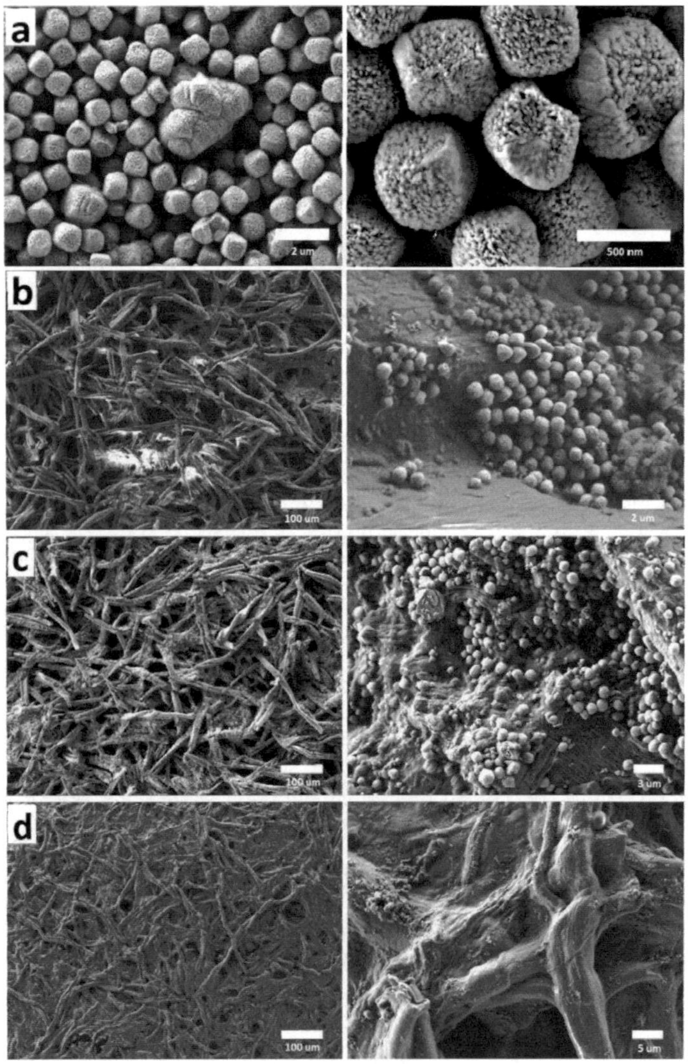

Figure 2. Low and high resolution SEM images of the sample (a) without cotton containing 0.495 g glycine and of the samples with 0.400 g cotton containing (b) 0.495 g, (c) 0.247 g, and (d) 0 g glycine.

With the addition of cotton, the cubic hematite particles with an average size of 0.712μm adhered on the surface of the fibers as shown in figure 2b. The cotton fibers have diameters and lengths of about 20 μm and 200 μm respectively. However, the cotton fibers are not uniformly covered with hematite.

For the sample prepared with 0.247 g glycine, the cotton fibers are also covered with hematite particles. Based on the SEM images (see figure 2c), more hematite particles are deposited on the surface at lower glycine concentration. The size of these particles ranges from 0.684 μm to 1.520 μm with an average size of 1.032 μm. It is larger compared to the average size of the previous sample, which is only 0.712 μm. But these particles are still much bigger compared to the average particle size of 80 - 90 nm that was reported using same synthesis method [5]. With over 100 particles measured, the standard deviation is about 14.77% of the average diameter indicating wide size distribution. Without addition of glycine, only cotton fibers are observed as seen in figure 2d. This agrees with the XRD pattern in figure 1b, which only show peaks due to cellulose I of the cotton.

The addition of glycine is crucial in preparing hematite particles. The mixing of $FeCl_3$ and glycine in deionized water may result to dissolution and hydrolysis forming iron cations and glycine in acidic (Gly^+: $^+H_3NCH_2$-COOH) and zwitterionic ($Gly^{\pm}$:^+H_3N-CH_2-COO^-) forms [14, 20]. Glycine acts as a strong chelating agent which may form stable complexes with the iron cations. These complexes become unstable at higher temperature, releasing the iron cations in the aqueous solution followed by nucleation and crystal growth upon hydrothermal process [5]. It has been reported that positively charged protonated glycine interacts with $Fe(OH)_3$, which is a product of $FeCl_3$ hydrolysis, via electrostatic and hydrogen bond [21]. Hematite particles are then formed by a two-phase transformation upon hydrothermal treatment [21], $Fe(OH)_3 \rightarrow$ -FeOOH $\rightarrow \alpha$-Fe_2O_3.

Glycine delays the growth of crystals due to the formation of the said complexes [5]. Low concentration of glycine would result to inefficient control of the particle size. Nuclei formed at early stages tend to outgrow each other since there is lesser glycine that delays growth, as suggested by the presence of larger hematite particles of the sample with lower glycine content (figure 2c). The nuclei formed at the later stages will no longer grow since the precursor is already depleted resulting to a wide size distribution. But without glycine, hematite cannot be formed. Perhaps, $Fe(OH)_3$ get strongly bonded to the cotton cellulose and is not able to release even at elevated temperature, hindering the formation of α-Fe_2O_3.

The three samples with the added cotton were subjected to thermal analysis. The TG and DTA thermograms for the sample with 0.495 g glycine are presented in figure 3a. There are three significant stages of mass reduction at temperature ranges of 25 - 260.9 °C, 260.9 - 351.6 °C and 351.6 - 542.8 °C. The first stage has two distinct levels. Firstly, a small decrease is observed until 74.73 °C that reduce the mass to 94.86%. This corresponds to the dehydration of cellulose. Secondly, a gradual mass loss occurs until 260.9 °C, effectively decreasing the mass to 88.91%. This is probably due to the elimination of physically or chemically adsorbed water. In the second stage, an abrupt decline in the mass to about 27.68% is determined at 351.6 °C. This is attributed to the oxidative decomposition of cotton which involves formation of combustible gases of carbonyl and carboxyl groups (e.g. aldehydes, ketones, ethers, etc.), evolution of gases (e.g. CO and CO2) and formation of carbonaceous residue or char [22]. In the final stage, the mass continues to decrease until it attains a stable mass of 15.2% around 542.8 °C. This is due to the decomposition of the previously formed residue [23].

A decreasing DTA curve indicates lowering of the sample's temperature with respect to the reference. Conversely, an increasing DTA curve means a rising relative temperature. The endotherms at 52.35 °C and 324.9 °C are brought by the absorbance of heat due to the evaporation of water and carbonization of cotton respectively. The peak at 351.6 °C signifies an increased temperature due to the burning of char.

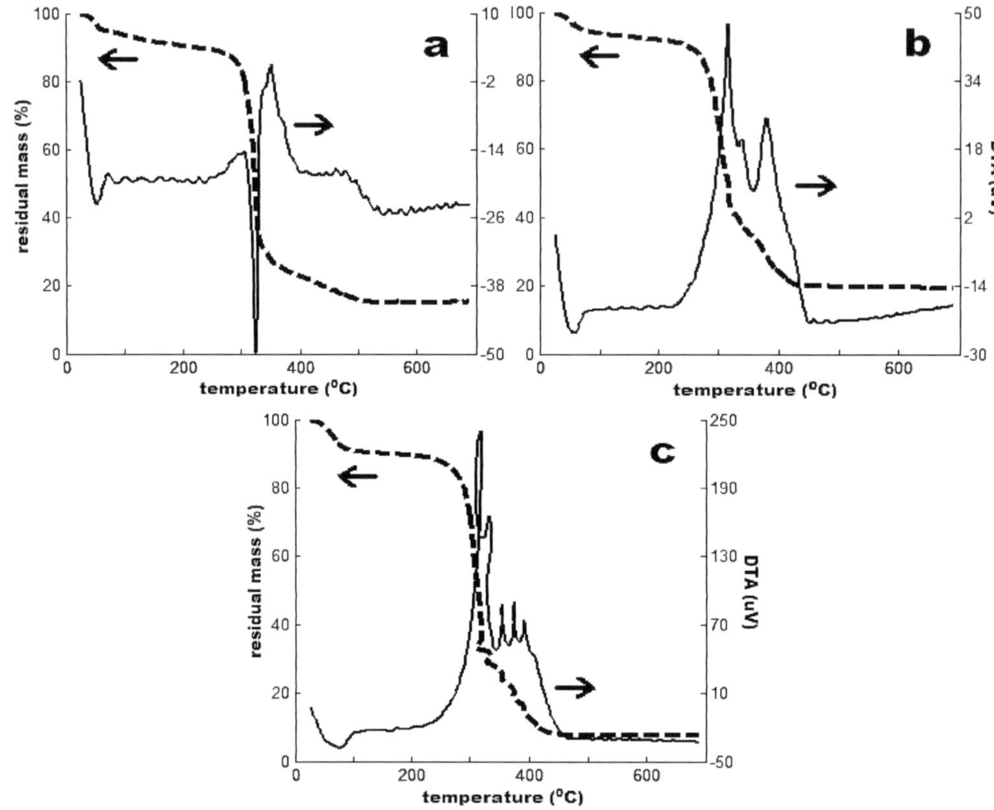

Figure 3. Thermal behavior of the samples with 0.400 g cotton and (a) 0.495 g, (b) 0.247 g, and (c) 0 g glycine.

On the other hand, from the thermogram of the sample prepared with 0.247 g of glycine (see figure 3b), a sudden drop occurs reducing the mass to 50.48% at the end of 315.7 °C. Then a multi-step reduction in mass occurs until it achieves a constant mass of 19.85% at about 451 °C. Its DTA curve shows an endotherm at 56.43 °C, which is due to the dehydration of cotton. An exothermic peak centered at 315.7 °C is observed and attributed to the oxidation of cotton. It is followed by two other peaks at 338.8 °C and 378.8 °C, which are basically the burning of the residue.

For the sample without glycine (see figure 3c), there is an immediate loss in mass eliminating 9.06% at the end of 100 °C. The mass is approximately stable until 217.15 °C, where it starts undergoing sharp decrease. Only 41.87% of the mass is left around 318.9 °C. From there, a multi-stage reduction in mass occurs where it ends with a stable mass of 7.77% around 460 °C. From the DTA thermogram, the endotherm brought by the removal of adsorbed water is broader and is shifted to a higher temperature of 76.9 °C. The shifting of the first endothermic peak and its broadening, as the amount of glycine is reduced, suggest that more water is adsorbed by the cellulose. On the other hand, the exothermic peak, which is associated with the burning of cotton, remains almost constant at 317.5 °C. The appearance of four additional peaks implies that unreacted precursor derivatives adsorbed by the cotton (i.e. $Fe(OH)_3$) underwent successive transformations beyond 315 °C.

Hematite is known for its stability even at very high temperatures [2, 5]. By assuming that cotton was completely removed by the burning, we suggest that the weight percentage of the hematite would be approximately 15% because the sample with 0.495 g glycine is found to be mainly composed of cellulose and hematite. The sample with 0.247 g glycine would have higher hematite content of around 20%. The results from the TG curves agree with the observations from the SEM showing more hematite particles deposited in the sample with lower amount of glycine. Finally, the sample containing no glycine generated 7.77% of stable mass. It is relatively higher than the residue when pure cotton is burnt, which is less than 2% as reported [22]. This implies that cellulose fibers might have absorbed significant amount of iron precursors, which eventually undergo transformations and remains as residual ash.

4. Conclusions

Successful deposition of hematite microcubes on cellulose was carried out via one-step hydrothermal process using anhydrous $FeCl_3$ as Fe(III) precursor and glycine as chelating agent respectively. The effect of varying amount of glycine was investigated. The addition of 0.495 g of glycine to 50 ml of 0.1 M FeCl3 solution with 0.400 g of cotton resulted to hematite-deposited cellulose having ~15% hematite content. The deposited hematite particles have an average size of 0.712 μm, with cubic structure having truncated edges and sponge-like surface morphology. The reduction of glycine to 0.247 g improved the amount of hematite to ~20% by weight, covering more surfaces of the cotton fibers with hematite particles. However, the particles have wide size distribution due to the reduced formation of complexes that controls crystal growth. Without glycine, no hematite was formed. It was also observed that cotton fibers absorbed significant amount of amorphous materials, probably iron precursors that were not able to transform to hematite after the hydrothermal treatment.

References

[1] Yin C Y, Minakshi M, Ralph D E, Jiang Z T, Xie Z and Guo H 2011 *J. Alloys Compd.* **509** 9821
[2] Huang M, Qin M, Chen P, Jia B, Chen Z, Li R, Liu Z and Qu X 2016 *Ceram. Int.* **42** 10380
[3] Cao Z, Qin M, Jia B, Gu Y, Chen P, Volinsky A A and Qu X 2015 *Ceram. Int.* **41** 2806
[4] Hao H, Sun D, Xu Y, Liu P, Zhang G, Sun Y and Gao D 2016 *J. Colloid Interface Sci.* **462** 315
[5] Chaudhari N K, Kim M S, Bae T S and Yu J S 2013 *Electrochim. Acta* **114** 60
[6] Kim H J, Choi K I, Pan A, Kim I D, Kim H R, Kim K M, Na C W, Cao G and Lee J H 2011 *J. Mater. Chem.* **21** 6549
[7] Pawar R C, Choi D H and Lee C S 2015 *Int. J. Hydrog. Energy* **40** 767
[8] Xu C, Wang Y, Zhao P, Chen H and Liu Y 2015 *Mater. Lett.* **159** 64
[9] Ruan C, Wang J, Gao M and Zhao G M 2016 *Mater. Chem. Phys.* **183** 158
[10] LaTempa T J, Feng X, Paulose M and Grimes C A 2009 *J. Phys. Chem. C* **113** 16293
[11] Darezereshki E, Bakhtiari F, Alizadeh M, Vakylabad A B and Ranjbar M 2012 *Mater. Sci. Semicond. Process.* **15** 91
[12] Bang J H and Suslick K S 2007 *J. Am. Chem. Soc.* **129** 2242
[13] Nazari M, Ghasemi N, Maddah H and Motlagh M M 2014 *J. Nanostruc. Chem.* **4** 1
[14] Casale A, De Robertis A, De Stefano C, Gianguzza A, Patanè G, Rigano C and Sammartano S 1995 *Thermochim. Acta* **255** 109
[15] Zhao W, Chen H, Li Y, Li L, Lang M and Shi J 2008 *Adv. Funct. Mater.* **18** 2780
[16] Neves M C, Neto C P, and Trindade T 2012 *J. Nanosci. Nanotechnol.* **12** 6817
[17] Sousa J J, Sousa A, Moura M J, Podczeck F and Newton J M 2002 *Int. J. Pharm.* **233** 111
[18] Blake R L, Hessevick R E, Zoltai T and Finger L W 1966 *Am. Mineral.* **51** 123
[19] Ford E N J, Mendon S K, Thames S F, Rawlins J W 2010 *J. Eng. Fiber Fabr.* **5** 10
[20] Prevot V, Caperaa N, Taviot-Guého C and Forano C 2009 *Cryst. Growth Des.* **9** 3646
[21] Biju C S, Raja D H and Padiyan D P 2014 *Chem. Phys. Lett.* **610** 103
[22] Xing W, Jie G, Song L, Hu S, Lv X, Wang X and Hu Y 2011 *Thermochim. Acta* **513** 75
[23] Muralidhara K S and Sreenivasan S 2010 *J. Sci. Ind. Res.* **69** 879

IOP Conf. Series: Journal of Physics: Conf. Series **985** (2018) 012028 doi:10.1088/1742-6596/985/1/012028

Structure modification of natural zeolite for waste removal application

W B Widayatno

Research Center for Physics, Indonesian Institute of Sciences, Tangerang Selatan, Indonesia

E-mail: wahy012@lipi.go.id

Abstract. Tremendous industrialization in the last century has led to the generation of huge amount of waste. One of the recent hot research topics is utilizing any advance materials and methods for waste removal. Natural zeolite as an inexpensive porous material with a high abundance holds a key for efficient waste removal owing to its high surface area. However, the microporous structure of natural zeolite hinders the adsorption of waste with a bigger molecular size. In addition, the recovery of natural zeolite after waste adsorption into its pores should also be considered for continuous utilization of this material. In this study, the porosity of natural zeolite from Tasikmalaya, Indonesia, was hydrothermally-modified in a Teflon-lined autoclave filled with certain pore directing agent such as distilled water, KOH, and NH_4OH to obtain hierarchical pore structure. After proper drying process, the as-treated natural zeolite is impregnated with iron cation and heat-treated at specified temperature to get Fe-embedded zeolite structure. XRD observation is carried out to ensure the formation of magnetic phase within the zeolite pores. The analysis results show the formation of maghemite phase (γ-Fe_2O_3) within the zeolite pore structure.

1. Introduction

Vast industrialization in the last decades has brought various emerging environmental problems around the globe. One of the big issue to be concerned is the water quality which is getting worse due to the contamination of heavy metals, oil spills, solid and liquid wastes, as well as bacteria. The socio-economic impacts caused by the degradation of water quality, such as the losses of agricultural production, the damage of tourism potential, the decline of public health quality, the ecosystem injury, and the potential of biological extinction, have been studied by many researchers [1 - 4]. To date, many studies and technological approaches have been also carried out to diminish the problem sources, as well as to suppress the occurred effects [3, 5 - 8].

Utilizing porous materials as an adsorbent to capture the pollutant from the water is one popular approach for improving the water quality [6, 8 - 11]. The pollutant is expected to be trapped within the pores of the materials herein. Some studies demonstrate the powerful performance of this trapping mechanism by developing new kind of porous materials [12 - 16]. Nevertheless, some issues such as the separation difficulty of the adsorbent from the solution, the complex synthesis process, and the indispensable production cost, need to be addressed in order to meet the economic value [9].

This study was conducted to address the above problems. Natural zeolite from Tasikmalaya area in Indonesia, was utilized and structurally-modified to obtain the maghemite phase (γ-Fe_2O_3). Previous studies elaborated the utilization of natural zeolite for removal of heavy metals from water [17-19]. However, their works do not include the insertion of magnetic phase. It is expected from this study that the magnetic properties of this phase can facilitate easier separation of the adsorbent from the solution.

2. Materials and methods

Natural zeolite from Tasikmalaya region in West Java, Indonesia, was used as a parent material. In order to study the effect of calcination environment on the formation of maghemite phase, the parent (non-treated) material was subjected into 5 wt.% iron nitrate solution prior to calcination in air and inert (N_2) gas stream, both at 500 °C for 2h. The parent and as-calcined samples were then marked as parent, non-treatment (NT), and non-treatment with gas (NTG), respectively.

The pore structure modification of the zeolite was carried out in a Teflon-lined autoclave filled with different pore directing agents (PDA), i.e. distilled water, 1M KOH, and 1M NH_4OH, at 200 °C for 24 h. Herein, the zeolite was hanged above the PDAs solution instead of immersed into it, to ease the collection process while avoiding the uncontrolled structural change of the zeolite. Figure 1 illustrates the arrangement during the modification process. In the case of base-treated samples, the collected samples were washed using distilled water until neutral. The neutralized samples were then dried overnight in a drying oven at 110 °C. The dried samples were then also subjected into 5 wt.% iron nitrate solution and calcined at 500 °C for 2 h to obtain the magnetic phase. For ease, the parent and treated zeolites were marked as parent, Z-H_2O, Z-KOH, and Z-NH_4OH, respectively.

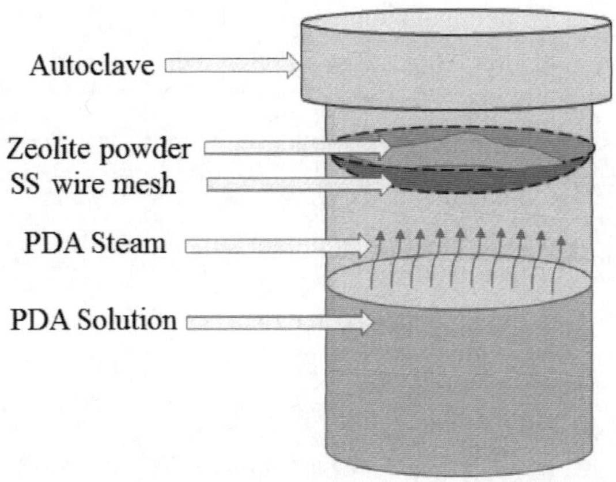

Figure 1. Setup arrangement for modification process.

The structure of parent, NT, NTG, and treated zeolites were compared by X-ray Diffraction (Smartlab, Rigaku, Japan). The morphology and pore distribution of the samples were qualitatively examined by observing Field Effect Scanning Electron Microscope (FE-SEM JIB-4610F, JEOL, Japan) images.

3. Results and discussion

3.1. Effect of calcination environment

While unexpected due to its low magnetic properties, hematite (α-Fe_2O_3) is the stable form of iron oxide up to around 600 °C in air. Thus, the strict control of oxygen amount is necessary to obtain iron oxide phase with a better magnetic properties (FeO, Fe3O4, γ-Fe_2O_3, etc.). Some previous studies utilized careful impregnation with lower post-treatment temperature to avoid the formation of hematite [7, 20]. Herein, the elevated post-treatment temperature with different environment was used in order to observe its effect on the formation of iron oxide phase.

Figure 2 shows the XRD profile of parent zeolite as well as calcined Fe-impregnated zeolite (without prior treatment) in air and N_2 gas stream, denoted as NT and NTG, respectively. The strong maghemite peak at (220) was obtained by utilizing N_2 gas stream. Meanwhile, the weak maghemite peak at (440) was obtained without gas stream. In addition, some unreacted iron phase was observed. This phenomenon was not observed when using N_2 stream, which is likely due to the limited reaction of N_2 and residual nitrate from the precursor (equation 1). This reaction reduces the reaction of nitrate and oxygen, as well as lets some oxygen finish the oxidation of iron phase.

$$2Fe(NO_3)_3 + 1.25N_2 \xrightarrow{heat} \gamma\text{-}Fe_2O_3 + 6.5NO_2 + 2NO \quad (1)$$

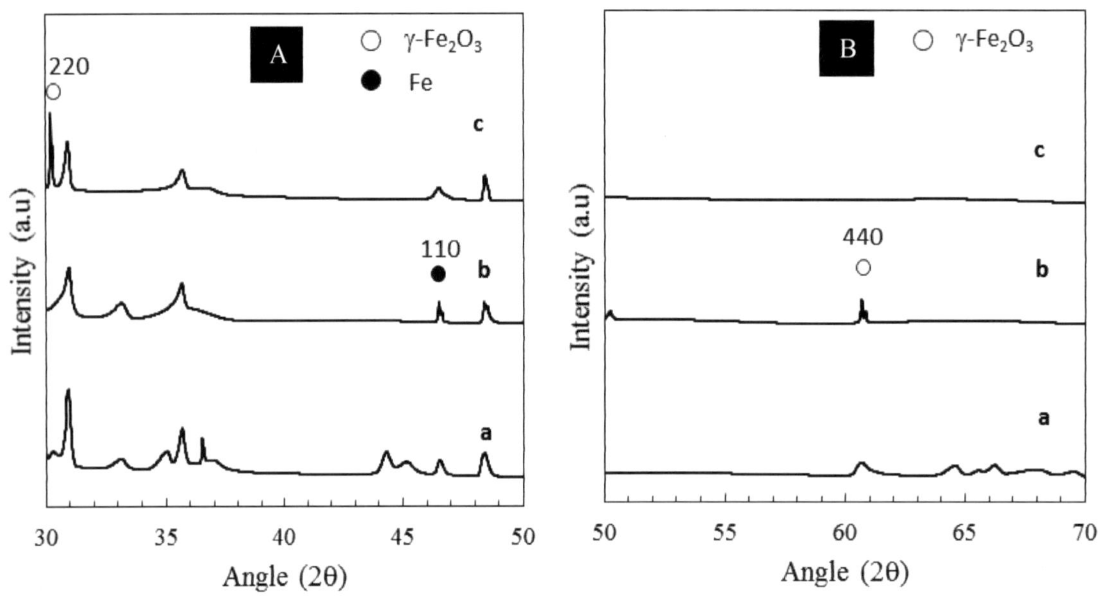

Figure 2. X-Ray diffraction profile of parent (a), NT (b), and NTG (c) zeolites in different observation angle range.

3.2. Effect of zeolite treatment

Figure 3 depicts the XRD profile of parent and Fe-impregnated zeolite (NT, Z-H_2O, Z-KOH, and Z-NH_4OH) after calcination at 500 °C for 2 h in air. It can be seen that prior treatment of zeolite in a water, KOH, as well as NH_4OH steam provides the similar result with that of NTG zeolite. The expected maghemite phase can be obtained easily. Moreover, there is no trace of unreacted iron phase within those samples, providing cheaper alternative for getting pure maghemite phase without expensive continuous gas stream.

Z-H_2O sample possess maghemite phase with many crystal orientation. The multi-crystal orientation was also observed for Z-KOH, with less amount of crystal orientation compared with that of Z-H_2O. Interestingly, by treating zeolite with NH_4OH steam, one can obtain single-orientation maghemite phase. This is likely due to the nature of NH_4OH as a "mild" pore structure-directing agent which can provide homogeneous pore structure as a host for the growth of maghemite crystal.

Figure 4 shows the electron micrograph comparison of Z-H_2O, Z-KOH, and Z-NH_4OH, respectively. It is clear that the grain morphology of Z-H_2O as well as Z-KOH is qualitatively rough compared with that of Z-NH_4OH. In addition, the morphology is quite heterogeneous with some spotted agglomerated particles. On the contrary, the Z-NH_4OH shows more homogeneous distribution with a much longer rod-like particles shape. This is in line with the XRD result which demonstrates the strong correlation between the smooth and more homogeneous morphology with that of the formation of single-orientation maghemite phase.

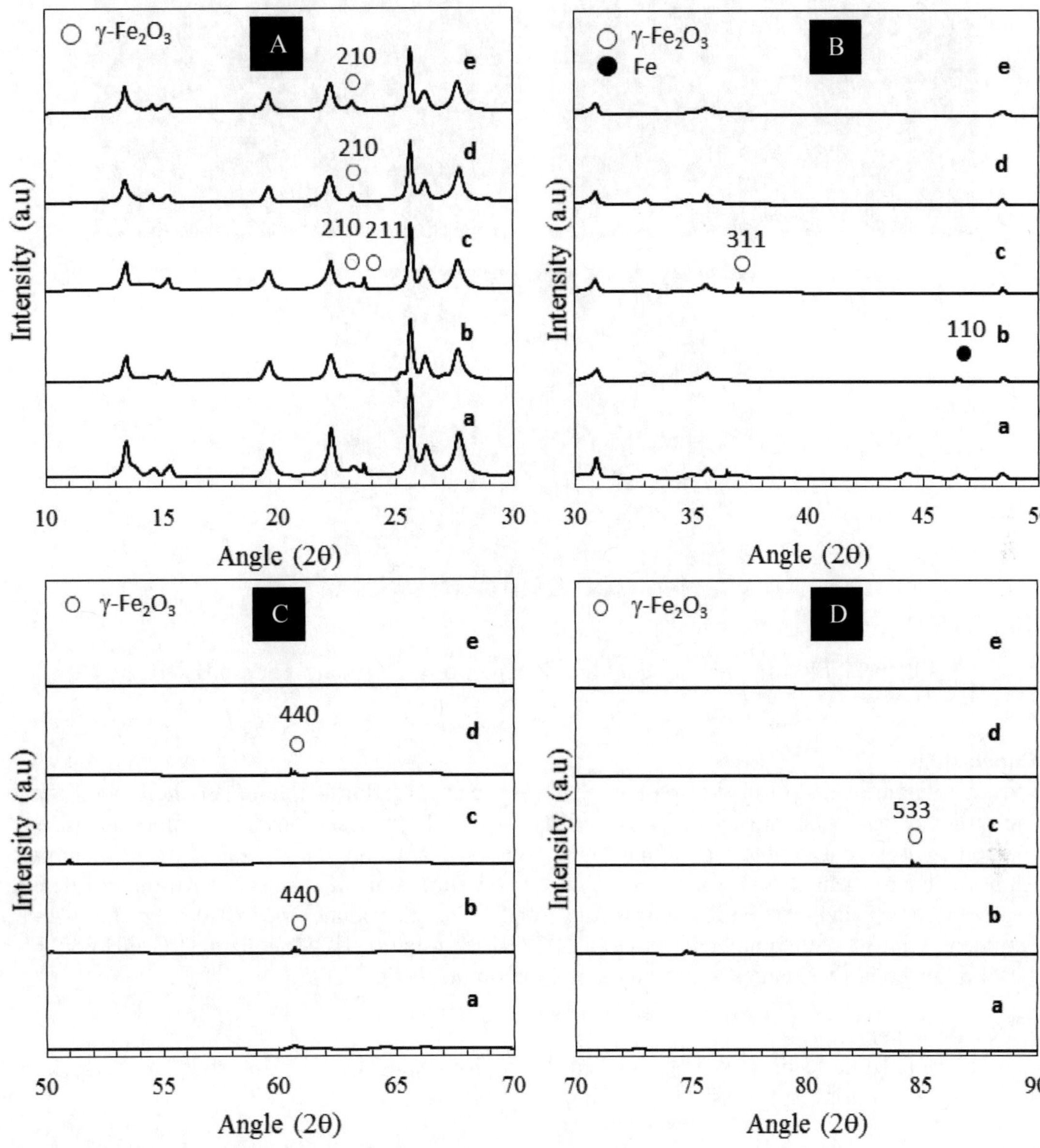

Figure 3. X-Ray diffraction profile of parent (a), NT (b), Z-H_2O (c), Z-KOH (d), and Z-NH_4OH (e) zeolites in different observation angle range.

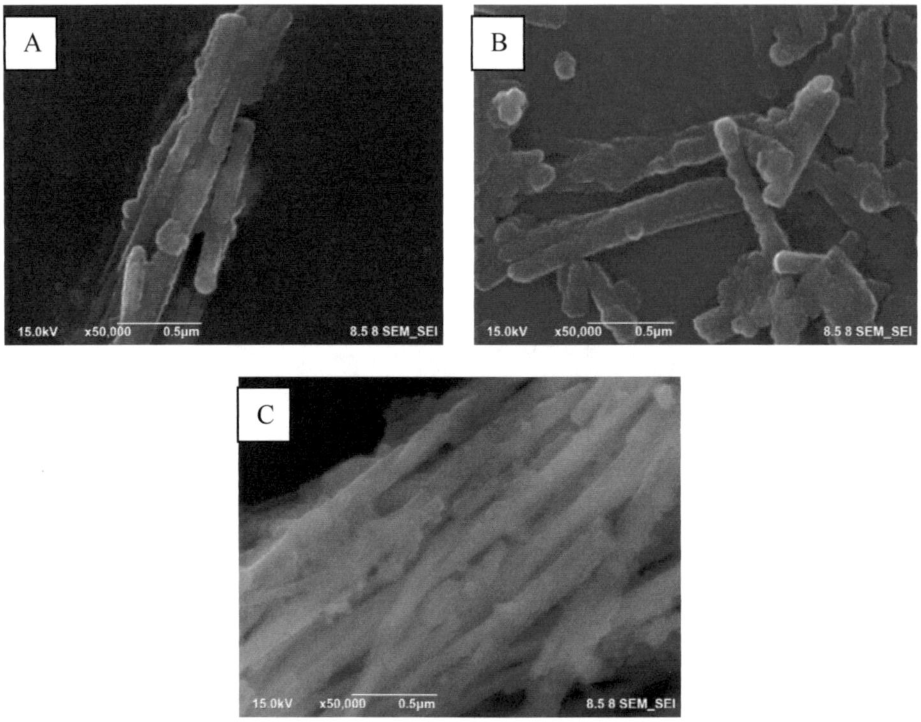

Figure 4. Electron micrograph of (a) Z-H_2O, (b) Z-KOH, and (c) Z-NH_4OH.

4. Conclusions

The structure modification of natural zeolite from Tasikmalaya, Indonesia, to obtain the Fe-embedded zeolite structure has been carried out successfully. XRD observation shows the important role of calcination environment to obtain pure maghemite phase. Meanwhile, the utilization of different pore structure-directing agent demonstrates its effect on the formation of single-orientation maghemite phase with more homogeneous morphology and size distribution. Considering the lower environmental impact and commercial price of NH_4OH than that of KOH, utilizing NH_4OH could be the best option for further studies to get single-orientation magnetic zeolite.

Acknowledgments

The author would like to thank Ms. Galih Padmasari from State Islamic University of Yogyakarta (UIN – Yogyakarta) for helpful assistance during the experiment.

References

[1] Peters N E and Meybeck M 2000 *Water Int.* **25** 185
[2] Reddy V R and Behera B 2006 *Ecol. Econom.* **58** 520
[3] Chang S E, Stone J, Demes K and Piscitelli M 2014 *Ecol. Soc.* **19** 26
[4] Beyer J, Trannum H C, Bakke T, Hodson P V and Collier T K 2016 *Marine Poll. Bull.* **110** 28
[5] Muyibi S A, Ambali A R and Eissa G S 2008 *Water Resour. Manag.* **22** 485
[6] Adebajo M O, Frost R L, Kloprogge J T, Carmody O and Kokot S 2003 *J. Porous Mater.* **10** 159
[7] Oliveira L C A, Petkowicz D I, Smaniotto A and Pergher S B C 2004 *Water Res.* **38** 3699
[8] Zhang W, Liu N, Cao Y, Lin X, Liu Y and Feng L 2017 *Adv. Mater. Interfaces* **4** 1600029
[9] Mehta D, Mazumdar S and Singh S K 2015 *Journal of Water Process Engineering* **7** 244

[10] Misaelides P 2011 *Microporous Mesoporous Mater.* **144** 15
[11] Shi W, Shao H, Li H, Shao M and Du S 2009 *J. Hazard. Mater.* **170** 1
[12] Dabwan A H A, Imai D, Kaneco S, Senmatsu I, Nakahama K, Katsumata H, Suzuki T and Ohta K 2008 *J. Environ. Sci.* **20** 172
[13] Amin M N, Kaneco S, Kato T, Katsumata H, Suzuki T and Ohta K 2008 *Chemosphere* **70** 511
[14] Alsbaiee A, Smith B J, Xiao L, Ling Y, Helbling D E and Dichtel W R 2016 *Nature* **529** 190
[15] Yang R X, Wang T T and Deng W Q 2015 *Scientific Reports* **5** 10155
[16] Wang J, Zhang P, Liang B, Liu Y, Xu T, Wang L, Cao B and Pan K 2016 *ACS Appl. Mater. Interfaces* **8** 6211
[17] Erdem E, Karapinar N and Donat R M 2004 *J. Colloid Interface Sci.* **280** 309
[18] Wingenfelder U, Hansen C, Furrer G and Schulin R 2005 *Environ. Sci. Technol.* **39** 4606
[19] Taamneh Y and Sharadqah S 2017 *Appl. Water Sci.* **7** 2021
[20] Gutiérrez M, Escudey M, Escrig J, Denardin J C, Altbir D, Fabris J D, Cavalcante L C D and García-González M T 2010 *Clays Clay Miner.* **58** 589

Synthesis of polymer hybrid latex poly(methyl methacrylate-co-butyl acrylate) with organo montmorillonite via miniemulsion polymerization method for barrier paper

J Chanra[1,], E Budianto[2] and B Soegijono[3]

[1] Graduate Material Science Program, Department of Physics University of Indonesia, Indonesia
[2] Department of Chemistry University of Indonesia, Indonesia
[3] Department of Physics University of Indonesia, Indonesia

E-mail: naufal@ui.ac.id

Abstract. Hybrid polymer latex based on combination of organic-inorganic materials, poly(methyl methacrylate-co-butyl acrylate) (PMMBA) and organo-montmorillonite (OMMT) were synthesized via miniemulsion polymerization technique. Modification of montmorillonite (MMT) through the incorporation of myristyltrimethylammonium bromide (MTAB) into the clay's interlayer spaces were investigated by Small-Angle X-ray Scattering (SAXS), Fourier Transform Infrared Spectroscopy (FTIR), Thermogravimetric Analysis (TGA) and Transmission Electron Microscopy (TEM). Barrier property and thermal stability of polymer latex film sample were investigated through its Water Vapor Transmission Rate (WVTR) and Thermogravimetric Analysis (TGA). The results indicated that addition of OMMT as filler in PMMBA increased the barrier property and thermal stability of the latex film. Addition of 8.0% (wt) OMMT increased the barrier property and thermal stability. Miniemusion polymerization process with higher addition (>8.0 wt%) of OMMT resulting in high latex viscosity, particle size, and high amount of coagulum. The utilization of this hybrid polymer could benefits paper and board industries to produce high quality barrier paper for food packaging.

1. Introduction

During the past decades, polymer science has entered the field of organic-inorganic polymer hybrids due to its remarkable barrier property, high cation exchange capacity, high platelet aspect ratio and good thermal stability[1]. Organoclays form an important type of modified clay material; clays such as montmorillonite (MMT) are widely used in various areas of sciences due to its ability to be modified with organic molecules [2].This modification can be carried out by exchanging the clay inter layer cations by organic cations which transform from organophobic to organophilic materials and also increase the basal spacing of the clay layers[3]. The use of OMMT in polymers offers improvement of its barrier property, thermal and physical stability. Modification of MMT by quaternary ammonium salts were used as intercalated agents and the intergallery distance $d(001)$ plane of the clay is relatively small, and the intergallery environment is hydrophilic[4]. In this study, MMT was modified using myristyltrimehtyl ammonium bromide (MTAB).

Numerous approach and methodologies have been developed to synthesize inorganic-organic polymer hybrids. Among various polymerization techniques, the hetero-phase polymerization system, particularly the miniemulsion polymerization technique, is widely used. Miniemulsion polymeriation is a type of polymerization which takes place in each of the stabilized monomer nano-droplets [5]. The stable nano-droplets (50-500 nm) are dispersed in the continous phase at the aid of high shear and an effective co-stabilizer to retard Ostwald ripening. The monomer transferring through the water phase to micelles and micelles does not take place as droplets themselves act as the loci of nucleation and latex particle droplets of 1:1 are ideal to copy of original droplets [6]. Every droplet behaves as independent reaction and polymerization inside droplets follows as suspension polymerization behavior.

Thermal stability of polymeric materials is an important keys to be considered; the properties may be profoundly affected by the presence of particular sequences of co-monomers or small proportions of additives. Co-monomer such as methyl methacrylic acid (MMA) as well as acrylic acid (AA) lead to increased thermal stability of co-polymers[7]. Thermogravimetry provides rapid method for determining thermal decomposition and thermal hazards of uncharacterized material or mixture (ASTM E2550-17). The scope of the test method describes the assessment of material thermal stability through determination of the temperature at which the material start to decompose or react and the extent of the mass change using thermogravimetry and is applicable over the temperature range from ambient to 800 °C [8].

Theoretical approaches on the barrier properties of polymer hybrids, treats fillers (clays) as impermeable non-overlapping particles, and it is assumed no permeability changes in the polymer matrix [9 - 11]. Permeability of the hybrid polymer will be smaller than permeability of the matrix (assuming the penetrant path does not cross fillers), this path of tortuosity was defined by Nielsen [12]. The key challenge is to obtain an effective dispersion and exfoliation of MMT in polymer matrix in order to obtained well aligned high aspect ratio particles for tortuous diffusion pathway for the improvement of barrier properties, these nanofillers can block gases and increase the tortuosity[13].

The hybrid OMMT/PMMBA latex with different OMMT content were prepared by miniemulsion polymerization method. Addition of OMMT as inorganic filler were expected to improve barrier properties and thermal stability of the hybrid polymer PMMBA latex. These properties could be used as barrier paper material in food packaging.

2. Experiment

2.1. Materials

Styrene, n-butyl acrylate, methyl methacrylic acid (MMA), azobisisobutyronitrile (AIBN) and hexadecane (HD) was supplied by Sigma Aldrich and used as received. Nanoclay montmorillonite (MMT) with CEC 125-130 meq/100g was supplied by FCC, disponil 103 (SDS) by BASF, and myristyltrimethyl ammonium bromide (MTAB) by Pioneer Chemicals, China.

2.2. Surface modification of MMT by ion-exchange

MMT (1.0 to its CEC) [14] was dispersed with stirrer in 750 ml distilled water. The mixture was stirred for 2 hours. MTAB (5.0 g) was dissolved in 250 ml distilled water and stirred for 2 hours. This solution was addded drop-wise to MMT clay dispersion under stirring for around 2 hours under regulated drop-wise regulation. The MMT particles were separated by centrifugation (5000 rpm, 10 minutes) and washed with large amount of distilled water until no bromide can be detected by silver nitrate.

2.3. Synthesis of polymer hybrid with miniemulsion polymerization

Oil phase containing hexadecane (co-stabilizer), AIBN (initiator), styrene, n-butyl acrylate, MMA monomers, and required OMMT (0 - 10 wt.%) was sonicated for 20 minutes with 90% amplitude using sonicator (Sonics 750 VCX - Sonics U.S.A). The aqueous phase containing disponil 103 and water were prepared by dissolving disponil 103 in distilled water at room temperature. The oil phase containing monomers and organoclay was poured into aqueous phase and the mixture was cooled in an icebath. After vigorous stirring for 3 minutes, the solution was sonicated for another 5 minutes with amplitude of 50%. The resultant hybrid miniemulsion was transfered into a three-neck round-bottom flask, and temperature was raised to 78 °C for 5 hours for polymerization. Latex film were prepared by casting the latex on plane glass. Samples were allowed to air dry for 24 hours, and after which were dried in an oven for another 24 hours.

3. Results and discussions

3.1. Modification of nanoclay montmorillonite

X-ray diffraction study was carried out using Small Angle x-ray Scattering (SAXS) with Cu-Kα radiation (λ = 1.5406 nm). The diffraction was scanned in the range from 2° to 30° at a scan rate of 1° min^{-1}. As shown in figure 1, MMT has an interlayer spacing of d = 121.677 nm corresponding to 2θ=7.2659° at d(001) reflection, and the surface modified MMT clay has shifted to d = 250.427 nm corresponding to 2θ=3.5282°. This confirmed that the MMT is intercalated between the layers as shown in figure 1.

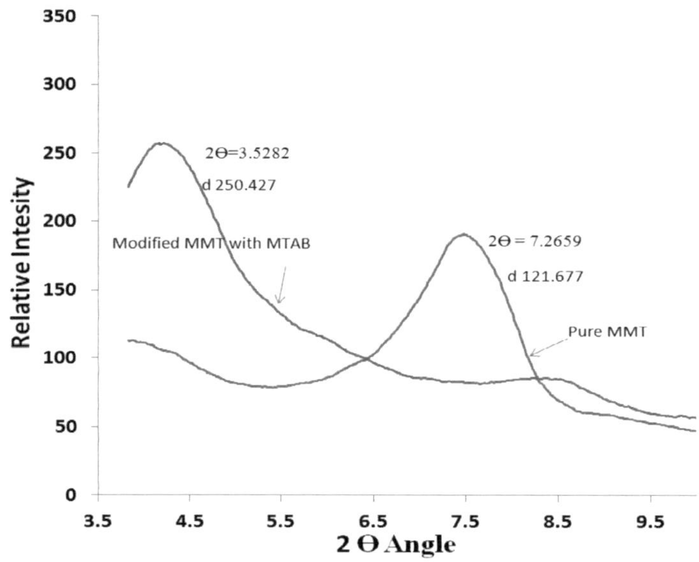

Figure 1. SAXS picture of pristine MMT (blue line) and modified MMT with MTAB (red line).

In figure 2 (FTIR graph), we could observe bands at 3000-3600 cm^{-1} , 1647 cm^{-1}, and 1000 cm^{-1}, correspond to bands of MMT (blue line). The broad peaks from 3000-3600 cm^{-1} are attributed to the adsorbed moisture during MMT modification. Peak at 1000 cm^{-1} indicates the presence of Si-O-Si groups in the octahedral sheets of MMT. Symmetrical and assymmetrical vibrations of -CH2- from alkyl chains of MTAB appear at peaks 2935 cm^{-1} and 2863 cm^{-1}. An Assymmetrical C-H vibration of alkyl ammonium appears to shift at peak 1493 cm^{-1}. The peak is attributed to attachment of MTAB surfactant on MMT surface. Peak at 1647 cm^{-1} is attributed to the coupling of water with octahedral cations and H-O-H bending vibrations. TEM images shows intercalation of MMT with MTAB (figure 3). It indicates that alkyl chains have pseudotrimolecular layer (20.0 - 21.0 Å) with several layers have been deformed.

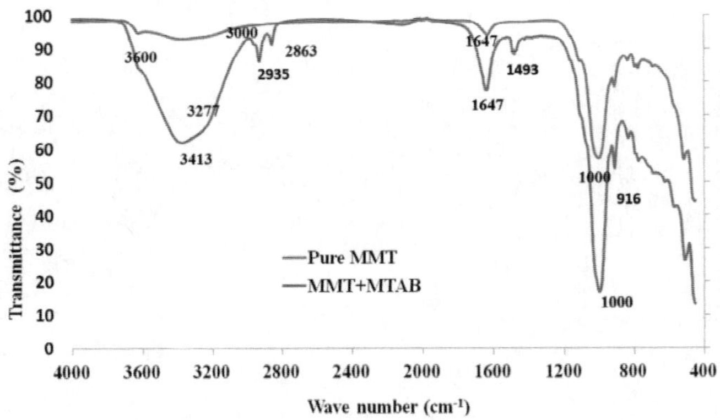

Figure 2. Comparison of FTIR spectra of unmodified (blue line) and modified MMT (red line).

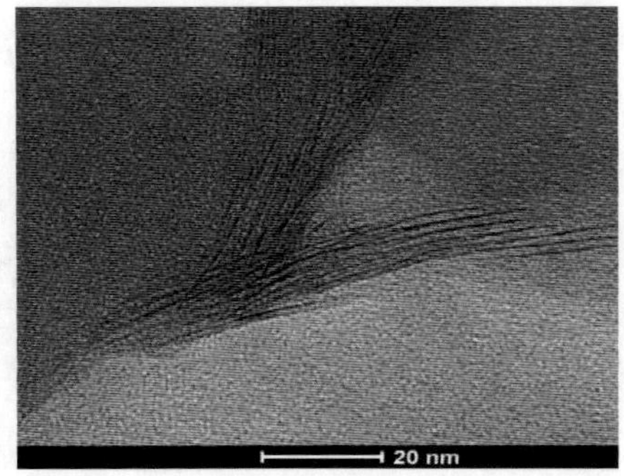

Figure 3. TEM image of modified MMT with MTAB.

3.2. Characterization of PMMBA in the presence of OMMT

MMT modified with MTAB in the PMMBA suspension was homogenous solution after sonication from addition of 2.0 - 8.0 wt.%. However, an addition of more than 8.0 wt.% created high amount of gel in monomers solution.

3.2.1. PMMBA/OMMT polymer hybrid viscosity

PMMBA Latex with addition of OMMT from miniemulsion polymerization was measured by Broekfield Viscosimeter RVDV, using spindle No.2 and speed 100 rpm at room temperature. Fiqure 4 shows an addition of OMMT will increase latex viscosity. An addition at higher OMMT level (> 8.0 wt%) created gel or coagulum during polymerization and limited workability. Latex PSD also increased from 0.4 μm to 0.8 μm as shown in figure 5 following an increase of latex viscosity.

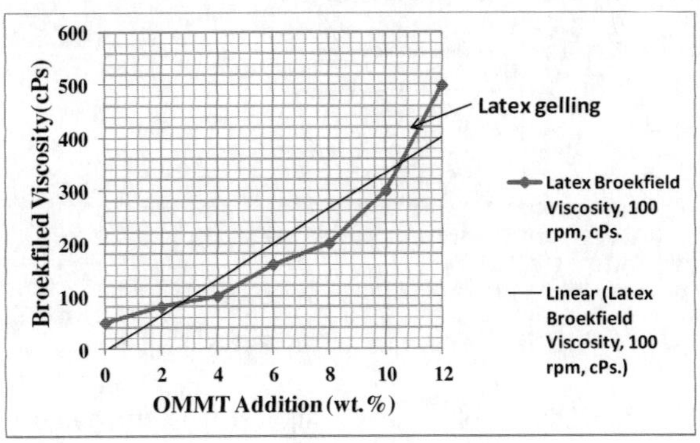

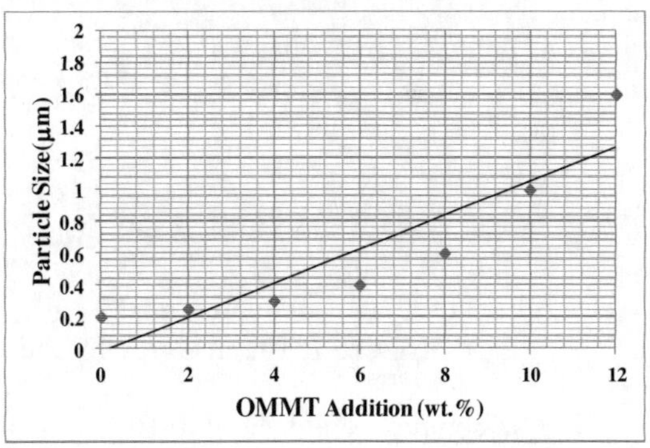

Figure 4. Latex viscosity vs OMMT addition.

Figure 5. Particle Size vs OMMT addition.

3.2.2. PMMBA/OMMT latex barrier and thermal stabilities

The standard method for WVTR testing is described by ASTM E96-08. This method measures how much moisture vapor to pass the materials in 24 hours. WVTR for food packaging generally can be accepted at 1cm^3/m^2/d for oxygen and 1g/m^2/d for water [15]. These levels of barrier are sufficient for food packaging application. However, further reducing of oxygen and water are needed. Using polymer latex with OMMT as filler could offer a defect free and better barrier property. In figure 6, an addition of OMMT starting 2.0 wt.% level could reduce WVTR value from 3.0 g/m^2/d to around 2.4 g/m^2/d. Up to 8.0 wt.% is recomended in order to achieve less than 1 g/m^2/d WVTR value, or similar to PE coated barrier paper.

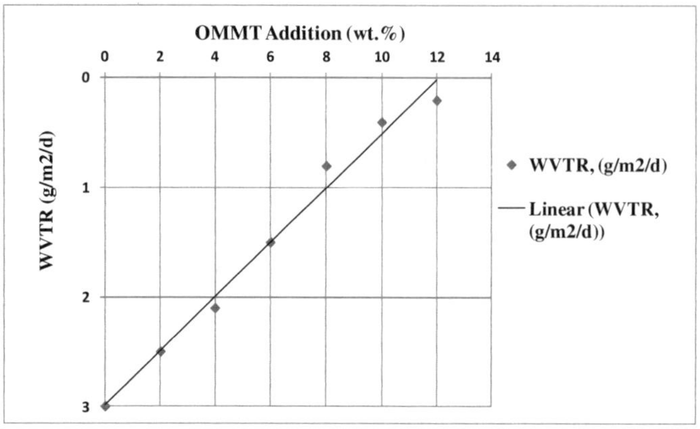

Figure 6. WVTR of PMMBA Latex Film.

The thermal stability of PMMBA/OMMT nanocomposites was analyzed by TGA, where the sample weight loss due to heating and monitored as a function of temperature ramp. Fiqure 7 shows the thermogram of PMMBA latex with different addition level of organoclay. The graph shows the decomposition behaviour of the materials as the temperature increased from 90 °C to 800 °C. We observed an increase of residual weight at temperature 600 °C, in all addition level of nanoclay.

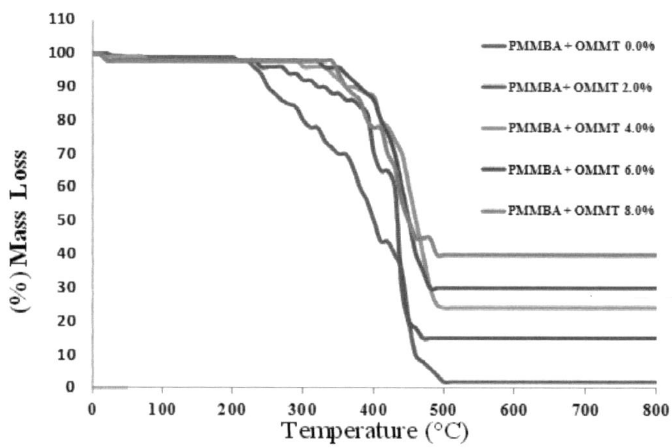

Figure 7. Thermal gravimetric curves of PMMBA and nanocomposites.

The decomposition of the PMMBA was observed at 200 - 300 °C. We suggest it is attributed to the themal degradation of MTAB molecules grafted on the MMT platelets. Improvement of thermal stability is not correlated linearly with addition level of organoclay. Several factors such as degree of nanoclay dispersion in the polymer matrix, morphology and type of interaction at the interface between the clay platelets and polymer matrix need to be considered [16]. Improvement in thermal stability of polymer matrices filled with organoclay acts as mass transport barrier and thermal insulator. And it is attributed to hindered out-diffusion of the volatile decomposition products, which is a direct result observed in the decrease in permeability and is usually observed in exfoliated polymer clay nanocomposite [17].

4. Conclusion

The use of miniemulsion polymerization was found to be an effective method to synthesize polymer hybrid latex poly(methyl methacrylate-co-butyl acrylate) with organo-montmorillonite as inorganic filler. Organo-montmorillonite are well intercalated with MTAB surfactant with 1.0 to its CEC. Basal spacing of SAXS give details arrangement of surfactant in the organoclay. Results from TEM analysis shows pseudotrimmolecular layer arrangement in the interlayers. Synthesis of PMMBA Latex with OMMT as inorganic filler with miniemulsion polymerization method at 8.0 wt.% addition level resulting an excellent barrier property and thermal stability. Addition at higher level of OMMT (> 8.0 wt.%) will cause high Broekfield viscosity, particle size (0.4 - 0.8 μm) and also high amount of coagulum/gelling on the latex.

Acknowledgments
This research is supported financially by "*Hibah Publikasi Internasional Terindeks untuk Tugas Akhir Mahasiswa University of Indonesia*" through letter of decree no.631/UN2.R3.1/HKP.05.00/2017.

References

[1] Xi Y, Ding Z, He H and Frost R L 2005 *Spectrochim. Acta A* **61** 515
[2] Tiwari R R, Khilar K C and Natarajan U 2008 *Appl. Clay Sci.* **38** 203
[3] Giannelis E P 1996 *Adv. Mater.* **8** 29
[4] Powell C E and Beall G W 2006 *Curr. Opin. Solid State Mater. Sci.* **10** 73
[5] Mirzataheria M, Mahtabani A, Lotfalieic AH and Mirzataherid M 2014 *Jordan Journal of Chemistry* **9** 229
[6] Ugelstad J, El-Aasser MS and Vanderhoff J W 1973 *J. Polym. Sci. C Polym. Lett.* **11** 503
[7] Chern C S and Liou Y C 1999 *J. Polym. Sci. A* 37 2537
[8] ASTM E2550-17 2017 *Standard Test Method for Thermal Stability by Thermogravimetry* (West Conshohocken-PA: ASTM International)
[9] Cussler E L, Hughes S E, Ward W J III and Aris R 1988 *J. Membr. Sci.* **38** 161
[10] Fredrickson G H and Bicerano J 1999 *J. Chem. Phys.* **110** 2181
[11] Bharadwaj R K 2001 *Macromolecules* **34** 9189
[12] Nielsen L E 1967 *J. Macromol. Sci. A* **1** 929
[13] Feldman D 2013 *J. Macromol. Sci. Pure and Appl. Chem.* **50** 441
[14] Nuntiya A, Sompech S, Aukkaravittayapun S and Pumchusak J 2008 *Chiang Mai J. Sci.* **35** 199
[15] Stefan J 1878 *Sitzungsber Akad. Wiss. Wien.* **77** 371
[16] Choi Y S, Choi M H, Wang K H, Kim S O, Kim Y K and Chung I J 2001 *Macromolecules* **34** 8
[17] Valandro S R, Loobandro P C, Poli A L, Horn M A Jr, Neumann M G and Cavalheiro C C S 2014 *Mat. Res.* **17** 265

Facile preparation of porous carbon from coffee bean waste using low temperature solvothermal method

L A Al Baroroh[1], D Fitria[1], M I Amal[2], A S Wismogroho[1] and W B Widayatno[1]

[1]Research Center for Physics, Indonesian Institute of Sciences, Tangerang Selatan, Indonesia
[2]Research Center for Metallurgy and Materials, Indonesian Institute of Sciences, Tangerang Selatan, Indonesia

E-mail: aisyahlinaab@gmail.com

Abstract. In this study, porous carbon made from coffee bean waste (CBW) was carbonized at 500 °C, 600 °C, and 700 °C to find effective temperature. It is verified from the IR spectrum that carbonization process at certain temperature can effectively break cellulose bonding and make aromatics functional group while preserving its carbon structure. The TG-DTA curve shows four stages of decomposition process and confirms most effective carbonization temperature. Activation process of as-carbonized CBW was carried out using solvothermal method in KOH and NH_4OH steam environment at 200 °C with variation of 30%, 40%, and 50% solvothermal volume. Scanning electron micrographs reveals significant increase of porosity on the carbon surface and differences of structural pores between the variations. The results show the possible potential of utilizing low temperature-solvothermal method for nanoporous carbon material.

1. Introduction

Nanoporous carbon is an activated carbon that has high porosity range. There are three ranges on the porous carbon, which are micropore (< 2 nm), mesopore (2-50 nm), and macropore (> 50 nm). Nanoporous carbon is widely used in device applications energy storage [1], catalyst [2], and water purification process [3] due to its high surface area. Porous carbon can be synthesized from several sources such as resin, coal, and biomass. Selection of porous carbon source from biomass is rated more easily because it is available in natural source, renewable, and relatively low cost. There are some commonly biomass used in synthesizing porous carbon such as coconut shell [4, 5], peanut shells [6, 7], bamboo [8, 9], coffee waste [10, 11], and tea waste [12, 13]. Basically, process of synthesizing porous carbon consists of two stages, which are carbonization and followed by activation [1]. Carbonization process is a process to synthesize carbon from raw materials and activation is a process of pores establishment. Activation process can be divided into two types: physical and chemical [14]. Physical activation involves controlling temperature, pressure, and gasification process. Chemical activation involves role of chemical compounds. Unfortunately, physical and chemical activation still consume large amount of energy in the process of temperature rise (> 800 °C). The usage of chemical compounds such as strong acids and strong bases can produce hazardous waste to the environment. Based on these facts, we need to find a method that can synthesize porous carbon by low-energy methods and environmentally friendly.

According to Statistics Indonesia (BPS) data, there is at least 600,000 ton of coffee produced in Indonesia which also means great number of waste produced as a consequence [15]. Coffee bean waste (CBW) must be processed to reduce waste and raise its utility. Considering this fact, this study explains a method to synthesize porous carbon from CBW. Porous carbon material from CBW was synthesized using simple heating carbonization method and low temperature solvothermal activation (200 °C) in KOH steam. Pressure variations from different KOH solution volume of 30%, 40% and 50% are subsequently investigated. KOH is a compound commonly used in porous carbon synthesis due to its excellent performance in high surface activated carbon manufacture [16, 17]. As a comparison, porous carbon activation using NH_4OH compounds is relatively similar but more environmentally friendly. Both KOH and NH_4OH activation results will be compared for carbon structure formation using scanning electron microscope.

2. Materials and methods

2.1. Carbonization of biomass and characterizations
CBW is carbonized using simple heating method under air atmospher. Carbonization process was carried out with heat flow of 5 °C/min, holding time of 1 hour, and temperature variations of 500 °C, 600 °C, and 700 °C to find an optimum value. A simple heating method was selected in order to find the most stable carbon form. The carbonization process must be able to break cellulose down and convert lignin to aromatic [18].

Mass changes during carbonization process which was investigated using TG-DTA testing (STA PT1600, Linseis, Germany). Transmitted IR spectrum was also investigated where samples were prepared using KBr pellets at 400 - 4000 cm^{-1} range (Nicolet iS10, Thermo Scientific, USA).

2.2. Carbon activation
Produced CBW carbon from simple heating carbonization process was milled to obtain smooth carbon powder. Furthermore, the carbon powder was activated through solvothermal activation. In order to create pores inside the material with low temperature, the solvothermal activation was selected by considering that steam energy and pressure formed inside a vacuum autoclave container will help to force the ion from the solution to scrape and pierce to create pores inside the material. Activation process started with 2 mg of carbon mixed with 0.5 M KOH and 0.5 M NH4OH solution. During the experiment, volume variations were made into 30%, 40%, and 50% of solvothermal cylinder container as illustrated in figure 1. Selected volume variations were based on the volume of solvothermal container by considering pressures that might affect pore structure [19].

Activation process was carried out through solvothermal method at 200 °C for 24 hours. Then, activated carbon was washed and dried at 200 °C. The porous structure of each variation was observed using SEM (Multi beam JIB 4610F, JEOL, Japan). Results and Discussion

3. Results and discussion

3.1. Effect of temperature on carbon structure
Figure 2 shows the curve of TG-DTA test results from CBW. The test was arranged with heating rate of 5 °C/min under normal air condition. From the figure, it can be seen that there are four decomposition processes from CBW [20]. The first process occurs at room temperature up to approximately 150 °C. This process signifies the loss of water vapour in the sample with weight loss of about 11.07%. The next process occurs at temperature of about 150 °C to 330 °C with weight loss of about 43.9%. At this temperature there is volatile loss in the sample due to hemicelluloses decomposition in biomass. Furthermore, in the temperature range of 330 °C to 500 °C, a weight loss of 15.8% indicates decomposition of cellulose and lignin. At temperatures above 500 °C there is no significant mass decline. It indicates that temperature of 500 °C can be selected as an effective temperature of carbonized coffee bean waste. Total relative mass change is about 70.77%.

Previous study showed different TG-DTA curve, in which they obtained TG-DTA curve showing only two steps of weight loss [20]. Stage one is moisture loss and stage two is volatile loss. In our study, CBW material used in this research shows four stages of weight loss. The difference might be caused by several factors such as coffee bean varieties, coffee bean content, different treatment received by coffee beans, and the different treatments received by the coffee waste.

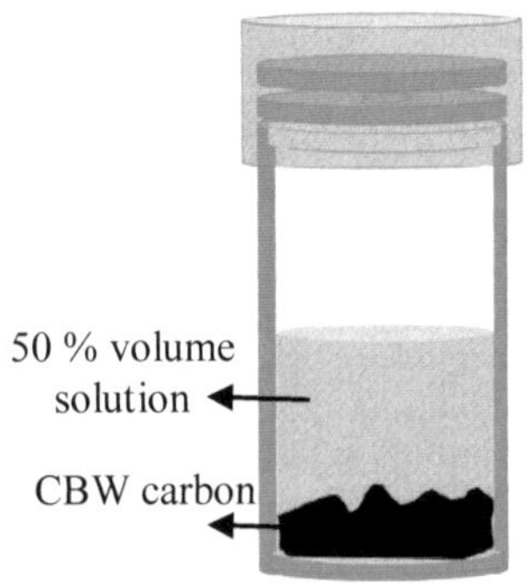

Figure 1. Illustration of solvothermal method on porous carbon activation.

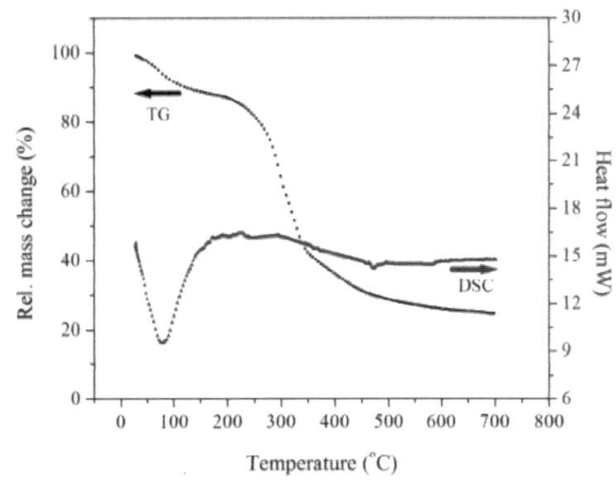

Figure 2. TG-DTA curve of coffee bean waste at room temperature up to 700 °C.

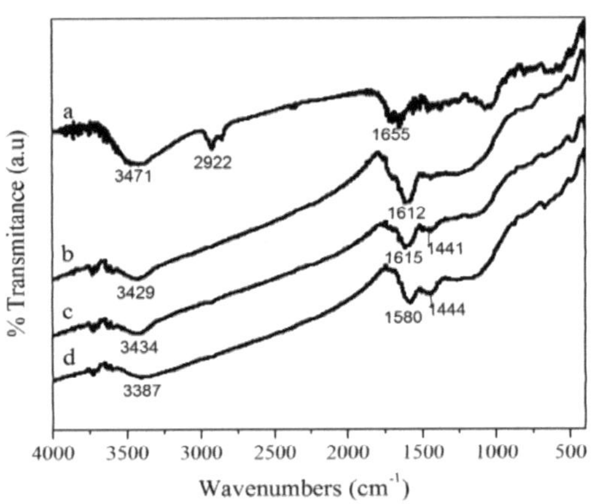

Figure 3. FT-IR spectra of (a) CBW, carbonized CBW at (b) 500 °C, (c) 600 °C, and (d) 700 °C.

Table 1. Functional group dan type of bond of CBW and CBW carbon [22].

Wave numbers (cm^{-1})	Functional group	Type of bond
3434	Hydroxyl group	O-H
3429	Hydroxyl group	O-H
3471	Hydroxyl group	O-H
3387	Hydroxyl group	O-H
2922	Alkanes group	C-H
1655	Alkenes group	C=C
1615	Aromatics double bond	C=C
1612	Aromatics double bond	C=C
1580	Aromatics double bond	C=C
1444	Single bond	C-H
1441	Single bond	C-H

From TG-DTA curve, carbonization temperature has been determined into 500 °C, 600 °C, and 700 °C. Carbonized CBW at these temperature are then analysed using FTIR to obtain IR spectra. Figure 3 shows the IR spectrum of CBW and carbon CBW with carbonization temperature of 500, 600, and 700 °C. The curve (a) displays IR spectrum of pure CBW. O-H bond is identified by the appearance of peak with widened intensity at bands 3200 - 3600 cm^{-1}. At band 2922 cm^{-1}, a peak appears indicating an existence of C-H bond while at 1655 cm^{-1} it may indicate an existence of alkene bond (table 1). Three carbonized CBW spectra (b-d) may show a presence of O-H groups, which is characterized by the appearance of wide peak intensity at bands of 3200 - 3600 cm^{-1}. At band of approximately 1500-1600 cm^{-1} a peak appears signifying aromatic formation. A change of lignin to aromatic becomes an important part because it signifies a strong structure and stable hexagonal carbon framework [21]. The IR spectrum of 500 °C carbonization temperature shows an aromatic formation, so that this temperature is considered more effective and requires less energy for the synthesis of stable carbon from coffee bean waste.

3.2. Analysis of activated carbon structural pores

CBW carbon with carbonized temperature of 500 °C and under free air conditions with a heating rate of 5 °C/minute have an amorphous structure as shown in figure 4a. It is shown that carbon particles seemed to have many cracks with rough surface. At greater magnification it is seen that crevices are observed with various gaps. This indicates that at temperature of 500 °C, organic materials have evolved from the carbon structure leaving crevices.

Figures 4(b-d) show SEM images of solvothermal activated carbon in KOH steam at 200 °C for 24 hours. Figure 4b shows the results of 30% KOH volume of solvothermal container. In figure 4b, it appears that pores have formed on rough carbon surfaces. At greater magnification, the pores formation have varying pore sizes but these pores are still classified as macropore.

Figure 4c displays the image of 40% KOH volume of solvothermal container. It exhibits more rough surface structure and more gap than activated carbon with 30% KOH volume. At greater magnification it can be seen that more pores are observed on carbon surface as well as variation in pore size. However, the pore size has mean size in range of macropore. Figure 4d shows an image of 50% KOH volume of solvothermal container. This variation is amorphous-like, rough, and has many crevices. At greater magnification, large amount of pores are observed with varying sizes in the macropore range.

Figure 4e displays image of 50% NH_4OH volume of solvothermal container. Surface of the particles formed in this variation has similar structure as amorphous, rough, and has many gaps. At larger magnification, it shows that formed pores have more quantity and size variation across the macropore range. Results of these SEM images show contradict behaviour to the image shown by previous studies [10]. In the previous study, it appears that CBW carbon has a structure that resembles graphite while in this study carbonized CBW before activation has an amorphous structure, has many crevice and cracks. The difference relates to the loss of material composition when carbonization takes place. This result can be indicated from the difference of TG-DTA curve. Different stages of mass loss will affect the morphology of formed carbon. In this study, the stages of relative mass change occur in four stages indicating many volatiles that escape and leave crevice and crack in the carbon structure.

Figure 5 shows pore size distribution curves per 100 μm^2 of CBW carbon of 50% KOH and 50% NH_4OH variation. KOH curve shows smaller line gradient compared to the NH_4OH curve. In the picked pore it was shown that the solvotermal activation using KOH has a higher uniformity level than that of activation using NH4OH. However, both activations show that pore distribution still looks dominant in the macropore range of >50 nm, while on NH_4OH activation curve more pores appear in mesopore range <50 nm.

4. Conclusions

CBW carbonization at a temperature of 500 °C under normal air with 5 °C/minute of heating rate is an effective temperature in steering stable and strong carbon framework. Low temperature of solvothermal activation under steam base particularly KOH and NH_4OH becomes a potential method of steering porous carbon. In variation of 50% volume KOH of solvothermal container, the pores formation are relatively more uniform than that of variation of 50% NH_4OH. However, dominant pore distribution range in the macropore and mesoporous range is found in the variation of 50% NH_4OH. In the future research, it will be good to confirm the surface area value in each variation and it is expected to conduct the research with variation on solvent concentration of solvothermal method.

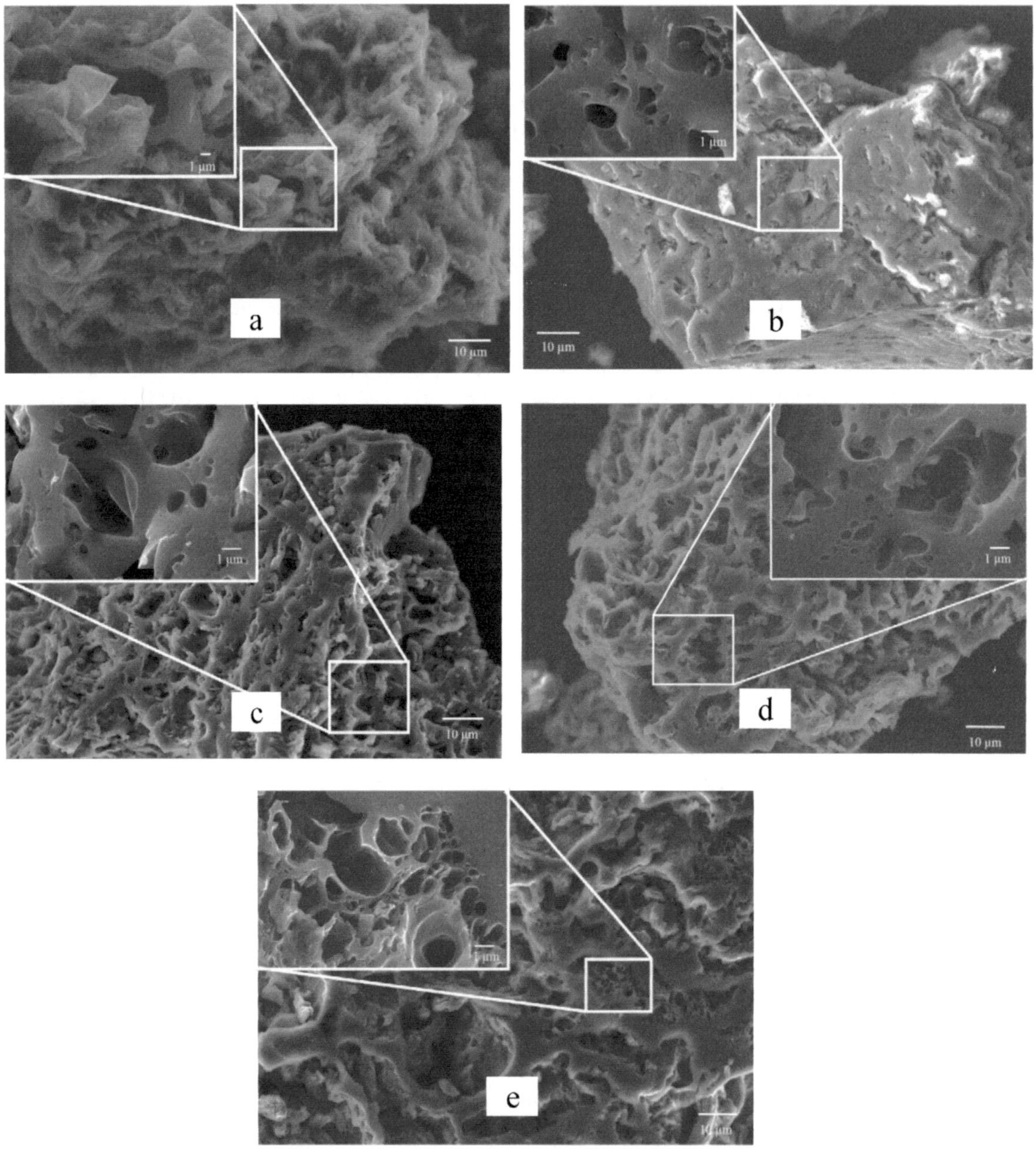

Figure 4. Scanning electron micrographs of (a) CBW carbon of 500 °C carbonization under normal air with 5 °C/minute of heating rate. Activated carbon of 0.5 M KOH with variation (b) 30%, (c) 40% and, (d) 50%, and (e) 0.5 M NH_4OH with 50% of volume at solvothermal activation.

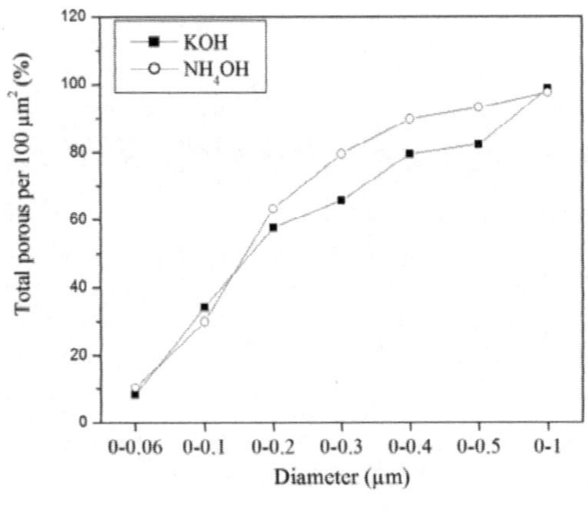

Figure 5. Cumulative distribution of pore size per 100 μm^2 of 50% KOH and 50% NH_4OH CBW porous carbon sample.

Acknowledgments

This study is the part of "Program INSINAS Riset Pratama Kemitraan" (04/INS-2/PPK/E/E4/2017) funded by Ministry of Research, Technology, and Higher Education of Republik Indonesia The author also would like to thank Dr. Memoria Rosi, Dr. Ismudiati Puri Handayani from Telkom University, Bandung, Indonesia, for helpful discussion during the analysis and Anif Mega from Institut Teknologi 10 November for the cooperation during the experiment.

References

[1] Pandolfo A G and Hollenkamp A F 2006 *J. Power Sources* **157** 11
[2] Bashkova S, Baker F S, Wu X, Armstrong T R and Schwartz V 2007 *Carbon* **45** 1354
[3] Oren Y 2008 *Desalination* **228** 10
[4] Laine J and Yunes S 1992 *Carbon* **30** 601
[5] Gratuito M K B, Panyathanmaporn T, Chumnanklang R A, Sirinuntawittaya N and Dutta A 2008 *Bioresour. Technol.* **99** 4887
[6] Ahmadpour A and Do D D 1997 *Carbon* **35** 1723
[7] Hayashi J, Horikawa T, Takeda I, Muroyama K and Ani F N 2002 *Carbon* **40** 2381
[8] Liu Q S, Zheng T and Guo L 2010 *Ind. Crops Prod.* **31** 233
[9] Ahmad A A and Hameed B H 2010 *J. Hazar. Mater.* **173** 487
[10] Boonamnuayvitaya V, Chaiya C and Tanthapanichakoon W 2004 *J. Chem. Eng. Jpn.* **37** 1504
[11] Rufford T E, Jurcakova D H, Zhu Z and Lu G Q 2008 *Electrochem. Commun.* **10** 1594
[12] Auta M and Hameed B H 2011 *Chem. Eng. J.* **171** 502
[13] Gundogdu A, Duran C, Senturk H B, Soylak M, Ozdes D, Serencam H and Imamoglu M 2012 *J. Chem. Eng. Data* **57** 2733
[14] Gao Z, Zhang Y, Song N and Li X 2017 *Mater. Res. Lett.* **5** 69
[15] Directorate General of Estate Crops, Ministry of Agriculture 2015 *Tree crop estate statistics of Indonesia 2014-2016 coffee* (Jakarta: Directorate general of estate crops, Ministry of Agriculture) p 3
[16] Otowa T, Nojima Y and Miyazaki T 1997 *Carbon* **35** 1315
[17] Sreńscek-Nazzal J, Kamińska W, Michalkiewicz B and Koren Z C 2013 *Ind. Crops Prod.* **47** 153
[18] Stoeckli H F 1990 *Carbon* **28** 1
[19] Funke A and Ziegler F 2010 *Biofuels. Bioprod. Bior.* **4** 160
[20] C. Saka 2012 *J. Anal. Appl. Pyrolysis* **95** 21
[21] Viswanathan B, Neel P I and Varadarajan T K 2009 *Methods of activation and specific applications of carbon material* (Chennai: National Centre for Catalys Research)
[22] Skoog D A, Holler F J and Crouch S R 2007 *Principles of instrumental analysis* 6th ed. (California: Thomson, Brooks/Cole) p 461-63

Identification of potential hydrophobic properties of carbon layer from the coffee bean waste

D Fitria[1], L A Al Baroroh[1], F Destyorini[1], W B Widayatno[1], M I Amal[2] and A S Wismogroho[1]

[1]Research Center for Physics, Indonesian Institute of Sciences, Tangerang Selatan, Indonesia
[2] Research Center for Metallurgy and Materials, Indonesian Institute of Sciences, Tangerang Selatan, Indonesia

E-mail: fdytta24@gmail.com

Abstract. The significant increase of waste due to vast development of human civilization and industrialization has plunged humanity into various environmental issues. Nowadays, the concern on waste handling and conversion into more valuable material has become one of hot research topics. Biomass waste has great abundance with various types that can be utilized for many applications such as landfill, recycled-material, adsorbent, separation, catalysis, and so on. In this study, coffee bean waste (CBW) was used as a source to produce hydrophobic layer. The CBW was converted into amorphous carbon using simple carbonization method at 500 °C, dispersed in acetic acid and then mixed with polyvinyl alcohol (PVA) at low temperature heating. In order to investigate effects of composition on hydrophobicity properties, ratio of carbon and PVA was varied. In addition, acetic acid was used to evaluate effect of dispersant on hydrophobic properties. SEM analysis reveals unique morphology of carbon layer. The measurement of contact angle demonstrates that this unique morphology possesses comparable hydrophobicity with that of some well-known materials. Fourier transform infrared spectroscopy (FTIR) analysis confirms the effect of PVA bonding and carbon layer on its hydrophobicity.

1. Introduction

Million tons of coffee beans are produced and consumed by people and health industry each year and most of it is discarded as coffee bean waste. Coffee bean waste (CBW) is mainly composed of cellulose and lignocellulose which contains many heteroatoms such as carbon, oxygen, and nitrogen [1, 2]. Biomass waste is a promising material to be converted as carbon precursor due to its abundance and long term storage potential [3, 4]. Developing carbon precursor from CBW via carbonization may lead to economical, environmental, and efficient benefits. The weight loss of pyrolysed CBW at the temperature of 600 °C is very slow, so carbonization in low thermal process (500 °C) has attracted researchers' interest [5].

Most carbon-based material is generally hydrophobic which is purely based on aromatic, non-polar and sheet structure, so that interaction with extremely polar molecules such as water is very weak [6]. Hydrophobic surface has value of water contact angle of $90° < \theta < 150°$. Surfaces with hydrophobic properties have many applications such as catalyst, anti-corrosion coating, textiles, and self-cleaning [7 - 9]. Unfortunately, materials with hydrophobic properties are relatively expensive and limited.

Therefore, the development of carbon from CBW as hydrophobic material is in great demand because it is very economical and it triggers unlimited development.

In this study, a potential hydrophobic property of CBW with variation of binder is identified. CBW was carbonized at low temperature to obtain carbon precursor which was then dispersed with acetic acid. Carbon layer was fabricated by using a simple coating method on an aluminum plate where PVA was used as the binder. To see the effect of acetic acid on the hydrophobic properties, the carbon layer was also fabricated without using dispersant.

2. Materials and methods

2.1. Carbon layer preparation and characterization

0.24 g CBW was ground until it passed through 100 meshes. CBW was carbonized at 500 °C and held for 1 hour. 20 ml of technical acetic acid with pH of 4 was added to the carbon. The mixture was sonicated for 30 min then dried at 90 °C. In different glasses, 0.03 g of polyvinyl alcohol (PVA) was dissolved homogeneously into distilled water. The dispersed carbon was added to the PVA solution while stirred with stirrer and heated to 90 °C until the mixture turned into slurry. The slurry was casted on aluminium plate using simple coating method. The formed layer was dried at room temperature. PVA and CBW with mass ratio of 1:6, 1:10, and 1:12 were made as other variation. The effect of dispersants was evaluated by fabricating samples with an addition of acetic acid and sonication process. Synthesis of carbon layers using commercial activated carbon with a ratio of 1:8 was used as control.

Functional groups of samples were evaluated using Fourier transform infrared spectroscopy (FTIR) by Nicolet iS10. Samples were characterized by Scanning Electron Microscopy (SEM) to identify surface morphology and justify the existence of PVA in the samples.

2.2. *Contact angle measurement*

The contact angle measurements were carried out using SLR Camera fitted to macro lens (Canon). 25 μl distilled water was delivered from micropipettes to form sessile drop on the substrate. The measurement was supported by ImageJ software. The average angle was measured four times by averaging left and right angle of the droplets on the substrate.

$$Contact\ Angle = \frac{\alpha^{*}+\alpha^{**}}{2} \tag{1}$$

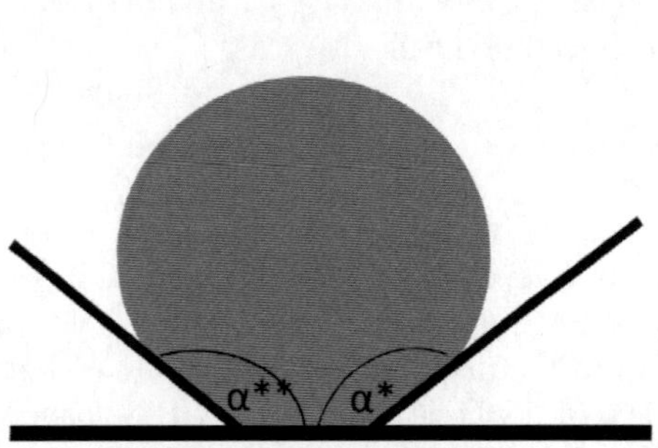

Figure. 1 Contact Angle Ilustration

Figure 1. Contact Angle Ilustration.

3. Results and discussion

3.1. Functional groups of carbon layer

FTIR spectra of the samples are presented in figures 2 and 3. Figure 2 shows a comparison of the use of acetic acid as a dispersant and the interaction response with PVA in carbon layer formation. In figures 2a and 2b, CBW 500 °C spectrum has significant peak at 1400 - 1600 cm^{-1} corresponding to the stretching vibration of the C-C bond (in ring) and around 2500 – 3500 cm^{-1} attributed to the stretching vibration of the hydrogen-bonded OH groups [10]. It indicates carbonized CBW is not pure due to the existence of other functional groups beside aromatics. Dispersed CBW has peak with strong intensity at 1402 cm^{-1} and 1262 cm^{-1} that are identified as C-H bond and C-O bond formed by interaction of acetic acid at the surface of CBW (figure 2c). Those functional groups will enhance the interaction of CBW and PVA to increase hydrophobic properties of carbon layer. In addition, strong peak at 1500 - 1700 cm^{-1} shows intensity increase compared to CBW 500 °C. It is because acetic acid successfully functionalized C-C bond in surface of CBW [11]. The peak of carbon layer at 1616 cm^{-1} is C-C bond as can be seen in figures 2d and 2e. It provides a shift and intensity decrease due to the effect of interaction between π-bond in the CBW surface and open double bond on PVA [12].

The FTIR spectra of Carbon layer with PVA variation are presented in figure 3. Carbon layer 1:6, 1:10, and 1:12 have strong intensity around 1400 cm^{-1} while carbon layer 1:8 is weak due to strong interaction between CBW and PVA.

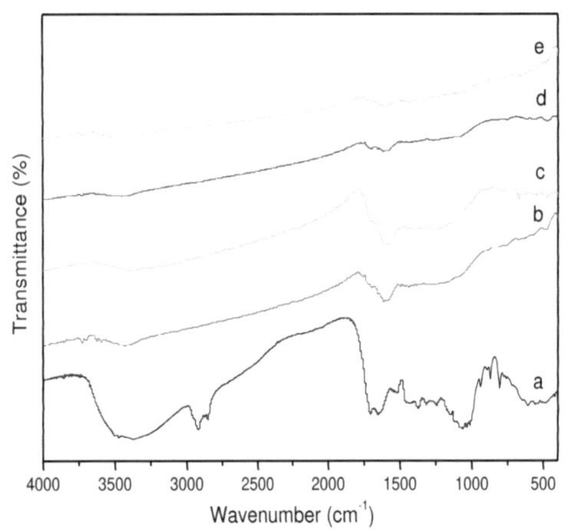

Figure 2. FTIR spectra (a) CBW, (b) CBW 500 °C, (c) dispersed carbon CBW with acetic acid, (d) carbon layer 1:8 with dispersant (acetic acid), and (e) carbon layer without dispersant.

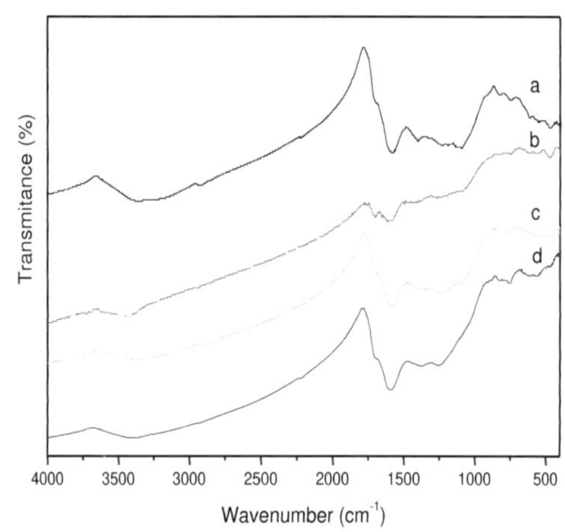

Figure 3. FTIR spectra of carbon layer with PVA variation (a) 1:6, (b) 1:8, (c) 1:10, and (d) 1:12.

3.2. Morphology of carbon layer

Figure 4 shows SEM images of carbon layer made from dispersed CBW and non dispersed CBW, raw material, and the activated carbon. SEM images show that carbon layer from non dispersed CBW has greater surface roughness compare to the dispersed CBW in spite of both samples having impurities (undissolved PVA). The particles of carbon layer from CBW have aggregate structure due to agglomeration with PVA binder as shown in figures 4a and 4b. Carbon layer from activated carbon clearly reveals greater surface roughness compare to CBW carbon based due to uniform size and purity of activated carbon (see figures 4c and 4e). Another phase is also identified in CBW 500 °C image besides carbon phase as observed from figure 4d.

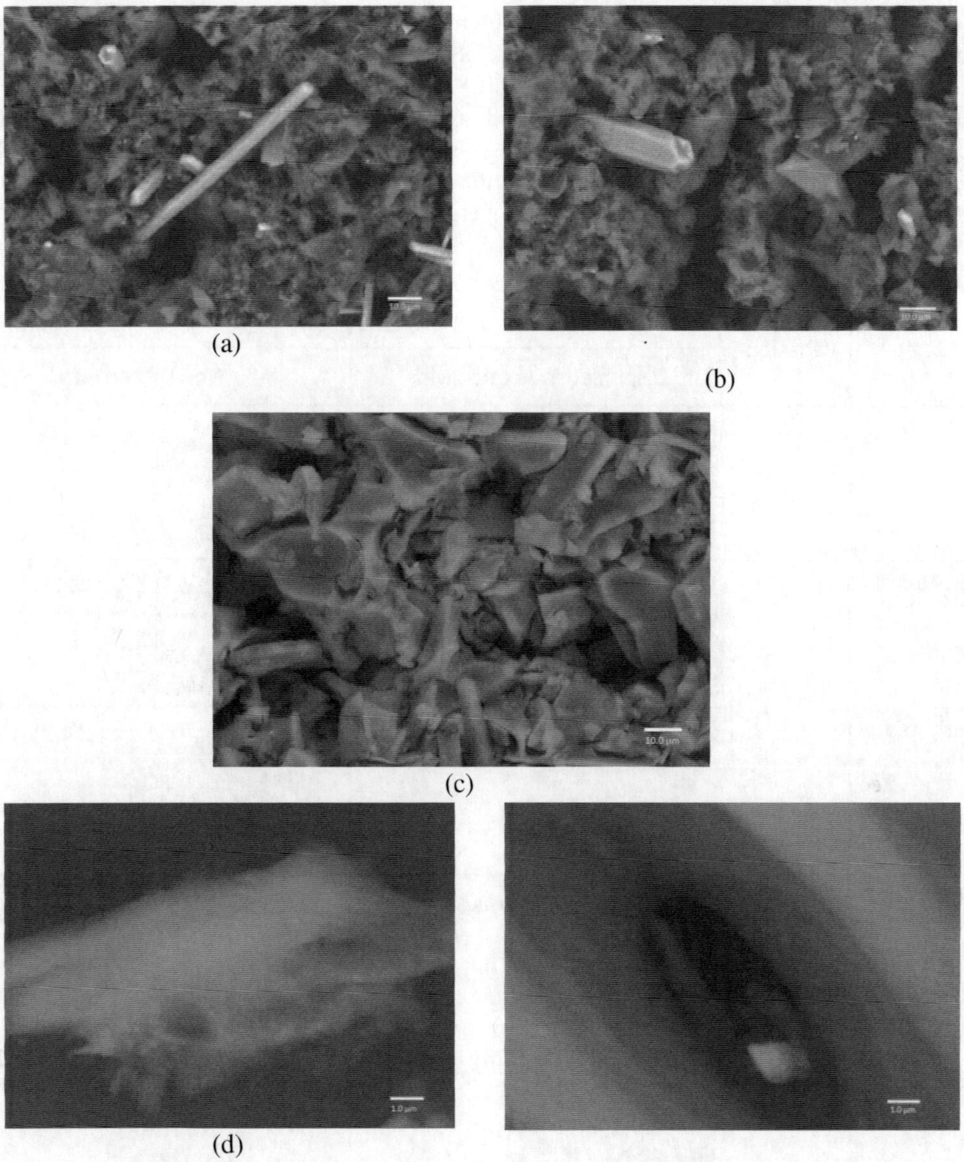

Figure 4. SEM Images of (a) carbon layer 1:8 from dispersed CBW, (b) carbon layer 1:8 from non dispersed CBW, (c) carbon layer 1:8 from activated carbon, (d) CBW 500 °C, and (e) activated carbon.

3.3. Contact angle of carbon layer

Carbon layers were fabricated using simple coating method with weight percentage (wt.%) variation of PVA binder. The effect of acetic acid as dispersant and PVA as binder on carbon layer are presented in table 1. Contact angle of carbon layers from CBW 500 °C are larger than 90°. The results suggest that CBW 500 °C has potential hydrophobic properties and it is comparable with commercial activated carbon. Carbon layer from CBW 500 °C has lower contact angle than carbon layer from commercial activated carbon (114.334°) because CBW 500 °C has hydrophilic functional groups on its surface as detected using FTIR [13] which decrease the potential hydrophobic properties [14]. Table 1 shows that acetic acid affects the contact angle. Carbon layer from dispersed CBW has lower contact angle compared to carbon layer from non dispersed CBW because acetic acid increases agglomeration with the results that decrease the roughness as provided by SEM images. The highest contact angle of carbon layer from non dispersed CBW is 117.414° with variation of PVA 1:10, whereas carbon layer 1:10 from dispersed CBW is 108.527°. Contact angle of carbon layer 1:10 with or without dispersed carbon is better than 1:6, 1:18, and 1:12 variation. It suggests that the optimal weight percentage ratio of PVA to CBW is 1:10. Contact angle increases with increasing of PVA binder because PVA has high permeability that decreases potential hydrophobic properties of carbon layer [11]. However, only potential hydrophobic properties of CBW is detected in the measurement.

Table 1. Contact angle measurement..

Name of Sampel	Variation	Contact Angle Measurement	
		Dispersed in Acetic Acid	Non Dispersed
PVA : CBW 500°C	1:6	 104.502°	 112.734°
PVA : CBW 500°C	1:8	 105.935°	 108.719°
PVA : Activated Carbon		-	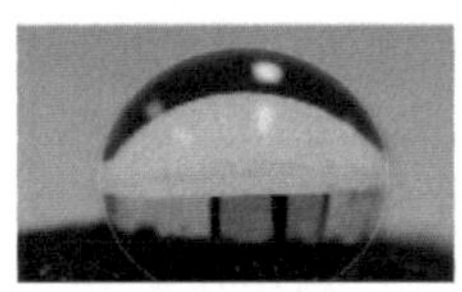 114.334°
PVA : CBW 500°C	1:10	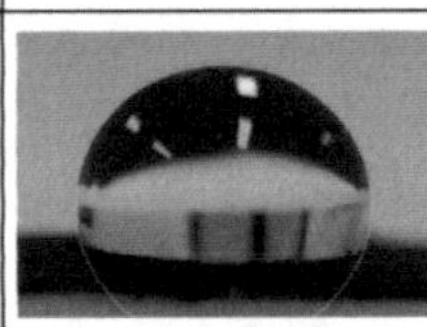 108.527°	 117.414°
PVA : CBW 500°C	1:12	 107.661°	 111.262°

4. Conclusion

Carbon layer from CBW with and without dispersion of acetic acid have been successfully fabricated using simple coating method. Potential hydrophobic propertiees of CBW 500 °C was successfully optimized using PVA binder. Acetic acid as dispersant controlled the contact angle due to interaction with PVA, but unfortunately decreased the contact angle when compared to non dispersed CBW. The optimal PVA variation (wt.%) for fabrication carbon layer from CBW 500 °C was 1:10 (117.414°) due to its suitable interaction with open double bond on PVA.

Acknowledgments
The author would like to thank LIPI for the fully supported by the "Program Riset Unggulan LIPI 2017" during the experiment.

References

[1] Yun Y S, Park M H, Hong S J, Lee M E, Park Y W and Jin H J 2015 *ACS Appl. Mater. Interfaces* **7** 3684
[2] Unni S M, George L, Bhange S N, Devi R N and Kurungot S 2016 *RSC Adv.* **6** 82103
[3] Park M H, Yun Y S, Cho S Y, Kim N R and Jin H J 2016 *Carbon Lett.* **19** 66
[4] Wang C H, Wen W C, Hsu H C and Yao B Y 2016 *Adv. Powder Technol.* **27** 1387
[5] Chen N, Ren J, Ye Z, Xu Q, Liu J and Sun S 2016 *Bioresour. Technol.* **221** 534
[6] Munz M, Giusca C E, Myers-Ward R L, Gaskill D K and Kazakova O 2015 *ACS Nano* **9** 8401
[7] Liang S, Kang Y, Tiraferri A, Giannelis E P, Huang X and Elimelech M 2013 *ACS Appl. Mater. Interfaces* **5** 6694
[8] Kowalczyk D, Brzeziński S, Makowski T and Fortuniak W 2015 *Appl. Surf. Sci.* **357** 1007
[9] Mohamed A M A, Abdullah A M and Younan N A 2015 *Arab. J. Chem.* **8** 749
[10] Skoog D A, Holler F J and Nieman T A 1998 *Principles of Instrumental Analysis* (Philadelphia : Saunders College Pub)
[11] Shameli A and Ameri E 2017 *Chem. Eng. J.* **309** 381
[12] Alghunaim N S 2016 *Results in Physics* **6** 456
[13] Lai D, Wei Y, Zou L, Xu Y and Lu H 2015 *Prog. Nat. Sci. Mater. Int.* **25** 445
[14] Asthana A, Maitra T, Büchel R, Tiwari M K and Poulikakos D 2014 *ACS Appl. Mater. Interfaces* **6** 8859

IOP Conf. Series: Journal of Physics: Conf. Series **985** (2018) 012032 doi:10.1088/1742-6596/985/1/012032

Thermal behaviour and corrosion resistance of nano-ZnO/polyurethane film

E Virgawati and B Soegijono

Department of Physics, University of Indonesia, Indonesia

E-mail: naufal@ui.ac.id

Abstract. Hybrid materials Nano-ZnO/polyurethane film was prepared with different zinc oxide (ZnO) content in polyurethane as a matrix. The film was deposited on low carbon steel plate using high volume low pressure (HVLP) method. To observe thermal behaviour of the film, the sample was investigated using thermogravimetric analysis (TGA) and differential scanning calorimetry (DSC). Fourier transform infrared spectroscopy (FTIR) was used to see whether any chemical reaction of ZnO in polyurethane occured. TGA and FTIR results showed that the decomposition temperature shifted to a higher point and the chemical reaction of zinc oxide in polyurethane occurred. The surface morphology changed and the corrosion resistance increased with an increase of ZnO content

1. Introduction

Polyurethanes (PUs) are commonly used in many applications such as adhesives, coatings, fibers, foams, paint, rubbers, thermoplastic elastomers, and composites [1 – 4]. Monomers can be flexibly chosen from many varieties of diisocyanates, polyester and chain extenders, as well as various types of molecular architectures that provide PUs with good properties such as abrasion resistance, mechanical properties, toughness, and chemical resistance [5 – 7]. However, there is still room to improve properties of PUs with emphasis on their mechanical properties and hydrophobicity. These can be achieved by varying microstructures of PUs or incorporating inorganic fillers. For example, organic-inorganic nanocomposites were developed to combine the required properties of PUs and also those of inorganic fillers [8 – 13] such as clay, silica, ZnO, and other nano-sized layered silicates. Significant improvements in performances such as mechanical properties, thermal stability have been achieved.

Present work deals with synthesis of nano zinc oxide-polyurethane composite with different nano-ZnO (1, 3 and 5 wt.%) content. Nano ZnO was produced by sol gel method. ZnO nanoparticles, which have high surface to volume ratio and more chemically reactive, were expected to modify the properties such as thermal behaviour and corrosion resistance of polyurethane as a matrix.

2. Experimental procedure

PUs was obtained from commercial product. The composition of the Polyurethane are diisocyanate (MDI), MDI based Polyisocuanate, 4,4'-Diphenylmethane, aluminum, aromatic solvent blend and isobutyl acetate. Figure 1 shows the urethane groups –NH–(C=O)–O– linking the molecular units. Nano-zinc oxide was synthesized from Zn $(CH_3COO)_2.2H_2O$ by sol-gel method. An appropriate amount of zinc acetate $Zn(CH_3COO)_2.2H_2O$ were dissolved in ethanol. The resulting solution was stirred at 80 °C for 2 hours until gel was formed. The obtained gel was then dried in an oven at 150 °C for 1 hour and then grounded with mortar and pestle to obtain fine powder. The powder was then sintered at 600 °C for 2 hours [13]. The composite were prepared by stirring polyurethane solution with various content of ZnO 1, 3 and 5 wt.%. The solution was deposited on low carbon steel plate using high volume low pressure (HVLP) method. The obtained films were taken from the substrate to be analyzed.

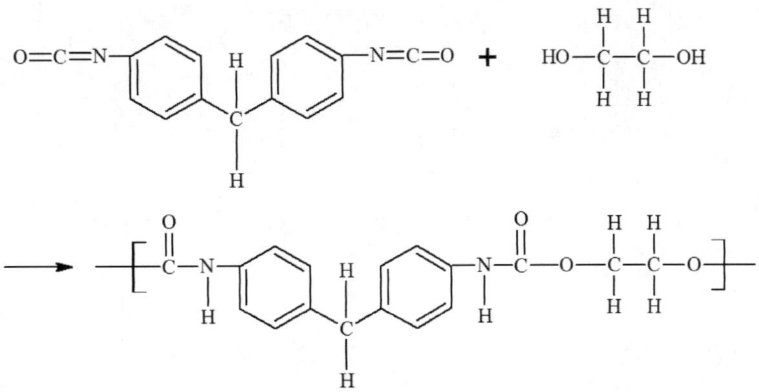

Figure 1. The urethane groups –NH–(C=O)–O– link the molecular units.

Thermogravimetry (TGA) was used to measure the weight loss of the nano-ZnO-polyurethane composite under nitrogen atmosphere. Samples were heated from room temperature to 800 °C at heating rate of 10 °C/min. Differential Scanning Calorimetry (DSC) was used to obtain information how the samples interact with calor at rate of 10 °C/min from room temperature to 800 °C. We did not use heating and cooling method to test the thermal stability due to permanent decomposition of the samples and limitation of the instruments. The samples were also investigated using Fourier transform infrared spectroscopy (FTIR) in the range of 300 - 4000 cm^{-1}. The method involves transmittance measurements. The resolution was set to 2 cm^{-1}, so that the spectra will be obtained at intervals of approximately 1 cm^{-1}. Morphology of the samples was observed by optical microscope and electron microscope. The samples were also subjected to salt spray according to ASTM-B117 for 72 hours. The samples were placed in closed testing chamber and exposed by fog or mist of a salt water (5% NaCl) solution. Information of the samples' corrosion resistance was obtained by observing its morphology.

3. Results and discussion

3.1. Thermogravimetry (TGA) analysis

TGA provides information about thermal stability of a material. The TGA curves of ZnO-polyurethane composite are shown in figure 2. It shows weight loss at temperature from room temperature to 800 °C. Onset decomposition temperature increases with increasing nano ZnO content. Temperature changes from 250 °C, 270 °C, 280 °C, and 291 °C with an increase in ZnO content. It suggests that the thermal stability of polyurethane also increases with increasing ZnO content. The end-set temperature remains the same, i.e. 424 °C, for various nano-ZnO content. It can be suggested that addition of ZnO cause an increases in thermal stability. PZnO 1 wt.% and PZnO 3 wt.% are more stable than PU (pure polyurethane without addition of ZnO) and PZnO 5 wt.%. A decrease in thermal stability may be due to agglomeration of ZnO in the matrix.

3.2. Differential Scanning Calorimetry (DSC)

DSC provides information heat absorb (endotherm) or loss (exotherm) during heating or cooling. The samples were heated in nitrogen atmosphere at heating rate 10 °C/min. Figure 3 shows DSC curve of four composite samples. It shows that during samples decomposition, heats released or absorbed are not significant. Samples PU and PZnO 5 wt.% show sudden exothermic heat for decomposition. It shows that PZnO 1 wt.% and PZnO 3 wt.% have more thermal stability than PU and PZnO 5 wt.%. Therefore, this evidence is consistent with results from TGA analysis.

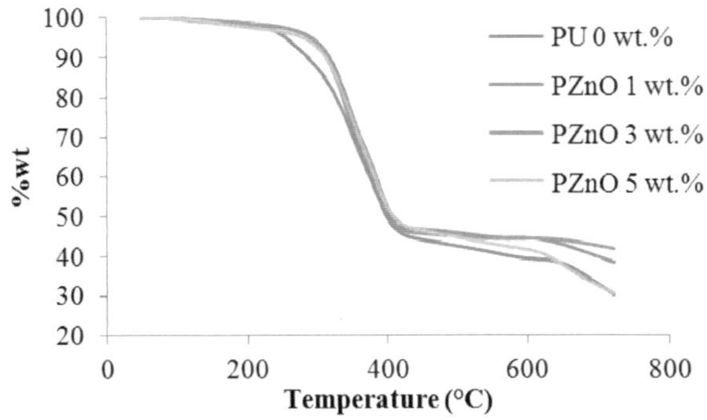

Figure 2. TGA curve of PU with various ZnO content (0, 1, 3 and 5 wt.%).

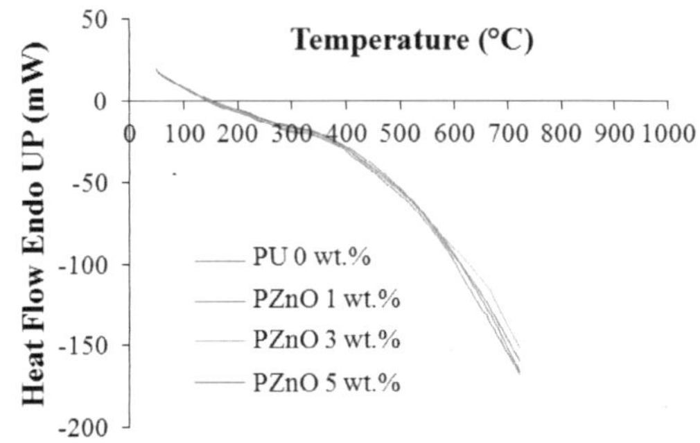

Figure 3. DSC curve of polyurethane without and with ZnO (0, 1, 3, and 5 wt.%) content.

3.3. Fourier transform infrared spectroscopy (FTIR) analysis
FTIR analysis was used to determine the functional groups of the samples. Figure 4 shows transmittance variation of polyurethane without and with ZnO. Stretching frequency for OH functional group is observed at 3350 cm^{-1} [8]. The peaks at 1598 and 1460 cm^{-1} are due to stretching frequency of C=C in the aliphatic chain and aromatic ring, respectively. The spectra show band around 2164 cm^{-1} which is for NCO group. The bands appearing at 1168 cm^{-1} and 1039 cm^{-1} correspond to Si-O-O stretching frequency and Al-O-Al stretching frequency, respectively [9]. According to the literature, the most affected functional groups by the ZnO filler are the N-H and C=O band at wave number 3521 cm^{-1} and 3358 cm^{-1} [10]. Samples PU and PZnO 5 wt.% have less intense transmittance than samples PZnO 1%wt and PZnO 3%wt. These trends are in line with DSC and TGA characterization.

3.4. Morphology and salt spray analysis
Figure 5 shows morphology of the nano ZnO powder used in preparing ZnO-polyurethane composite. The shape of the particle is rod-like with 30 nm in diameter and 200 nm in length. Due to nanoscale size of the particle, the powder have high surface to volume ratio which lead to more chemically reactive.

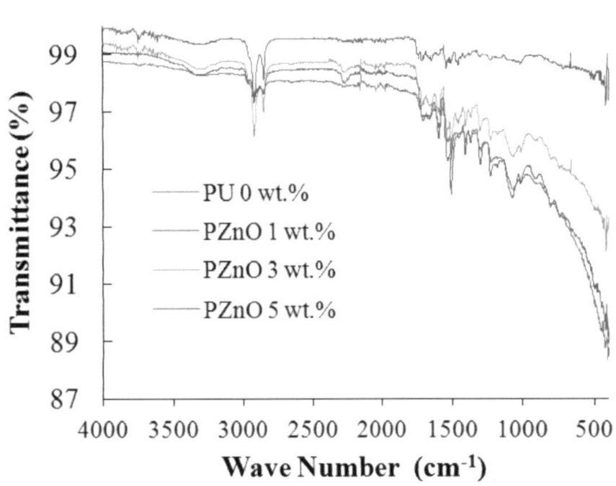

Figure 4. FTIR curve for composite with different ZnO content.

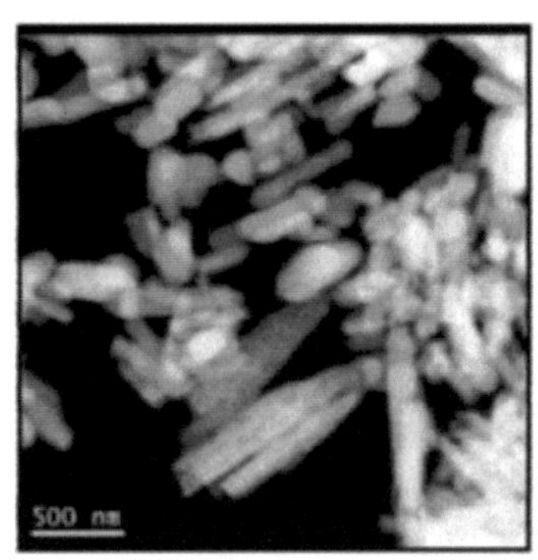

Figure 5. TEM image of ZnO powder.

Figures 6 and 7 show surface morphology of the samples before and after salt spray test. It is shown that samples PZnO 1 wt.% and PZnO 3 wt.% have smoother surface morphology compared to PU and PZnO 5 wt.% (figure 6). Rougher surface was found in samples with higher ZnO content due to agglomeration of ZnO. After salt spray test for 72 hours, sample surface shows different roughness compared to initial condition (Figure 7). Sample roughness indicated that samples PZnO 1 wt.% and PZnO 3 wt.% have better corrosion resistance than samples PU and PZnO 5 wt.%. The smooth morphology in PZnO 1 wt.% and PZnO 3 wt.% can be related to interaction of ZnO with polyurethane which were shown in FTIR pattern.

PU 0 wt.% 400x PZnO 1 wt.% 400x PZnO 3 wt.% 400x PZnO 5 wt.% 400x

Figure 6. Surface Morphology of ZnO-Polyurethane composite with different ZnO content before salt spray test.

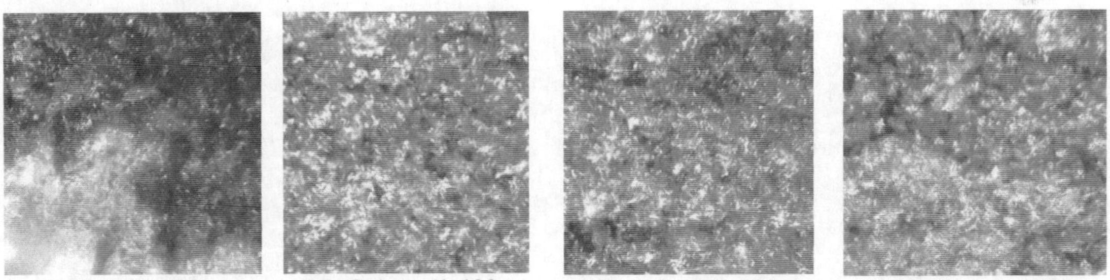

PU 0 wt.% 400x PZnO 1 wt.% 400x PZnO 3 wt.% 400x PZnO 5 wt.% 400x

Figure 7. Surface Morphology of ZnO-Polyurethane composite with different ZnO content after corrosion.

4. Conclusion

Nano ZnO/polyurethane composite with different ZnO content have been successfully prepared. Samples PZnO 1%wt and PZnO 3%wt seem to have higher thermal stability and corrosion resistance than samples PU and PZnO 5 wt.%. There may be chemical interaction of ZnO and polyurethane due to nano-scale size of ZnO. The composite ZnO and polyurethane may be used as a coating material to protect material from corrosion.

Acknowledgments

This research is supported financially by "Hibah Publikasi Internasional Terindeks untuk Tugas Akhir Mahasiswa University of Indonesia" through letter of decree no.631/UN2.R3.1/HKP.05.00/2017.

References

[1] Zhang Q, Hu J and Gong S 2011 *J. Appl. Polym. Sci.* **122** 3064
[2] Jóźwiak A B, Kielty C M and Black R A 2008 *J. Mater. Chem.* **18** 2240
[3] Osman M A, Mittal V, Morbidelli M and Suter U W 2003 *Macromolecules* **36** 9851
[4] Ihata O, Kayaki Y, Ikariya T 2004 *Angew. Chem. Int. Ed.* **43** 717

[5] Buma P, Ramrattan N N, van Tienen T G and Veth R P H 2004 *Biomaterials* **25** 1523
[6] Yeganeh H and Moeini H R 2007 *High Perform. Polym.* **19** 113
[7] Zhang C, Zhang X, Dai J and Bai C 2008 *Prog. Org. Coat.* **63** 238
[8] Malay O, Oguz O, Kosak C, Yilgor E, Yilgor I and Menceloglu Y Z 2013 *Polymer* **54** 5310
[9] Fan Q, Fang J, Chen Q and Yu X 1999 *J. Appl. Polym. Sci.* **74** 2552
[10] Okkema A Z, Fabrizius D J, Grasel T G, Cooper S L and Zdrahala R J 1989 *Biomaterials* **10** 23
[11] Jeon H T, Jang M K, Kim B K and Kim K H 2007 *Colloids Surf. A* **302** 559
[12] Fadillah L, Soegijono B, Budiawanti S and Mudzakir I 2017 *AIP Conf. Proc.* **1862** 030053
[13] Suharno, Soegijono B, Budiawanti S and Fadillah L 2017 *J. Phys.: Conf. Ser.* **817** 012022

IOP Conf. Series: Journal of Physics: Conf. Series **985** (2018) 012033 doi:10.1088/1742-6596/985/1/012033

Characterization of dross and its recovery by sulphuric acid leaching

S A Rahmani[1], A Meidianto[1], M I Amal[2], A S Wismogroho[1] and W B Widayatno[1]

[1] Research Center for Physics, Indonesian Institute of Sciences, Tangerang Selatan, Indonesia
[2] Research Center for Metallurgy and Materials, Indonesian Institute of Sciences, Tangerang Selatan, Indonesia

E-mail: wahyubw@gmail.com

Abstract. This paper reports the characterization of dross from galvanizing process and its recovery using acidic leaching method. The diffraction profile of dross showed identical peaks with that of ZnO. The X-ray Fluorescence (XRF) analysis identified the content of following metals: Zn, Fe, Mn, Ga, Co, and W. The thermal behaviour examination revealed the existence of some volatiles within the initial sample. The acidic leaching at various concentrations of sulphuric acid was conducted to determine the optimum concentration for zinc recovery and the highest yield of zinc sulphate. It is concluded that the optimum concentration of H_2SO_4 for this kind of dross is 4 M with 71.9% yield of $ZnSO_4$. The result of leaching process was confirmed by infrared spectrum, where various absorptions corresponding to SO_4^{2-} and Zn-O bands were observed.

1. Introduction

In general, zinc can be obtained from primary source (i.e. natural mineral rocks) and secondary source (e.g. industrial waste such as galvanizing, smelting, and casting industries). Herein, the secondary source could be in the form of zinc ash, zinc dross, galvanizer dross, electric arc furnace dust (EAF Dust), and zinc flux. Regarding the content of metallic zinc, the galvanization process of metal sheet or metal tube can produce waste (in form of zinc ash, dross, and flux) with high zinc metallic content, e.g. 60 - 85% [1]. On the contrary, EAF Dust has low amount of remaining zinc species [2, 3]. Meanwhile, the highest content of zinc can be obtained from galvanizer dross with Zn and Fe content of 96% and 4%, respectively [4]. Those wastes can be recovered in order to obtain the alternative metal sources and to reduce environmental pollution.

Recently, study on waste recovery focuses on the development of novel and effective recovery process which is also environmental friendly. Basically, this can be carried out using either pyrometallurgy or hydrometallurgy approach. Pyrometallurgy uses thermal treatment to initiate chemical and physical transformations of metals species during recovery process. Unfortunately, this process requires high energy, costs, and pollution due to some mass loss during combustion stage [5, 6]. On the other hand, hydrometallurgy approach uses aqueous or organic solution which involves chemical reaction during the process [7]. This approach also generates some liquid pollutants. However, the handling is still much easier than that of the pollutant generated by pyrometallurgy approach. To date, various studies of hydrometallurgy approach have been conducted using various reagents such as hydrochloric acid [8], sulphuric acid [9], acetic acid [10], and sodium hydroxide with addition of hydrogen peroxide [11]. Yet, broad variation of waste sources still urges specific study of a recovery process to get industrially-optimized results.

Previous study conducted in The Warren Spring Laboratory of the UK concluded that zinc recovery from EAF Dust can be carried out using sulphuric acid with recovery percentage of 85 - 90% and 80% at pH of 2 and 3 - 4, respectively [12]. Nesbitt and Xue also reported that the use of sulphuric acid is best for metal recovery such as zinc, copper, and lead [13]. In another study, Nii and Hisamatsu found that sodium hydroxide can also be used for the recovery process. Yet, this lead to some difficulties during separation process [14].

This study is conducted to investigate the characteristic of the obtained dross and determine the laboratory-optimized recovery process of zinc as an initial basis for further scale-up recovery process. The study is expected to give some insights on the optimized recovery process of zinc from dross, especially from vehicle galvanizing industries.

2. Materials and methods

2.1. Dross preparation and characterization

Dross was obtained from vehicle galvanizing industries. Particularly, the material was produced during the burning process in electric arc furnace at temperatures more than 1400 °C under an oxidizing atmosphere. The initial structure and composition of the material was determined using X-Ray Diffraction (XRD) and X-Ray Fluorescence (XRF), respectively. Meanwhile, the thermal behaviour of the material was investigated using coupled Thermogravimetry - Differential Scanning Calorimetry (TG-DSC) analysis.

Based on the result of XRF analysis, the obtained dross contains zinc and iron of 96.25% and 3.60%, respectively. The remaining content, on the other hand, consists of trace metals such as gallium, manganese, tungsten, and cobalt, as described in table 1. The main impurity is iron, which could be in the form of hematite (Fe_2O_3), magnetite (Fe_3O_4), or zinc ferrite ($ZnFe_2O_4$).

To support the XRF results, XRD characterization was also conducted. As depicted in figure 1, the prominent peaks at 31.770°, 34.436°, 36.25°, 47.537°, 56.589°, 62.842°, and 67.944° are clearly observed, which is identical with the pattern of ZnO. This is in line with the XRF result, where the dominant species is Zn species in oxide form. However, another unidentified metal phase was also observed. This can be attributed to the poor crystallinity of the phase or the insignificant amount of the phase so it cannot reach the detection limit of XRD. It is then confirmed that the main species of the dross is zinc oxide.

Figure 2 shows the TG-DSC analysis result of dross under nitrogen atmosphere. At first, the water content of the sample evaporates as the temperature increases to 200 °C. The process is followed by the evaporation of volatiles as the temperature increases to 350 °C. Above this temperature, the first stage decomposition of dross begins, which is denoted by the increase of the heat flow. At this stage, the mass loss exceeds 12%. At elevated temperature (around 900 °C), the melting phase started to occur. In addition, the increase of mass can be observed at about 1200 °C, which is likely due to the atmospheric reaction of the sample.

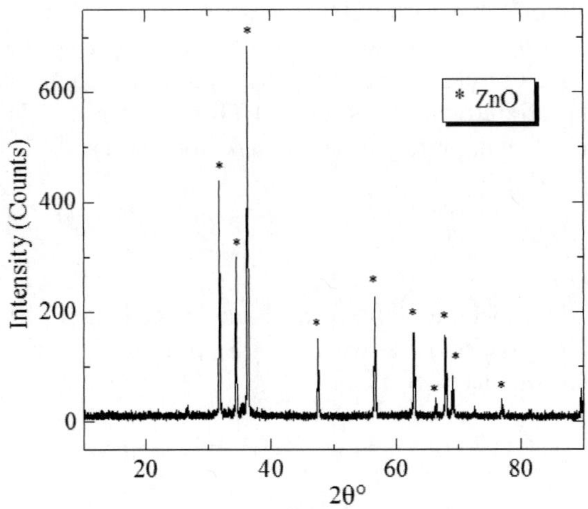

Figure 1. XRD patterns of dross.

Table 1. Chemical compositions of dross.

Elements	Weight (%)
Zn	96.255
Fe	3.605
Ga	0.037
Mn	0.037
W	0.054
Co	0.012

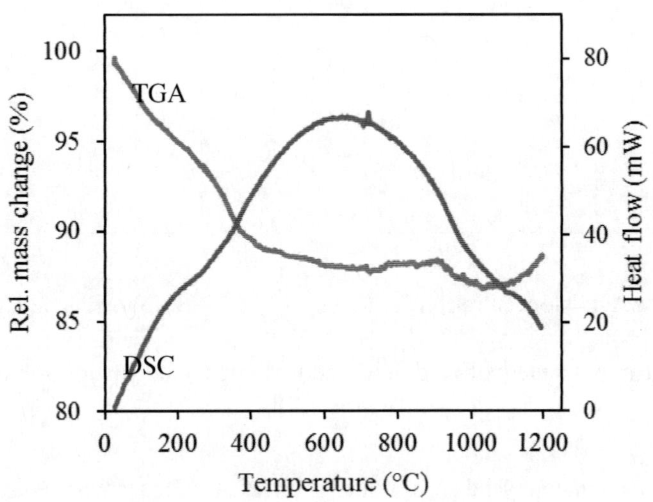

Figure 2. DSC-TGA curves of dross.

2.2. Acidic leaching and iron precipitation

At first, the as-received dross was washed using hot deionised water to dissolve the water-soluble impurities (e.g. Cl^- , OH^-). The recovery process was carried out in two steps, which are acidic leaching using sulphuric acid and iron precipitation using ammonia. The high-grade sulphuric acid (Merck, purity 95-97%) was used to prepare 2 M, 4 M, 6 M, and 8 M sulphuric acid solutions for acidic leaching. The leaching step was conducted by dissolving 2 g of dross into sulphuric acid with a certain concentration at 60 °C for 2 hours, with solid to liquid ratio of 10:1. The leaching reaction can be described as follows [15]:

$$ZnO_{(s)} + H_2SO_{4\,(aq)} \rightarrow ZnSO_{4\,(aq)} + H_2O_{(l)} \quad (1)$$

$$Fe_2O_{3\,(s)} + 3H_2SO_{4\,(aq)} \rightarrow Fe_2(SO_4)_{3\,(aq)} + 3H_2O_{(l)} \quad (2)$$

After solid-liquid separation, the remaining leaching solution was characterized using Atomic Absorption Spectroscopy (AAS). The iron species was precipitated by adding ammonia into the leaching solution until pH of 9 is reached and ammonium jarosite $NH_4[Fe(SO_4)_2(OH)_6]$ is formed. After separating the ammonium jarosite, the solution that contains zinc sulphate was heated up until the water and NH_3 gas vaporized, while leaving out the clear zinc sulphate crystal. The obtained crystal was analyzed using Fourier Transform Infrared Spectroscopy (FTIR) analysis. FTIR measurement was conducted using KBr pellet in the wavelength region of 400 cm^{-1} to 4000 cm^{-1}.

3. Results and discussion

3.1. Atomic Absorption Spectroscopy (AAS) analysis

The effect of sulphuric acid concentration on the yield of either zinc or iron was investigated using AAS analysis. Figure 3 reveals that highest yield of zinc and iron (i.e. 19600 ppm and 796 ppm, respectively) can be obtained when concentration of 4 M was used.

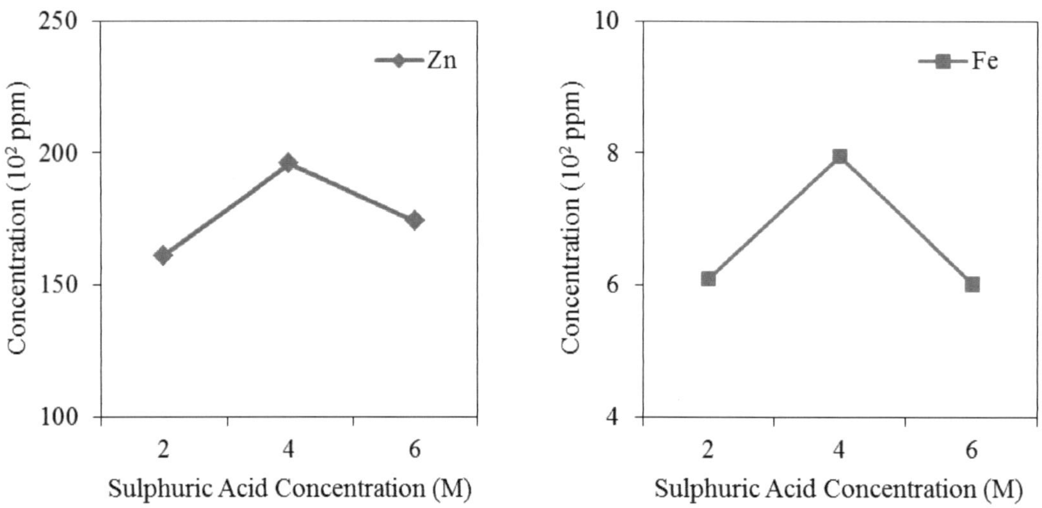

Figure 3. Content of zinc (left) and iron (right) in leaching solution.

3.2. Fe precipitation

The formation of zinc sulphate crystal from liquid is difficult to occur when iron is still present in the leaching solution. Thus, iron precipitation must be carried out beforehand. Iron can be precipitated by adding ammonia and iron sulphate into the leaching solution to initiate the formation of jarosite. First, ammonia will react with the excess of sulphuric acid as follows:

$$2NH_4OH_{(aq)} + H_2SO_{4\,(aq)} \rightarrow (NH_4)_2SO_{4\,(aq)} + 2\,H_2O_{(aq)} \quad (3)$$

Ammonium sulphate is involved in iron precipitation after the completion of leaching reaction. At low concentration of sulphuric acid, the hematite and magnetite could not dissolve and remain as the residue. When the concentration is increased, zinc ferrite dissolves easily as the bond between zinc oxide and hematite breaks. After addition of ammonia, the formed iron sulphate is then reacted with ammonium sulphate, forming ammonium jarosite. Below is the detail of the reaction:

$$3\,ZnO_{(s)} + 3\,Fe_2O_{3\,(s)} + 12\,H_2SO_{4\,(aq)} \rightarrow 3\,ZnSO_{4\,(aq)} + 3\,Fe_2(SO_4)_{3\,(aq)} + 12\,H_2O_{(l)} \quad (4)$$

$$3\,Fe_2(SO_4)_{3\,(aq)} + (NH_4)_2SO_{4\,(aq)} + 6\,H_2O_{(l)} \rightarrow 2\,NH_4[Fe(SO_4)_2(OH)_6]_{(s)} + 6\,H_2SO_{4\,(aq)} \quad (5)$$

Giving overall reaction:

$$3ZnO_{(s)}+3\,Fe_2O_{3(s)}+6\,H_2SO_{4(aq)}+(NH_4)_2SO_{4(aq)}\rightarrow 3ZnSO_{4(aq)}+2NH_4[Fe(SO_4)_2(OH)_6]_{(s)}+6\,H_2O_{(l)} \quad (6)$$

IOP Conf. Series: Journal of Physics: Conf. Series **985** (2018) 012033 doi:10.1088/1742-6596/985/1/012033

The obtained zinc sulphate solution is then crystallized by evaporating the water at room temperature to get zinc sulphate crystal.

3.3. Fourier Transform Infrared Spectroscopy (FTIR) analysis
Figure 4 depicts the absorption of various functional groups of the zinc sulphate crystal. There is no single trace of N-H bond, which implies that the obtained zinc sulphate crystals are NH_4^+ ions free. The broad vibration band of symmetric stretching mode of –OH group from water molecule was observed at 3134.99 cm^{-1}. The wavelength region of 400 – 600 cm^{-1} is the range for metal oxide vibration bands. Hence, the band observed at 460.07 cm^{-1} is expected to be Zn-O vibration. The bands observed from 1000 – 400 cm^{-1} are expected to be SO_4^{2-} ion vibrations. The bands observed at 618.22 and 978.99 cm^{-1} are attributed to the υ^2 and υ^3 of SO_4^{2-} mode, respectively, while the bands observed at 1108.53 and 1397.99 cm^{-1} are attributed to the υ^1 of SO_4^{2-} mode [16, 17].

3.4. Effect of concentration of sulphuric acid
Figure 5 plots the effect of sulphuric acid concentration on the yield of zinc sulphate. The %yield is calculated as follows:

$$\%\ Yield = \frac{Actual\ Yield}{Theoretical\ Yield}\ x\ 100\% \qquad [7]$$

The concentration of 4 M gives the highest yield of zinc sulphate crystal (i.e. 2.73 g or 71.9% of initial mass). This result is in line with the AAS result which estimates the optimum concentration of sulphuric acid for zinc recovery to be 4 M. The crystallization of zinc sulphate was expected to reach more than 80% after iron precipitation. This could be attributed to the insufficient crystallization time of zinc sulphate. Another issue is the possibility of precipitation of some dissolved zinc into ammonium jarosite. Hence, factors affecting the optimum crystallization time should also be investigated for further study.

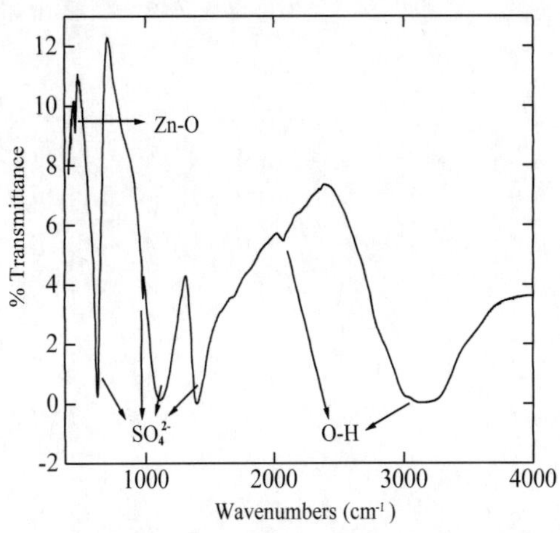

Figure 4. FTIR spectrum of zinc sulphate crystal.

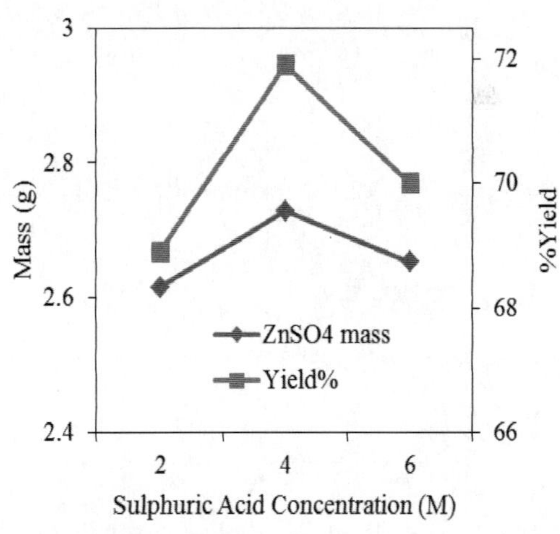

Figure 5. Mass and %yield of zinc sulphate crystal.

4. Conclusions
The main compound of the dross is zinc oxide with iron as the major impurities. The dross can be recovered using sulphuric acid with optimum concentration of 4M, wherein the yield exceeds 71.9% (in the form of zinc sulphate crystal).

IOP Conf. Series: Journal of Physics: Conf. Series **985** (2018) 012033 doi:10.1088/1742-6596/985/1/012033

References

[1] Dvořák P and Jandová J 2005 *Hydrometallurgy* **77** 29
[2] Caravaca C, Cobo A and Alguacil F J 1994 *Resour. Conserv. Recycl.* **10** 35
[3] Martins F M, do Reis Neto J M and da Cunha C J 2008 *J. Hazard. Mater.* **154** 417
[4] Jha M K, Kumar V and Singh R J 2001 *Resour. Conserv. Recycl.* **33** 1
[5] Sayilgan E, Kukrer T, Civelekoglu G, Ferella F, Akcil A, Veglio F and Kitis M 2009 *Hydrometallurgy* **97** 158
[6] Jadhav U U and Hocheng H 2012 *J. Achievements Mater. Manuf. Eng.* **54** 159
[7] Kinoshita T, Akita S, Kobayashi N, Nii S, Kawaizumi F and Takahashi K 2003 *Hydrometallurgy* **69** 73
[8] Yoshida T 2003 *Mater. Trans.* **44** 2489
[9] Ahmed I M, Nayl A A and Daoud J A 2016 *J. Saudi Chem. Soc.* 20 S280
[10] Nagib S and Inoue K 2000 *Hydrometallurgy* **56** 269
[11] Dutra A J B, Paiva P R P and Tavares L M 2006 *Miner. Eng.* **19** 478
[12] Pearson D 1981 Recovery of zinc from metallurgical dusts and fumes *Process and Fundamental Considerations of Selected Hydrometallurgical Systems* ed M C Kuhn (New York: Society of Mining Engineers of AIME) pp 153–168
[13] Nesbitt C C and Xue S 1995 *Conf. Treatment and minimization of heavy metals containing wastes* 12-16 Feb (Las Vegas, USA) 12-16 Feb p 43–56
[14] Nii K and Hisamatsu Y 1965 *Trans. Nat. Res. Inst. Met.* **7** 9
[15] Leclerc N, Meux E and Lecuire J M 2003 *Hydrometallurgy* **70** 175
[16] Miller F A and Wilkins C H 1952 *Anal. Chem.* **24** 1253
[17] Saha J K and Podder J 2011 *J. Bangladesh Acad. Sci.* **35** 203

Influence of layer number and sintering on ferroelectric behaviour of barium titanate films prepared with sol-gel method

S Budiawanti[1,2], B Soegijono[1] and I Mudzakir[2]

[1]Department of Physics, Indonesia University, Depok, Indonesia
[2]Faculty of Teacher Training and Education, Sebelas Maret University, Solo, Indonesia

E-mail: awanty77@yahoo.com

Abstract. We have studied ferroelectric behaviour of barium titanate ($BaTiO_3$) thin films prepared with sol-gel method. $BaTiO_3$ solutions were deposited onto silicon substrates with different numbers of layers via spin coating method. The obtained samples were sintered at 700 °C and 800 °C. Structure and ferroelectric properties of the samples were investigated using X-ray diffraction (XRD) and Sawyer–Tower circuit. As observed from XRD patterns, number of layers affected the films structure. Results also showed that crystallographic system of $BaTiO_3$ thin films was characterized by a tetragonal structure. Maximum polarization increased with an increase in layer number. Spontaneous polarization was also observed increasing with an increase in temperature.

1. Introduction

Barium titanate ($BaTiO_3$), which has a tetragonal structure at room temperature, is considered a good perovskite ferroelectric material. This material has found application as a multilayer ceramic capacitor and as a component of ferroelectric memory devices. The wide usage of $BaTiO_3$ is attributed to its high dielectric constant, ferroelectric characteristics, and piezoelectric properties, in which cations and anions are displaced in opposite directions of the polar axis [1 - 4].

Ferroelectric films have been fabricated through various deposition techniques, such as the hydrothermal method[5], molecular beam epitaxy, chemical vapor deposition[6], electrophoretic deposition[7], laser chemical vapor deposition[8, 9], and sol-gel processing[10, 11]. Among these techniques, sol-gel processing has attracted considerable interest due to its low cost and ease of integration with semiconductor silicon (Si) technology [12].Through this method, $BaTiO_3$ films can be integrated with ferromagnetic films, thereby enabling the electrical or magnetic tuning of these microwave materials [13 - 15]. The main advantages of sol-gel processing are that it presents good homogeneity, ability for chemical composition control, reducing sintering temperature for crystallization, and good control over the formation of nano-sized particles [12 - 16]. Using sol-gel method and sintering temperature 700-800 °C, BaTiO3 become crystalline with crystallite in the nano-sized range (<100 nm). Sintering time and temperature determine crystallinity, crystallite size, and compactness of $BaTiO_3$ films. It is therefore important to consider various time and temperature of sintering process [2].

Studies on single or multilayer thin film have made significant progress than on the bulk $BaTiO_3$. Single or multilayer BaTiO3 thin films have different properties than bulk materials. Noticeable differences are Pockels effect or linear electro-optic and dielectric properties in $BaTiO_3$ thin films compared to $BaTiO_3$ bulk crystals [5, 6]. By investigating single or multilayer thin film deposition method and its properties, it is possible to synthesize thin films with certain properties.

In this study, sol-gel method was used to prepare $BaTiO_3$ multilayer thin films. Effects of layer number, annealing temperature, and annealing time on structural, morphological, and ferroelectric behaviors of the thin films were investigated.

2. Experimental procedures

Barium acetate and titanium (IV) isopropoxide $[Ti((CH_3)_2CHO)_4]$ were used as precursors for fabrication of $BaTiO_3$ thin films. Glacial acetic acid, 2-methoxyethanol ($CH_3OCH_2CH_2OH$), and ethylene glycol were used as solvents. First, barium acetate was dissolved in acetic acid. The solution was continuously stirred at 70 °C for 60 min to obtain a clear and homogeneous solution and then was diluted with $CH_3OCH_2CH_2OH$. An equimolar amount of $Ti((CH_3)_2CHO)_4$ was added to the solution at a molar ratio of Ba:Ti = 1:1 under continuous stirring. The solution was continuously stirred at 90 °C for 2 h and was filtered using microfiber glass filter paper.

The obtained precursor solution was transparent and bright yellow. After the precursor solution was aged for 24 h, it was deposited onto a silicon (Si) substrate by spin coating at 3000 rpm for 30 s. Following the deposition, the films were pyrolyzed for 10 min at 200 °C in air. These steps were repeated several times (1, 5, 7) with reference to film thickness. Finally, the films were sintered in an air atmosphere furnace to crystallized and compacted at 700 °C and 800 °C for 3 h.

To investigate structural properties of the films, X-ray diffraction (XRD) was carried out using a Bruker D8 ADVANCE diffractometer with Cu-Kα radiation at λ= 1.547 Å. From the XRD patterns, crystallite size could be determined using HighScore Plus software. Morphology of the films surfaces was also investigated using scanning electron microscopy (SEM) FEI Inspect F50. Hysteresis loops, which are indicators of ferroelectric properties, were measured using a Sawyer–Tower circuit standard model.

3. Results and discussion

3.1. Crystal structure and crystallite size

Figure 1 shows XRD patterns of $BaTiO_3$ films with various layer numbers (1, 5, and 7 layers). We did not present XRD patterns from the films with 3 layers because the data could not be read. XRD patterns with 1, 5, and 7 layers show quite similar. Multilayer $BaTiO_3$ have different properties to bulk $BaTiO_3$ [3, 6]. All films were heat-treated at 800 °C in air for 3 h. Peaks of XRD patterns were successfully indexed with tetragonal structure of $BaTiO_3$ for all samples. No other phases were observed in the XRD patterns. We observed that with increasing layer number, intensity of major diffraction peaks increases indicating that crystallization of $BaTiO_3$ nanostructure improves at high layer numbers. The most intense peak (110) was found at 2θ = 31.57°. Peaks at 2θ= 22.17°, 2θ = 38.89°, 2θ = 45.44°, 2θ = 51.03°, and 2θ = 56.17° were (101), (021), (202), (113), and (122), respectively. We observed that XRD peaks corresponding to the (021) and (122) planes split into two peaks due to formation of a tetragonal structure, which is important for ferroelectricity.

Average crystallite sizes of $BaTiO_3$ samples, which were calculated using HighScore Plus, were 47, 33, and 23 nm at layer number 1, 5, and 7, respectively. It is shown that crystallite size of all samples are in the range of nano-sized (less than 100 nm).

Figure 2 shows XRD patterns of $BaTiO_3$ samples sintered at 700 °C and 800 °C for 3 h. We found XRD patterns of the film sintered at 700 °C still shows small peaks corresponding to barium carbonate–titanium dioxide. But, a well-controlled composition, single-phase perovskite structure was formed in the film sintered at 800 °C. Figure 3 shows XRD patterns of the $BaTiO_3$ samples sintered at 800 °C for 3 and 4 h. We suggest that intensity of major diffraction peaks increased with increasing sintering time. It may indicate that a long sintering time improves crystallization of $BaTiO_3$ nanostructure. From figures 1, 2 and 3, full width at half maximum (FWHM) changes with number of layers, annealed temperature and annealed time. Broadening occurs due to smaller crystallite size in the range of nano scale (<100 nm).

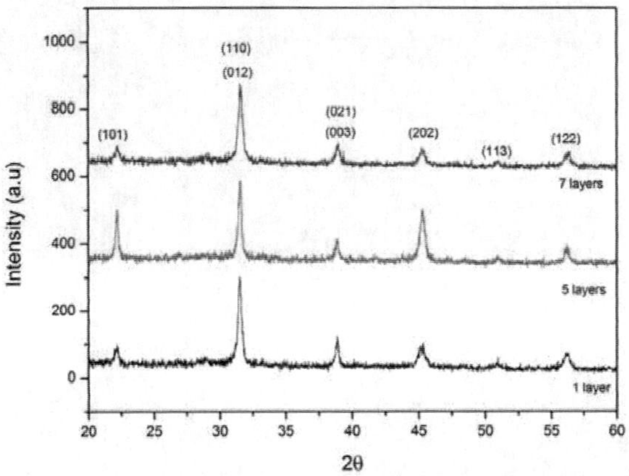

Figure 1. XRD patterns of $BaTiO_3$ films at various layer numbers.

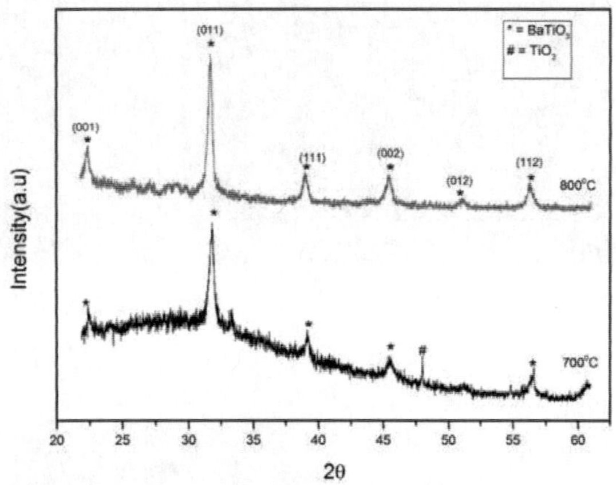

Figure 2. XRD patterns of 5 layers $BaTiO_3$ films sintered at 700 °C and 800 °C.

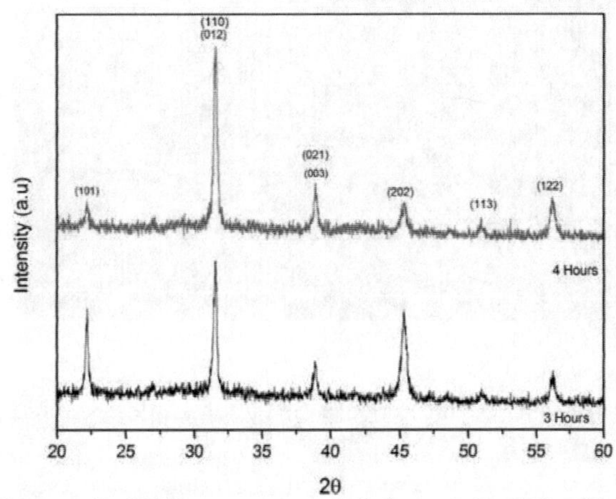

Figure 3. XRD patterns of 5 layers $BaTiO_3$ films sintered at 800°C for 3 and 4 h.

3.2. Morphology

SEM images for a single phase $BaTiO_3$ with five layers sintered at 800 °C at various sintering times are shown in figure 4. Figure 4a shows that $BaTiO_3$ thin film deposited on Si (100) substrate and sintered at 800°C for 3 h exhibits cracks. $BaTiO_3$ thin film sintered at 800 °C for 4 h exhibits a crack-free and uniform morphology (figure 4b). $BaTiO_3$ thin film sintered at 700 °C is not presented because it is not a single phase. The films sintered for 3 and 4 h have average grain sizes of 33 and 48 nm, respectively. We suggest a significant increase in grain size occurs with rising sintering time, as shown in the SEM cross-section images in figure 4c. Average thickness of $BaTiO_3$ with five layers is 2.9 μm.

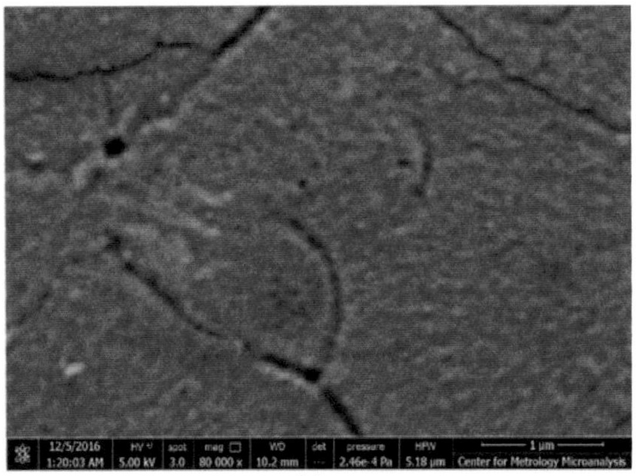

(a)

(b)

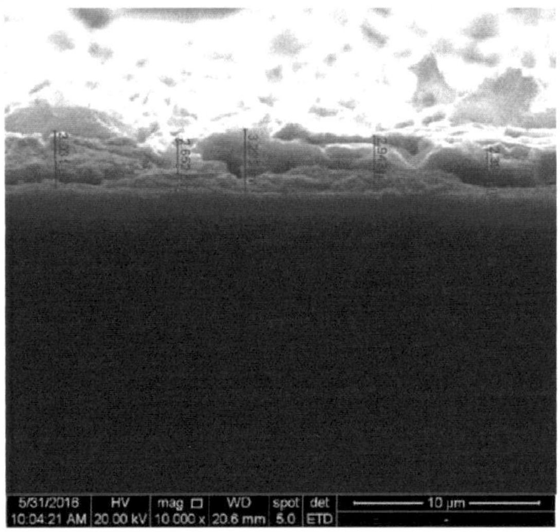

(c)

Figure 4. SEM images of five layers single phase $BaTiO_3$ samples sintered at 800 °C for (a) 3 h and (b) 4 h. (c) Cross-section of $BaTiO_3$.

3.3. Polarization-electric (P-E) hysteresis loops

Figure 5 shows curves of room-temperature *P–E* loops of $BaTiO_3$ films with various layer numbers. We observed that value of maximum polarization increases increasing layer number. However, the area spanned by a loop decreases. This behavior may be due to the particle size of the samples [17-18]. A previous study reported that spontaneous polarization (*Ps*) and coercive field (*Ec*) in a $BaTiO_3$ single crystal are $Ps = 26$ C/cm^2 and $Ec = 1.5$ kV/cm [3]. The low *Ps* and high *Ec* values of the films in the current work might be due to small grain size of the films and clamping of the films to the substrates [17]. Room-temperature *P–E* loops of $BaTiO_3$ films sintered at 700 °C and 800 °C are presented in figure 6. We found that these films exhibit almost similar *Ps* and *Ec* values because both films have almost similar particle size. Figure 7 shows room-temperature *P–E* loops of $BaTiO_3$ films sintered at 800 °C for 3 and 4 h. We observed that remanent polarization increases when sintering time increases, whereas *Ec* decreased. Decrease in *Ec* might be caused by an increase in particle size as sintering time increases [18]. Homogeneity increases with sintering time and this may lead to soften the materials and then decreases *Ec*. In other word, crystallinity degree increases or FWHM decreases. The highest polarization value of 3.32 μC/cm^2 was generated when $BaTiO_3$ film have seven layers and sintered at 800°C for 3 h. The curve-hysteresis loops are well-identified as corresponding to ferroelectric behavior.

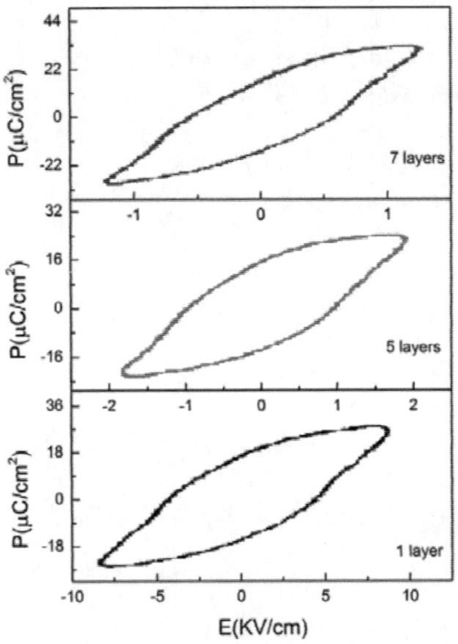

Figure 5. Room-temperature *P–E* loops of $BaTiO_3$ films at various layer numbers.

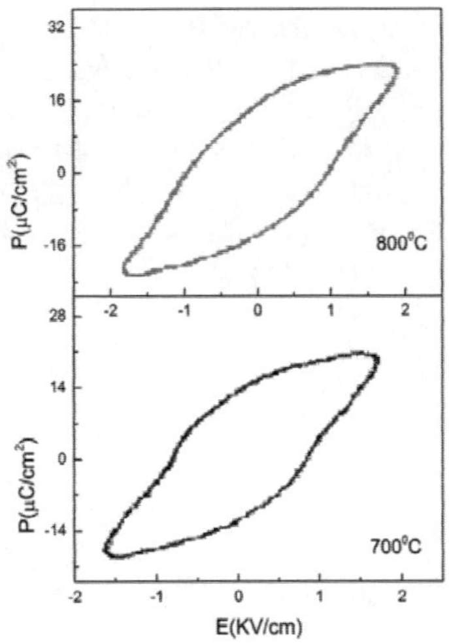

Figure 6. Room-temperature *P–E* loops of five layers $BaTiO_3$ films sintered at 700 °C and 800 °C.

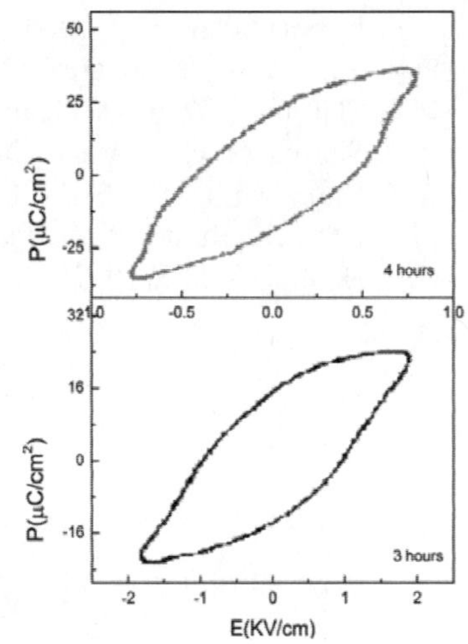

Figure 7. Room-temperature *P–E* loops of five layers $BaTiO_3$ films sintered at 800 °C for 3 and 4 h.

4. Conclusion

Effects of layer number, sintering temperature, and sintering time on structural, morphological, and ferroelectric properties of $BaTiO_3$ thin films were investigated. XRD patterns indicate that increasing layer number, sintering temperature, and sintering time may cause an increase in crystallization. SEM images show that increasing sintering time may also increase crystallite size. SEM images also show that sample annealed for 4 h at 800 °C had a crack-free surface. Results of *P–E* loop measurements shows that value of maximum polarization increases with increase of layer number. The highest polarization of 3.32 $\mu C/cm^2$ was achieved with $BaTiO_3$ film has seven layers and sintered at 800 °C for 3 h.

Acknowledgments

This research is supported financially by Hibah Publikasi Internasional Terindeks untuk Tugas Akhir Mahasiswa Universitas Indonesia through letter of decree no. 631/UN2.R3.1/HKP05.00/2017 and Hibah Disertasi Doktor Universitas Sebelas Maret through letter of decree no. 623/UN27.21/PP/2017.

References

[1] Srinivas A, Gopalan R and Chandrasekharan V 2009 *Solid State Commun.* **149** 367
[2] Cai W, Fu C, Gao J, Guo Q, Deng X and Zhang C 2011 *Physica B* **406** 3583
[3] Ertuğ B 2013 *Am. J. Eng. Res* **2** 1
[4] Chávez E, Fuentes S, Zarate R A and Padilla-campos L 2010 *J. Mol. Struct.* **984** 131
[5] Kobayashi Y, Saito H, Kinoshita T, Nagao D and Konno M 2011 *Thin Solid Films* **519** 1971
[6] Kormondy K J *et al.* 2017 *Nanotechnology* **28** 075706
[7] Jian G, Zhou D, Yang J, Shao H, Xue F and Fu Q 2013 *J. Eur. Ceram. Soc.* **33** 1155
[8] Ito A, Guo D, Tu R and Goto T 2012 *Mater. Chem. Phys.* **133** 398

[9] Guo D, Ito A, Tu R and Goto T 2013 *J. Asian Ceram. Soc.* **1** 197
[10] Sharma S, Tomar M, Puri N K and Gupta V 2014 *Procedia Eng.* **87** 1172
[11] Sharma S, Sharma A, Tomar M, Puri N K and Gupta V 2014 *Procedia Eng.* **87** 1067
[12] Bassiri-Gharb N, Bastani Y and Bernal A 2014 *Chem. Soc. Rev.* **43** 2125
[13] Suastiyanti D, Soegijono B and Hikam M 2013 *Advanced Materials Research* **789** 118
[14] Sharma S, Tomar M, Kumar A, Puri N K and Gupta V 2014 *Physica B* **448** 125
[15] Sharma N, Gaur A and Gaur U K 2014 *Ceram. Int.* **40** 16441
[16] Budiawanti S and Soegijono B 2016 *AIP Conf. Proc.* **1729** 020020
[17] Das S K, Rout P P, Pradhan S K and Roul B K 2012 *J. Adv. Ceram* **1** 241
[18] Hoshina T 2013 *J. Ceram. Soc. Jpn.* **121** 156

IOP Conf. Series: Journal of Physics: Conf. Series **985** (2018) 012035 doi:10.1088/1742-6596/985/1/012035

Thermal behaviour properties and corrosion resistance of organoclay/polyurethane film

O Kurniawan and B Soegijono

Graduate Materials Science Program, Department of Physics, Universitas Indonesia, Depok, Indonesia

E-mail: naufal@ui.ac.id

Abstract. Organoclay/polyurethane film composite was prepared by adding organoclay with different content (1, 3, and 5 wt.%) in polyurethane as a matrix. TGA and DSC showed decomposition temperature shifted to a lower point as organoclay content change. FT-IR spectra showed chemical bonding of organoclay and polyurethane as a matrix, which means that the bonding between filler and matrix occured and the composite was stronger but less bonding occur in composite with 5 wt.% organoclay. The corrosion resistance overall increased with the increasing organoclay content. Composite with 5 wt.% organoclay had more thermal stability and corrosion resistance may probably due to exfoliation of organoclay.

1. Introduction

Among all the available nano fillers, those derived from layered silicates (nanoclays) are the most studied for polymer nanocomposites. By adding small amounts of nanoclay into polymer, it is possible to achieve consistent increments of some properties such as elastic modulus and strength, flame retardancy, lower gas permeability, improved solvent, and heat resistance [1-3].

Thermal stability is defined by specific temperature or temperature time limit within which the material can be used without excessive loss of its properties. Thermal decomposition of polyurethane (PU) has been known as the complex heterogeneous [3]. The thermal decomposition of PU (their degradation attributed to absorbed thermal energy) and corrosion resistance are important phenomenon from both fundamental and industrial applications. The understanding of degradation processes allows determination of optimum conditions for designing PU in order to obtain high-performance polymer materials.

Organoclay such as montmorillonite have been researched for more than 10 years as a filler for polymer based composite [4-5]. Exchanging cations present in clay by Na^+ could allow water to increase interlayer spacing up to several hundred angstrom [6-7]. Modifications of montmorillonite by means of organic modifiers such as alkylammonium to improve the compatibility between hydrophilic organo-montmorillonite and hydrophobic polyurethanes are often used [8, 9]. These organic chains having positively charged ends can be bonded to the surface of the negatively charged silicate layers resulting in an increase of the interlayer gap [10-11].

Composite on polyurethane and montmorillonite particles have been widely investigated [8,10] and frequently enhanced advanced properties such as mechanical, water vapor permeability, water permeability properties due to exfoliation dispersion, and high surface area [11-14].

The presents paper will discuss the characterization of nanocomposite clay-polyurethane. Polyurethane was obtained from commercial product with aluminum addition. Addition of aluminum differs it with others PU. Thermal stability and corrosion resistance were studied with different content of organoclay. With unique properties of organoclay, we expect that the thermal stability and corrosion resistance changed due to different content of organoclay.

2. Experimental procedure

PU was obtained from commercial product with aluminum addition. The composition of the PU are diisocyanate (MDI), MDI based polyisocyanate, 4.4'– diphenylmethane, aluminum, aromatic solvent blend, and isobutyl acetate. Figure 1 shows the urethane groups –NH–(C=O)–O– link the molecular units.

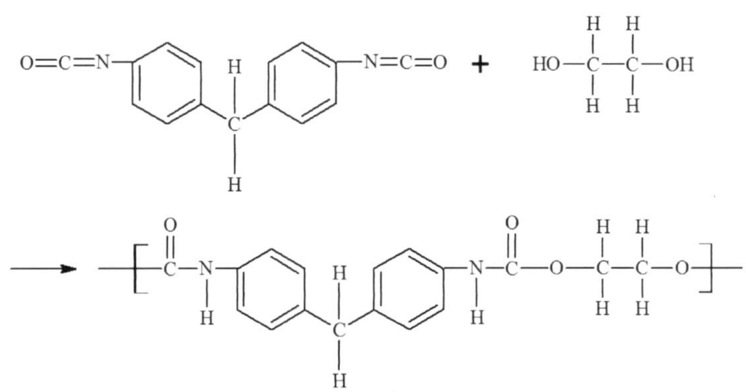

Figure 1. The urethane groups –NH–(C=O)–O– link the molecular units.

Organoclay was procured from commercial product. The composite were prepared by stirring PU solution with various content of clay: 1 wt.% (PClay1), 3 wt.% (PClay3), and 5 wt.% (PClay5). Then composite were deposited on low carbon steel plate using High Volume Low Pressure (HVLP) method. Then the films were taken from the substrate to be characterized. Thermogravimetry (TGA) was used to measure the weight loss of the clay-polyurethane composite under nitrogen atmosphere. Samples were heated from 30 °C room temperature to 800 °C at heating rate 10 °C/min. Differential Scanning Calorimetry was used to obtained information heat needed or loss during heating or cooling rate of 10 °C/min from 30 °C temperature to 800 °C. The salt spray according to ASTM B-117 test was also used for 72 hours to provide information of corrosion resistance. The samples were also investigated by FTIR spectrometer in the wave number range of 300 – 4000 cm^{-1}. Morphology of the samples was investigated by optical microscope before and after salt spray test.

3. Results and discussion

3.1. TGA analysis

The study of the decomposition of PU is often difficult. PU degrades with the formation of various gaseous products and a number of decomposition steps are typically observed in thermogravimetric analysis (TGA). TGA also provide information of lost mass during heating as function of temperature of the samples. Figure 2 shows TGA curve as a function of time. According to figure 2 there are two step of degradation. The first (onset temperature) is round 260 – 290 °C and the other stage (end-set temperature) is between 420 – 450 °C. The first stage is major stage which the sharp degradation occurred. In the second stage, the composite with different content clay displayed higher thermal resistance than pure PU. The smallest lost mass is PU and the lost mass increases with increasing clay content. It seems that the onset temperature become sharp and slightly shifting to the higher temperature with increasing clay content, but the end-set temperature remain more or less unaltered.

IOP Conf. Series: Journal of Physics: Conf. Series **985** (2018) 012035 doi:10.1088/1742-6596/985/1/012035

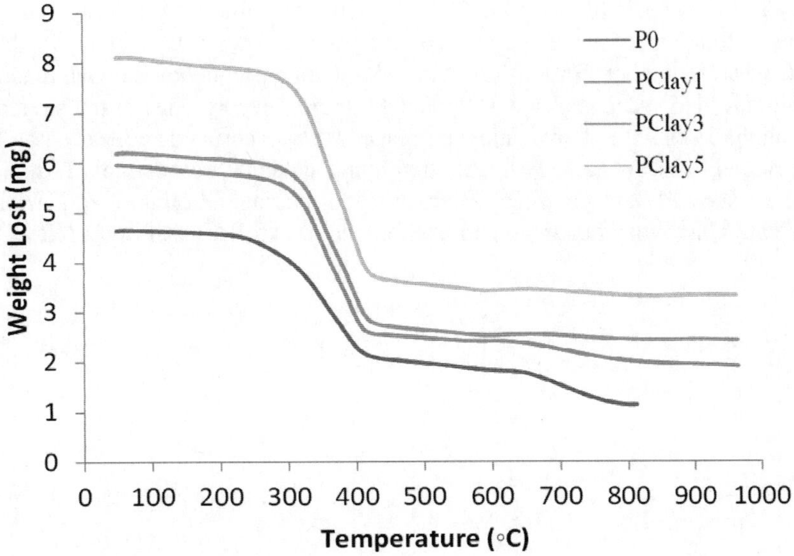

Figure 2. TGA curve of the samples clay/PU with various content of clay. These are pure PU (P0), 1 wt.% (PClay1), 3 wt.% (PClay3), and 5 wt.% (PClay5).

3.2. DSC analysis

Figure 3 displays heat needed or loss versus temperature of the samples. Samples P0, PClay1, and PClay3 suddenly decreased in heat flow around 800 °C, but PClay5 was better in thermal stability. Some researchers have shown that thermal-degradation behavior of PU and its composite at high temperatures may provide a fingerprint of the materials that may not only inform us the material characteristics or the filler type but also its processing and quality of the end products [15]. The complexity of the PU's or its composite's thermal stability and degradation drives researchers to do further study.

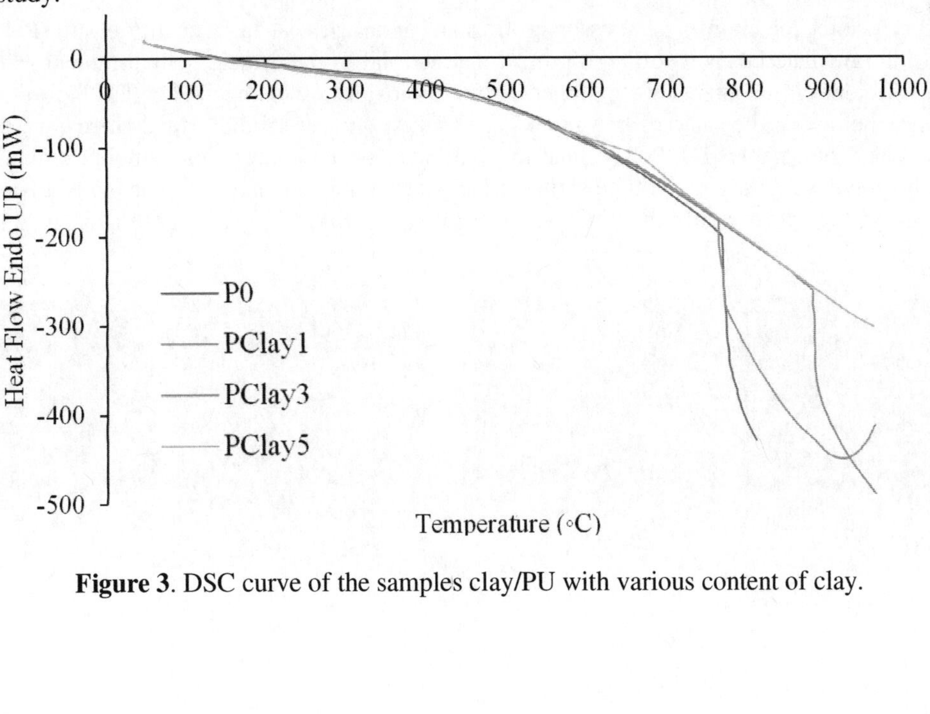

Figure 3. DSC curve of the samples clay/PU with various content of clay.

3.3. FTIR analysis

Figure 4 shows FTIR patterns of the pure PU (P0) and its addition with various clay content. The transmittance as a function of wave number of PU with 5 wt.% clay (PClay5) showed the best transmittance for wave number 500 – 4000 cm^{-1}. Stretching frequency of OH functional group is observed at 3349 cm^{-1} [15]. The peaks at 1596 and 1460 cm^{-1} occured due to the stretching frequency of C=C in the aliphatic chain and aromatic ring respectively. The bands appearing at 1167 cm^{-1} and 1041 cm^{-1} correspond to Si-O-O and Al-O-Al stretching frequency, respectively [16]. The addition of 5wt.% organoclay showed less chemical interaction comparing to others samples. But the better thermal stability of PClay5 may occure due to phenomena of exfoliation of organoclay [17].

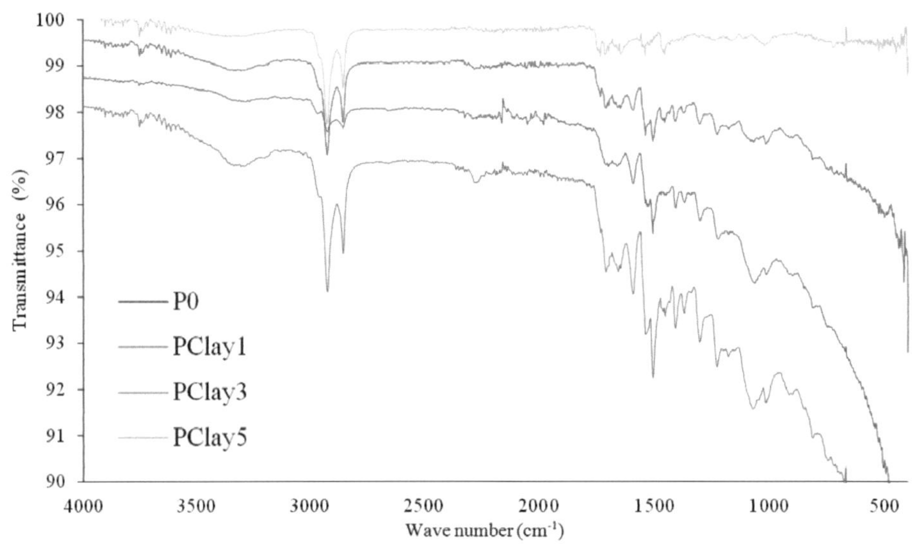

Figure 4. FTIR curve of the samples clay/PU with various content of clay.

3.4. Morphology and salt spray analysis

Durable corrosion protection is very essential in many industries. A high quality elastic PU coating system using raw materials is required to protect the material from corrosion environtment. Therefore, PU with different clay contents were subjected for corrosion test using salt spray. Figure 5 shows surface morphology of the samples before and after salt spray test to understand corrosion resistance of PU coating. The samples P0, PClay1 and PClay3 have surface morphology smoother than PClay5 before salt spray test. After salt spray test, the surface morphology changed. We observed that samples PClay1 and PClay3 are rougher than PClay5. We suggested that PClay5 is more resistance than the others.

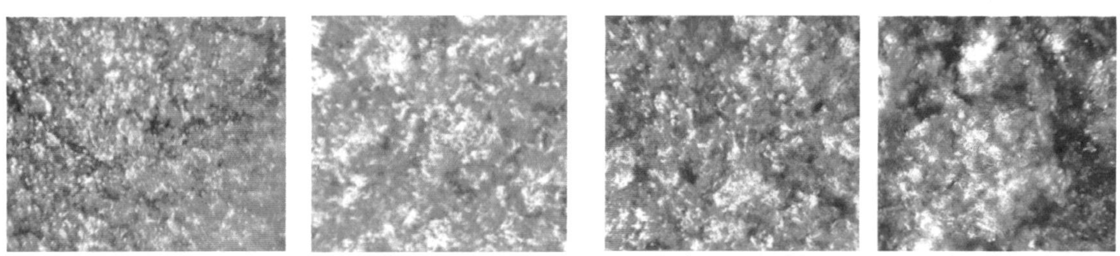

P0 400 X PClay1 400 X PClay3 400 X PClay5 400 X

(a)

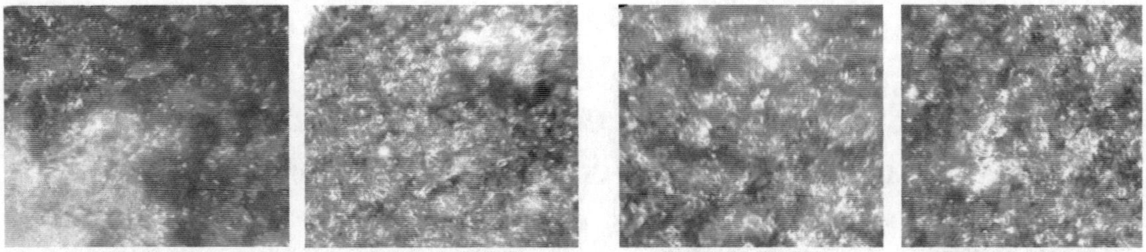

(b)

Figure 5. Surface Morphology of clay/PU composite with different clay content (a) before and (b) after salt spray.

4. Conclusion

Organoclay/PU composite with different clay content have been investigated. PClay5 have more thermal stability and good corrosion resistance than PO, PClay1 and PClay3. It revealed that addition of clay modify the properties of pure PU thus it was suitable for coating materials. The exfoliation of organoclay in PU matrix indicated a good interaction between organoclay and PU.

Acknowledgment
This research is financially supported by grant for international publication at University of Indonesia through letter no.630/UN2.R3.1/HKP.05.00/2017

References

[1] Yusoh K, Jin J and Song M 2010 *Prog. Org. Coat.* **67** 220
[2] Naguib H F, Aziz M S A and Saad G R 2013 *J. Ind. Eng. Chem.* **19** 56
[3] Song L, Hu Y, Tang Y, Zhang R, Chen Z and Fan W 2005 *Polym. Degrad. Stab.* **87** 111
[4] Peponi L, Puglia D, Torre L, Valentini L and Kenny J M 2014 *Mater. Sci. Eng.* **85** 1
[5] Fathurrohman M I, Soegijono B, Budianto E, Rohman S and Ramadhan A 2015 *Macromol. Symp.* **353** 62
[6] Ayres E, Orefice R L and Sousa D 2006 *Macromol. Symp.* **245** 330
[7] Barick A K and Tripathy D K 2011 *Appl. Clay Sci.* **52** 312
[8] Cheng A, Wu S, Jiang D, Wu F and Shen J 2006 *Colloid. Polym. Sci.* **284** 1057
[9] Han B, Cheng A, Ji G, Wu S and Shen J 2004 *J. Appl. Polym. Sci.* **91** 2536
[10] Kim G M, Lee D H, Hoffmann B, Kressler J and Stöppelmann G 2001 *Polymer* **42** 1095
[11] Malay O, Oguz O, Kosak C, Yilgor E, Yilgor I and Menceloglu Y Z 2013 *Polymer* **54** 5310
[12] Rahman M M, Kim H D and Lee W K 2008 *J. Appl. Polym. Sci.* **110** 3697
[13] Salahuddin N, Enein S A A, Selim A and Dien O S E 2010 *Appl. Clay Sci.* **47** 242
[14] Tien Y I and Wei K H 2001 *Macromolecules* **34** 9045
[15] Prime R B, Burns J M, Karmin M L, Moy C H and Tu H B 1988 *J. Coat. Technol.* **60** 55
[16] Fan Q, Fang J, Chen Q and Yu X 1999 *J. Appl. Polym. Sci.* **74** 2552
[17] Fathurrohman M I, Soegijono B, and Budianto E 2013 *J. Mater. Sci. Eng. B* **3** 575

IOP Conf. Series: Journal of Physics: Conf. Series **985** (2018) 012036 doi:10.1088/1742-6596/985/1/012036

Synthesis of alumina ceramic encapsulation for self-healing materials on thermal barrier coating

O P Golim[1], N Prastomo[1], H Izzudin[2], S Hastuty[3], R Sundawa[2], E Sugiarti[2] and K A Z Thosin[2]

[1]Engineering Physics Department, Surya University, Tangerang, Banten, 15810 Indonesia
[2]Research Center for Physics, Indonesian Institute of Sciences, Komplek Puspiptek, Serpong, Tangerang Selatan 14314, Indonesia
[3]Mechanical Engineering Department, Universitas Pertamina, South Jakarta, Jakarta, 12220 Indonesia

E-mail: obert.golim@gmail.com

Abstract. Durability of Thermal Barrier Coating or TBC can be optimized by inducing Self-Healing capabilities with intermetallic materials $MoSi_2$. Nevertheless, high temperature operation causes the self-healing materials to become oxidized and lose its healing capabilities. Therefore, a method to introduce ceramic encapsulation for $MoSi_2$ is needed to protect it from early oxidation. The encapsulation process is synthesized through a simple precipitation method with colloidal aluminum hydroxide as precursor and variations on calcination process. Semi-quantitative analysis on the synthesized sample is done by using X-ray diffraction (XRD) method. Meanwhile, qualitative analysis on the morphology of the encapsulation was carried out by using Scanning Electron Microscope (SEM) and Field Emission Scanning Electron Microscope (FESEM) equipped with dual Focus Ion Beam (FIB). The result of the experiment shows that calcination process significantly affects the final characteristic of encapsulation. The optimum encapsulation process was synthesized by colloidal aluminum hydroxide as a precursor, with a double step calcination process in low pressure until 900 °C.

1. Introduction

Thermal Barrier Coating or TBC is a type of coating that is commonly applied on substrate used on a high temperature, such as super alloys on turbine gas engine. It enables the substrate to work on a higher temperature [1]. The main structure of TBC is composed of two main parts, top coat and bond coat. The top coat is an oxide layer that has high temperature resistance and the bond coat connects the oxide layer to the substrate. During operation, oxygen may diffuse through the oxide layer and reacts with the bond coat to form another layer which was called the *thermally grown oxide* (TGO) layer, that has a very distinctive coefficient of thermal expansion with the oxide layer on top of it [2]. Extreme cyclic operating condition may cause stress to occur due to the difference in thermal expansion coefficient and induced crack on TBC which later would cause the coating to peel off, leaving the substrate to be unprotected [3].

Experiment had shown that self-healing properties induced on the TBC may help close up cracks appeared and extend its lifetime. The self-healing mechanisms proposed occurred whenever cracks formed on the TBC, causing the self-healing material to be exposed with oxygen and formed an oxide materials to heal the crack [4].

Due to its extreme operating conditions, there are several limitations for the healing material that can be used. First, the healing material must have a solid phase to prevent any significant difference on the coefficient of thermal expansion between the healing materials and the TBC. Second, it must be able to form a healing product that has similar properties to the TBC. Literatures show that materials composed of early transition metal and group A element exhibit properties fulfil the criteria specified above. Materials used on past experiment, $MoSi_2$, was also included in this group. Nonetheless, materials from this group will undergo an oxidative decomposition on high temperature environment, thus reducing its healing capabilities [4-7].

Therefore, a specific method was needed to protect the self-healing materials from early oxidation. A study has shown that self-healing materials are isolated from the outside medium through an encapsulation process. These encapsulations prevent contained self-healing materials to undergo early reaction, and enables the healing process to occur when the capsule was ruptured [5]. However, there are several criteria that must be fulfilled for the encapsulation, which is to be oxygen impenetrable and have a similar coefficient of thermal expansion with the TBC. Literatures had shown that alumina encapsulation was the best type of encapsulation that able to fulfil those criteria [4, 6].

Several researches had shown that alumina encapsulation can be synthesized through several methods. However, the simplest method to produce an alumina encapsulation was through the precipitation method [5, 6, 8]. This method consists of two main steps, the first one is the formation of simple coating through precipitation method. The second step is the formation of oxide encapsulation layer through calcination process [8,9].

A study had shown that synthesis of alumina encapsulation for self-healing material $MoSi_2$ was possible using aluminium chloride as precursor. However, experiment results show the healing materials undergo premature oxidation that cause it to lose some of it healing properties during the encapsulation process [6].

This experiment was done to introduce a better method for synthesizing encapsulation for self-healing material $MoSi_2$ through the precipitation method with modification on the precursor used and the calcination method.

2. Methodology

2.1. Materials

The self-healing materials encapsulated on this experiment was molybdenum disilicide (Japan New Metals, Co. Ltd), and the materials used for the alumina encapsulation were aluminium hydroxide (Taimei Chemicals), nitric acid (65%, Merck), ammonium hydroxide (28%, Tedia), and locally produced distilled water.

2.2. Encapsulation procedure

The procedure of the encapsulation started with synthesizing the precursor by diluting aluminium hydroxide powder into nitric acid to form a colloid. Then the self-healing material was dispersed into the colloid with mol ratio of 1:10. On these experiment, a 5M precursor was used. The dispersion process was done with magnetic stirring and ultrasonication for 30 minutes. Next, the mixture was hydrolysed by addition of diluted ammonium hydroxide 25% slowly until the pH of mixture reach 9. While being titrated with ammonium hydroxide, the mixture was stirred with magnetic stirrer to prevent agglomeration of particles. The mixture was then aged for more than 15 hours at room temperature to settle the precipitate. After aging process, the sample was then filtrated with a filtration paper while being washed with distilled water at 80 °C to help clean the sample from salt residue from the hydrolyzation. Then, the sample was dried on low vacuum condition (10mbar) at 100 °C. Low vacuum condition was used to prevent premature oxidation of the self-healing material. The last step was calcining the pre-encapsulated $MoSi_2$ with a vacuum furnace. On this experiment, there are 4 types of calcination processed used: direct calcination process into 900 °C, step calcination process into 900 °C, direct calcination process into 1000 °C, and step calcination process into 1000 °C. Low-pressure condition was held at 15 Pa to prevent any premature oxidation of the self-healing material.

2.3. Characterization

The samples synthesized on this experiment were characterized with x-ray diffraction (XRD) and scanning electron microscope (SEM), before and after the calcination process. The XRD used to characterize crystalline structure of the sample using RIGAKU X-ray diffraction instrument in 2theta range 20°- 80° with a step size of 0.002°. The phase of sample obtained through Rietveld analysis profile fitting. SEM images were obtained on HITACHI SEM along with energy dispersive x-ray spectroscopy (EDS) to characterize the morphology of the encapsulated self-healing particle. The beam energy used was 20kV and the beam current was set into 86.1μA. EDS analysis was using the point analysis method with molybdenum, silicon, aluminium and oxygen element.

Sample with the best result was then characterized again using JEOL multi-beam field effect scanning electron microscope (FESEM) equipped with focus ion beam (FIB) to obtain the cross-section view of an encapsulated particle before and after the calcination process. The cross section of the sample was obtained by milling an encapsulated self-healing particle with Gallium ion beam under low vacuum condition. To gain the morphology of encapsulation, EDS mapping was used on molybdenum, silicon, aluminium, and oxygen element.

Prior to electron microscopy the samples were treated with thin carbon coating to prevent sample charging.

3. Results and discussions

3.1. Precursor synthesis

Aluminium hydroxide ($Al(OH)_3$) was diluted by nitric acid, considering its high solubility in strong acid or base solution [9]. However, some parts of the diluted aluminium hydroxide will react with nitric acid to form aluminium nitrate through equation 1 [10]. Therefore, the hydrolyzation process is needed for the formation of aluminium hydroxide precipitate that will encapsulate the self-healing particle. Hydrolyzation process will occur according to the chemical reaction on equation 2 [11].

$$Al(OH)_{3_{(s)}} + 3HNO_{3_{(aq)}} \rightarrow Al(NO_3)_{3_{(s)}} + 3H_2O_{(aq)} \quad (1)$$

$$Al(NO_3)_{3_{(l)}} + 3NH_4OH_{(aq)} \rightarrow Al(OH)_{3_{(s)}} + 3NH_4NO_{3_{(aq)}} \quad (2)$$

During the hydrolyzation process, the precipitates interacts with the self-healing material to form an encapsulation. This interaction mainly caused by the difference in surface potential of $MoSi_2$ and aluminum hydroxide. Past studies have shown that the zeta potential of $MoSi_2$ at pH 9 was around -80 mV, while the zeta potential of aluminum hydroxide under the same conditions are around -35 mV [12, 13]. Even though both particle has a negative surface charge, the difference in the magnitude of the potential itself cause an interaction between both particles. It is stated that particle with charge that is more positive are attracted to particle with negative charges. Therefore, it can be assumed that aluminum hydroxide particle will be attracted to $MoSi_2$ particle to form an encapsulation.

3.2. Encapsulation result before calcination process

Alumina encapsulation will form through the calcination process of boehmite. Boehmite itself will form on the drying process that evaporates water according to this below reaction [14].

$$Al(OH)_{3_{(l)}} \rightarrow AlOOH_{(s)} + H_2O_{(g)} \quad (3)$$

Figure 1a shows the boehmite encapsulated $MoSi_2$ after the drying process. Aside from the coated material, there could be observed some agglomerates of the boehmites particle that forms on the surface of the self-healing material. These agglomerates of boehmite particle formed when the precipitates failed to form encapsulation and agglomerates with other precipitates.

The XRD result on figure 2 shows that phase diagram of the coated self-healing material is dominated by boehmite. It is caused by the large number of the aluminium hydroxide precipitate which later formed the boehmites particle according to equation 3.

3.3. Encapsulation result after calcination process

After the calcination process, the boehmite encapsulation of $MoSi_2$ particle undergo a significant change that can be shown from the SEM and XRD results. The surface morphology of encapsulated $MoSi_2$ particle can be shown through the SEM images on figure 1b. Through the EDS point analysis, it is known that particle surfaces contain aluminium element, which means that the encapsulation was still intact.

The effect of calcination process variations to the encapsulation can be seen from the XRD result in figure 2. Boehmite encapsulation undergo a change into alumina encapsulation according to equation 4 [15]. However, it is also shown that when the calcination temperature increased from the 900 °C to 1000 °C, Mullite and Mo_5Si_3 are formed, which means that the self-healing material undergo a premature oxidation according to equation 5 and 6 [6, 16]. The premature oxidations are caused by the rapid crystallization of alumina on higher temperature, which later caused the encapsulation to peel off from the surface and fail to protect the self-healing materials.

$$AlOOH \xrightarrow{500-700\,^0C} \gamma - alumina \xrightarrow{>900\,^0C} \theta - alumina \xrightarrow{>1100\,^0C} \alpha - alumina \quad (4)$$

$$7\,MoSi_{2(s)} + \frac{7}{5}xO_2 \xrightarrow{800-1000\,^0C} \frac{x}{5}\,Mo_5Si_{3_{(s)}} + \frac{7}{5}SiO_{2_{(s)}} + (1-x)\,MoSi_{2(s)} \quad (5)$$

$$\frac{7}{5}\,xSiO_{2_{(s)}} + \frac{21}{10}x\gamma,\eta,\delta Al_2O_{3_{(s)}} \xrightarrow{1000\,^0C} \frac{7}{5}x\,(3Al_2O_3.2SiO_2)_{(s)} \quad (6)$$

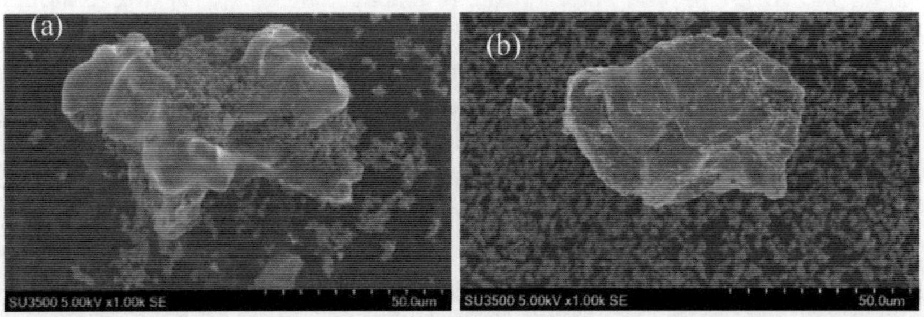

Figure 1. SEM images of coated $MoSi_2$ particle (a) before calcination (b) after step calcination to 900 ° C.

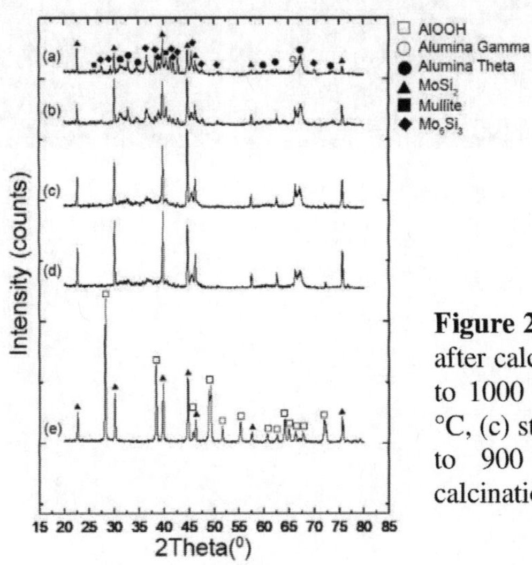

Figure 2. XRD result of sample after calcination method (a) step to 1000 °C, (b) direct to 1000 °C, (c) step to 900 °C, (d) direct to 900 °C, and (e) before calcination process.

The XRD results contradicts the result of an experiment, where calcination up to 1000 °C are proven to better [6]. On this experiment, it can be observed that calcination process up to 900 °C does not produce a significant amount of Mullite through the XRD analysis even though there are faint peaks of Molybdenum Silicide (Mo_5Si_3), which proves $MoSi_2$ does not undergo a lot of premature oxidation. Apparently, this result could be achieved while doing the calcination process under low pressure condition. Medium vacuum condition of 15 Pa has enabled oxygen from ambient air to be limited, therefore oxidations of self-healing materials may be prevented [17]. The large alumina phases on XRD result was caused by the large amount of boehmite particle from the drying process. Boehmite that does not form an encapsulation still undergo phase change due to the heat treatment process to form alumina crystal as shown in figure 1b.

3.4. Morphology of encapsulation by FESEM

The best encapsulated self-healing materials obtained through calcination up to 900 °C with step process, was characterized again using FESEM. Figure 3a and b shows characterization result of encapsulated $MoSi_2$ before calcination process. According to the observation, previous statement on the existence of aluminium hydroxide precipitate on the surface of encapsulated particle can be proven. It is shown through the mapping of aluminium in figure 3b that some of the boehmites that does not form encapsulation, agglomerated and some others form encapsulation on the surface of the material. While on figure 3c it can be observed that oxygen is also located on the aluminium position. Both mapping image shows that there is a thin encapsulation that covers the self-healing material composed of aluminium and oxygen. However, due to the thickness of the encapsulation is lesser than the depth of penetration of EDS analysis, the mapping is dominated by molybdenum element on top of the material surfaces (red-coloured).

Figure 4 shows the result of characterization of encapsulated $MoSi_2$ after calcination process. According to the mapping, the surfaces are still covered with a faint of aluminium and oxygen element, which correlates with previous statement that after the calcination process alumina coating was still intact with the self-healing particle. It can also be observed that minimum number of oxygen is located on the materials, which proves that premature oxidation of the self-healing material is reduced. The cracks observed in SEM images from figure 4a were formed when the cross-section process using destructive ion gallium beam was done.

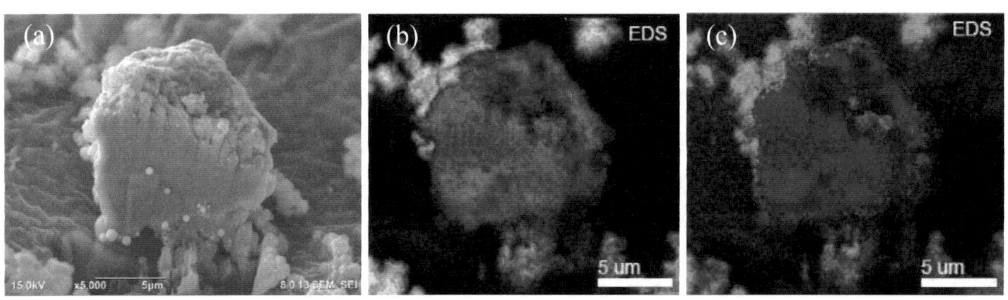

Figure 3. (a) SEM images of cross sectioned coated $MoSi_2$ particle, (b) EDS mapping of aluminum (blue) and molybdenum (red), and (c) EDS mapping of oxygen (green) and molybdenum (red) before calcination.

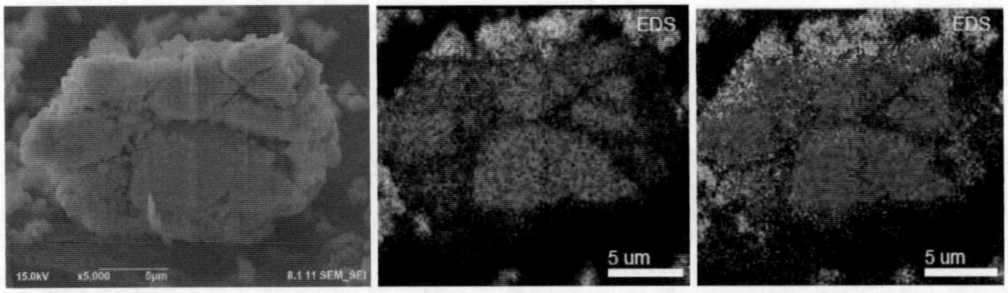

Figure 4. (a) SEM images of cross sectioned coated $MoSi_2$ particle, (b) EDS mapping of aluminum (purple) and molybdenum (red)), and (c) EDS mapping of oxygen (yellow) and molybdenum (red) after calcination.

4. Conclusions

Encapsulation of self-healing material $MoSi_2$ was successfully synthesized through the proposed precipitation method with colloidal aluminium hydroxide as precursor. Through the experiments, it is shown calcination processes are proven to affect the final properties of encapsulation greatly. Step calcination until 900 °C under low pressure condition proved to be the most effective calcination processes in forming the alumina encapsulation with reduction in premature oxidation of the self-healing material.

References

[1] Sampath S, Schulz U, Jarligo M O and Kuroda S 2012 *MRS Bull.* **37** 903
[2] Clarke D R, Oechsner M and Padture N P 2012 *MRS Bull.* **37** 891
[3] van der Zwaag S 2010 *Bull. Pol. Ac.: Tech.* **58** 227
[4] Sloof W G, Turteltaub S, Carabat A L, Derelioglu Z, Ponnusami S and Song G M 2015 Self healing materials - pioneering research in the Netherlands Crack healing in yttria stabilized zirconia thermal barrier coatings (Netherlands: Delft University Press) pp 219-227
[5] Ollier R, Rodriguez E and Alvarez V 2013 *Compos. Part A: Appl. Sci. Manuf.* **48** 137
[6] Carabat A L, van der Zwaag S and Sloof W G 2015 *J. Am. Ceram. Soc.* **98** 2609
[7] Kötteritzsch J, Schubert U S and Hager M D 2013 *Nanotechnol. Rev.* **2** 699
[8] Chavda V P, Soniwala M M and Chavda J R 2012 *Crit. Rev. Pharm. Sci.* **1** 15
[9] Tregubenko V Yu, Udras I E, Drozdov V A and Belyi A S 2011 *Russ. J. Appl. Chem.* **84** 9
[10] Levin O V, Sidel'kovskaya V G, Aliev R R and Leshcheva E A 1997 *Chem. Technol. Fuels Oils* **33** 97
[11] Morgado E Jr, Lam Y L and Nazar L F 1997 *J. Colloid Interface Sci.* **188** 257
[12] Yan Y, Li W, Yang J, Zheng A, Liu F, Feng X and Sparks D L 2014 *Environ. Sci. Technol.* **48** 6735
[13] Sandlin M S, Butt D P, Taylor T N and Petrovic J J 1997 *J. Mater. Sci. Lett.* **16** 1336
[14] Chiou Y H, Tsai M T and Shih H C 1994 *J. Mater. Sci.* **29** 2378
[15] Santos P S, Santos H S and Toledo S P 2000 *Mat. Res.* **3** 104
[16] Zhang H A and Gu S Y 2011 *Advanced Material Research* **284-286** 1630
[17] Shigematsu T and Nakanishi N 1990 *Solid State Ion.* **40** 799

Thermal properties and corrosion resistance of organoclay /epoxy resin film

M Baiquni[1] and B Soegijono[2]

[1]Graduate Materials Science Program, Dept. of Physics University of Indonesia
[2]Dept. of Physics University of Indonesia

E-mail: naufal@ui.ac.id

Abstract. Hybrid materials organoclay/epoxy resin films were prepared by varying organoclay content in epoxy resin as a matrix. The film were investigated by thermogravimetric analysis (TGA), differential scanning calorimetry (DSC), Fourier transform infrared spectroscopy (FT-IR), scanning electron microscopy (SEM), and thermal conductivity. TGA and FT-IR results confirmed that the melting temperature shifted to a lower point. The thermal conductivity and corrosion resistant generally increase with increasing organoclay content. The changes on these properties may due to cross link between organoclay and epoxy.

1. Introduction

Because of the properties of epoxy resin such as chemical stability, good adhesion, and excellent mechanical properties, epoxy resin have been often used as coating, adhesives, laminates, casting material [1-6], and as matrix for composite materials [7-8]. However, researcher are still searching for enhancing the properties of epoxy resin for many applications such as heat resistance, moisture resistance, material for printed board, liquid crystal displays, and plasma display panel.

By incorporating inorganic materials such as organoclay in epoxy resin, rubber or other matrix, the properties of the composite improved [9-11]. However, the role of organoclay in the enhancement of the properties is not well explained yet. The processing of the composite determined its properties, such as the dispersion of clay in epoxy resin matrix. In most cases, many researcher show that well-dispersed and exfoliated clay could enhance the properties of the polymer matrix [12–15].

Composite with more than one compound of filler are interesting, such as organoclay and rock-wool with epoxy resin as a matrix. In this paper, composite epoxy resin and different content of organoclay as a filler have been studied to understand the role of organoclay on thermal stability and corrosion resistance.

2. Experimental procedure

A commercially available two-part epoxy thermoset were used. Epoxy resin is based on a diglycidyl ether of bisphenol A/epichlorohydrin (DGEBA), while Hardener (EPH 555) is a cycloaliphatic amine. MMT organoclay functionalized with an octadecyl ammonium salt was purchased from Nanocor. Epoxy was mixed with various content of organoclay of 1, 3, and 5 phr (parts per hundred resin) and stirred at 300 rpm for 24 h at 27 °C in an Erlenmeyer flask. Hardener EPH555 was added to the mixture and it was highly degassed in a vacuum until the bubble come out from the solution. The samples were molded and cured at room temperature for 24 h.

Thermal stability of the samples was characterized by thermogravimetric analysis (TGA) and differential scanning calorimetry which was conducted from 30 to 800 °C in nitrogen atmosphere. The heating rate and nitrogen flow rate were 10 °C/min and 0.1 ml^3/min, respectively. The composite were subjected to salt spray test according to ASTM B-117 for 72 h to obtain information of its corrosion resistance. Corrosion resistance was obtained by observing morphology of the samples using electron microscope instead of mass difference method.

3. Results and discussion

3.1. TGA Analysis

Thermal stability information of the samples can be obtained from TGA pattern. Figure 1 shows TGA curve of organoclay/epoxy resin composite with different content of organoclay at temperature 35 to 800 °C. The onset decomposition temperature of epoxy resin is different with various content of clay. This indicates that thermal stability is shifting. There are at least three stages of decomposition as shown in figure 1.

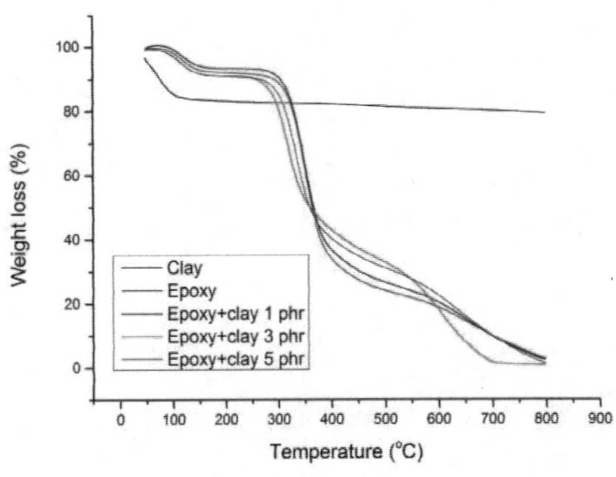

Figure 1. TGA pattern of organoclay/epoxy resin with different content of clay.

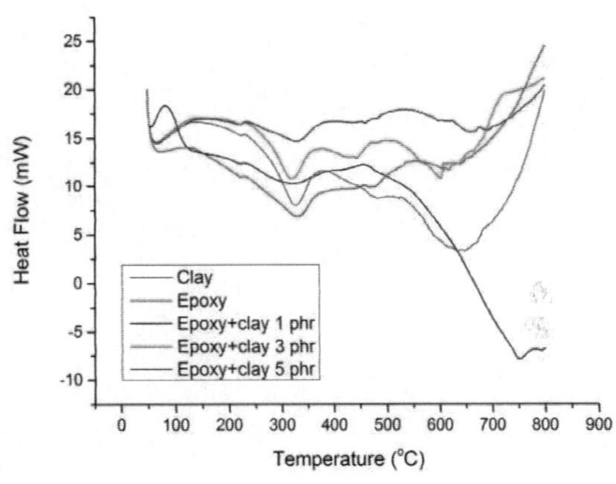

Figure 2. DSC pattern of organoclay/epoxy resin with different content of clay.

The onset and end set temperature of the four samples for three stages decomposition are each different. It can be concluded that the addition of organoclay influenced the thermal properties. Sample with 1 phr clay, below 325 °C, show the smallest weight loss, but above 325 °C the TGA curve shows gradual weight loss of all composites. It is also noted from the TGA that increase in weight percentage of nano-clay in epoxy resin shows reduction in weight loss of material at high temperature [16]. The sample with 1 phr clay showed optimum thermal stability.

3.2. DSC analysis

Figure 2 shows DSC curve of clay/epoxy composite with different clay content. Generally exothermic peaks involving weight loss, no weight change, and weight gain are known to be a result from combustion, crystallization, and oxidation, respectively. The same goes with endothermic peaks involving weight loss indicate decomposition or dehydration [16]. In the present study, the DSC curve indicate and exothermic peak around 250 °C and endothermic peaks around 300 °C and 600 °C, accompanied by weight loss of material. This indicates that decomposition of material occurs at temperature above 300 °C. The latest is probably due to clay content (5 phr). Addition of 1 phr clay showed the least heat needed for decomposition during heating process from room temperature to 800 °C. The optimum thermal stability of the samples is the composite with 1 phr clay content. Increasing amount of organoclay in the samples (3 and 5 phr) did not show good thermal stability probably due to agglomeration of organoclay, or poor dispersion in the samples.

3.3. FTIR analysis

Figure 3 describes FTIR curve of organoclay/epoxy resin with different content of clay. From the curve, it seems that the influence of clay addition on the absorption behavior is not quite clear.

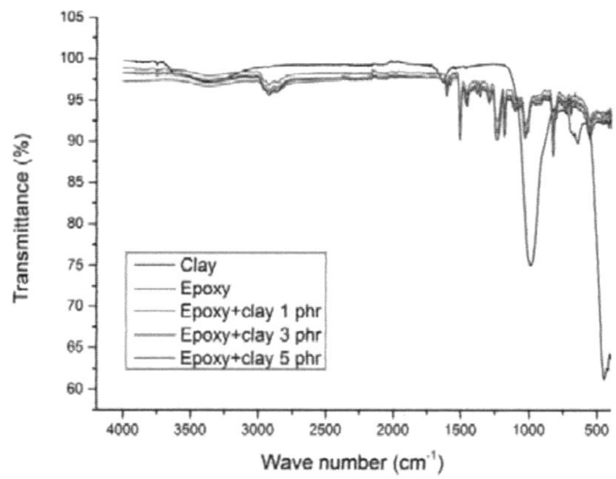

Figure 3. FTIR curve of Organoclay/epoxy resin with different content of clay.

Table 1 shows detailed wave number absorption of the samples. The wave number 3389.93 coming from the clay is affected and slightly shifting to higher wave number 3361.17 when mixed with epoxy resin. However, wave number 447.05 is shifting to the right (464.96) and then back to the left (459.39). Wave number 1383.16 appears only in the sample with addition of 1 phr clay. Addition of organoclay to the epoxy resin slightly affects infrared absorption. The shift of wave number might indicate a tighter structure and stronger bonds compared to neat epoxy. A shift toward high wave numbers also indicates decreased cross-linking for any nanoclay contents [17]. Figure 3 has clearly shown the optimum absorption of the sample, e.g. with 1 phr clay.

IOP Conf. Series: Journal of Physics: Conf. Series **985** (2018) 012037 doi:10.1088/1742-6596/985/1/012037

Table 1.The absorption wave number of organoclay/epoxy resin with different content of clay.

clay neat	epoxy neat	epoxy clay 1 phr	epoxy clay 3 phr	epoxy clay5 phr	Notes
3389.93	3351.35	3358.25	3354.53	3361.70	Amides bond, alcohol bond
	2923.42	2922.87	2923.05	2923.04	sp3 C-H
1630.78	1606.79	1606.86	1606.76	1606.85	alkenes bond
		1581.73			amides N-H bond
	1508.16	1508.04	1508.13	1508.10	nitro compound
	1456.63	1456.20	1456.70	1456.58	alkane (methylen group)
		1383.16			Aldehyde bond
	1361.79	1361.74	1361.51	1364.51	sulfoamide bond (S=O)
	1294.96	1295.01	1294.90	1294.81	Alcyl bond (C-O)
	1235.01	1235.10	1235.30	1235.01	Amine bond
	1181.22	1181.22	1181.25	1181.22	tertiary alcohol bond
	1105.88	1105.89			secondary alcohol bond
	1034.26	1033.31	1033.65	1032.22	
988.18					alkene bond (monosubtituted)
	826.79	826.73	826.87	828.69	alkene bond (trisubtituted)
	735.10	734.84	734.53	734.53	alkene bond (Cis)
	697.42	697.39	697.37	697.46	alkene bond (cis)
673.16					
647.97					
	556.76	558.79	558.51	555.71	
447.05		458.99	464.96	459.39	
419.60		419.06			
402.96					

3.4. Thermal conductivity

Table 2 shows thermal conductivity of organoclay-resin composite. The measurement was taken 8 times for each sample. From table 2, thermal conductivity of the samples increases with increasing organoclay content. This increase is due to addition of clay as a filler, and the fact that they have higher thermal conductivity than epoxy.

Table2. Thermal conductivity of organoclay-resin composite (watt/m K).

No	Epoxy neat	Epoxy+Clay 1 phr	Epoxy+Clay 3 phr	Epoxy+Clay 5 phr
1	0.2553	0.2619	0.2553	0.2627
2	0.2516	0.2767	0.2677	0.3411
3	0.2519	0.2667	0.2745	0.3341
4	0.2369	0.2642	0.2540	0.2241
5	0.2452	0.2627	0.2681	0.3103
6	0.2310	0.2439	0.2407	0.2788
7	0.2301	0.2552	0.2445	0.2626
8	0.2134	0.2238	0.2526	0.2706
AVG	0.2394	0.2569	0.2572	0.2855

The increase in average thermal conductivity may be due to cross link and also thermal conductivity of organoclay which is higher than epoxy.

3.5. Corrosion test

Corrosion resistance test of the samples were conducted by salt spray method for 72 hours. Figure 4 show the surface morphology after corrosion test with 0, 1, and 3 phr clay addition.

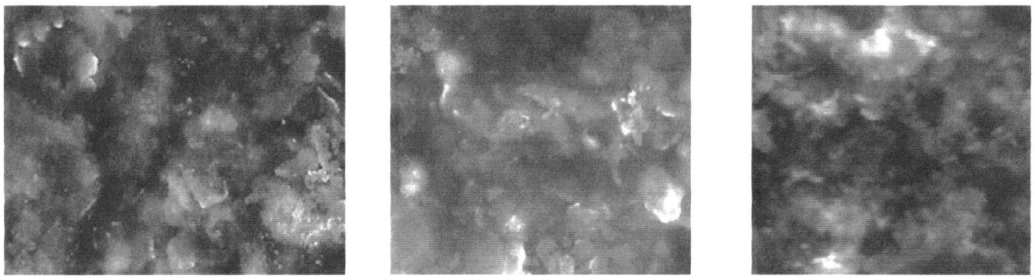

(a) (b) (c)

Figure 4. Surface morphology of organoclay/epoxy resin with (a) 0, (b) 1 and (c) 3 phr clay after salt spray test for 72 hours (magnification 1000x)

Different surface morphology provides evidence that addition of clay influences how epoxy resin changes its corrosion resistance. All of these may due to interaction between clay and epoxy as matrix. From figure 4, it is clearly shown that the addition of 1 phr organoclay has shown the best corrosion resistance.

4. Conclusion

Composite organoclay/epoxy resins with different content of clay have been well prepared. Thermal stability of the samples with 1 phr clay content shows the best performance. Increasing organoclay content in the composite organoclay/epoxy resin also increases the thermal conductivity. Corrosion resistance of composite change with addition of organoclay 1 phr shows the best result. Overall, the influence of

IOP Conf. Series: Journal of Physics: Conf. Series **985** (2018) 012037 doi:10.1088/1742-6596/985/1/012037

organoclay on the properties of clay/epoxy resin composite are clearly seen with 1 phr organoclay. These may due to crosslink of clay and epoxy as a matrix

Acknowledgment

This research is supported financially by "Hibah Publikasi Internasional Terindeks untuk Tugas Akhir Mahasiswa University of Indonesia" through letter of decree no.630/UN2.R3.1/HKP.05.00/2017.

References

[1] Fan Q C, Wu C Z and Zhang L J 2008 *Technol. Res. Dev.* **23** 17
[2] Teh P L, Jaafar M and Akil H M 2008 *Polym. Adv. Technol.* **19** 308
[3] Koerner H, Hampton E, Dean D, Turgut Z, Drummy L, Mirau P and Vaia R A 2005 *Chem. Mater.* **17** 1990
[4] Wang D, Jiang M T, Zheng D M, Cheng B and Huang Z 2006 *Journal of Materials Engineering* **1** 417
[5] Zhang L and Wei Y B 2005 *Journal of Chemical Engineering of Chinese Universities* **19** 720
[6] Zhang D T 2002 *Thermoset. Resin* **17** 44
[7] Chen G X, Kim H S, Shim J H and Yoon J S 2005 *Macromolecules.* **38** 3738
[8] Tamura K, Yokoyama S, Pascua C S and Yamada H 2008 *Chem. Mater.* **20** 2242
[9] Fathurrohman M I, Soegijono B and Budianto E 2015 *Macromol. Symp.* **353** 62
[10] Hussain M, Varley R J, Mathys Z 2004 *J. Appl. Polym. Sci.* **91** 1233
[11] Zang Y, Xu W, Liu G, Qiu D and Su S 2009 *J. Appl. Polym. Sci.* **111** 813
[12] Lee J Y, Shim M J and Kim S W 1997 *Mater. Chem. Phys.* **48** 36
[13] Miyagawa H, Rich M J and Drzal L T 2004 *J. Polym. Sci. B* **42** 4384
[14] Zheng Y, Chonung K, Wang G, Wei P and Jiang P 2009 *J. Appl. Polym. Sci.* **111** 917
[15] Stone G C and Boulter E A 2004 Electrical Insulation for Rotating Machines *Design, Evaluation, Aging, Testing, and Repair* Chapter 1 (Canada: IEEE Press) pp 173–214
[16] Sarathi R, Sahu RK and Rajeshkumar P 2007 *Mater. Sci. Eng. A* **445-446** 567
[17] Nyczyk A, Paluszkiewicz C, Hasik M, Cypryk M and Pospiech P 2012 *Vib. Spectrosc.* **59** 1

IOP Conf. Series: Journal of Physics: Conf. Series **985** (2018) 012038 doi:10.1088/1742-6596/985/1/012038

Development of $MoSi_2$ coating with Al doping by using high energy milling method

C M S Simanjuntak[1], S Hastuty[2], H Izzuddin[3], R Sundawa[3], T Sudiro[3], A Sukarto[3] and K A Z Thosin[3]

[1]Engineering Physics Department, Surya University, Tangerang, Banten 15810 Indonesia
[2]Mechanical Engineering Department, Universitas Pertamina, South Jakarta, Jakarta, 12220 Indonesia
[3]Research Center for Physics, Indonesian Institute of Sciences, Komplek Puspitek, Serpong, Tangerang Selatan 14314, Indonesia

E-mail: hubby.izzuddin@lipi.go.id

Abstract. $MoSi_2$ is well known as a material for high temperature application because it has high oxidation and corrosion resistance. The aim of this research is to develop $MoSi_2$ coating with Al doping on Stainless Steel 316 (SS316) substrate using High-Energy Milling method. Aluminium is added to the coating as a dopant to increase formation of $MoSi_2$ coating layer on the substrate. The variations used here based on the concentrations of doping Al (at.%) and duration of milling. Results show that the $MoSi_2$ coatings with variations of 30 and 50 at.% of Al doping and 3 and 6 hours of milling times were successfully coated on the surface of SS 316 using the high-energy milling method. The most optimum coating result after oxidation test at 1100 °C for 100 hours is shown by $MoSi_2$-30%Al with 3 hours of milling times. From the oxidation results, the Al doping into $MoSi_2$ coating was able to increase the oxidation resistance of the SS 316 substrate.

1. Introduction

$MoSi_2$ is a promising high-temperature material due to its density (6.31 g / cm^3) and can operate at high temperatures because $MoSi_2$ has high melting point of 2020 °C. In addition, $MoSi_2$ has a good oxidation resistance at high temperatures. Through thin-layer formation of SiO_2, $MoSi_2$ could protect materials from continuous oxidation [1]. Some coating methods have been successfully applied for $MoSi_2$ coatings, such as pack siliconizing, plasma spraying, and chemical vapor deposition [2-4].

For powder materials, mechanical alloying with high-energy milling is one of the most efficient techniques in synthesizing nanomaterials and preparing nanocrystalline alloying. In previous studies, it had been demonstrated that mechanical alloying could also be used in coating processes such as Ni-Al coating [5-7], Fe-Al coating [8-9], and Cr-Al coating [10]. During mechanical alloying with ball milling process, the ball mills together with powder particles collide with the substrate surface with repeated high energy impact resulted in the powder deposition on its surface [11].

Aluminum was a metal element that showed great interest to the oxygen. This was due to the formation of alumina (Al_2O_3) in high temperature. The Al_2O_3 oxide layer became a good protective layer on the surface of the substrate [12]. The addition of Al for coating process affected significantly on the resulted coating formation [8-9].

This research was conducted with the aim to study factors that influence coating process of $MoSi_2$ on Stainless Steel 316 substrate using high-energy milling method with variation of milling time process and concentration of Al (at.%) as doping. In addition, coating on the SS 316 substrate would be tested for oxidation at 1100 °C, which then would be analyzed for its oxidation resistance.

2. Experimental Method

2.1. Material and Coating Process

$MoSi_2$ and Al powder were used as a coating material for Al-$MoSi_2$ coating. Stainless Steel 316 plates with dimension of 10×10×3 mm^3 were used as a substrate. Prior to the coating process using high-energy milling method, SS 316 substrates were polished using carbide papers up to 1200 grit. The SS 316 substrates were subsequently cleaned using ultrasonic bath to remove impurities from the surface of the substrate.

The high-energy milling process was carried out in the air atmosphere using shaker mill with vials and stainless steel ball mills. The Al-$MoSi_2$ powder mixture was prepared by mixing $MoSi_2$ powder with Al. Initially, $MoSi_2$ and Al powder with composition of 30 at.% and 50 at.% Al were mixed using shaker mill for 1 hour with 9:1 ball to powder ratio. For the coating process, the SS 316 substrate was milled together with ball mills and mixture of Al-$MoSi_2$ in a vial with variations of 3 and 6 hours of milling times.

2.2. Oxidation Test

The substrate dimension and weight were measured before the oxidation test. The cyclic oxidation test was performed for all coated substrates under air atmosphere with temperature of 1100 °C for 100 hours in a muffle furnace. For each cycle of 20 hours of oxidation time, the samples were removed from the furnace, cooled down to room temperature, and the mass gain was measured, without the spalled oxidation. This cycle was repeated for 5 times until it reached a total of 100 hours of oxidation test.

2.3. Characterization

The coated samples were characterized using optical microscope, X-ray diffraction (XRD), and scanning electron microscope equipped with energy dispersive X-ray analysis (SEM-EDX). The optical microscope was used to view and make sure the samples were free from scratches before the sample was observed using the SEM-EDX. XRD characterization was used to determine phase formation in the deposited $MoSi_2$-Al layer on the surface of SS 316. SEM-EDX characterization was used to observe sample morphology at micro scale and distribution of the detected elements in the sample.

3. Results and discussions

3.1. Coating results with high-energy milling method

XRD characterization results shows that $MoSi_2$ coating with Al doping was successfully formed on the surface of SS 316 substrates after 3 and 6 hours of milling time. It could be seen in figure 1 that all diffraction peaks of substrate surfaces after milling process mostly consisted of $MoSi_2$, and some Mo_3SAl and FeCr phases. The detected FeCr phases are main compositions of the 316 substrates. From the XRD results, Al on this coating was found in the form of Mo_3Al [13].

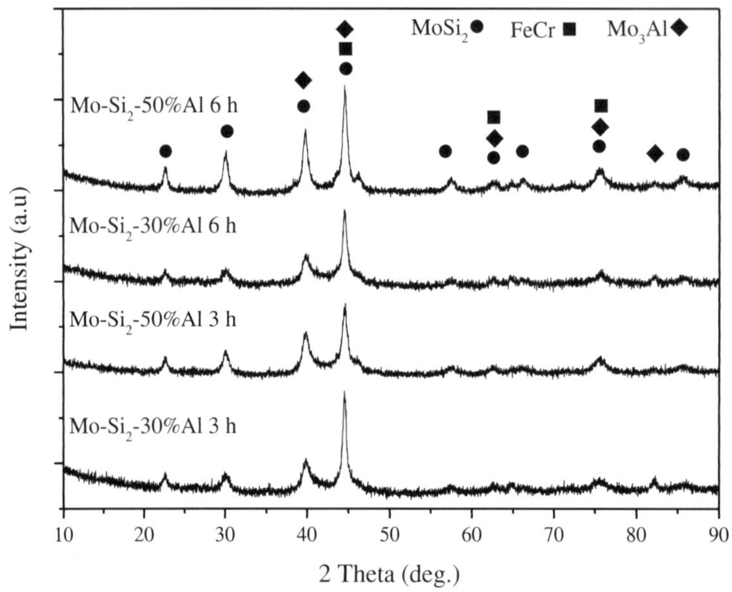

Figure 1. XRD pattern of $MoSi_2$-Al coating on the 316 substrates after 3 and 6 hours of milling time.

In order to support the XRD results, the cross section of the sample surface was studied using SEM. SEM images on figure 2 reveals that the $MoSi_2$-Al had been successfully deposited on the surface of SS 316, but with some porous and rough surfaces. $MoSi_2$-30%Al coating layer is less porous and rough than that of $MoSi_2$-50%Al coating. The average coating thickness of $MoSi_2$-30Al and $MoSi_2$-50Al was 25.69 µm and 38.32 µm, respectively. The coating layers of $MoSi_2$-Al formed were rough and clearly distinguishable between the coating layer and the substrate. The coating layer was formed by repeated collisions occurring between ball mills, mixed $MoSi_2$-Al powder particles, and the SS 316 substrate.

Figure 2. SEM images of the cross-sectional 3 h milling $MoSi_2$-Al coating (a) with 50% at.% Al (b) 30% at.% Al.

3.2. Oxidation test

Figure 3 shows the mass change of the SS 316 substrates with $MoSi_2$-Al coating after oxidation test at 1100 °C in air. From all samples, the oxidation rate in the $MoSi_2$-50%Al - 3h sample has relatively the lowest mass decrease compared to the other samples, followed by the $MoSi_2$-30%Al - 3h sample.

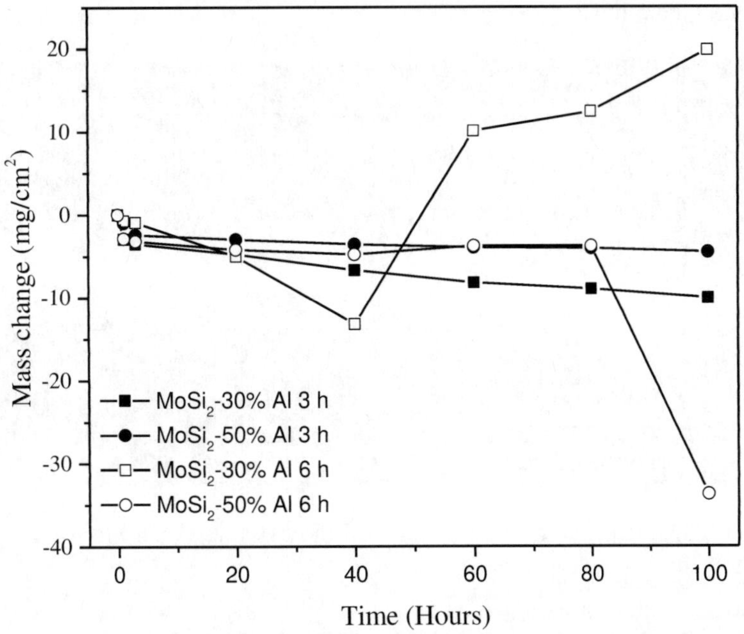

Figure 3. Mass gain after oxidation at 1100 °C in air of the SS 316 substrates with $MoSi_2$-Al coating.

At beginning of the oxidation test until 40 hours, all samples had mass loss due to the instability of the MoO_3 formed during the oxidation process and the volatile of MoO_3 that made a way for oxygen to reacted with the substrate [14].

After 40 hours, samples with 3 hours milling time were able to minimize the mass loss unlike samples with 6 hours of milling time that showed irregular trends. For the $MoSi_2$-50%Al 6h sample, the mass reduction was relatively low and stable until 80 hours of oxidation test. However, after 80 hours the sample mass decreased significantly. As the $MoSi_2$-30%Al - 6h sample, after 40 hours, the sample mass increased significantly until the end of oxidation test. Although the sample mass increased, figure 4 shows that sample surface began to crack after 60 hours of oxidation process.

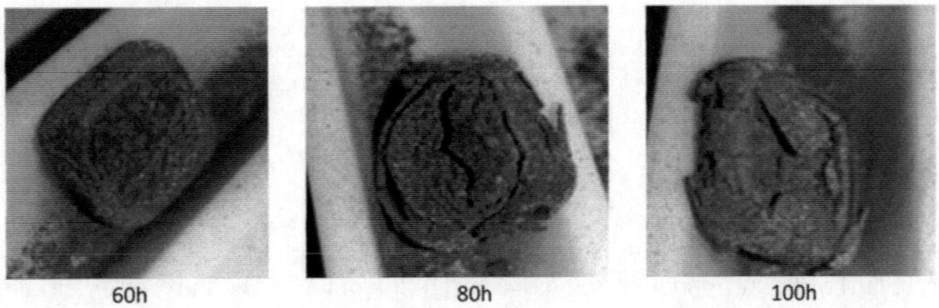

Figure 4. $MoSi_2$-30%Al-6h coating layer changing during oxidation at 1100 °C.

The surface morphologies with EDS mapping of a cross section of the 316 substrate with $MoSi_2$-50%Al-3h after oxidation test at 1100 °C for 100 hours is shown on figure 5. After 100 hours of oxidation test, the $MoSi_2$-50%Al-3h surface consisted of 3 layers. Layer 1 is identified as a thick layer of mixed SiO_2 and Al_2O_3 (point 1). Layer 2 consisted of thin layer of Al_2O_3 (point 2). Meanwhile, at layer 3, thin internal oxidation of mixed $MoSi_2$ – Al is formed (point 3). The detailed composition of each layer is shown in table 1.

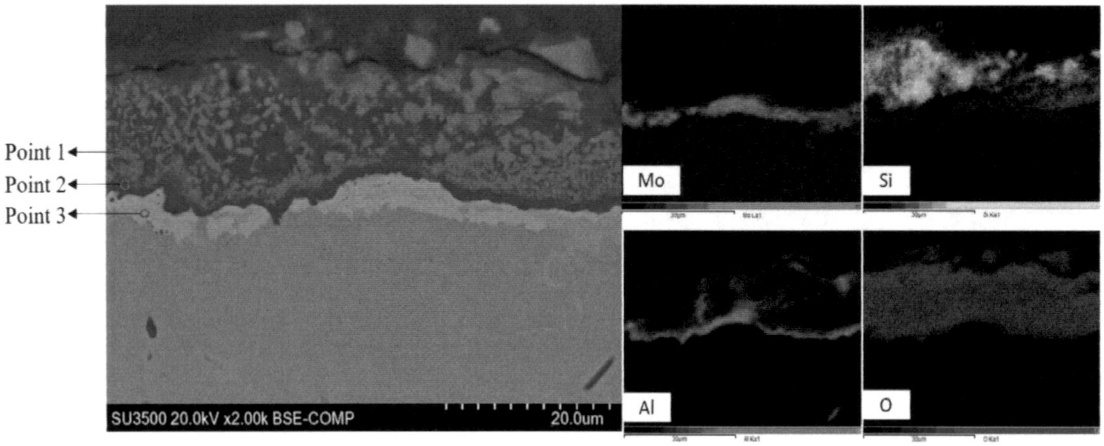

Figure 5. SEM and EDX elemental mapping images of the cross-sectional $MoSi_2$-50%Al -3h coating after oxidation test at 1100 °C.

Table 1. Element compositions from the EDX elemental analysis of $MoSi_2$-50%Al - 3h after oxidation.

Elements	Point 1 (at.%)	Point 2 (at.%)	Point 3 (at.%)
Al	3,862	25,498	4,418
Mo	-	-	48,886
Si	13,345	1,252	23,506
O	82,74	73,251	23,118

Formation of double SiO_2 and Al_2O_3 layers on the substrate surface with $MoSi_2$-50%Al-3h coating during oxidation test enables the sample to minimize the mass loss during the oxidation test. However, since the layer 1 is not formed as a completely closed layer, this correlates with the oxidation results that the substrate had still mass loss during oxidation test. Thus, the substrate had relied on the thin protective Al_2O_3 layer against the oxidation attack. This happened because $MoSi_2$ has high coefficient thermal expansion mismatch [15]. With the high-energy milling process, this affected more with the oxidation results of coated $MoSi_2$-Al coating substrates. This explains the results of oxidation test of the substrates with 6 hours of milling times.

4. Conclusions

The $MoSi_2$ coatings with variations of 30 and 50 at. % of Al doping and 3 and 6 hours of milling times were successfully coated on the surface of SS 316 using high-energy milling method. Coating thickness increased with increasing atomic concentration from Al. The most optimum coating result after oxidation test at 1100 °Cfor 100 hours is $MoSi_2$-30%Al with 3 hours of milling times. From the oxidation results, the Al doping into $MoSi_2$ coating was able to increase the oxidation resistance of the SS 316 substrate.

Acknowledgments
The author gratefully acknowledges the support from Research Center for Physics of Indonesian Institute of Sciences.

References

[1] Liu Z D, Hou S X, Liu D Y, Zhao L P, Li B and Liu J J 2008 *Surf. Coat. Technol.* **202** 2917
[2] Tiwari R, Herman H and Sampath S 1992 *Mater. Sci. Eng.* **155** 95
[3] West G A and Beeson K W 1989 *J. Electrochem. Soc.* **135** 1752
[4] Yoon J K, Lee K H, Kim G H, Lee J K, Doh J M and Hong K T 2004 *J. Electrochem. Soc.* **151** B309
[5] Zadorozhnyy V Y, Shahzad A, Pavlov M D, Kozak D S, Chirkov A M, Zagrebin D S, Khasenova R S, Komarov S V and Kaloshkin S D 2017 *J. Alloys Compd.* **707** 351
[6] Chen C, Feng X and Shen Y 2017 *Surf. Coat. Technol.* **321** 8
[7] Saba F, Kabiri E, Khaki J V and Sabzevar M H 2016 *Powder Technol.* **288** 76
[8] Aryanto D, Sudiro T and Wismogroho A S 2016 *AIP Conf. Proc.* **1725** 020008
[9] Aryanto D, Wismogroho A S and Sudiro T 2016 *J. Phys.: Conf. Ser.* **739** 012132
[10] Hia A I J, Sudiro T, Aryanto D and Sebayang K 2016 *J. Phys.: Conf. Ser.* **739** 012131
[11] Romankov S, Sha W, Kaloshkin S D and Kaevitser K 2006 *Surf. Coat. Technol.* **201** 3235
[12] Sheasby P and Pinner R 2001 *The Surface Treatment and Finishing of Aluminium and Its Alloys* 6th Edition (Ohio: ASM International) pp 584-596
[13] Materials Data on AlMo3 (SG:223) by Materials Project https://materialsproject.org/docs/calculations (last accessed on september 2015)
[14] Yan J, Xu H, Zhang H and Tang S 2009 *Rare Metals* **28** 418
[15] Cook J, Khan A, Lee E and Mahapatra R 1992 *Mater. Sci. Eng. A* **155** 183

The reactive element effect of yttrium and yttrium silicon on high temperature oxidation of NiCrAl coating

S Ramandhany[1,2], E Sugiarti[1], R D Desiati[1], E Martides[3], E Junianto[3], B Prawara[3], A Sukarto[1] and A Tjahjono[2]

[1] Laboratory of High Temperature Materials and Coatings, Research Centre for Physics, Indonesian Institute of Sciences (LIPI), Serpong 15314, Indonesia
[2] Physics Division, Faculty of Science and Technology, State Islamic University of Jakarta 15411, Indonesia
[3] Research Center for Electrical Power and Mechatronics, Indonesian Institute of Sciences (LIPI), Komplek LIPI Jl. Sangkuriang, Building 20, Bandung, West Java 40135, Indonesia

E-mail: enis002@lipi.go.id

Abstract. The microstructure formed on the bond coat affects the oxidation resistance, particularly the formation of a protective oxide layer. The adhesion of bond coat and TGO increased significantly by addition of reactive element. In the present work, the effect of yttrium and yttrium silicon as reactive element (RE) on NiCrAl coating was investigated. The NiCrAl (without RE) and NiCrAlX (X:Y or YSi) bond coating were deposited on Hastelloy C-276 substrate by High Velocity Oxygen Fuel (HVOF) method. Isothermal oxidation was carried out at 1000 °C for 100 hours. The results showed that the addition of RE could prevent the breakaway oxidation. Therefore, the coating with reactive element were more protective against high temperature oxidation. Furthermore, the oxidation rate of NiCrAlY coating was lower than NiCrAlYSi coating with the total mass change was ±2.394 mg/cm^2 after 100 hours of oxidation. The thickness of oxide scale was approximately 1.18 μm consisting of duplex oxide scale of spinel $NiCr_2O_4$ in outer scale and protective α-Al_2O_3 in inner scale.

1. Introduction

Hastelloy X as a common Ni based superalloy is mostly used in gas turbines that operate at elevated temperature. Ni-based superalloy substrate has good mechanical properties at high temperature. The microstructure superalloy composes of high stable austenite containing significant amounts of alloying elements such as Cr, Mo, Al, Ti, Fe, and C [1]. It is well known that coating on the surface of Ni-based superalloy is required to improve the oxidation resistance. Thermal Barrier Coating (TBC) is widely applied in gas turbine. This system consists of a bond coat, a Thermally Grown Oxide (TGO), and a ceramic top coat. Bond coat with formula of MCrAlY (M: Ni, Co, Fe, or both of these three, Y: reactive element) provides a good protection against high temperature oxidation [2, 3]. Many studies have stated that oxidation resistance occurs due to the TGO conditions. A continuous, adherence, and slow-growing of TGO protects high temperature components against oxidation [4].

Naumenko et al. studied that the addition of reactive element (e.g. Y, Si, Hf, Zr and La) could increase the TGO adherence [4]. Each reactive element has a different effect on TGO performance. Yttrium promotes formation of oxide pegs at interface between oxide scale and the underlying coating, and prevents scale rumpling [5,6]. In addition, silicon can produce protective oxide scale by formation

IOP Conf. Series: Journal of Physics: Conf. Series **985** (2018) 012039 doi:10.1088/1742-6596/985/1/012039

of SiO_2. Protective scale of silicon dioxide has a good oxidation resistance as well as alumina in high temperature. Therefore, a lot of studies of silicon addition on bond coat has been conducted for Ni-based supperalloy [7-9]. However, the comparison of oxidation resistance in high temperature between NiCrAl with Y and YSi has not yet explored.

In the present study, the effect of Y and YSi addition on NiCrAl bond coat under oxidation test in air at elevated temperature of 1000 °C for 100 hours was investigated. The microstructure of bond coat with and without yttrium and yttrium-silicon provided information on the effect of Y and YSi reactive elements addition on NiCrAl coating that deposited on Hastelloy C-276.

2. Experimental procedures

Hastelloy C-276 substrate with composition balNi-15Mo-14Cr-4Fe-3W-2.5Co was cut into dimensions of 15×15×1.6 mm. The substrate was polished with a silicon carbide sand paper of #100 until final finishing of #1200. NiCrAlX coating powder was prepared by mechanical milling for 36 hours. The milling process produced an agglomerated particle with an average particle size of 130 μm. Table 1 shows the composition of coating powder.

Table 1. Composition of coating powder.

Coating	Composition (wt.%)				
Powder	Ni	Cr	Al	Y	Si
NiCrAlSi	Bal	24	7		-
NiCrAlY	Bal	24	7	0.4	-
NiCrAlYSi	Bal	24	7	0.4	0.4

Table 2. The parameters of HVOF coating.

N_2 pressure	4.8 bar
O_2 pressure	8.2 bar
Propane pressure	5.2 bar
Air Screw Compressor	4 bar
Fuel gas	60 lt/min

NiCrAlX was deposited using High Velocity Oxy Fuel (HVOF) method. This technology exhibits good performance on high adhesivity, low porosity, and high density [9]. Table 2 shows the parameter used in HVOF thermal spraying method. Afterwards, the coated samples were exposed to heat treatment in a vacuum (20 Pa) at 1100 °C for 4 hours in order to increase the adhesion of coating to the substrate. Isothermal oxidation was carried out at 1000 °C for 100 hours in air condition.

The heat-treated and oxidized samples were analyzed by X-Ray diffractometer (Rigaku Smart Lab Type) to identify phases. Surface morphology and micro-structural characterizations were carried out using a Field Emission Scanning Electron Microscope (FE-SEM) equipped with Energy Dispersive Spectroscopy (EDS) (JEOL JIB4610F).

3. Results and discussion

3.1. Isothermal oxidation

Figure 1 shows the mass change of NiCrAlX coating. The mass change was grouped into 3 stages as shown in table 3. The result shows that increasing in mass gain at initial stage of oxidation is due to the rapid formation of meta-stable oxide on the coating surface. Mass steady indicated low formation of oxides and less diffusion of oxygen to form oxides. Result in table 3 reveals that the mass steady of NiCrAlY specimen was lower than NiCrAl and NiCrAlYSi samples. According to figure 1, the scale spallation only appeared on NiCrAl of oxidized sample after 31 hours exposure and it indicated the failure protective scheme. It is also verified by the appearance of photograph sample which indicated the scale spalling occurred only on NiCrAl specimen as shown in figure 2. It is concluded that NiCrAlY sample had the best oxidation resistance due to the lowest mass change after 100 hours of oxidation exposure.

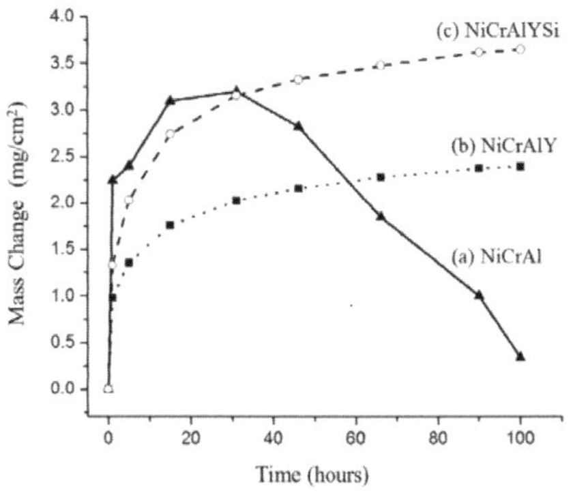

Figure 1. Mass change after isothermal oxidation in 1000 °C of three types coated sample (a) NiCrAl, (b) NiCrAlY, and (c) NiCrAlYSi.

Table 3. Three stages of mass change.

Stage	Mass Change (mg/cm^2)		
	NiCrAl	NiCrAlY	NiCrAlYSi
mass gain	3.2019	2.0247	3.1556
mass steady/loss	-2.86	0.369	0.493
total	0.342	2.394	3.649

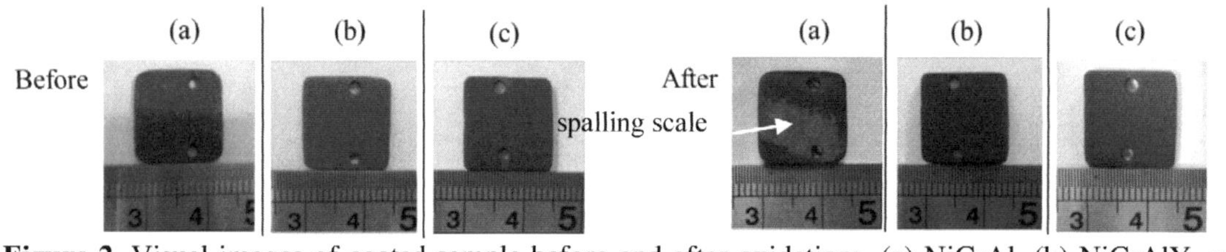

Figure 2. Visual images of coated sample before and after oxidation: (a) NiCrAl, (b) NiCrAlY, and (c) NiCrAlYSi.

3.2. Surface morphology of oxide scale

Figure 3 shows the surface morphology of three types of coated samples before and after oxidation. It revealed that surface samples without reactive elements contained unmelted particles, whereas the samples with the addition of reactive element did not contain unmelted particles (figure 3 top). The surface morphology changes gradually during oxidation process due to the interaction with oxygen. Consequently, a typical cauliflower-like structure with varying sizes were observed on the surface of all oxidized samples [10].

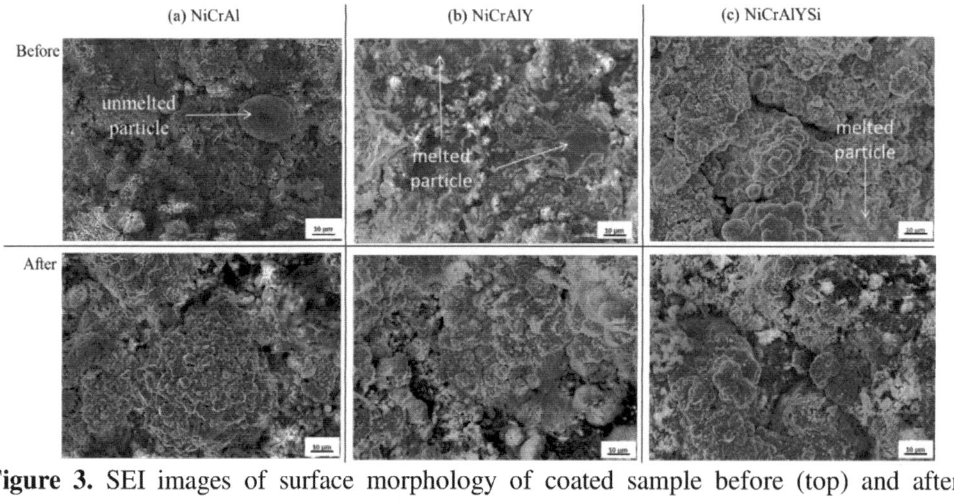

Figure 3. SEI images of surface morphology of coated sample before (top) and after oxidation (bottom): (a) NiCrAl, (b) NiCrAlY, and (c) NiCrAlYSi.

3.3. Microstructure of oxide scale

Figure 4 shows the cross sectional structure of coated sample before and after oxidation. Coating thicknesses are measured above 200 µm. The sample consists of two layers which is secondary reaction zone (SRZ) and NiCrAl/+RE coating layer. Precipitates are formed in SRZ layer between substrate and coating layer due to the low solubility of Mo and Cr elements from Hastelloy substrate [10,11]. These precipitates improve the creep resistance of the substrate due to their grain boundary pinning action [1]. However, excessive quantities of SRZ formation decreases the mechanical properties of the substrate, thus it leads to unwanted phase formation [10,11]. The precipitate zones in NiCrAl decompose an amount of chromium and will cause spalling on the bond coat. On the contrary, in NiCrAlYSi sample, the secondary reaction zone is lower than the other two samples of NiCrAl and NiCrAlY due to the existence of silicon. Silicon as a reactive element could prevent outward diffusion of refractory elements of Mo from the substrate [7]. Therefore, a refractory precipitate was just observed in NiCrAl and NiCrAlY, which was indicated by bright gray area in the vicinity of coating interface.

Furthermore, a refractory precipitate is remained in NiCrAl sample after oxidation. It might aggravates internal stress at the coating/substrate interface which could result in mechanical failure of the coating system [3]. This result corresponds to the mass loss as shown in figure 1 for NiCrAl sample. However, the spalling area is not occurred on entire of oxidized sample. Therefore, the coating and oxide layer could observe as shown in figure 4a. Spallation of coating layer was also due to the excessive intensity of the oxide interface in the NiCrAl sample, which worsens the internal stress and causes TB failure [3]. In the case of NiCrAlY and NiCrAlYSi samples, oxide is observed not only on the top surface but also in coating layer indicated as black area in figure 4.

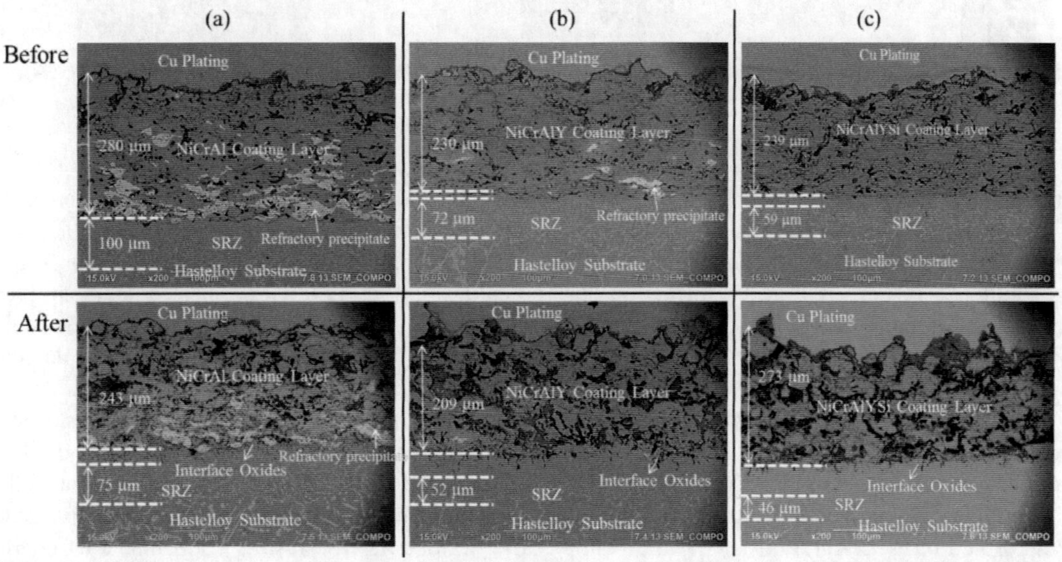

Figure 4. BSE images of cross sectional structure of coated sample before (top) and after oxidation (bottom): (a) NiCrAl, (b) NiCrAlY and (c) NiCrAlYSi.

Figure 5 shows cross sectional structure of oxide scale on the top layer (TGO) after 100 hours oxidation. The thickness of the oxide scale is about 1.03 µm, 1.18 µm, 1.17 µm on the top layer of NiCrAl, NiCrAlY, and NiCrAlYSi, respectively. According to the elemental mapping, the oxide scale consists of Cr, Ni, Al, and O constituents. This oxide scale is formed with duplex scale of spinel structure in the outer scale and stable alumina in the inner scale as indicated by arrow marked in figure 5. The detrimental mixed oxides are found to be preferred sites of crack nucleation, leading to premature TBC failure on the bond coat for a long time oxidation exposure [13]. Rumpling, continuous, and thin TGO layer are found in NiCrAl sample as shown in figure 5a. The rumpling of TGO can initiate the failure of coating system [6] thus it is also confirmed that NiCrAl sample has less oxidation resistance. Moreover, thick and continuous TGO layer is observed in NiCrAlY sample. Mixed oxide is randomly distributed on the outer scale. Elemental mapping analysis also indicates that high intensity of aluminium and oxygen covered the NiCrAlY sample figure 5b bottom . It is assumed that stabilized alumina was formed in TGO layer. In contrast, a quite thick mixed oxide and discontinuous TGO layer is observed in NiCrAlYSi sample. Silicon addition in this coating sample could hardly prevent the formation of thick mixed oxide.

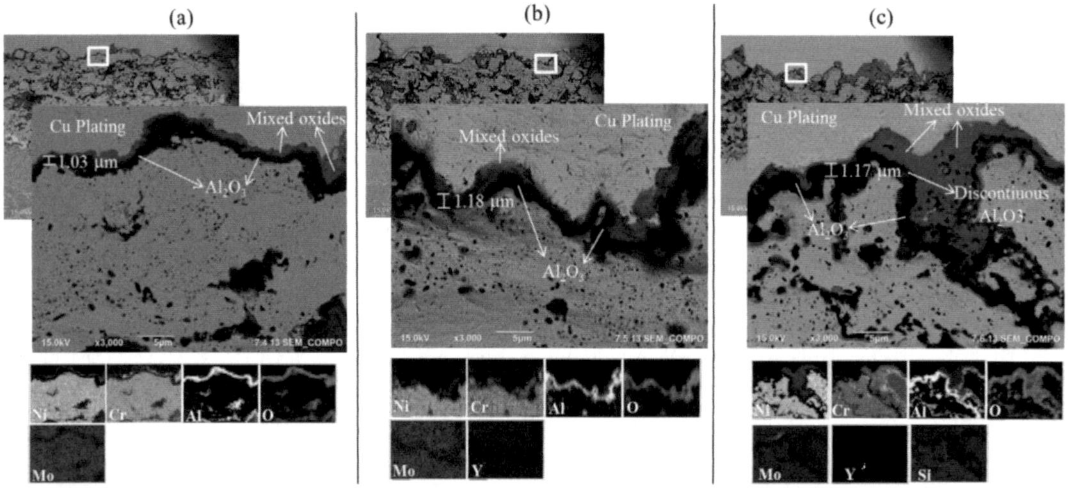

Figure 5. Enlargement images on top layer of coated sample after 100 hours oxidation: (a) NiCrAl, (b) NiCrAlY, and (c) NiCrAlYSi.

3.4. Phase formation

Figure 6 shows the XRD patterns before and after oxidation of NiCrAl (green), NiCrAlY (red), and NiCrAlYSi (blue) coated samples. The results showed that γ-Ni and γ'-Ni_3Al with cubic structure formed in all coated samples. The phase of γ-Ni increase mechanical properties [14] whereas γ'-Ni_3Al contributes to promote the Al_2O_3 formation during heating period [11]. On the heat treated sample, a hexagonal structure of Cr_2O_3 only observed in the sample without reactive element. The stable alumina (α-Al_2O_3) with rhombohedral structure was formed on NiCrAlY and NiCrAlYSi sample while NiCrAlYSi and NiCrAl sample identified meta-stable alumina θ-Al_2O_3 (monoclinic structure).

$NiCr_2O_4$ as a spinel oxide was formed in all oxidized samples and it confirmed the existence of mixed oxide with spinel structure in the outer scale of figure 5. The expected phase on oxidized sample was protective oxide. However, on NiCrAl sample after oxidation could not be identified as protective oxide, unlike on the spallation part such as meta-stable and stable alumina. Protective oxide was formed on NiCrAlY and NiCrAlYSi sample after oxidation. The results confirmed that coating with reactive element has good oxidation protective than coating without reactive element.

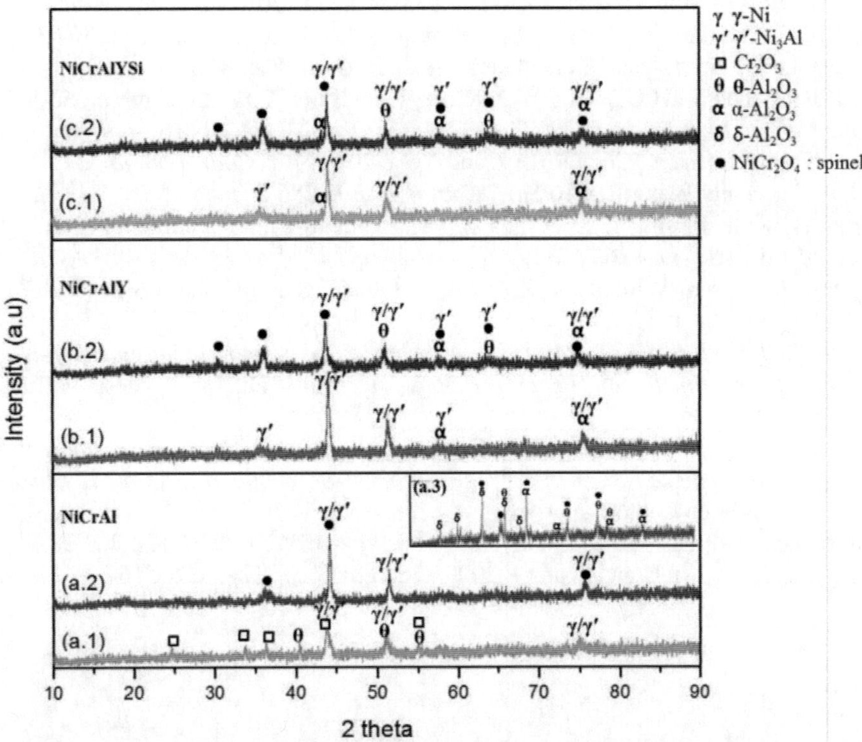

Figure 6. XRD patterns of the coated sample: (a.1) before (a.2) after oxidation and (a.3) spallation part of NiCrAl, (b.1) before and (b.2) after oxidation of NiCrAlY, (c.1) before and (c.2) after oxidation of NiCrAlYSi.

4. Conclusions

NiCrAlX coating was successfully developed on Hastelloy C-276 substrate by HVOF thermal spraying method. The following conclusions can be drawn from the results:

(1) Yttrium addition has improved the oxidation resistance of NiCrAl coatings.

(2) Silicon addition could prevent outward diffusion of Mo from the substrate and it increased the adherence of bond coat interface. However, silicon hardly suppresses the formation of thick spinel oxide.

(3) TBC failure was indicated by spallation of coating layer in NiCrAl oxidized sample which consisting of meta-stable and stable alumina after 31 hours oxidation.

(4) Meta-stable θ-Al_2O_3 and stable α-Al_2O_3 were formed in NiCrAlY and NiCrAlYSi oxidized sample, however a continuous of stable α -Al_2O_3 was only formed in NiCrAlY coated sample.

(5) NiCrAlY had the best oxidation resistance with total mass change of ±2.394 mg/cm^2 after 100 hours of oxidation.

Acknowledgments

The authors would like to thank to Kegiatan Unggulan LIPI, INSINAS IRPI grant No. 04/INS-2/PPK/E/E4/2017 and 2016 Toray Science Technology Research Grant for financial support of this work.

References

[1] Osorio JD, Toro A and Hernandes-Ortiz JP 2012 *Dyna* **79** 149

[2] Wang H, Liu Y, Ning X, Wang Q, Wang F and Chen D 2017 *Powder Metall.* **60** 15
[3] Liu YZ, Hu XB, Zheng SJ, Zhu YL, Wei H and Ma XL 2015 *Mater Des.* **80** 63
[4] Naumenko D, Pint BA and Quadakkers WJ 2016 *Oxid. Met.* **86** 1
[5] Ma J, Jiang S M, Li H Q, Wang W X, Gong J and Sun C 2011 *Corros. Sci.* **53** 1417
[6] Wang J, Chen M, Yang L, Liu L, Zhu S, Wang F and Meng G 2016 *Appl. Surf. Sci.* **366** 245
[7] Cao J, Zhang J, Hua Y, Chen R, Li Z and Ye Y 2017 *Surf. Coat. Technol.* **311** 19
[8] Ren P, Zhu S and Wang F 2016 *Surf. Coat. Technol.* **286** 262
[9] Sarjas H, Kulu P, Juhani K, Viljus M, Matikainen V and Vuoristo P 2016 *Proc. Estonian Academy of Sciences* **65** 101
[10] Sugiarti E, Zaini K A, Sundawa R, Wang Y, Ohnuki S and Hayashi S 2017 *J. Phys.: Conf. Ser.* **817** 012067
[11] Mu N 2007 *High temperature oxidation behavior of [gamma]Ni+[gamma]'-Ni3Al alloys and coatings modified with Pt and reactive elements* Doctoral Thesis (Ames: Iowa State University) p 35-6
[12] Locci I E, Western C, Mackay R A, Garg A and Ritzert F 2004 *Successful Surface Treatments for Reducing Instabilities in Advanced Nickel-Base Superalloys for Turbine Blade* (Ohio: Glenn Research Center) pp 1-21
[13] Daroonparvar M, Hussain M S and Yajid M A M 2012 *Appl. Surf. Sci.* **261** 287
[14] Yang X, Peng X, Xu C and Wang F 2009 *J. Electrochem. Soc.* **156** C167

IOP Conf. Series: Journal of Physics: Conf. Series **985** (2018) 012040 doi:10.1088/1742-6596/985/1/012040

Ni-Al phase transformation of dual layer coating prepared by pack cementation and electrodeposition

A Afandi, E Sugiarti, R Ekaputra, T Sudiro and K A Z Thosin

Center of Physics, Indonesian Institute of Science (LIPI), Serpong, 45134, Indonesia

E-mail: kema001@lipi.go.id

Abstract. In this work, Fe-Cr alloys were coated via Aluminum (Al) pack cementation, followed by Nickel (Ni) electrodeposition. The process of pack cementation was done with mixing powders of Al, Al_2O_3 and NH_4Cl with weight percentage of 15%, 85%, and 5% respectively. To control successful Al diffusion to the substrate, pack cementation was conducted for 7 hours with two holding temperatures treatment at 400 °C for 4 hours, and 800 ° C hours for 2 hours. Subsequently, the electrodeposition of Ni was applied with the solution consisting of $NiSO_4$, H_3BO_3, and $NiCl_2$. The samples were placed in the cathode, and then dipped in the solutions, while Ni plate used as anode. Successfully the samples were coated by dual Al-Ni layers, the samples were slowly heat treated at 900 °C for 10 hours. The inter-diffusion of Al and Ni were characterized with SEM/EDX to investigate the distribution of the elements. Mechanical properties of the coated substrates were analyzed with Hardness Vickers (HV). It was found the hardness of the substrate increased significantly, from originally 255 HV to the 1177 HV after pack cementation. The hardness of the substrates has decreased to 641 HV after Ni plating, but subsequent heat treatment has been able to increase the hardness to 842 HV. This phenomenon can be correlated to the inward Al diffusion, and outward Fe, Cr diffusion. The formation of intermetallic compounds due to Al inward and Fe, Cr outward diffusion were discussed in details.

1. Introduction

Nickel aluminide (NiAl) has been considered as a promising alloys for high temperature applications after exhibiting high melting point, high corrosion and oxidation resistance, and high hardness and ductility [1, 2]. There are at least five stable intermettalic compounds found from NiAl alloys namely $NiAl_3$, Ni_2Al_3, NiAl, Ni_2Al_5, and Ni_3Al. The formation of those intermetallic compounds corresponds respectively with the atomic compositions. Among mentioned compounds, NiAl and Ni_3Al are considered to having desirable properties, and respectively known as β and γ' phases. Despite that NiAl and Ni_3Al have been reported to have better oxidation resistance, high specific strength, combined with high hardness, and high electrical conductivity, the application of those compounds are still limited due to poor ductility at room temperature [3].

In corporating NiAl compounds on the surface of the engineering alloys such as stainless steels has been becoming a major investigation recently. Stainless steels are widely used for many industrial applications, especially for the alloys with Fe-Cr as major constituents. Fe-Cr exhibits good oxidation resistant, however with the increasing operating temperature, the oxidation rate increase significantly mainly caused by water vapors [4-6]. In this paper, Fe-Cr with chemical compositions shown in table 1 is used as the substrate. The aim of the study is investigating the diffusion characteristic of the constituent elements, especially Fe, Cr, Ni, and Al in the substrate due to dual Ni-Al coatings. This paper also identifies the hardening mechanism of the substrate after Al pack cementation, and different hardening mechanism after Ni electroplating and heat treatment.

Table 1. Chemical composition of Fe-Cr alloys as the substrate (in atomic %)

Fe	C	Cr	Ni	Mo	Mn	Co	Si	Other as Impurities (P, S)
balanced	<0.08	16-18	10-14	2-4	2	1	Max 1	Max 1

2. Experimental procedures

The samples were cut to the dimension of 9.3×20.2×3 mm, and then the surface were ground with sandpapers of 1000 SiC grit to remove the scratches and oxidation on it. The sample was further cleaned with acetone and dried up with hot air. There were 4 samples kept in Al_2O_3 crucible, and then dipped with pack cementation powder consisting of Al, Al_20_3, and NH_4Cl, respectively with 15%, 85%, and 5% weight ratio. The crucible was closed and sealed with concrete and then kept at the room temperature for 2 hours allowing calcination to occur. The vacuum furnace was prepared with the heating rate of 5 °C/min. When temperature inside the vacuum reached 400 °C, the temperature was hold for 4h. Subsequently the temperature was increased until 800 °C with the same heating rate and then holds it again for 2h. The sample was kept in the vacuum furnace until it cooled down.

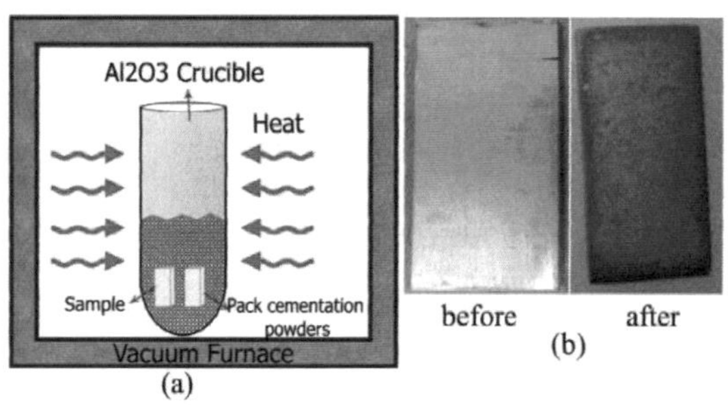

Figure 1. (a) Illustration of pack cementation process. The samples were dipped in the mixing powders, heat it for 7 hours with two holding temperature at 400 °C for 4 h, and 800 °C for 2 h. (b) The appearance of the sample before and after pack cementation.

After the sample was aluminized, the Ni electroplating was prepared with the solution of $NiSO_4$, $NiCl_2$, H_3BO_3, having weight portion of 165 gr, 22.5 gr, and 20 gr respectively. The sample was put on the cathode, and the nickel plate was on the anode. The current density for the Ni plating was 20 mA/cm^2. The heat treatment was done for 10 to allow complete reaction between Al and Ni. The microstructures were investigated with Scanning Electron Microscopy (SEM) Hitachi SU-3500 equipped with Energy Dispersive X-ray (EDX) to investigate the element distributions. The micro hardness was conducted with automatic LECO LM 100AT to measure substrate, surface after pack cementation and heat treatment.

3. Results and discussion

The microstructures of the sample after pack cementation and Ni-plating and heat treatment are shown in figure 2. Figures 2a and 2b, respectively, correspond to the after Al pack cementation and Ni electroplating and heat treatment for 10 h. The grains produced by Ni plating and heat treatment are finer than the one produced by pack cementation. The hardness of the sample was measured and we obtained 1177 HV after pack cementation, which is very hard as compared to the original substrate at around 255 HV (figure 3). The hardness after pack cementation strongly indicates the formation of Fe-Al intermetallic compounds. Intermetallic compounds of Fe-Al were summarized by Potesser et al [7], among which the most hardness of the compounds found from Al rich to form Fe_2Al_5 with orthorhombic crystal structure and the hardness may reached over 1000 HV [8-9]. The hardness of the sample after Ni-plating and subsequent heat-treatment for 10 h were measured at 641 HV and 842 HV, respectively. The cross-section of the substrate after two coatings can be seen in figure 4. The thickness of the coating from Al pack cementation is 124 μm, while the double coating layers of additionally Ni-plating has the thickness of 153 μm.

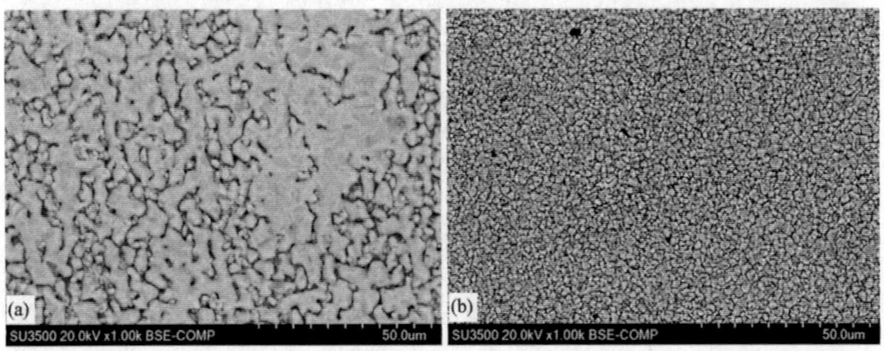

Figure 2. (a) The microstructure of the substrate after pack cementation indicates strongly the diffusion process, (b) the microstructure of the substrate after Ni plating and heat treatment

There is inter-diffusion zone (IDZ) obtained from Al pack cementation, which the thickness is 25.1 μm figure 4b. Above IDZ is Al matrix with intermetallic compounds act as the solid solution hardener, which is identified as white flakes. Further coating with Ni has produced additional layer with thickness of about 21 μm, with second IDZ thickness is about 11.7μm. From the cross section of the substrate, there are cracks formed in the interface between Al matrix and the outer-zone, near the surface. The cracks are pronounced significantly after Ni-plating and heat treatment, which could be influenced by the different of the thermal expansion coefficient.

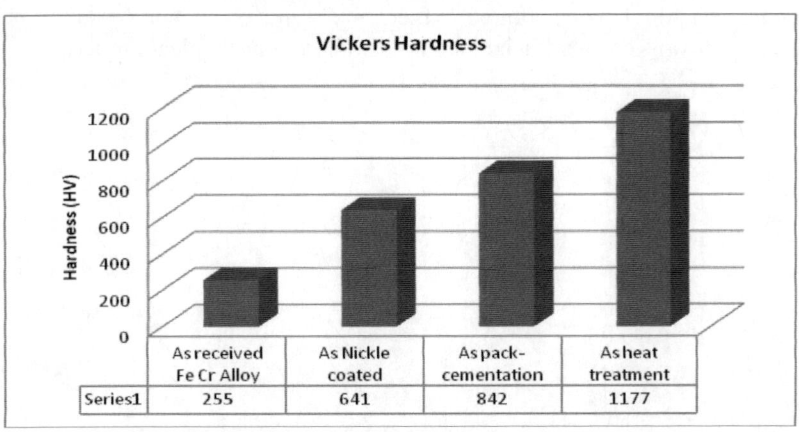

Figure 3. The hardness of the sample is 1177 HV after pack cementation, which is very hard as compared to the original substrate at around 255 HV.

IOP Conf. Series: Journal of Physics: Conf. Series **985** (2018) 012040 doi:10.1088/1742-6596/985/1/012040

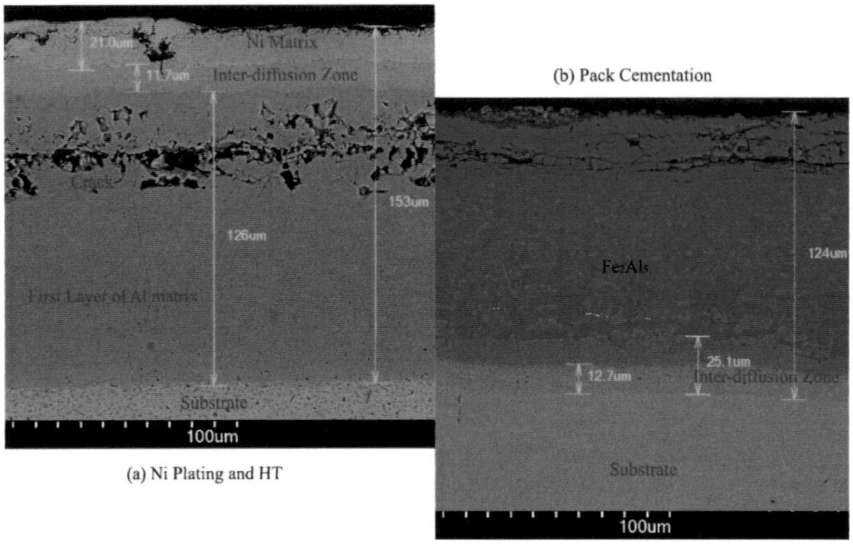

Figure 4. (a) The cross section of the substrate after Al-pack cementation, and Ni electroplating plus heat treatment. The thickness of the coating layers is 153 μm. (b) The cross section of the substrate after Al pack cementation, the thickness of the layer is 124 μm.

The mapping of element diffusions is shown in figure 5 for the substrate with Al pack cementation. The specific interest is observing the distribution of Fe, Al, and Cr through the cross-sectional substrate. In general, Al is found on the top of substrate, while Fe and Cr are concentrated mainly in substrate with few atoms are distributed in the top layer. Interesting to note, the inward diffusion of Al is not seen from figure 5b. This indicates that IDZ has been formed to prevent any Al diffusion to the substrate.

Further mapping analysis is carried out from the cross section of the substrate after dual coatings, which produce significant cracks between Al and Ni matrixes as shown in figure 6a. The mapping of element Oxygen (O) is strongly concentrated in the crack region as shown in figure 6b. This indicates the large portion of porosity is created, which could disrupt the mechanical properties of the substrate.

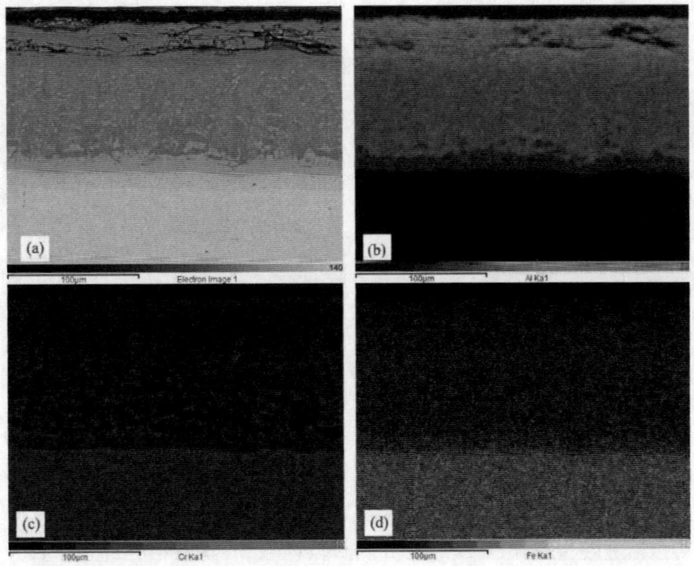

Figure 5. The elemental mapping from the cross section of the substrate (a), indicating Al rich on the top layer, while the Fe and Cr are distributed mainly in the substrate as shown in (c) and (d).

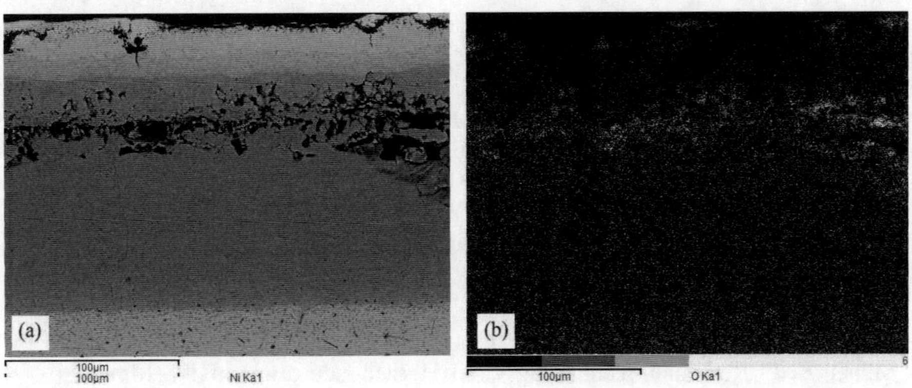

Figure 6. (a) The mapping of the cross section of the substrate, and (b) the porosity is created in the cracks area that was produced from the Al pack cementation.

Further mapping analysis is carried out to observe the diffusion behavior of the atoms after dual coatings, as shown in figure 7. Element of Cr and Fe are mainly distributed on the substrate up to the Al coating. Interesting to note, there are no Fe and Cr outward diffusion through the IDZ that is formed between Al and Ni matrix. The hardness was measured after Ni-plating and heat treatment in the range of 600-850 HV, which correspond well to the hardness of NiAl intermetallic compounds [10]. Further Ni deposition via electroplating has produced IDZ, which could prohibit the outward diffusion of Fe and Cr. This is very important finding because any outward elemental diffusion will create microstructure instabilities of the substrate, as it has been proved by Watson et al. and Locci eat al. in different materials [10-12].

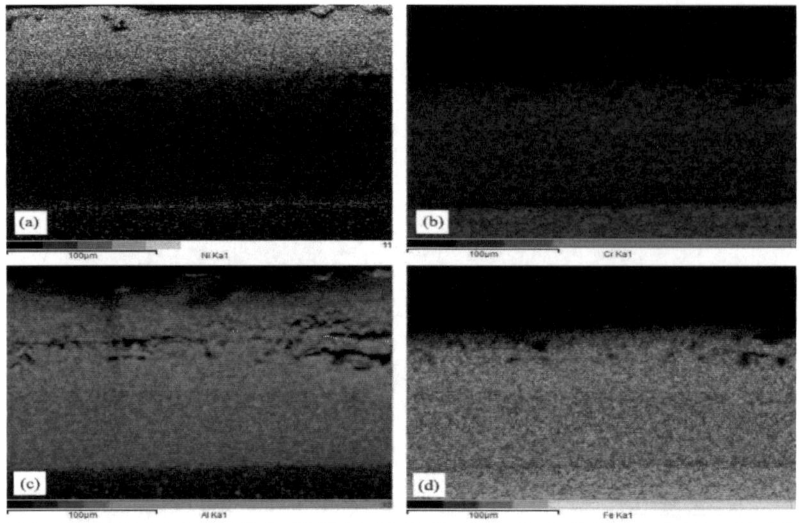

Figure 7. (a) Elemental mapping of Ni is concentrated in the top layer, while Cr (b) and Fe (d) are distributed on the substrate and Al matrix (c).

The formation of Fe-Al intermetallic compounds could act as the diffusion layer if the substrate is exposed at elevated temperature and Oxygen environment. It is expected that Al_2O_3 layer will be formed, and Al could act as a container to prevent outward diffusions of the elements from the matrix. Further analysis is needed to statistically investigate solid solutions and distribution of every element in the sample.

4. Conclusions

From this work, it can be concluded that Al pack cementation is effective to increase the hardness of the substrate. The hardening mechanism can be correlated to the formation of solid solutions such as Fe_2Al_5, which is strongly identified by high micro-hardness that reach 1077 HV. The second layer produced by Ni electrodeposition could act as the barrier for the Fe and Cr outward diffusions. The hardness of the substrate after dual coatings is reduced to the 844 HV. This investigation is a first step to develop protective layer Al_2O_3 that could improve oxidation resistance of Fe-Cr alloys.

Acknowledgments

This investigation was carried out with financial support from the Indonesian Institute of Sciences LIPI and Minsitry of Reseaech, Technology and Higher Education of the Republic of Indonesia (Development of ODS Steel INSINAS fiscal year 2017).

References

[1] Liang X and Wang X 2016 *Mater. Des.* **112** 519
[2] Deevi S C and Sikka V K 1996 *Intermetallics* **4** 357
[3] Guo J T, Sheng L Y, Xie Y, Zhang Z X, Ovcharenko V E and Ye H Q 2011 *Intermetallics* **19** 137
[4] Saunders S R J, Monteiro M and Rizzo F 2008 *Prog. Mater. Sci.* **53** 775
[5] Zhu C, Li P and Wu X Y 2016 *Ceram. Int.* **42** 7708
[6] Tortorelli P F and Natesan K 1998 *Mater. Sci. Eng. A* **258** 115
[7] Potesser M, Schoeberl T, Antrekowitsch H and Bruckner J 2006 *Conf. Proc. EPD Congress* 167
[8] Tomida S and Nakata I 2003 *Surf. Coat. Technol.* **174** 559
[9] Rathod M J and Katsuna M 2004 *Weld. J.* ***83*** 16s
[10] Krasnowski M, Gierlotka S and Kulik T 2013 *Intermetallics* **42** 35

[11] Watson W S, Schaeffer J C and Murphy W H 1996 A new type of microstructure instability in supperalloy-SRZ *Supperalloys* ed R D Kissinger et al. (Warrendal : The Minerals, Metals and Materials Society)

[12] Locci I E, Western C, Mackay R A, Garg A and Ritzert F 2004 *Successful Surface Treatments for Reducing Instabilities in Advanced Nickel-Base Superalloys for Turbine Blade* (Ohio: Glenn Research Center)

IOP Conf. Series: Journal of Physics: Conf. Series **985** (2018) 012041 doi:10.1088/1742-6596/985/1/012041

Characterization of corrosion resistant on NiCoCr coating layer exposed to 5%NaCl

E Sugiarti, R Sundawa, R D Desiati and K A Zaini

Laboratory of High Temperature Materials and Coatings, Research Centre for Physics, Indonesian Institute of Sciences (LIPI), Serpong 15314, Indonesia

E-mail: enis002@lipi.go.id

Abstract. Highly corrosion resistant of carbon steel coated NiCoCr was applied in corrosive of marine environtment. Carbon steel coated NiCoCr was prepared by a two step technique of NiCo electro-deposition and Cr pack cementation. The samples were exposed to 5 wt.% NaCl for 48 and 168 hours. The microstructure and corrosion product were characterized by scanning electron microscopy (SEM), X-ray diffraction (XRD), and transmission electron microscopy (TEM). The corrosion resistance of carbon steel coated NiCoCr was found to be better than that of carbon steel substrate without coating. The results showed the microstructure of 48 h corroded sample has duplex layer composed of inner α-(Ni,Co), α-Cr and outer Cr_2O_3, while a quite thin and continues protective oxide of Cr_2O_3 was observed in outer layer of 168 h corroded sample. The formation of oxide scale rich in Cr_2O_3 has contributed for the better corrosion resistance of carbon steel coated NiCoCr, whereas the formation of non protective oxide of iron might caused low corrosion resistance of carbon steel substrate.

1. Introduction

Carbon steel is widely applied in the construction sector and is part of mechanical component in marine sector. Although carbon steel has excellent mechanical properties and low cost, ingress of aggressive environment in marine atmosphere can lead to the formation of unprotect surface due to the formation of hydrate iron oxide (rust, $Fe_2O_3.H_2$) [1]. Therefore, protection against extreme condition is required in order to maintain carbon steel performance. Carbon steel can be protected by applying metallic coating of resistant elements, such as nickel based coating. Nickel forms high melting and corrosion resistance phase [2-3]. Sharifi et al. investigated the corrosion resistance of Ni-Co coatings electrodeposited on steel substrate in sulphate bath [4]. The research revealed that the addition of cobalt with low content in nickel possesses good adhesion and increases the corrosion resistance of steel substrate due to the formation of a passive layer film of NiO or Ni(OH)2 on the surface. The corrosion resistance of Ni-1%Co was measured 3.75 mpy after polarization test in 3.5%NaCl solution.

A 3.5%NaCl solution is used frequently as synthetic seawater to simulate the actual condition in marine atmosphere. However it is known that a 3.5%NaCl is more aggressive toward carbon steel than natural seawater [5]. In order to prolong the lifetime of Ni-Co coatings on carbon steel substrate, chromium is selected as alloying element or supplementary coating layer. Chromium shows good corrosion resistance in chlorine and in high temperature condition by forming Cr_2O_3 as a protective oxide scale [6]. Fernando et al examined the adhesive strength and corrosion resistance of Cr based coatings on carbon steel substrate. In that research, four types of samples was prepared, i.e, FeCr, FeCoCr, FeCoCrNi and FeCrNi alloys by arc sprayed coatings. In both cases, FeCoCr had the best

IOP Conf. Series: Journal of Physics: Conf. Series **985** (2018) 012041 doi:10.1088/1742-6596/985/1/012041

corrosion resistance in 5%NaCl solution using salt spray method [7]. Further, NiCoCr coatings on carbon steel have exhibited good corrosion resistance after immersion test using 10%HCl as described in our previous study [8]. It revealed that the corrosion rate of uncoated carbon steel was 7.5 mm/y, which was almost 11 times greater than carbon steel coated NiCoCr with the corrosion rate of 0.67 mm/y.

According to the recent results, the present study investigated the corrosion resistance of duplex coating layer based on Ni-Co-Cr constituent using 5%NaCl. Electro-deposition and pack cementation techniques are applied to deposit NiCo as a first layer and chromium layer as a second layer, respectively. These two techniques are easiest and cheapest processes to deposit a coating layer and improve the corrosion resistance of carbon steel [9-10]. Thus, in this study, the micro-structural characterization and corrosion behavior of carbon steel coated NiCoCr is presented.

2. Experimental procedures

A rectangular shape with the dimension of 15×10×1.5 mm from a carbon steel plate was used as a substrate. A substrate was mechanically polished by using emery paper #100, #220, #400, #800, #1000, and #1200. After degreasing and washing with distilled water, the substrate was dried with hot air. Initially, Ni-5%Co was electro-deposited by using Ni-strike solution for 30 s followed by Ni-5%Co watt solution for 120 min with a constant current density of 20 mA/Cm^2. The composition of Ni-strike and Ni-5%Co watt solutions have been describe elsewhere [11]. After electroplating process, the sample was rinsed and dried before weighing. The second coating process is pack cementation of chromium or is well known as chromizing process. The pack powder was mixture of Cr, Al_2O_3 (inert filler) and NH_4Cl (activator) with the concentration (wt.%) of 30, 60, and 10, respectively. The pack was performed at temperature of 800 °C for 10 hours in Ar atmosphere.

The coating thickness was estimated by measuring the weight before and after the chromizing process. Corrosion test was carried out by an immersion test using 5%NaCl solution for 48 and 168 h. The mass loss was measured after exposure period using a balance with a sensitivity of 10^{-5} g. Corrosion rate in mills per year (mpy) was calculated from the weight loss formula by assuming that the substrate was uniformly corroded over the entire surface [11]. A surface morphology, cross-sectional microstructure and elemental distribution of the samples before and after corrosion test were characterized by FE-SEM (JEOL, JIB4610). Phase formed on the sample was identified by XRD (Smartlab, Rigaku) with Cu-Kα radiation operated at 40 kV and 30 mA. A nanostructural characteristic was observed by 200 kV TEM (JEOL, JEM 2010F) equipped with an energy dispersive X-ray spectroscope (EDS).

3. Results and discuccions

3.1. Corrosion resistance

Figure 1 shows the mass loss of carbon steel with and without NiCoCr coating in 5%NaCl solution. The results indicated that the corrosion resistance of the coated sample is significantly higher than that of uncoated carbon steel not only for 48 hours but also 168 hours of exposure time. The corrosion rate of uncoated carbon steel was measured to be approximately 8.016 and 15.560 mpy for 48 and 168 hours immersion time, respectively. The corrosion rate increased more than two times with increasing immersion time. In addition, the corrosion rate of carbon steel coated NiCoCr was very low with a value of ~0.025 mpy and ~0.282 mpy for duration of 48 and 168 hours, respectively. The result showed the corrosion rate was 11 times lower than longer immersion time. It also revealed that the corrosion rate was significantly increased almost 55 times than uncoated carbon steel for duration of 168 hours. Thus, NiCoCr coating layer could act as a protective barrier against chloride ion attack into substrate which originated from 5%NaCl solution. The result was corresponded to sharifi.et.al results which showed that NiCo coating increased the corrosion resistance better than Ni pure coating [4]. Further, addition of chromiun was increased corrosion resistance significantly in chlorine atmosphere [6].

IOP Conf. Series: Journal of Physics: Conf. Series **985** (2018) 012041 doi:10.1088/1742-6596/985/1/012041

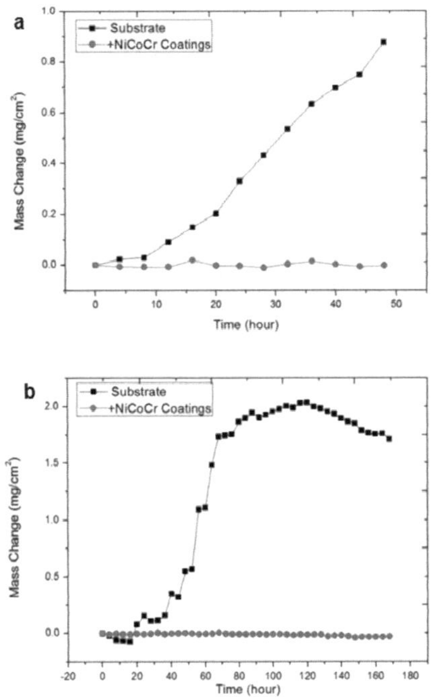

Figure 1. Corrosion performance of carbon steel without and with NiCoCr coating immersed in 0.5%NaCl during (a) 48 h and (b) 168 h.

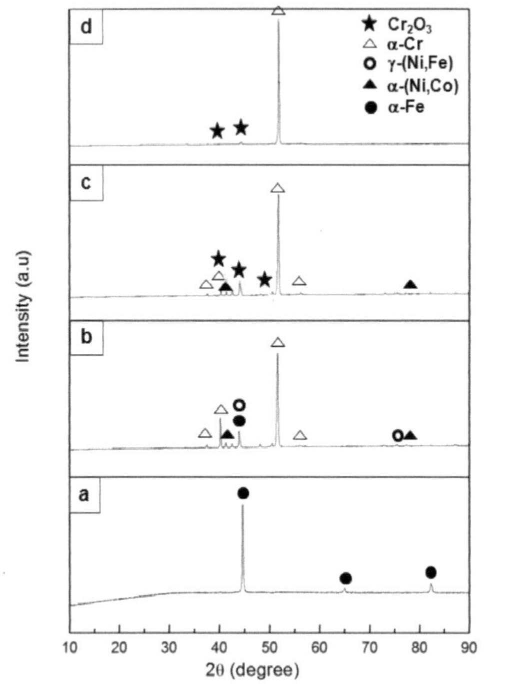

Figure 2. Phase identification using XRD analysis (a) carbon steel substrate, (b) NiCoCr coating, (c) 48 h, and (d) 168 h corroded samples.

3.2. X-ray diffraction analysis

Figure 2 shows the phase identification of carbon steel substrate, NiCoCr coating before and after corrosion for 48 h and 168 h exposure. The result indicated that the phases of as-coated sample mainly consist of γ-(Ni, Fe), α-(Ni, Co), and α-Cr with cubic structure. However, the oxide scale was identified to be mostly composed of α-Cr_2O_3 and a small amount of α-Cr. In addition, by exposing to longer corrosion time, the relative intensity of peaks for α-Cr decreased gradually, which means that more α-Cr_2O_3 was formed in 168 h corroded sample. It confirmed the previous result from corrosion performance that NiCoCr coating exhibited superior corrosion behavior due to the formation of protective oxide scale, α-Cr_2O_3, formed on the coating surface [6-8].

3.3. Surface morphology

Figure 3 shows surface morphology of NiCoCr coating sample before and after corrosion test for 48 and 168 h. it showed that the coated surface on carbon steel substrate was found to be irregular shape in nature. Moreover, figure 3b shows the irregular size fine grain with blocky structure was randomly separated on the entire surface of the scale. The continuous and very fine oxide particles were uniformly distributed on the surface of corroded sample after 168 h immersion time figure 3c. It confirmed that a well adherent and single protective oxide scale was homogenously distributed over the surface of 168 h corroded sample.

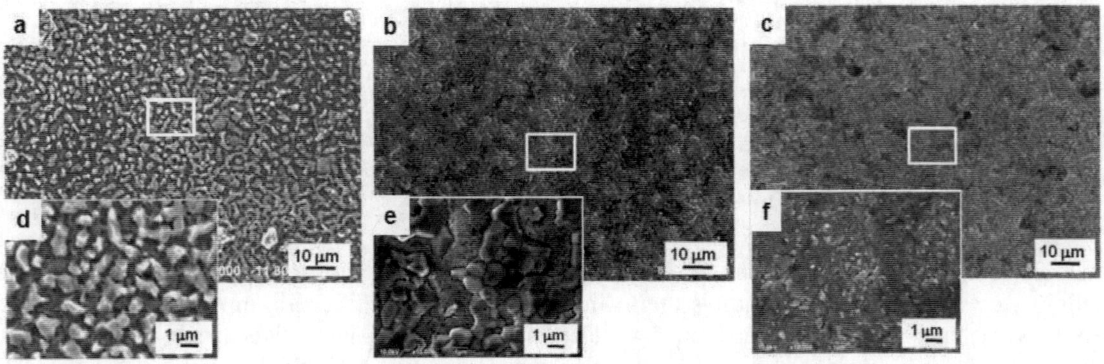

Figure 3. Surface morphology of (a) NiCoCr coating, (b) 48 h, and (c) 168 h corroded samples.

3.4. Cross-sectional morphology

The SEM method was employed to obtain the cross-sectional elemental mapping images of as-coated and corroded samples as shown in figure 4. The microstructural images show that typical two layers consisting NiCo layer and Cr layer were successfully formed on the samples. For as-coated sample, the homogenous layer was established with an approximate thickness of 32 µm and 5 µm for NiCo and Cr layer, respectively. The thickness of oxide scale was measured with an approximate of 5 µm and 1 µm for the corroded sample after 48 and 168 h exposure time, respectively.

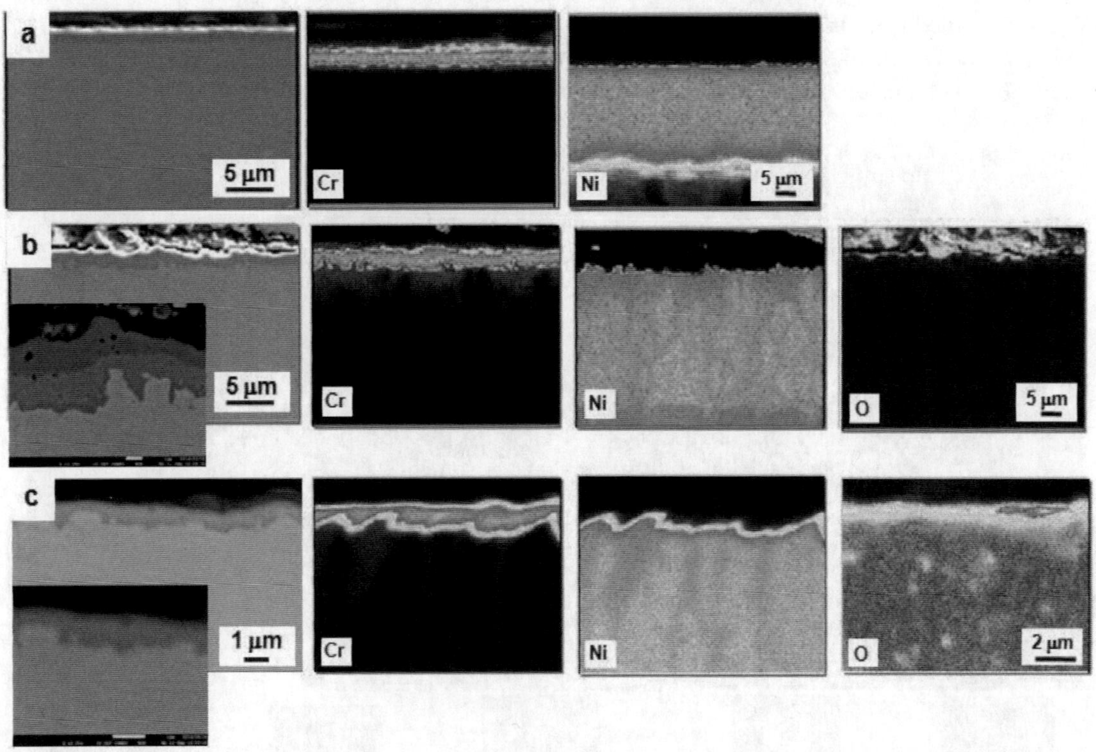

Figure 4. Cross-sectional micrographs (left) and corresponding elemental distribution (right): (a) NiCoCr coating, (b) 48 h, and (c) 168 h corroded samples.

Trace amounts of Ni and Cr are noticeably detected for as-coated samples figure 4 a-right. Different condition of coating can be distinguished in which strong intensity of Cr on the top layer of the coated sample after corrosion test. The corrosion test promotes the formation of corundum structure of Cr_2O_3 that protect the substrate from harsh environment. In addition, a duplex scale consisting of Ni-Co-Cr with high content in inner scale and Cr-O in outer scale was examined in 48 hours corroded sample. On the contrary, only single oxide scale with high amount of Cr-O element was detected in 168 hours corroded sample. It revealed that longer immersion time produced stable, dense, slow growing Cr_2O_3 protective oxide as identifid by XRD analysis in figure 2.

3.5. Nanostructure of protective scale

An XTEM analysis was performed to examine the detailed microstructure of the corroded layer. In addition, crystallographic technique is the main tool for phase identification in multiphase material. Thus, systematic phase identification by electron diffraction pattern indexation routines was performed. It has been clearly understood that the corroded sample has a thick oxide scale of about 2 μm. The high magnification image observed by SEM in figure 4b and c was only focused on the top surface of the coated layer. Therefore, XTEM sample was prepared in the area adjacent the metal/oxide interface of the corroded sample as shown in figure 5. The results show that Cr_2O_3 grains were seen to nucleate at the scale/alloy interface on the corroded sample figure 5-top. It described the nucleation of Cr_2O_3 occurs at the scale/alloy interface during oxidation of NiCoCr coating system. Good adherence was observed on the metal/oxide interface. The presence of the Cr_2O_3 phase was also identified by electron diffraction figure 5-bottom. Point analysis in area 1, 2, and 3 and its selected area electron diffraction (SAED) patterns of 48 hours of corroded sample indicated the formation of Cr_2O_3, α-Cr and α-(Ni,Co). The oxide of Cr_2O_3 forms a corundum type lattice with hexagonal structure having lattice parameter of a = 0.4961 nm and c = 1.360 nm [13]. Further, α-Cr and α-(Ni,Co) identified in 48 hours of corroded sample possessed cubic structure with lattice parameter of a = 0.291 nm and a = 0.352 nm, respectively [11]. In addition, only two phases of Cr_2O_3 and α-(Ni,Co) was identified on the top of oxide layer in 168 hours of corroded sample.

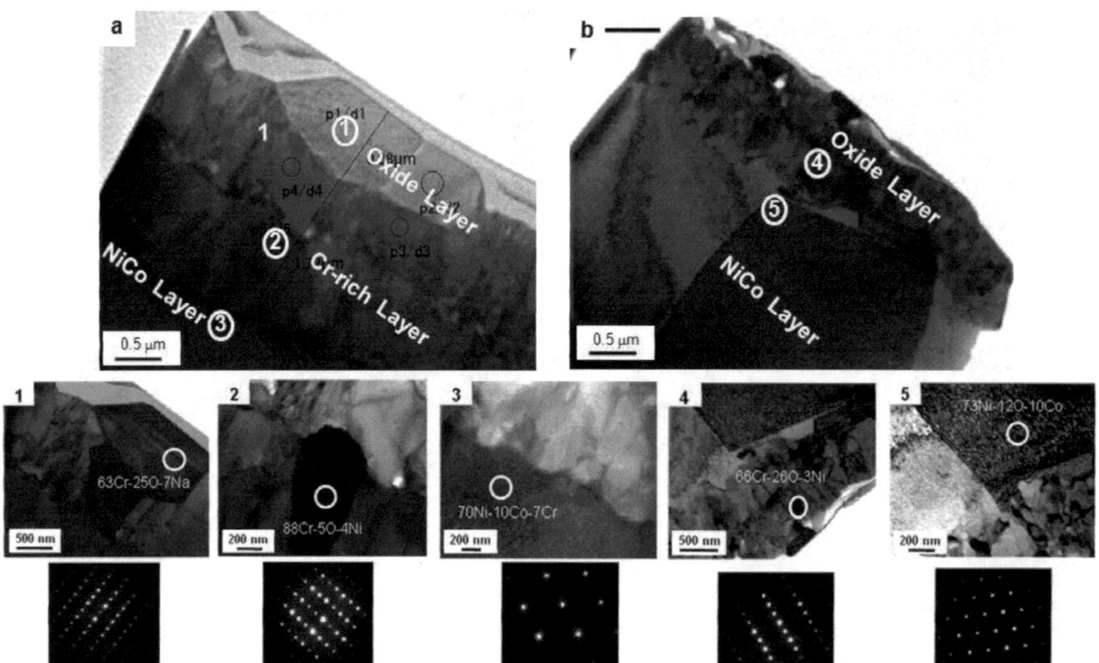

Figure 5. Bright field XTEM images and SAED patterns of (a) 48 h and (b) 168 h corroded samples. 1-2-3 indicates the magnified images obtained from circled area in top figure and SAED pattern obtained from the point EDS analysis area.

4. Conclusions
NiCoCr coatings were successfully developed by electroplating and pack cementation techniques. The corrosion rate of uncoated and NiCoCr coating sample was very low with a value of ~0.025 mpy and ~0.282 mpy for duration of 48 and 168 hours, respectively. It showed that the corrosion resistance of carbon steel coated NiCoCr was 55 times higher than uncoated sample. The formation of Cr_2O_3 on the surface of coating layer could significantly protect the substrate in the corrosive environment. The results revealed that longer immersion time produced stable, dense, slow growing Cr_2O_3 protective oxide scale over the surface of corroded sample.

Acknowledgments
The authors would like to thank to Kegiatan Unggulan LIPI, INSINAS IRPI grant No. 04/INS-2/PPK/E/E4/2017 and Toray Science Technology Research Grant for financial support of this work.

References
[1] Tawancy H M, Al-Hadhrami L M, Al-Yousef F K 2013 *Case Studies Engineering Failure Analysis* **1** 6
[2] Porcayo-Calderon J, Ramos-Hernandez J J, Mayén J, Porcayo-Palafox E, Pedraza-Basulto G K, Gonzalez-Rodriguez J G, and Martinez-Gomez L 2017 *Advances in Materials Science and Engineering* 8929873
[3] Podchernyaeva I A and Panasyuk A D 2000 *Powder Metall. Met. Ceram.* **39** 434
[4] Sharifi K and Ghorbani M 2014 *Bull. Mater. Sci.* **37** 713
[5] Moller H, Boshoff E T, and Froneman H 2006 *J. South African Inst. Min.Metall.* **106** 585
[6] Sahri M I, Othman N K, Samsu Z, and Daud A R 2014 *AIP Conf. Proc.* **1614** 152
[7] Antunes F J, Sá Brito V R S, Bastos I N and Costa H R M 2013 *Applied Adhesion Science* **1** 1
[8] Destyorini F, Eni S and Zaini K. A 2013 *Telaah* **31** 51
[9] Eni S, Zaini K A, Sundawa R, Wang Y, Ohnuki S and Hayashi S 2017 *J. Phys.: Conf. Ser.* **817** 012067
[10] Tsai L C, Sheu H H, Chen C C and Ger M D 2015 *Int. J. Electrochem. Sci.* **10** 317
[11] Eni S, Destyorini F, Zaini K A, Wang Y, Hashimoto N, Ohnuki S and Hayashi S 2015 *Surf. Coat. Tech.* **265** 68
[12] Oparaodu K O and Okpokwasili G C 2014 *International Journal of Environmental Bioremediation & Biodegradation* **2** 243
[13] Kiselev V F and Krylov O V 1989 *Adsoprtion and Catalysis on Transition Metals and Their Oxides* (New York: Springer-Verlag Berlin Heidelberg) chapter 3 pp 95-97
[14] Zhao P, Zhao H, Yu J, Zhang H, Gao H and Chen Q 2018 *Ceram. Int.* **44** 1356

IOP Conf. Series: Journal of Physics: Conf. Series **985** (2018) 012042 doi:10.1088/1742-6596/985/1/012042

Microstructure transformation of Cr-Al coating on carbon steel prepared by ball milling method as a function of tungsten doping

A S Wismogroho, T Sudiro, Didik A and Ciswandi

Research Center for Physics, Indonesian Institute of Sciences, Serpong 15314, Tangerang Selatan, Banten, Indonesia

E-mail: agus.sukarto.wismogroho@lipi.go.id

Abstract. In present work, Cr-Al coatings containing 0, 1, 2, 3, and 5% W have been prepared on the surface of low carbon steel by mechanical alloying technique. The composition of each powder was milled for 2 hour in a stainless steel crucible with a ball to powder ratio of 10:1. Afterward, the Cr-Al-W powder and substrate were mechanically alloyed in air for 1 hour. The heat treatment of coated samples was carried out at 800 °C in a vacuum furnace for 2 hour. In order to characterize the phase composition and microstructure of the coating before and after heat treatment, XRD and SEM-EDX were used. The analysis results reveal that the ball milling process induces the formation of homogeneous Cr-Al-W coating structure with a thickness of about 80 μm. The phase observation shows individual peaks of each starting elements, along with the occurrence of powder refinement and solid solution formation. After heat treatment, $AlCr_2$ and Al_8Cr_5 phases were formed. The addition of W accelerates the formation of $AlCr_2$, but inhibits the formation of Al_8Cr_5. The detail of the results was presented in this paper.

1. Introduction

Carbon steels have been widely used for wide range of applications due to its commercial consideration. However, it shows low oxidation resistance at high temperature [1]. In order to improve steel resistance against wear, oxidation, and corrosion at high temperatures, surface coatings can be applied on the surface of carbon steel. The Al-Cr coating is often used for corrosion resistance material due to its promising corrosion resistance to room temperature aqueous corrosion and hot corrosion of fused salt deposits [2,3].

In order to improve its properties, various studies based on Al-Cr coating have been conducted. The Al-Cr-Zr-N and Al-Cr-Co-Fe coatings were developed for corrosion resistance applications [4] and thermal barrier coating system [5], respectively. The composite ceramic coatings as $(Al,Cr)_2O_3$-Cr [6] and Al-Cr-N [7] were prepared to improve wear resistance. Currently, a mechanical alloying (MA) technique [8-12] has been used to deposit various coatings on the surface of materials. It was known that mechanical alloying is a new technique which can be utilized to produce intermetallic coatings on steel [8] and nanostructure coating on various metal surfaces [9].

In the previous study, we successfully prepared Al-Cr coating by a mechanical alloying technique [12]. Currently, to improve the hardness of Al-Cr coating, the addition of tungsten (W) was conducted to the Al-Cr coating by a mechanical alloying technique. Heat treatment was also performed in order

IOP Conf. Series: Journal of Physics: Conf. Series **985** (2018) 012042 doi:10.1088/1742-6596/985/1/012042

to refine the coating structure. The metallographic characterizations of the coating before and after heat treatment were studied by means of XRD and SEM-EDX.

2. Experimental details

Commercial low carbon steel with an approximate dimension of 10×8×3 mm^3 was used as substrate in this study. For coating deposition, the alloy surface was ground using SiC papers for up to 1200 grit. Prior to coating deposition, the specimen was ultrasonically cleaned in an aqueous solution and dried using hot flowing air.

The Cr_{50}-Al_{50}, $Cr_{49.5}$-$Al_{49.5}$-W, $Cr_{48.5}$ $Al_{48.5}$ W_3, $Cr_{47.5}$-$Al_{47.5}$-W_5 coatings were prepared using high purity Cr, Al, and W powders. Firstly, the coating powders were mixed in steel vial for 2 h using a shaker mill with powder to steel balls ratio of 1 to 10. The specimens were then subjected to mechanical alloying for 1 h.

The collisions of steel balls and powder on the surface of low carbon steel lead to deposition of coating powder on the steel surface. The heat treatment was carried out to improve the coating structure at 800 °C for 2 h in vacuum atmosphere of 5.6 Pa. The specimens were then allowed to cool down to room temperature in the same chamber. The phase compositions of the coating before and after heat treatment were analyzed by X-ray diffraction (SmartLab Rigaku) with Cu Kα radiation at 40 kV and 30 mA. The coating morphologies were characterized by means of scanning electron microscope equipped with energy dispersive X-ray spectrometer (SEM Hitachi SU 3500-EDX).

3. Results and discussion

The cross-sectional SE images of Cr_{50}-Al_{50}, $Cr_{49.5}$-$Al_{49.5}$-W, $Cr_{48.5}$-$Al_{48.5}$-W_3, and $Cr_{47.5}$-$Al_{47.5}$-W_5 coatings before heat treatment are shown in figures 1 respectively. It appears that the coatings were successfully deposited on the surface of low carbon steel and exhibits good adherence to the steel substrate. A relatively dense structure can be observed in the Cr_{50}-Al_{50}, $Cr_{49.5}$-$Al_{49.5}$-W, and $Cr_{48.5}$-$Al_{48.5}$-W_3 coatings. However, some micro cracks can be observed in the $Cr_{47.5}$-$Al_{47.5}$-W_5 coating. In addition, the cross-sectional images also show that the sample surface is rough.

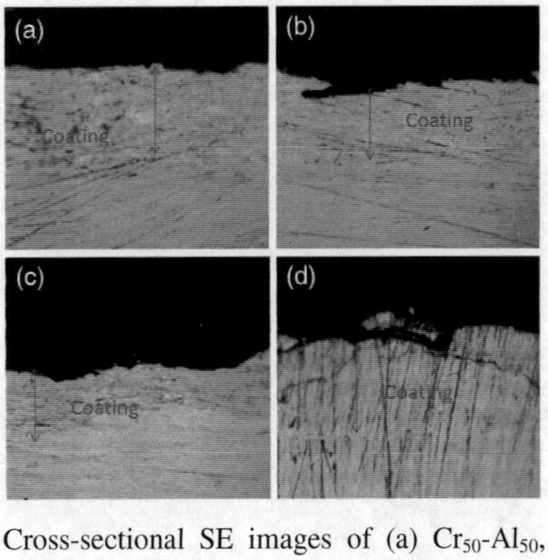

Figure 1. Cross-sectional SE images of (a) Cr_{50}-Al_{50}, (b) $Cr_{49.5}$-$Al_{49.5}$-W, (c) $Cr_{48.5}$-$Al_{48.5}$-W_3, and (d) $Cr_{47.5}$-$Al_{47.5}$-W_5 coatings before heat treatment.

Figure 2 shows the X-ray diffraction patterns of Cr_{50}-Al_{50}, $Cr_{49.5}$-$Al_{49.5}$-W, $Cr_{48.5}$-$Al_{48.5}$-W_3, and $Cr_{47.5}$-$Al_{47.5}$-W_5 coatings before heat treatment. The XRD analysis shows reflections of Cr and Al in the Cr_{50}-Al_{50} coatings as shown in figure 2a with no reflection peak of Fe from substrate of low carbon steel. Meanwhile in the $Cr_{49.5}$-$Al_{49.5}$-W, $Cr_{48.5}$-$Al_{48.5}$-W_3, and $Cr_{47.5}$-$Al_{47.5}$-W_5 coatings, Cr and W reflections can be clearly seen in Figures 2b, 2c, and 2d with relatively low diffraction peaks of Al.

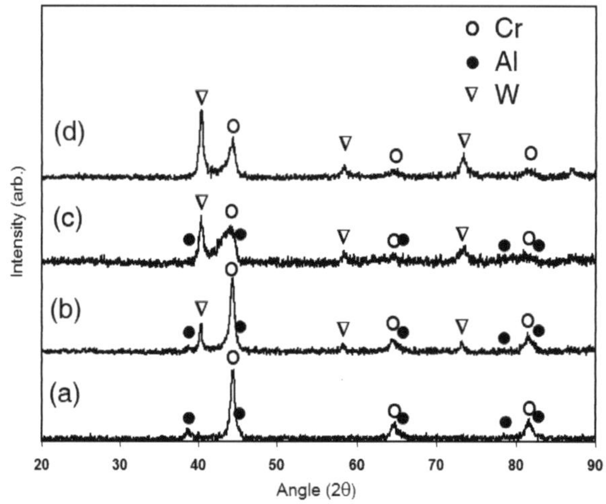

Figure 2. X-ray diffraction patterns of (a) Cr_{50}-Al_{50}, (b) Cr_{49}-Al_{49}-W, (c) $Cr_{48.5}$-$Al_{48.5}$-W_3 and (d) $Cr_{47.5}$-$Al_{47.5}$-W_5 coatings before heat treatment.

The cross-sectional BSE microstructures and corresponding EDX elemental maps of $Cr_{50}Al_{50}$ and $Cr_{48.5}Al_{48.5}W_3$ coatings before heat treatment are shown in figures 3. In the coating layer, the coatings are composed of grey particles of Cr, dark area of Al, and white particles of W. The Cr-particles in the $Cr_{48.5}Al_{48.5}W_3$ coating are smaller and more uniformly distributed compared to in the $Cr_{50}Al_{50}$ coating. In the $Cr_{48.5}Al_{48.5}W_3$ coating layer, as the distance from the substrate surface to coating surface increases, the particles become smaller and more uniform. Meanwhile, in the $Cr_{50}Al_{50}$ coating, the Cr particles seem to be less smooth. The thickness of coatings, as shown in figure 3, is about 80 μm.

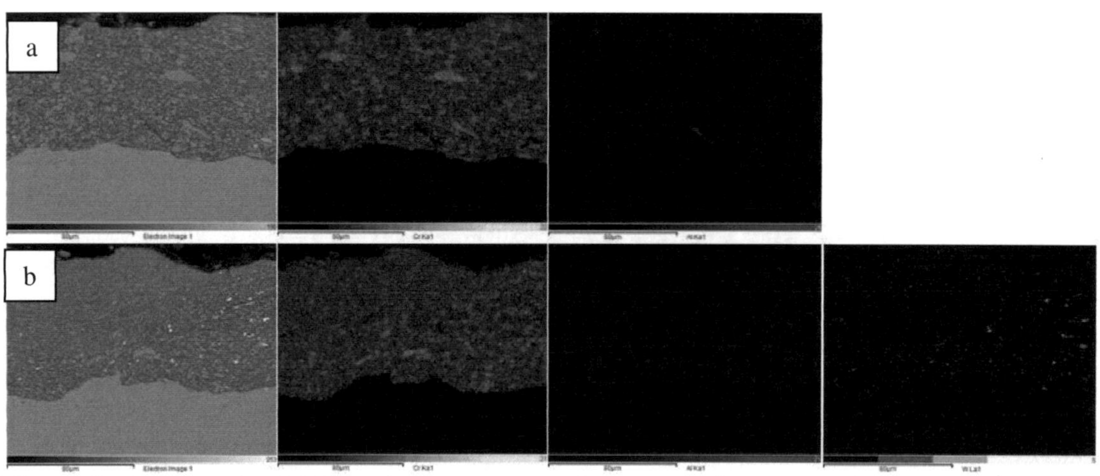

Figure 3. Cross-sectional BSE microstructures and corresponding EDX elemental maps of (a) $Cr_{50}Al_{50}$ and (b) $Cr_{48.5}Al_{48.5}W_3$ coatings before heat treatment.

The XRD results in figure 2 show that although Al has the same concentration as Cr, the Al peaks appear significantly reduced than Cr peaks. This indicates that during milling, Al particles are more susceptible to deformation than Cr particles. In comparison to the results presented in figures 3a and

3b, it can be seen that the Al rich area is deformed to be the matrix surrounding the Cr rich area. As the W concentration increases, the Cr diffraction peak intensities tend to decrease. The peak intensity of W is relatively high despite the coating containing a small W concentration. This shows that the addition of W accelerates the smoothing and solid solution formation between Cr-Al. Al atoms are dissolved into Cr lattice to form Cr (Al) solid solution. This process is likely to occur due to hardness differences between Al, Cr, and W which hardness are about 160-350, 1060, and 3430-4600 MPa, respectively. The addition of harder materials accelerates the formation of a solid solution of a softer mixture. On the other hand, the addition of 5 at.% W (figure 1d) leads to crack formation in the coating, suggesting that the coating becomes more brittle with the increase of W concentration. When the coating surface is hit with the milling ball, the cracks can be generated easily.

The cross-sectional SE images of Cr_{50}-Al_{50}, $Cr_{49.5}$-$Al_{49.5}$-W, $Cr_{48.5}$-$Al_{48.5}$-W_3, and $Cr_{47.5}$-$Al_{47.5}$-W_5 coatings after heat treatment at 800 °C for 2 h are shown in figures 4. The Cr_{50}-Al_{50}, $Cr_{49.5}$-$Al_{49.5}$-W, and $Cr_{48.5}$-$Al_{48.5}$-W_3 coatings exhibit good adherence to the steel substrate and a relatively dense structure can be observed. In the $Cr_{47.5}$-$Al_{47.5}$-W_5 coating, however, some voids and micro cracks can be observed.

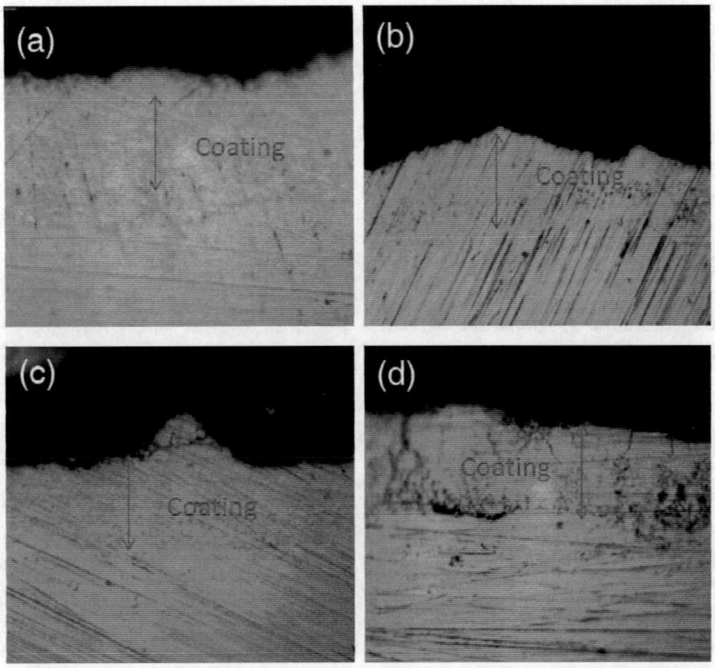

Figure 4. Optical cross-sectional images of (a) Cr_{50}-Al_{50}, (b) $Cr_{49.5}$-$Al_{49.5}$-W, (c) $Cr_{48.5}$-$Al_{48.5}$-W_3 and (d) $Cr_{47.5}$-$Al_{47.5}$-W_5 coatings after heat treatment.

The results of X-ray analysis of Cr_{50}-Al_{50}, $Cr_{49.5}$-$Al_{49.5}$-W, $Cr_{48.5}$-$Al_{48.5}$-W_3, and $Cr_{47.5}$-$Al_{47.5}$-W_5 coatings after heat treatment are shown in Figure 5. In Cr_{50}-Al_{50}, $Cr_{49.5}$-$Al_{49.5}$-W, and $Cr_{48.5}$-$Al_{48.5}$-W_3 coatings, intermetallic phases as $AlCr_2$ and Al_8Cr_5 are formed with $AlCr_2$ as the main phase. Meanwhile, in the $Cr_{49.5}$-$Al_{49.5}$-W and $Cr_{48.5}$-$Al_{48.5}$-W coatings, Cr reflection can still be observed. The $AlCr_2$ and W phases are observed in the $Cr_{47.5}$-$Al_{47.5}$-W_5 coating. XRD characterization reveals that as W concentration in Cr-Al coating increases, the diffraction peak of $AlCr_2$ increases and Al_8Cr_5 decreases. At 5 at.% W, however, the Al_8Cr_5 phase is not formed. This means that W addition inhibits Al_8Cr_5 phase formation.

IOP Conf. Series: Journal of Physics: Conf. Series **985** (2018) 012042 doi:10.1088/1742-6596/985/1/012042

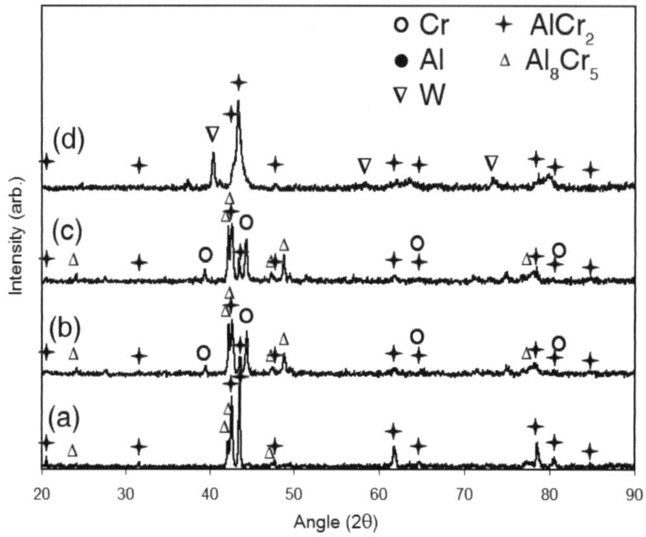

Figure 5. X-ray diffraction patterns of (a) Cr_{50}-Al_{50}, (b) Cr_{49}-Al_{49}-W, (c) $Cr_{48.5}$-$Al_{48.5}$-W_3 and (d) $Cr_{47.5}$-$Al_{47.5}$-W_5 coatings after heat treatment.

The cross-sectional BSE images of Cr_{50}-Al_{50} and $Cr_{48.5}$-$Al_{48.5}$-W_3 coatings after heat treatment at 800 °C for 2h are shown in figure 6. As shown in figure 6a, the Cr-Al coating exhibits good adherence to the steel substrate and a dense structure can be observed. The results of EDX elemental maps show that Cr and Al are uniformly distributed in the coating. However, in the Cr-Al coating interface, the coating becomes Cr-rich and Al-depleted. On the contrary, some micro cracks are observed in the $Cr_{48.5}$-$Al_{48.5}$-W_3 coating (see figure 6b). Similar to the Cr-Al coating, Cr and Al are distributed in the $Cr_{48.5}$-$Al_{48.5}$-W_3 coating with small white precipitates detected by EDX as W-rich area.

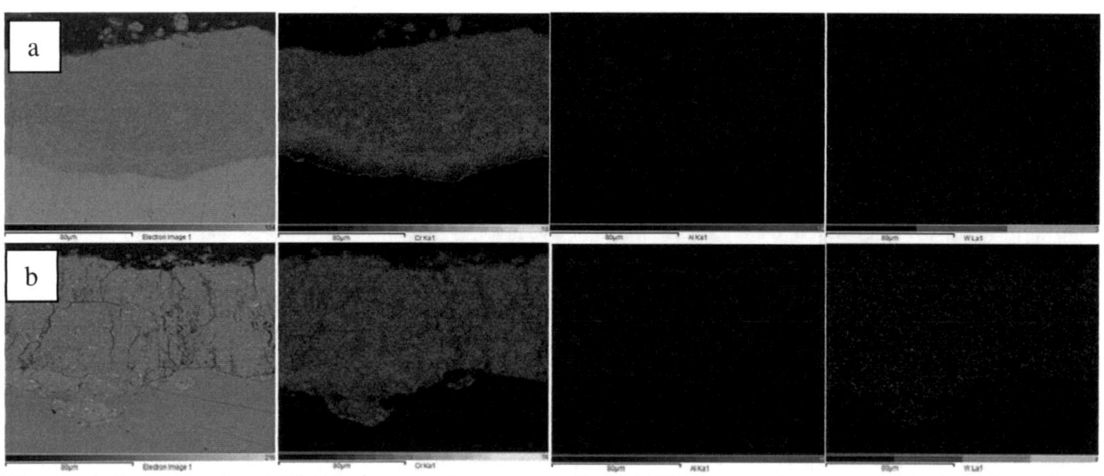

Figure 6. Cross-sectional BSE images and corresponding EDX elemental maps of (a) $Cr_{50}Al_{50}$ and (b) $Cr_{48.5}Al_{48.5}W_3$ coatings after heat treatment.

4. Conclusion

Four different compositions of Cr-Al coatings with the W addition as Cr_{50}-Al_{50}, $Cr_{49.5}$-$Al_{49.5}$-W, $Cr_{48.5}$-$Al_{48.5}$-W_3, and $Cr_{47.5}$-$Al_{47.5}$-W_5 have been prepared on low carbon steel by a mechanical alloying technique. The obtained results show a dense structure of Cr_{50}-Al_{50}, $Cr_{49.5}$-$Al_{49.5}$-W_1, $Cr_{48.5}$-$Al_{48.5}$-W_3 coatings. However, voids and microcracks are formed in the $Cr_{47.5}$-$Al_{47.5}$-W_5 coating. After heat treatment at 800 °C for 2h, the intermetallic phase of $AlCr_2$ and Al_8Cr_5 are formed in the Cr_{50}-Al_{50}, $Cr_{49.5}$-$Al_{49.5}$-W, $Cr_{48.5}$-$Al_{48.5}$-W_3 coatings. The high W concentration ($Cr_{47.5}$-$Al_{47.5}$-W_5) leads to suppression of Al_8Cr_5 formation.

Acknowledgements

This work was supported by Insinas Kemitraan (gel 2) 2017, "Development of Aluminum Metal Powder for Pigments and Hebel additive". The authors are grateful to the Laboratory of High Resistant Materials, Research Center for Physics-LIPI for providing research facilities.

References

[1] Wang B, Wu J, Zhang Y, Wu Z, Li Y and Xue W 2015 *Surf. Coat. Technol.* **269** 302
[2] Xu J, Chen Z Y, Tao J, Jiang S Y, Liu Z L and Xu Z 2009 *Sci. China, Ser. E* **52** 1225
[3] Zhang M, Chen C, Zhang G, Rao Y and Ling G 2013 *Physics Procedia* **50** 206
[4] Chen L, Liu Z Q, Xu Y X and Y Du 2015 *Surf. Coat. Technol.* **275** 289
[5] Wolf W, Schulz R, Savoie S, Bolfarini C, Kiminami C S and Botta W J 2017 *Surf. Coat. Technol.* **319** 241
[6] Dai X, Yan D, Yang Y, Chu Z, Chen X and Song J 2017 *Ceram. Int.* **43** 6340
[7] Yang B, Chen L, Chang K K, Pan W, Peng Y B, Du Y and Liu Y 2012 *Int. J. Refract. Metals Hard Mater.* **35** 235
[8] Mohammadnezhad M, Shamanian M and Enayati M H 2012 *Appl. Surf. Sci.* **263** 730
[9] Mohammadnezhad M, Shamanian M and Enayati M H and Salehi M 2013 *Surf. Coat. Technol.* **217** 64
[10] Canakci A, Erdemir F, Varol T and Ozkaya S 2013 *Powder Technology* **247** 24
[11] Romankov S, Sha W, Kaloshkin S.D and Kaevitser K 2006 *Surf. Coat. Technol.* **201** 3235
[12] Hia A I J, Sudiro T, Aryanto D and Sebayang K 2016 *J. Phys.: Conf. Ser.* **739** 012131

IOP Conf. Series: Journal of Physics: Conf. Series **985** (2018) 012043 doi:10.1088/1742-6596/985/1/012043

Effect of forming process by using external magnetic field of bonded magnet made from NdFeB flakes to microstructure and magnetic properties

Suprapedi[1], P Sardjono[1], Muljadi [1], N R Djauhari[1] and Ramlan[2]

[1]Research Center for Physics, Indonesian Institute of Sciences (LIPI), Kompleks Puspiptek, Serpong, Indonesia
[2]Physics Department, FMIPA Sriwijaya University, Palembang, Indonesia

E-mail: suprapu@gmail.com, ramlan@unsri.ac.id, muljadi2002@yahoo.com

Abstract. Research of fabricated bonded magnets NdFeB made from NdFeB flakes with variation of external magnetic field has been done. The materials preparation process begins with milling NdFeB flakes using High Energy Milling (HEM) for 60 minutes and mixing it with 5 wt % celuna binder and performing compaction to form pellet with a pressure of 40 Kgf /cm^2 and then applying external magnetic field (0, 2000, 5000, 8000 and 11000 Gauss). The pellet samples were then dried using vacuum dryer with temperature of 100 °C for 1 hour. Characterization includes bulk density, measurement of magnetic properties with gauss meter, and Vibrating Sample Magnetometer (VSM). From the characterization results the best value was obtained on the external magnetic field orientation of 8000 to 11000 Gauss with a density value of 5.38 g/cm^3, flux magnetic value of 465.9 – 467.1 Gauss, remanence value of 2.63–2.776 kGauss, and coercivity value of 1.905–1.925 kOe.

1. Introduction

High performance Nd-Fe-B-type bonded magnets have increasingly found new applications, such as the magnet in small motor (e.g. motor for power window), small generator, and component for flow meter [1, 2]. For preparation of bonded magnet, fine powder Nd-Fe-B-type, isotropic particles are required and they need to maintain the high permanent magnetic performance even when they are processed to fine particles [3]. The easiest way to prepare fine particles is probably crushing and milling of bulk or flakes of Nd-Fe-B-type precursor magnets. In addition to those materials, the HDDR-treated material may also be a promising candidate as a precursor material for fine particles. It is common, however, that the fine particles prepared by mechanical milling of the precursor materials lose significantly the coercivity [3, 4]. There are many sources of raw materials for making bonded magnet NdFeB such as ribbon flakes NdFeb, product from magnet quench with name code MQP-A, MQP-B, and etc.

The preparation of NdFeB powder is important step in manufacture of bonded magnet. The atmosphere of powder preparation process can influence the quality of powder, because NdFeB is easy to cause corrosion with oxygen to form oxide [5, 6]. There are some methods for making NdFeB fine powder, such as HDDR process method and mechanical alloying by milling process with hydrogen or argon gas atmosphere [7]. NdFeB magnet has high values of the remanence and coercivity, as well as relatively high curie temperatures (312 °C) [8]. This type of magnetic alloys are identified as suitable for research and further development of magnetic composite materials with polymer matrix, and therefore it is called bonded magnets [7, 8]. There are several different process routes for bonded magnet production. The compression molding is the most common technique utilized for materials with thermo setting polymer matrices. Advantages of using bonded composite materials include their simple technology, possibility of forming their final properties, low manufacturing costs because of no costly finishing, and low material losses resulting from the possibility of forming any shape [9].

There are two kinds of bonded magnet, such as isotropic bonded magnet and anisotropic bonded magnet. Isotropic bonded magnet is a material made at forming process that is done without using an external magnetic field. It means that the direction of the domain of the particles from this material still leads randomly. It is however different for anisotropic bonded magnet, which is made at forming process using an external magnetic field. As a result, the direction of the domain of the particles from this material will be oriented in one particular direction. In general, the magnetic properties of an isotropic are greater than that of isotropic magnets [10]. In this research, we will make an isotropic permanent bonded magnet using NdFeB flakes and celuna as binder. The purpose of this research is to know the effect of external magnetic field to the microstructure and its magnetic properties.

2. Experimental works

A bonded magnet is made from NdFeB flakes by using a celuna as binder. The NdFeB powder preparation step was done by using High Energy milling (HEM) machine for milling of NdFeB flakes for 60 minutes. The resulting powder was then dried at 80 °C by using a vacuum dryer. Then the fine powder was analyzed by using XRD to observe the crystal structure. The dry powder was then mixed with 5% of the celuna binder and was formed as pellets by using an external magnetic field (0, 2000, 5000, 8000 and 11000 Gauss) and a pressure of 40 kg/cm^2. The resulting pellets were dried at 100 °C by using a vacuum dryer for 2 hours, and then all samples were magnetized by using impulse magnetizer. Characterization was performed including measurement of bulk density, flux magnetic, hysteresis curve, and microstructure.

3. Results and discussion

Sample NdFeB powder after wet milling for 60 minutes was analyzed by using XRD-Rigaku; figure 1 shows XRD patterns. The analysis using XRD aims to observe the phases formed on the sample powder. The peaks matched with JCPDS cards number 79-1994 were marked with circle which show that all peaks in the figure 1 are $Nd_2Fe_{14}B$ phase.

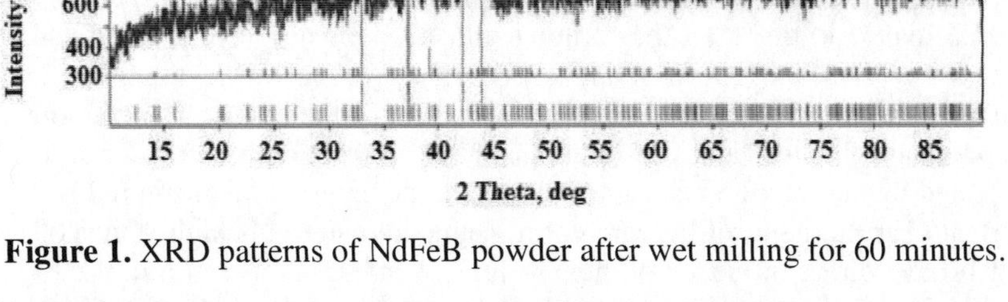

Figure 1. XRD patterns of NdFeB powder after wet milling for 60 minutes.

The fine powder was measured distribution of particle size by using laser particle size analyzer (PSA), figure 2 shows PSA result of sample after milling 60 minutes. The fine powder NdFeB has mean diameter about 21.29 μm after milling using HEM for 60 minutes and the PSA curve as shown in figure 2 has a narrow distribution curve, which means that the NdFeB powder has a homogeneous particle size.

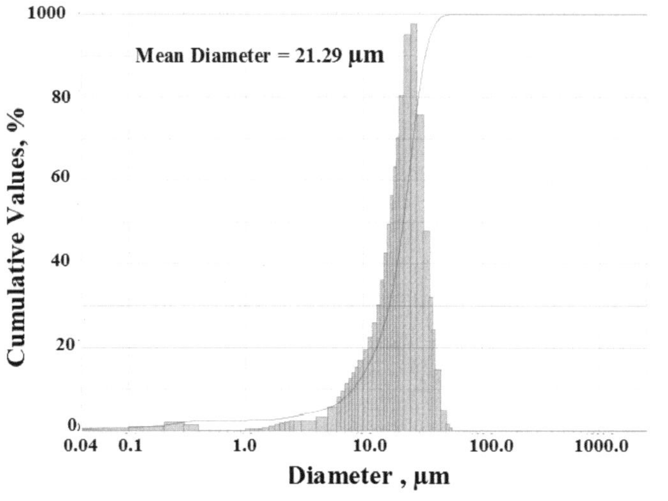

Figure 2. Distribution particle size of NdFeB powder after milling 60 minutes.

NdFeB fine powder was formed by using pressure at 40 kgf/cm^2 and external magnetic field was applied at 0, 2000, 5000, 8000, and 11000 Gauss. The resulted pellets were dried at 100 °C in vacuum dryer, then all samples were magnetized by using impulse magnetizer. Subsequently, bulk density and flux magnetic of the bonded magnet were measured for all samples and the results were shown at figure 3. The bulk density value of the bonded magnet NdFeB tends to be constant with an external magnetic field variation of about 5.13-5.14 g/cm^3. When compared with the theoretical density value of NdFeB that is about 6-7 g/cm^3 [11], then the density obtained from this experiment only reaches about 78.92–79.08% from theoretical density. The density of the samples made in this experiment has not yet reached a high density (> 90%), which is strongly influenced by the pressure of compaction used; the compression pressure of 40 kgf/cm^2 is probably not enough and the compaction pressure should be much higher. Nevertheless, the magnetic properties, especially the flux magnetic value, are greatly influenced by the value of the external magnetic field.

The highest flux magnetic value achieved is about 465.9–467.1 Gauss with applied external magnetic field 8000–11000 Gauss. When compared with other author's research results, where NdFeB permanent bonded magnets have magnetic flux values of 40-60% of the remanence value (Br) that should be 1000-2000 Gauss for NdFeB bonded magnets [12, 13]. This is due to the raw material factor used in the experiment, as seen from the XRD analysis (figure 1) where the intensity of the NdFeB peaks seems to be still too low. This is causing the flux magnetic to remain very low. The effect of providing external magnetic fields on the forming process does not have a significant effect on flux magnetic value.

The properties of magnetic materials are a manifestation of the hysteresis phenomena, where hard magnetic materials have greater values of hysteresis [14]. Magnetic properties of an isotropic bonded magnets are affected by the fraction of magnetic powder and external magnetic field used [11]. One of the most important characteristics of the rare-earth magnetic material is high values of remanence and coercivity, which have direct influence on high values of maximal energy product [9]. The result of hysteresis curve analysis for bonded magnet NdFeB as shown in figure 4 shows that the magnitude of the external magnetic field used affects the area of the hysteresis curve. The larger the external magnetic field, the VSM hysteresis curve tends to be slightly enlarged. The samples prepared by using external magnetic field pressing were measured using VSM, whereas the samples prepared without using external magnetic field pressing were not. Magnetic flux measurement results and hysteresis curve at figures 4a to 4d indicated that the magnetic flux value and hysteresis curves are almost the same for all samples. From the hysteresis curve shown in figure 4 properties of isotropic bonded magnet NdFeB such as remanence and coercivity can be obtained.

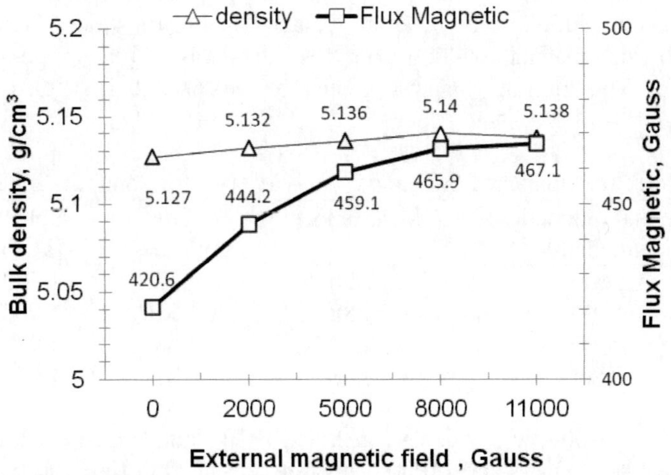

Figure 3. Bulk density value and flux magnetic value as function of external magnetic field.

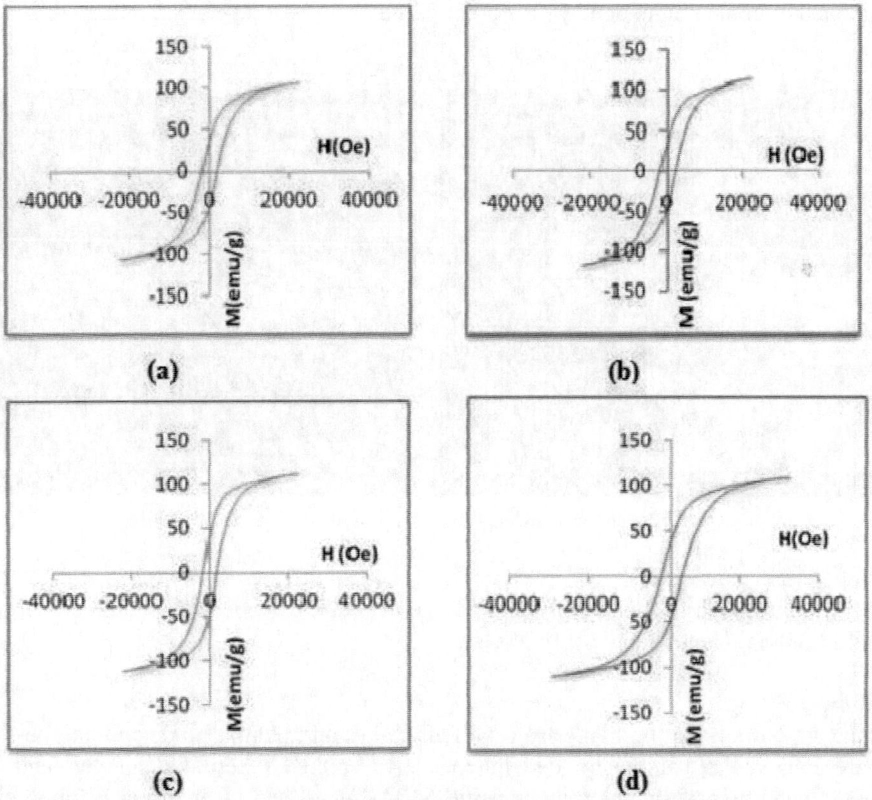

Figure 4. The hysteresis curve of an isotropic NdFeB bonded magnet samples made with external magnetic fields: (a) 2000 Gauss; (b) 5000 Gauss; (c) 8000 Gauss; (d) 11000 Gauss.

IOP Conf. Series: Journal of Physics: Conf. Series **985** (2018) 012043 doi:10.1088/1742-6596/985/1/012043

Table 1 shows the remanence and coercivity values of the isotropic bonded magnetic samples made from NdFeB flakes and the variations of external magnetic application. When compared to the magnet (Br, Hc) bonded magnetic NdFeB value of another author having Br about 6-8 kGauss and Hc about 4-10 kOe [7-9], the bonded magnet resulting from this research is still lower, possibly due to different raw materials result. The starting material needs to be heated first to obtain more perfect crystallization.

Table 1. Value of remanence (Mr) and coercivity (Hc) for bonded magnet NdFeB.

External Magnetic Field, Gauss	Mr (emu/g)	Mr (kGauss)	Hc (kOe)
2000	37.86	2.550	1.783
5000	38.88	2.570	1.903
8000	40.77	2.632	1.925
11000	42.68	2.776	1.905

SEM micrographs of isotropic bonded magnet Nd-Fe-B sample made with celuna binder and external magnetic field 8000 Gauss and 11000 Gauss are presented in figure 5. It shows that the shape and size of NdFeB particles is in homogeneous. The largest particles have size of about 4-5 μm, and there is an agglomeration of smaller particles of about 0.5-1 μm as shown in figures 5a and 5b. The celuna binder is not clearly visible and possibly envelops the entire surface of the particle. The effect of applied external magnetic field is not clearly seen, so the change of remanence value and coercivity in various applied external magnetic field is not too large.

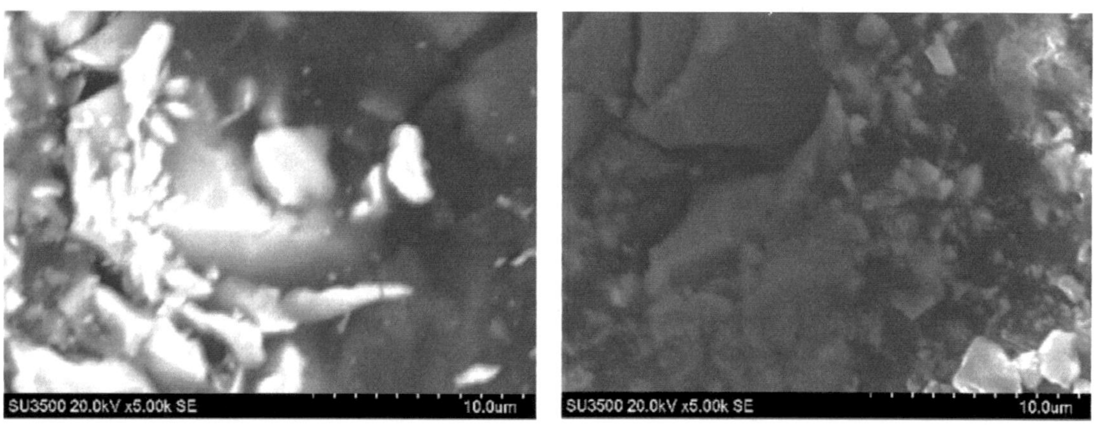

(a) (b)

Figure 5. SEM micrographs of anisotropi bonded magnet NdFeB made with celuna binder and extenal magnetic field (a) 8000 Gauss and (b) 11000 Gauss.

4. Conclusion

A bonded magnet NdFeB can be made from NdFeB flakes and celuna binder; the highest magnetic properties are achieved at sample bonded magnet with applied external magnetic field of 8000–11000 Gauss. Evaluation of microstructure using SEM cannot see clear effect of applied external magnetic field. The magnetic properties of resulted samples bonded magnet NdFeB are flux magnetic value = 465.9–467.1 Gauss, remanence value = 2.63–2.776 kGauss, and coercivity value =1.905–1.925 kOe.

Acknowledgments
Authors wishing to special work by SEM and VSM operators from Research Center for Physics-LIPI.

References
[1] Grujić A S, Lazić N L, Talijan N M, Spasojević V, Stajić-Trošić J T, Ćosović V R and Aleksić R 2010 *Acta Physica Polonica A* **117** 859
[2] Haavisto M, Tuominen S, Santa-Nokki T, Kankaanpää H, Paju M and Ruuskanen P 2014 *Advances in Materials Science and Engineering 760584*
[3] Coey J M D 2010 *Magnetism and Magnetic Materials* (Cambridge: Cambridge University Press)
[4] Schultz L, Wecker J and Hellstern E 1987 *J. Appl. Phys.* **61** 3583
[5] Gutfleisch O 2000 *J. Phys. D* **33** 157
[6] Zhao Z, Wang X, Zhang X, Ma Q, Liu Y, Li Y, Liu F and Wang G 2017 *Journal of Magnetics* **22** 60
[7] Brown D, Ma B M and Chen Z 2002 *J. Magn. Magn. Mater.* **248** 432
[8] Ma B M, Herchenroeder J W, Smith B, Suda M, Brown D N and Chen Z 2002 *J. Magn. Magn. Mater* **239** 418
[9] Dobrzanski L A, Drak M and Ziebowicz B 2007 *J. Mater. Process. Technol.* **191** 352
[10] Yang Y, Wu Q, Hu Y C, Zhang P Y and Ge H L 2016 *Journal of Magnetics* **21** 235
[11] Liu W, Cao L, Wu J and Li T 2003 *Mater. Trans.* **44** 2094
[12] Tao S, Lu X, Tian J, Qu X, Honkura Y, Mitarai H and Noguchi K 2009 *Rare Metals* **28** 245
[13] Sardjono P, Muljadi, Suprapedi, Sinuaji P, Ramlan and Gulo F 2017 *J. Phys.: Conf. Ser.* **817** 012052
[14] Stajić-Trošić J, Grujić A, Stijepović M Z, Nedeljković D, Stajčić A and Zak T 2014 *Zaštita Materijala* **55** 170

IOP Conf. Series: Journal of Physics: Conf. Series **985** (2018) 012044 doi:10.1088/1742-6596/985/1/012044

Effect of Fe-Mn addition on microstructure and magnetic properties of NdFeB magnetic powders

C Kurniawan[1], A S Purba[2], E A Setiadi[1], S Simbolon[1], A Warman[2] and P Sebayang[1]

[1] Research Center for Physics, Indonesian Institute of Science (LIPI), Puspiptek Office Area, Tangerang Selatan, Banten, Indonesia 15314
[2] Department of Physics, The Univercity of Sumatera Utara, Jl. Bioteknologi No. 1 USU, Medan, Sumatera Utara, Indonesia 20155

E-mail: cand002@lipi.go.id

Abstract. In this paper, the effect of Fe-Mn alloy addition on microstructures and magnetic properties of NdFeB magnetic powders was investigated. Varied Fe-Mn compositions of 1, 5, and 10 wt% were mixed with commercial NdFeB type MQA powders for 15 minutes using shaker mill. The characterizations were performed by powder density, PSA, XRD, SEM, and VSM. The Fe-Mn addition increased the powder density of NdFeB/Fe-Mn powders. On the other side, particle size distribution slightly decreased as the Fe-Mn composition increases. Magnetic properties of NdFeB/Fe-Mn powders changed with the increasing of Fe-Mn content. SEM analysis showed the particle size of NdFeB/Fe-Mn powder was smaller as the Fe-Mn composition increases. It showed that NdFeB/Fe-Mn particles have different size and shape for NdFeB and Fe-Mn particles separately. The optimum magnetic properties of NdFeB/Fe-Mn powder was achieved on the 5 wt% Fe-Mn composition with remanence M_r = 49.45 emu/g, coercivity H_c = 2.201 kOe, and energy product, BH_{max} = 2.15 MGOe.

1. Introduction

Nanocomposite magnets composed by magnetically hard and soft phases have attracted more attention due to its higher remanence, coercivity and energy product for permanent magnet development [1–4]. In order to enhance magnetic properties such as remanence (M_r/M_s > 0.5 for isotropic magnets) and fairly coercivity values, both hard and soft magnetic phases are in nanoscale range so that the adjacent grain are exchange coupled [5]. To optimize the exchange couple magnetization, some soft magnetic phases have been added by researchers for specific elements such as α-Fe, Zr, Mn, and MnBi [6–11].

The work from Li et al. [12] showed that the core-shell structure of nanocomposite system based NdFeB/α-Fe could increase both the remanence and coercivity, which leads to energy product of isotropic magnets exceeding 25 MGOe. The addition of FeB to $Nd_2Fe_{14}B$ structure also enhanced the coercivity field as reported by Mingxiang et al. [13]. Previous studies revealed that the addition of Mn to NdFeB could promote crystallization and enhance hard magnetic properties of materials [5]. On the other hand, the particle size of Mn doped NdFeB could affect the curie temperature, but lower the coercivity values. These effects were contributed by the decreasing of magnetocrystalline anisotropy [8].

In this paper, we reported and discussed the effect of Fe-Mn alloy addition to the NdFeB structure by mechanical mixing method. The magnetic properties from VSM measurements are reported. The microstructure and physical characteristics analyzed by XRD, SEM, PSA and density measurement are also discussed.

2. Experimental method

The commercial powders of NdFeB type MQA from Magnequench Inc. and Fe-Mn were utilized in this experiment. The powders were sieved using standard sieves until they passed #200 meshes. Varied composition of Fe-Mn from 1, 5, and 10 wt% was added to NdFeB powders and mixed with shaker mill for 15 minutes in toluene medium to prevent oxidation from environment. During mixing process, the powder to ball ratio was 1:10. After milling, the mixed slurry were dried in vacuum chamber (P = 10 mbar) at room temperature to obtain dried NdFeB/Fe-Mn powders.

The powder density was measured by pycnometer method using toluene as medium. Particle size distribution was analyzed by CILAS Particle Size Analyzer. SEM images were obtained by SEM Hitachi SU-3500. Magnetic phase analysis was performed by Smartlab Rigaku X-Ray Diffractometer using Cu-Kα radation. Magnetic properties were analyzed using Vibrating Sample Magnetometer type VSM250 from Dexing Magnet Ltd.

3. Results and discussions

The mean particle size and powder density measured by PSA and Pycnometer method are displayed in figure 1. It showed that the average value of particle size decreases as Fe-Mn content increases. This effect was contributed by the brittle characteristics of Fe-Mn alloys [14]. Therefore, the powder density of NdFeB/Fe-Mn is expected to increase as also displayed in figure 1. The results showed that the correlation of mean particle size and powder density is inversely proportional as the Fe-Mn content increases.

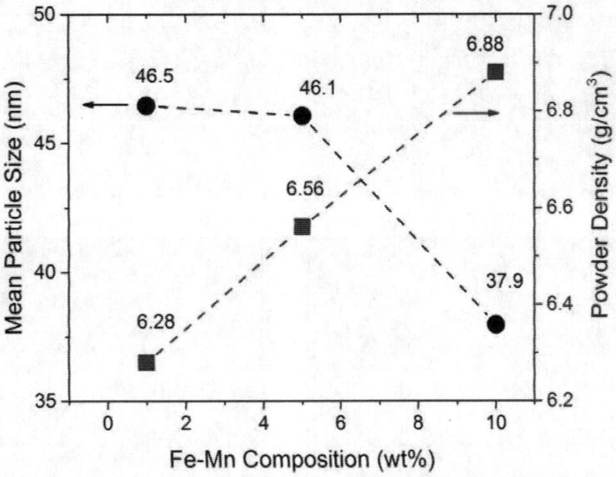

Figure 1. Mean particle size and powder density of NdFeB/Fe-Mn powders as a function of Fe-Mn composition.

The size distribution of NdFeB/Fe-Mn powders as measured by PSA is displayed in figure 2. It showed that the particle size distributions are narrower as the Fe-Mn content is increasing. Compared to the mean particle size values, the 50% cumulative diameter values in figure 2 are shifting to lower values.

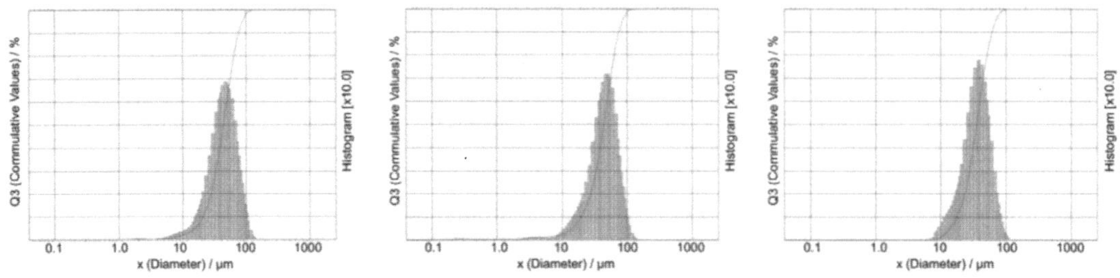

(a) (b) (c)

Figure 2. Size distribution of NdFeB/Fe-Mn powders measured by PSA with variation Fe-Mn composition of (a) 1 wt%, (b) 5 wt% and (c) 10 wt%.

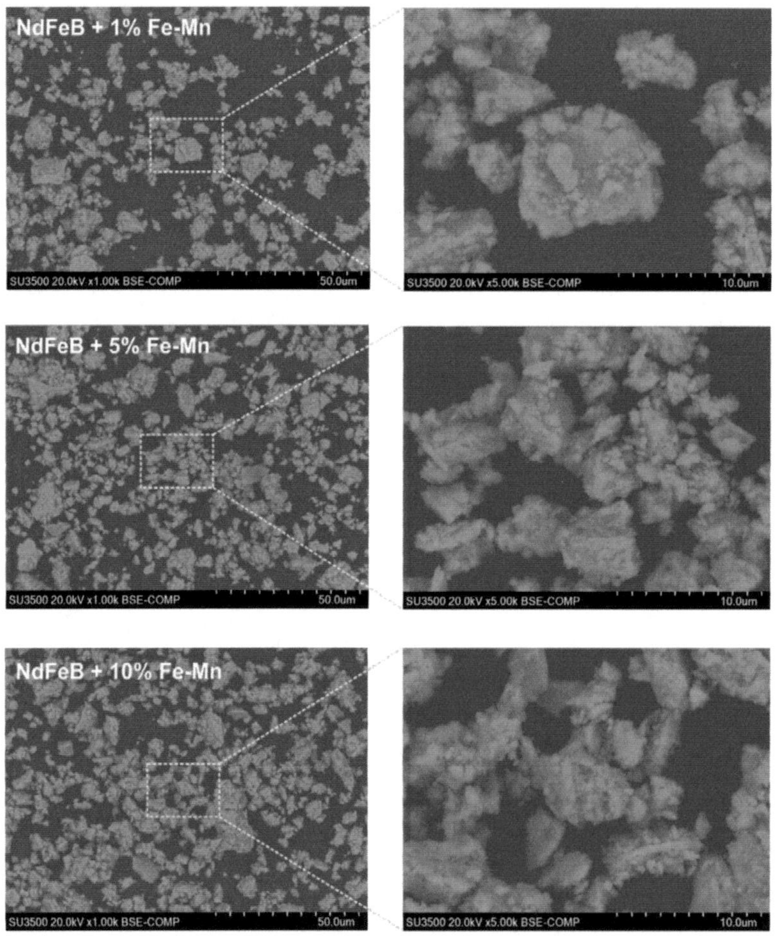

(a) (b)

Figure 3. SEM analysis of NdFeB/*x*Fe-Mn (x = 1, 5, 10 wt%) powder with different magnification of (a) 1.000X and (b) 5.000X.

The morphology and microstructure analysis was performed by SEM analysis as shown in figure 3. The results showed that the particle size of NdFeB/Fe-Mn powder is confirmed to be smaller as the Fe-Mn composition increases. It showed that the NdFeB/Fe-Mn particles have different size and shape for NdFeB and Fe-Mn particles separately. In addition, the NdFeB has larger size and rounded shape with particle size of 5–10 µm. The Fe-Mn has smaller size and platelet shape with particle size of 1–3 μm. These values are far lower than the PSA analysis average values due to sampling area method in SEM analysis.

The magnetic phases analysis performed by XRD is displayed in figure 4. It showed that the Fe-Mn alloy (sample a) is composed by Fe_3Mn_7, α-Fe, and MnO_2. On the other hand, pure NdFeB powder is composed by single phase $Nd_2Fe_{14}B$ (sample b). The increasing amount of Fe-Mn phases in NdFeB/Fe-Mn powders are shown by the increasing Fe-Mn phases peak intensity. However, the mixing process by shaker mill affects the peak broadening of $Nd_2Fe_{14}B$ structure. It indicates that the crystalline size decreases as the milling process conducted. The work of Chikhalia et al. stated that the milling process could deteriorate the crystalline structure and reduce the particle size [15]. Based on the Scherrer formula, the average crystalline size of the NdFeB/Fe-Mn powders is 22.8 nm. Compared to the unmilled sample, the average $Nd_2Fe_{14}B$ crystalline size is 35.6 nm.

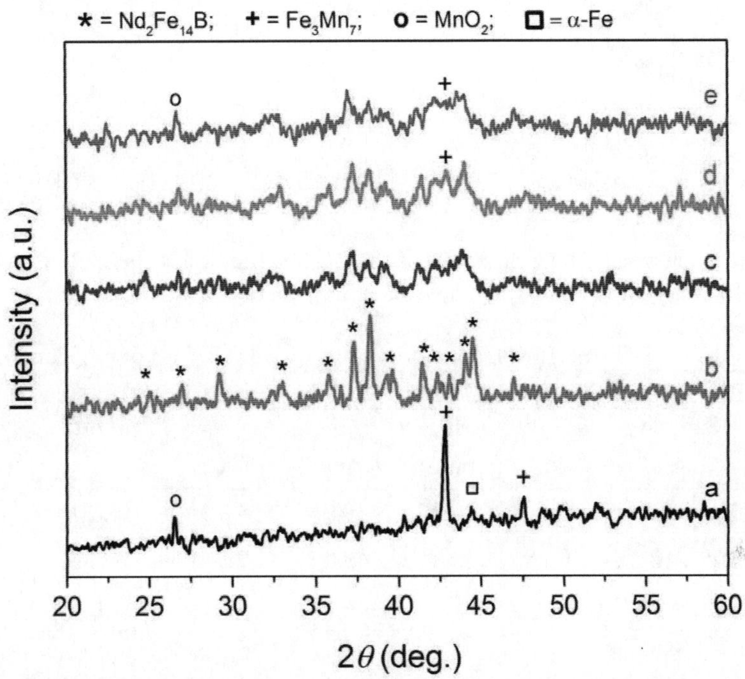

Figure 4. XRD curve of (a) pure Fe-Mn, (b) pure NdFeB and NdFeB/Fe-Mn powder with varied Fe-Mn composition of (c) 1 wt%, (d) 5 wt% and (e) 10 wt%.

The hysteresis curve analyzed by VSM is displayed in figure 5. It showed that the addition of Fe-Mn to NdFeB powders changed the magnetic behavior such as magnetic saturation (M_s), remanence (M_r), coercivity field (H_c), and energy product (BH_{max}). The magnetic remanences slightly decreased as the Fe-Mn content increases. However, the coercivity values have maximum point, which is obtained at 5 wt% Fe-Mn addition with M_r = 49.45 emu/g and H_c = 2.201 kOe. These properties lead to higher energy product, BH_{max} with value of 2.15 MGOe. The detailed magnetic properties data for all samples are presented in table 1.

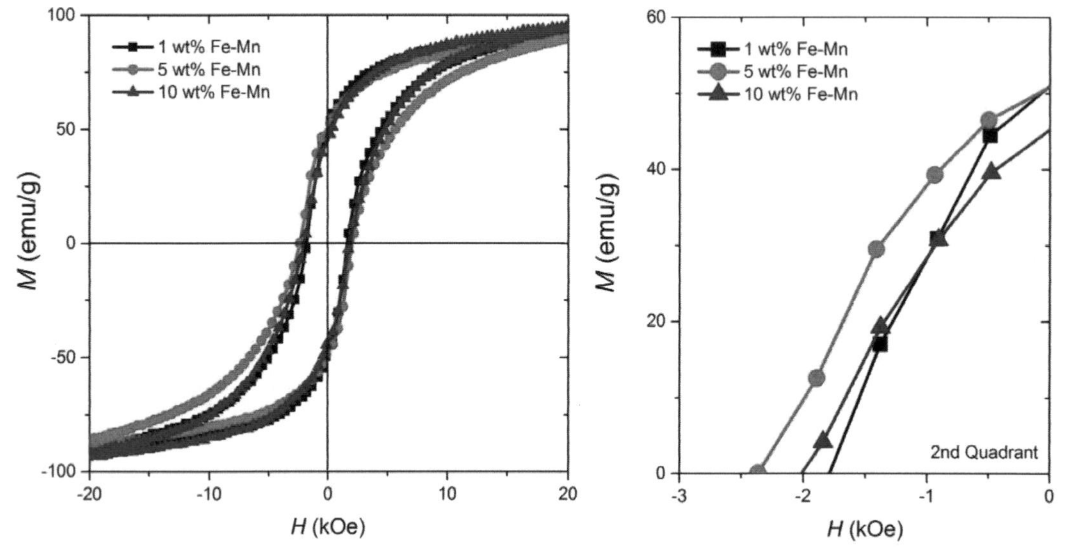

Figure 5. Hysteresis curve and magnification of the 2nd quadrant of NdFeB/Fe-Mn magnetic powders with varied Fe-Mn composition of 1 wt%, 5 wt%, and 10 wt%.

Table 1. Magnetic properties of NdFeB/Fe-Mn powder.

Sample	M_s (emu/g)	M_r (emu/g)	H_c (kOe)	M_r/M_s	BH_{max} (MGOe)
NdFeB/1 wt% Fe-Mn	94.12	49.94	1.706	0.531	1.63
NdFeB/5 wt% Fe-Mn	92.18	49.45	2.201	0.536	2.15
NdFeB/10 wt% Fe-Mn	96.22	44.07	1.911	0.458	1.65

4. Conclusion

In this work, the effect of Fe-Mn addition to NdFeB powders by mechanical mixing method has been investigated. The Fe-Mn addition increased the powder density of NdFeB/Fe-Mn powders. On the other side, the particle size distribution slightly decreased as the Fe-Mn composition increases. The SEM images showed that the particle size of NdFeB/Fe-Mn powder was smaller as the Fe-Mn composition increases. It showed that the NdFeB/Fe-Mn particles have different size and shape for NdFeB and Fe-Mn particles separately. Magnetic properties of NdFeB/Fe-Mn powders change as the increasing of Fe-Mn content. The optimum magnetic properties of NdFeB/Fe-Mn powder was achieved on the 5 wt% Fe-Mn composition with remanence M_r = 49.45 emu/g, coercivity H_c = 2.201 kOe, and energy product, BH_{max} = 2.15 MGOe.

Acknowledgement

This work is supported by Mandiri Research Program from Research Center for Physics, Indonesian Institute of Sciences (LIPI) of financial year 2017.

IOP Conf. Series: Journal of Physics: Conf. Series **985** (2018) 012044 doi:10.1088/1742-6596/985/1/012044

References

[1] Wang Y, You C, Wang J, Tian N, Lu Z and Ge L 2012 *Journal of Rare Earths* **30** 757
[2] Suryanarayana C and Al-Aqeeli N 2013 *Prog. Mater. Sci.* **58** 383
[3] Lou L, Hou F C, Wang Y N, Cheng Y, Li H L, Li W, Guo D F, Li X H and Zhang X Y 2014 *J. Magn. Magn. Mater.* **352** 45
[4] Soderžnik M et al. 2017 *Acta Mater.* **135** 68
[5] Xie G, Yin S, Zhang F, Lin P, Gu B, Lu M, Du Y and Yuan Z 2004 *Mater. Lett.* **58** 636
[6] Ahmad I, Davies H A and Kanwal M 2012 *J. Magn. Magn. Mater.* **324** 3971
[7] Liu Z W, Qian D Y, Zhao L Z, Zheng Z G, Gao X X and Ramanujan R V 2014 *J. Alloys Compd.* **606** 44
[8] Akdogan N G, Li W and Hadjipanayis G C 2014 *J. Nanopart. Res.* **16** 2797
[9] Han G, Su H, Gao R, Yu S, Kang S, Zhu M, Li W and Liu X B 2015 *Journal of Rare Earths* **33** 1303
[10] Zhang D T, Wang P F, Yue M, Liu W Q, Zhang J X, Sundararajan J A and Qiang Y 2016 *Rare Metals* **35** 471
[11] Ma Y L, Liu X B, Nguyen V V, Poudyal N, Yue M and Liu J P 2016 *J. Magn. Magn. Mater.* **411** 116
[12] Li H, Li X, Guo D, Lou L, Li W and Zhang X 2016 *Nano Lett.* **16** 5631
[13] Mingxiang P, Pengyue Z, Hongliang G, Hangfu Y and Qiong W 2012 *J. Magn. Magn. Mater.* **324** 2953
[14] Jones R T and Mills K. 2011 *Southern African Pyrometallurgy 2011 International Conference: [6-9 March 2011 Misty Hills, Muldersdrift* (Johannesburg: Southern African Institute of Mining and Metallurgy)
[15] Chikhalia V, Forbes R T, Storey R A and Ticehurst M 2006 *Eur. J. Pharm. Sci.* **27** 19

IOP Conf. Series: Journal of Physics: Conf. Series **985** (2018) 012045 doi:10.1088/1742-6596/985/1/012045

The effect of calcination temperature on the formation and magnetic properties of $ZnMn_2O_4$ spinel

B Hermanto, Ciswandi, F Afriani, D Aryanto and T Sudiro

Research Center for Physics, Indonesian Institute of Sciences, Kawasan Puspiptek Serpong, Tangerang Selatan 15310, Indonesia

E-mail: toto011@lipi.go.id

Abstract. The spinel based on transition-metal oxides has a typical composition of AB_2O_4. In this study, the $ZnMn_2O_4$ spinel was synthesized using a powder metallurgy technique. The Zn and Mn metallic powders with an atomic ratio of 1:2 were mechanically alloyed for 3 hours in aqueous solution. The mixed powder was then calcined in a muffle furnace at elevated temperature of 400, 500 and 600 °C. The X-ray Diffractometer (XRD) was used to evaluate the formation of a $ZnMn_2O_4$ spinel structure. The magnetic properties of the sample at varying calcination temperatures were characterized by a Vibrating Sample Magnetometer (VSM). The results show that the fraction of $ZnMn_2O_4$ spinel formation increases with the increase of calcination temperature. The calcination temperature also affects the magnetic properties of the samples.

1. Introduction

The spinel structure based on transition metal oxides having a general formula of AB_2O_4 (A and B = a transition metal but not in same element) is a promissing material for wide range applications. The one of an attractive oxide spinel is $ZnMn_2O_4$. Due to its thermal and electrochemical properties, the zinc manganese oxide spinel materials have been used in practical devices such as negative temperature coeficient (NTC) thermistor, catalyst, high temperature materials, cathode and anode electrode for battery, supercapacitor applications, and solar energy fields [1-11].

The various methods have been used to synthesize the $ZnMn_2O_4$ spinels as: sol-gel [12], hydrothermal [4, 8, 10, 13], polymer-pyrolysis [1], solvothermal reaction [2, 5, 9, 11], flame spray pyrolysis [3], reactive template route combined with a post-calcination process [4], oxalate decomposition [6], and solid state reaction [14]. Nevertheless, the properties of $ZnMn_2O_4$ oxide spinel is often affected by its purity and crystalline structure. Accordingly, the appropriate method and processing parameter used are important to obtain the high purity and cristallinity of oxide spinel.

In this study, we synthesized the zinc manganese oxide spinel from Zn and Mn metallic powders using a powder metallurgy technique. The effect of calcination temperature on the formation of $ZnMn_2O_4$ spinel was studied by X-ray diffractometer and magnetic properties was investigated by Vibrating Sample Magnetometer (VSM).

2. Experimental

2.1. Material preparation and synthesis
The $ZnMn_2O_4$ samples were prepared using Zn (≥95%) and Mn metallic powders (≥99%) from Merck and Sigma-Aldrich, respectively. Initially, the Zn-Mn powder mixture with an atomic ratio of 1:2 (at%) was prepared by a mechanical alloying technique. The powder was mixed under wet milling condition in steel vial using a shaker mill with a powder to ball ratio of 1:10 for 3 h. After powder mixing was conducted, the Zn-Mn powder mixture was dryed to evaporate the milling solution. In order to form the $ZnMn_2O_4$ oxide spinel, the mixture powder was then calcined in muffle furnace at elevated calcination temperature of 400, 500 and 600 °C in an air atmosphere. Prior to powder characterization, the calcined powder was crushed using a mortar.

2.2. Characterization examination
A Smartlab-Rigaku X-ray diffractometer with CuK α radiation was used to analyze the formation of $ZnMn_2O_4$ spinel phases after the powders were calcined at different temperature. The X-ray diffraction measurement was carried out in the angle range of 10° - 90°. In this study, we estimated the average crystallite size (*D*) using Debye Scherer's formula [15], as shown below:

$$D = \frac{0.9\lambda}{\beta \cos\theta} \tag{1}$$

The λ, β and θ represent the wavelength of X-ray used (1.541862 Å), the full width at half maximum (FWHM) diffraction peak of $ZnMn_2O_4$ and the Bragg's diffraction angle, respectively. While the lattice strain (ε) was calculated using the following formula [15],

$$\varepsilon = \frac{\beta}{4\tan\theta} \tag{2}$$

The magnetic properties of calcined powder was characterized using a Vibrating Sample Magnetometer Model VSM-250 Dexing Magnet Tech. Co., Ltd. at room temperature.

3. Results and discussion

3.1. Spinel formation at varying calcination temperature
Figure 1 shows the X-ray diffraction patterns of Zn, Mn powders as a starting material and Zn-Mn mixture after mechanical alloying. It can be seen that the starting material of the powder as shown in figure 1a and 1b is composed mainly by Zn and Mn phases, respectively. After mixing by a shaker mill for 3 h, X-ray diffraction analysis shows a main peak in the diffraction angle of 43.16° (figure 1c). This suggest that the Zn and Mn powders form the (Zn, Mn) solid solution. There are no other phase formed in the powder after mechanical milling. The ball and powder collison during milling enhance the complete mixing of the starting powder materials.

Figure 2 shows the XRD patterns of formed $ZnMn_2O_4$ oxide spinel synthesized at varying calcination temperature. The results reveal that small diffraction peaks of $ZnMn_2O_4$ can be observed at calcination temperature of 400 °C. The other diffraction peaks were identified as MnO and ZnO phases. Accordingly, it can be predicted that during calcination, Zn and Mn elements are preferentiallty oxidized to form its oxide as shown in equation 3 and 4. While, the formation of $ZnMn_2O_4$ oxide spinel may take place in the two ways according to equation 5 and 6.

$$2Zn + O_2 \rightarrow 2ZnO \tag{3}$$

$$2Mn + O_2 \rightarrow 2MnO \tag{4}$$

$$Zn + 2Mn + 2O_2 \rightarrow ZnMn_2O_4 \tag{5}$$

$$2ZnO + 4MnO + O_2 \rightarrow 2ZnMn_2O_4 \quad (6)$$

At higher calcination temperature of 500 and 600 °C, the formation of $ZnMn_2O_4$ oxide spinel tends to raise. The small amounts of MnO and ZnO phases are still observed at 500 °C. The $ZnMn_2O_4$ spinel was observed as a mojor phase after calcined at 600 °C for 18 h. Since the formation of ZnO and MnO decreses with the increase of calcination temperature, the formation of $ZnMn_2O_4$ is likely to occur due to the reaction between ZnO and MnO, as shown in equation 6.

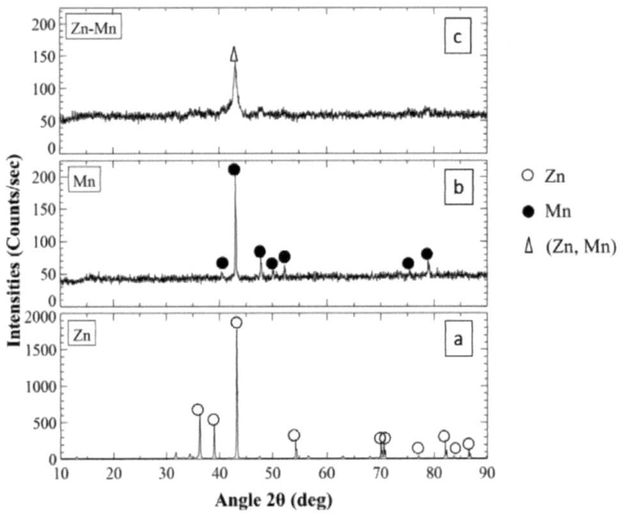

Figure 1. X-ray diffraction patterns of (a) Zn, (b) Mn starting materials, and (c) Zn-Mn powder after mechanical alloying.

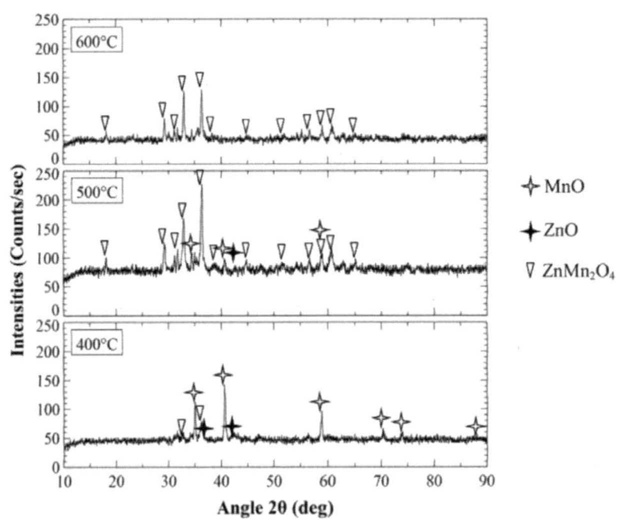

Figure 2. X-ray diffraction patterns of formed $ZnMn_2O_4$ prepared at varying calcination temperature of 400, 500 and 600 °C.

The crystallite size of synthesized powder was estimated using a Scherer formula based on the Full Width Half Maximum (FWHM) of X-Ray diffraction data. According to the results as presented in table 1, the cristallite size of $ZnMn_2O_4$ increases with an increasing calcination temperature. It is about

14 - 28 nm. It seems that the cristallite size of $ZnMn_2O_4$ spinel tends to grow with the increase of calcination temperature. On the contrary, the lattice strain is gradually decreased by increasing calcination temperature.

3.2. Effect of calcination temperature on the magnetic properties

Figure 3 shows the magnetization curves of formed $ZnMn_2O_4$ synthesized at elevated calcination temperatures of 400, 500 and 600 °C. The corresponding magnetic properties of the magnetization measurement are listed in table 1. Initially when the external field was applied to the sample, the hysteresis curve looks like it has not been saturated. But the line between magnetization and demagnetization has a coincided. This indicates that the measurement has undergone magnetization saturation state. These results correspond to the magnetization measurement of $ZnMn_2O_4$ studied by Song et al. [5]. The magnetization curve first shows an extremely increases as the external field increases approximately a hundred Oersteds, and then a slow increase with the increases of external field. In the present study, the maximum external field which can be applied is up to 22 kOe (2.2 Tesla). The magnetization curve was possible to rise slowly with increasing external magnetic field when it was compared to the previous study [5] which was measured for up to 7 T.

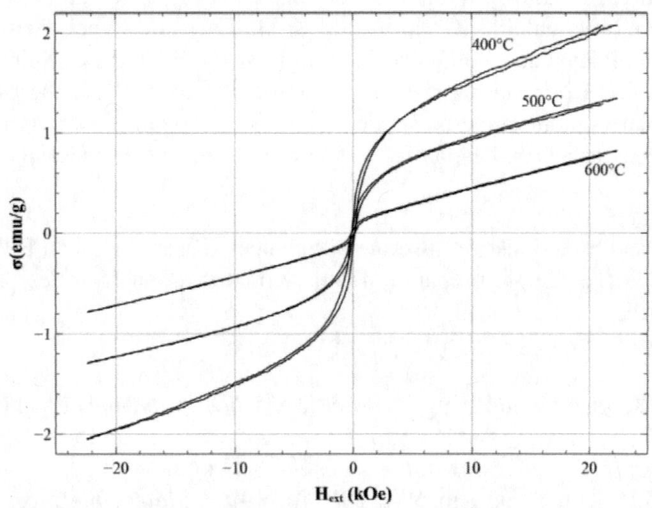

Figure 3. Hysteresis curve of formed $ZnMn_2O_4$ prepared at elevated calcinations temperatures of 400 °C, 500 °C and 600 °C.

As can be seen in table 1, the saturation magnetization of the samples decreases from 2.08 to 0.82 emu/g and for the remanance magnetization is decreased from 0.14 to 0.03 emu/g when the calcination temperature increases from 400 to 600 °C. Meanwhile, the coercivity of the sample is varied between 120–133 Oe. It was known that Mn [16] and Zn [17] as a starting materials in this study are paramagnetic and diamagnetic, respectively. The Zn oxide (ZnO) and Mn oxide (MnO) exhibit diamagnetic [18] and antiferromagnetic [19], respectively. The table 1 shows that the formed $ZnMn_2O_4$ prepared at varying calcinations temperature has a low magnetic saturation, remanence and coercivity. Obviously from figure 2, the calcination temperature of 600 °C leads to the formation of high purity $ZnMn_2O_4$ spinel oxide, while at 400 and 500 °C, the small amount of MnO and ZnO can still be found. In addition, table 1 indicates that the crystallite sizes of $ZnMn_2O_4$ are also depended on the calcinations temperature. The aforesaid evidences strongly suggest that the magnetic properties of synthesized sample are affected by the phase composition and crystallite size. At calcination temperature of 600 °C, the synthesized material consisting of $ZnMn_2O_4$ spinel shows antiferromagnetic behavior [20] with low magnetic saturation, remanence and coercivity. Thus, according to the magnetic properties clasification as reported in Ref. [21], the $ZnMn_2O_4$ oxide spinel can be classified as semi-hard magnetic (jHc between 25 to 700 Oe).

Table 1. Crystallite size, lattice strain and magnetic properties of formed $ZnMn_2O_4$ prepared at varying calcination temperature.

Calcination temperature T (°C)	Crystallite size D (nm)	Lattice strain ε [%]	Saturation magnetization σ_s (emu/g)	Remanence magnetization σ_r (emu/g)	Coercivity jHc (Oe)
400	15	0.763	2.08	0.14	133
500	25	0.444	1.34	0.07	120
600	29	0.386	0.82	0.03	124

4. Conclusions

The $ZnMn_2O_4$ spinel was sucsesfully synthesed using powder metalurgy technique. The phase formation and stability of $ZnMn_2O_4$ spinel were affected by calcination temperatures. At low calcination temperature (400 and 500 °C), instead of $ZnMn_2O_4$ oxide spinel, the phase characterization indicates the formation of ZnO and MnO phases. A high purity of $ZnMn_2O_4$ spinel was mostly formed after calcined at 600 °C. The crystallite size of the sample increases while the lattice strain decreases with the increase of calcination temperature. Few hundred Oerstead in coercivity of $ZnMn_2O_4$ oxide spinel reveals that this material can be classified as semi-hard magnetic material.

Acknowledgments
This work was supported by Indonesian Institute of Sciences (Riset Unggulan LIPI 2017). The authors gratefully acknowledge the sample preparation and characterization facilities provided by Research Center for Physics-LIPI.

References

[1] Yang Y, Zhao Y, Xiao L and Zhang L 2008 *Electrochem. Commun.* **10** 1117
[2] Xiao L, Yang Y, Yin J, Li Q and Zhang L 2009 *J. Power Sources* **194** 1089
[3] Choi S H and Kang Y C 2013 *Int. J. Electrochem. Sci.* **8** 6281
[4] Kim J G, Lee S H, Kim Y and Kim W B 2013 *ACS Appl. Mater. Interfaces.* **5** 11321
[5] Song M S, Cho Y J, Yoon D Y, Nahm S, Oh S H, Woo K, Ko J M and Cho W I 2014 *Electrochim. Acta* **137** 266
[6] Knight J C, Therese S and Manthiram A 2015 *J. Mater. Chem. A.* **3** 21077
[7] Zhang N, Cheng F, Liu Y, Zhao Q, Lei K, Chen C, Liu X and Chen J 2016 *J. Am. Chem. Soc.* **138** 12894
[8] Guo N, Wei X Q, Deng X L and Xu X J 2015 *Appl. Surf. Sci.* **356** 1127
[9] Zhao L, Li X and Zhao J 2013 *Appl. Surf. Sci.* **268** 274
[10] Courtel F M, Abu-Lebdeh Y and Davidson I J 2012 *Electrochim. Acta* **71** 123
[11] Zeng X, Shi L, Li L, Yang J, Cheng X and Gao M 2015 *RSC Adv.* **5** 70379
[12] Peng H and Wu T 2009 *Appl. Phys. Lett.* **95** 152106
[13] Zhang X D, Wu S Z, Zang J, Li D and Zhang Z D 2007 *J. Phys. Chem. Solids* **68** 1583
[14] Peiteado M, Caballero A C and Makovec D 2007 *J. Solid State Chem.* **180** 2459
[15] Aryanto D, Jannah W N, Masturi, Sudiro T, Wismogroho A S, Sebayang P, Sugianto and Marwoto P 2017 *J. Phys.: Conf. Ser.* **817** 012025
[16] Yeom T H and Lee S H 2013 *Journal of Magnetics* **18** 221
[17] Upadhyay A et al. 2017 *Chem. Eur. J.* **23** 4903
[18] Gilliland S, Segura A, Sánchez-Royo J F, García L M, Bartolomé F, Sans J A, Martínez-Criado G and Jimenez-Villacorta F 2010 *J. Appl. Phys.* **108** 073922

[19] Chatterji T, Su Y, Iles G N, Lee Y C, Khandhar A P and Krishnan K M 2010 *J. Magn. Magn. Mater.* **322** 3333
[20] Huber S, Sofer Z, Nadherny L, Jankovsky O, Simek P, Sedmidubsky D and Marysko M 2013 *Ceramics – Silikaty* **57** 162
[21] Steve Constantinides *The Elements of Magnetics* https://www.magmatllc.com

The effect of temperature on synthesis of $MgFe_2O_4$ based on natural iron sand by Co-precipitation method as adsorbent Pb ion

E A Setiadi[1], M Yunus[1], N Nababan[2], S Simbolon[1], C Kurniawan[1], S Humaidi[2], P Sebayang[1] and M Ginting[1]

[1]Research Center for Physics, Indonesian Intitute of Sciences (LIPI), Puspiptek, Serpong, Tangerang Selatan, Banten, Indonesia
[2]Department of Physics, State University of North Sumatera, Jalan Bioteknologi No.1, Medan, Indonesia.

E-mail: eko.arief.setiadi@lipi.go.id

Abstract. Magnetic materials of magnesium ferrite ($MgFe_2O_4$) based on natural iron sand has been successfully prepared as an adsorbent of Pb ions. This material was synthesized by co-precipitation method with natural iron and $Mg(CH_3COO)_2.4H_2O$ as precursor powders which were mixed with 2.0 M NH_4OH at a synthesis temperature of 50, 70 and 90 °C. The material characterizations were carried out using X-Ray Diffraction (XRD), Scanning Electron Microscopy (SEM), and Vibrating Sample Magnetometer (VSM). The XRD pattern showed that $MgFe_2O_4$ has single phase and its crystallite size increased and the powder density decreased as the temperature increased. The optimum temperature for the synthesis of $MgFe_2O_4$ was found at 50 °C, when the saturation magnetization and coercivity were at the highest values of 51.73 emu/g and 308 Oe, respectively. The result also showed that $MgFe_2O_4$ as an adsorbent of Pb ions has adsorption capacity of 143.07mg/g and removal efficiency of 97.3%.

1. Introduction

Iron sand is a one of abundant natural mineral in Indonesia. Natural iron sand contains high iron (Fe) content and can be processed into magnetic material to have a high economic value. One of the magnetic materials prepared from natural iron sand is based on spinel ferrite. Spinel ferrite possess two different crystal structures: tetrahedral and octahedral. The chemical formula of the spinel ferrite is $M\text{-}Fe_2O_4$, where M is a divalent of metal ion such as Cu, Zn, Ni, Co, Fe, Mn, Mg or their mixtures. Spinel ferrite has widely known for its low coercivity as well as high permeability and resistance; but, coercivity is low when the particle size is in nanometer scale [1].

The magnesium ferrite ($MgFe_2O_4$) is one of spinel ferrites which have high saturation magnetization, Curie temperature and electrical resistivity [1]. It is also known for its unique chemical properties and thermal stability. The synthesis methods for $MgFe_2O_4$ production developed so far included sol-gel [2], hydrothermal [3], polyol, co-precipitation [4], combustion [5], and solvothermal [6]. The $MgFe_2O_4$ was categorized as an *n*-type semiconductor materials so it can be applied as adsorption, sensor, Magnetic Resonance Imaging (MRI) materials [7], magnetic separation, hyperthermia treatment [2], and drugs delivery [8, 9]. In its application as heavy metal ion adsorbents, optimum conditions will be obtained if the material has a small particle size, large absorbent surface area, high saturation, and low coercivity [10].

In this research, the $MgFe_2O_4$ from natural iron sand was synthesized by coprecipitation method that is advantageous for its relatively simple procedure at low temperatures. The synthesis temperature was varied at 50, 70 and 90 °C. $MgFe_2O_4$ properties were analyzed by X-Ray diffraction (XRD), scanning electron microscope (SEM), and vibrating sample magnetometer (VSM).

2. Experiment method

Iron sand powder from Deli Serdang River of North Sumatra was used as raw material The precursor solution was made from the mixture of iron sand (2.72 g) and magnesium acetate tetrahydrate, $Mg(CH_3COO)_2.4H_2O$ (5.91 g), dissolved in 40 mL HCl solution (5.1 M). The solution was stirred using magnetic stirrer at 500 rpm and then filtered with whatman filter paper (125 μm diameter). Subsequently, the all solution was added dropwise into 99 ml NH_4OH 2 M solution at temperatures of 50, 70, and 90 °C, named as T50, T70, and T90 samples. The solutions were continuously stirred for 120 minutes at 500 rpm to form dark precipitate. The precipitate was washed with distilled water to reach neutral pH and dried in the oven at 100 °C to evaporate the water. The resulted $MgFe_2O_4$ powder was tested using pycnometer, X-Ray diffraction (XRD - Rigaku SmartLab, wavelength λ = 1.5418 Å), scanning electron microscope (SEM - Hitachi 3500 SU), and vibrating sample magnetometer (VSM - Electromagnetic VSM250) for the analysis of powder density, phase, microstructure, and magnetic properties respectively. The Pb ion adsorption was carried out by mixing 50 mg of $MgFe_2O_4$ powder into a 25 mL Pb ion solution in water. The mixture was then shaken with a shaker mill for 30 minutes. Finally, the ion content of Pb before and after adsorption was tested using atomic absorption spectroscopy (AAS - Shimadzu AA6800).

3. Results and discussion

Figure 1 shows diffraction pattern of the $MgFe_2O_4$ samples synthesized with various temperatures. Based on XRD database (PDF card No. 04-102-1070), all data showed that the samples has a single phase of cubic crystal structure. This observation can be used to calculate the lattice parameter and the crystallite diameter using the following equations:

$$a^2 = \frac{\lambda^2}{4sin^2\theta}(h^2 + k^2 + l^2) \quad (1)$$

$$D = \frac{K\lambda}{Bcos\theta} \quad (2)$$

In which *D* is crystallite diameter, *a* is lattice parameter, *θ* is angle of diffraction, *h*, *k*, and *l* are miller index, *K* is a Scherrer constant (0.9), *λ* = wavelength of the x-ray, and *B* is full width at half maximum (FWHM). The calculated lattice parameter and crystallite size were shown in table 1.

Based on table 1, it can be seen that the calculated lattice parameters for all samples are the same with those for $MgFe_2O_4$ material (8.410 Å). This result strongly confirms that the synthesized powder was $MgFe_2O_4$ with crystalline structure. Furthermore, we can also see from table 1 that the crystallite diameter tends to increase with the increase of synthesis temperature, while the crystal structure and cell volume are not affected [11, 12]. From the calculation results, the smallest crystallite size of 27.87 nm is obtained with synthesis temperature of 50 °C (T50). Meanwhile, an increase in crystallite size is directly proportional to the number of pores and inversely to powder density [13].

Obtained hysteresis curve from VSM measurement for all samples are shown in figure 2, and their associated magnetic properties in table 2. Table 2 shows that the coercivity and remanence increases with an increase of synthesis temperatures. The magnetic properties of material depend on the composition, crystallinity, size, and phases [11]. Smaller particle size will tend to be single domain, so that the energy barrier will be smaller. As a result, the magnetic moment will be easily switched by the influence of the external magnetic field. When magnetic field is eliminated the magnetic moment will be more easily to return to its original position, so that the coercivity value tends to decrease that the optimum synthesis temperature is 50 °C with coercivity value of 382 Oe.

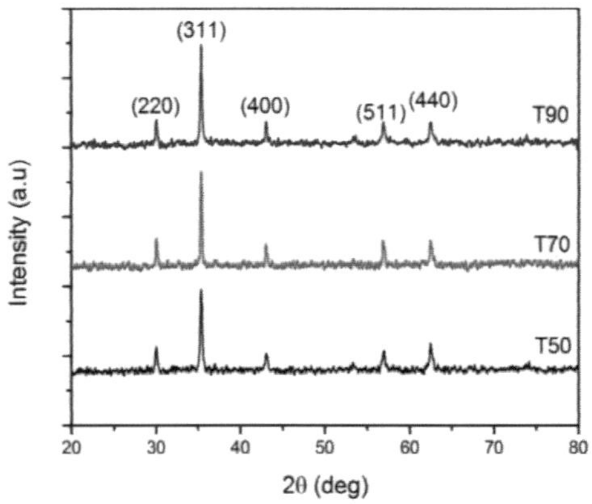

Figure 1. XRD diffraction pattern of $MgFe_2O_4$.

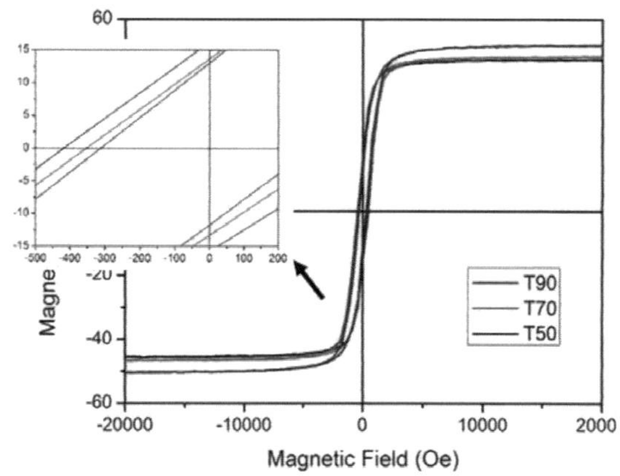

Figure 2. Hysteresis curves of $MgFe_2O_4$.

Table 1. Lattice parameter and crystallite size of $MgFe_2O_4$.

Sample	Powder density (g/cm^3)	Lattice parameter (Å)	Crytallite size (nm)
T50	4.677	8.411	27.87
T70	4.666	8.412	32.22
T90	4.482	8.411	34.30

Table 2. Magnetic properties of $MgFe_2O_4$.

Sample	Coercivity (Oe)	Saturation (emu/g)	Remanence (emu/g)
T50	308	51.73	13.24
T70	382	48.37	13.79
T90	416	47.41	16.48

SEM image and its EDX spectra of T50 sample are shown at figures 3a and 3b, respectively. From figure 3a, it can be observed that the particle size is ranged from 0.1 to 0.8 μm and the particle tends to agglomerate. EDX spectra as shown in figure 4b revealed that the T50 sample contains Mg, Fe, and O with the weight percentages of 0.6, 69.9, and 29.5 wt%.

We used obtained $MgFe_2O_4$ powder as Pb ion adsorbent by simply mixing the powder with an artificial waste solution made from Pb salt dissolved in water. The adsorption capacity (q) and removal efficiency (%R) of the powder can be calculated by the following equations:

$$q = \frac{C_o - C_e}{W} V \quad (3)$$

$$\%R = \frac{C_o - C_e}{C_o} 100 \quad (4)$$

where C_o is concentration of Pb ion before treatment, C_e is concentration of Pb ion after treatment, W is particle weight of $MgFe_2O_4$, and V is volume of waste solution. AAS analysis obtained that the adsorption capacity decreases with increasing synthetic temperature. The highest adsorption capacity was obtained for T50 samples at 143.07 mg/g (table 3) due to its small particle size and the largest surface area (table 3). From this result, we can conclude that T50 sample has a superparamagnetic property with high saturation and small coercivity [6, 10].

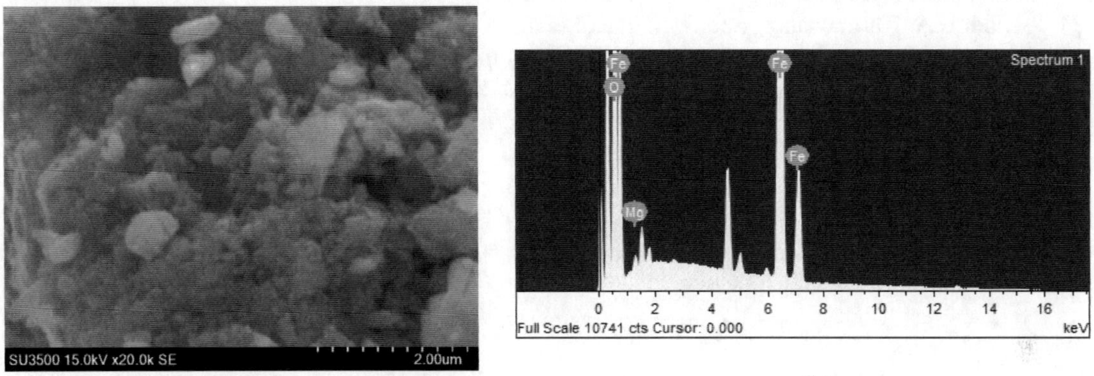

(a) (b)

Figure 3. (a) SEM image and (b) its correlated EDX spectra of T50.

Table 3. Adsorption properties of $MgFe_2O_4$.

Sample Name	Syntesize temperature (°C)	Concentration of Pb Ions, (ppm)		Adsorption capacity (mg/g)	Removal efficiency (%)
		Before treatment	After treatment		
T50	50	294	7.86	143.07	97.32
T70	70	294	34.70	129.65	88.20
T90	90	294	156.00	69.00	46 .93

4. Conclusion

Magnesium ferrite ($MgFe_2O_4$) as magnetic material has been successfully fabricated by co-precipitation method. XRD measurement revealed that the resulted $MgFe_2O_4$ powder has a single phase with spinel cubic structure. We also confirmed that the synthesis temperature was found to affect the crystallite diameter of the powder, where the smallest diameter was obtained for the sample synthesized at 50 °C. The magnetic properties of this material such as coercivity and magnetization saturation also confirmed this finding. The application of the powder as Pb ion adsorbent in the artificial waste water showed that the powder has a good adsorption capacity and high removal efficiency.

References

[1] Chen Q, Rondinone A J, Chakoumakos B C and Zhang Z J 1999 *J. Magn. Magn. Mater.* **194** 1
[2] Hussein S I, Elkady A S, Rashad M M, Mostafa A G and Megahid R M 2014 *J. Magn. Magn. Mater.* **379** 9
[3] Verma S, Joy P A, Khollam Y B, Potdar H S and Deshpande S B 2004 *Mater. Lett.* **58** 1092
[4] Shen Y F, Tang J, Wang Y, Ren Y and Zuo L 2009 *Bioresour. Technol.* **100** 4139
[5] Patil J Y, Khandekar M S, Mulla I S and Suryavanshi S S 2012 *Curr. Appl. Phys.* **12** 319

[6] Shen Y, Wu Y, Li X, Zhao Q and Hou Y 2013 *Mater. Lett.* **96** 85
[7] Bangale S V, Patil D R and Bamane S R 2011 *Arch. Appl. Sci. Res.* **3** 506
[8] Arruebo M, Fernández-Pacheco R, Ibarra M R and Santamaría J 2007 *Nano Today* **2** 22
[9] Wahajuddin and Arora S 2012 *Int. J. Nanomedicine* **7** 3445
[10] Setiadi E A, Sebayang P, Ginting M, Sari A Y, Kurniawan C, Saragih C S and Simamora P 2016 *J. Phys.: Conf. Ser.* **776** 012020
[11] Suharyadi E, Setiadi E A, Shabrina N, Kato T and Iwata S 2014 *Advanced Material Research* **896** 126
[12] Kim Y I, Kim D and Lee C S 2003 *Physica B* **337** 42
[13] Setiadi E A, Kurniawan C, Sebayang P and Ginting M 2017 *J. Phys.: Conf. Ser.* **817** 012054
[14] Goya G F, Fonseca F C and Jardim R F 2003 *J. Appl. Phys.* **93** 6531
[15] Lu L, Li J, Ng D H L, Yang P, Song P and Zuo M 2017 *J. Ind. Eng. Chem.* **46** 315
[16] Ghasemi E, Heydari A and Sillanpää M 2017 *Microchem. J.* **131** 51

IOP Conf. Series: Journal of Physics: Conf. Series **985** (2018) 012047 doi:10.1088/1742-6596/985/1/012047

Analysis of physical and magnetic properties of composite NdFeB bind with polyvinyl alcohol

Ramlan[1], P Sardjono[2], Muljadi[2], D Setiabudidaya[1] and F Gulo[3]

[1]Physics Department FMIPA, Sriwijaya University, Palembang Indonesia
[2]Research Center for Physics, Indonesian Institute of Sciences
[3]Department of Enviroment , Sriwijaya University, Palembang Indonesia

E-mail: ramlan@unsri.ac.id, priyo_sardjono@yahoo.com, fgulo@unsri.ac.id

Abstract. The composite magnet NdFeB has been made using magnetic powder MQP-B and polyvinyl alcohol (PVA) as the binder. The mixing compositions of raw materials used are: 95 wt% NdFeB - 5 wt% PVA, 92.5 wt% NdFeB - 7.5 wt% PVA, 90 wt.% NdFeB - 10 wt% PVA, and 87.5 wt% NdFeB - 12.5 wt% PVA. Both raw materials are weighed according to the composition, and then mixed until homogeneous. Furthermore, pellet forming was made using dry pressing at 50 kgf/cm^2 pressures and continued with drying at 100 °C and 10 mmbar for 4 hours. The characterization includes bulk density, hardness, compressive strength measurements, and magnetic properties testing. The characterization results show that the optimal composition of binder PVA is achieved at 5–7.5 wt% NdFeB composite magnet with following properties: bulk density = 5.21–5.25 g/cm^3, hardness = 302.17 - 304.32 Hv, compressive strength = 25.17–3.17 kgf/cm^2, magnetic flux = 1150-1170 Gauss, remanence = 70.90–74.97 emu/g or 4.7–5.0 kGauss, coercivity = 8.68–8.76 kOe, and energy product = 2.89–3.04 MgOe.

1. Introduction

The bonded or composite permanent magnets are technical products whose properties can be produced to suit a very wide variety of requirements. These materials have very important role as functional components within the wider spectra of contemporary devices in different industrial branches, as well as in the wider consumption. One of the most important applications of the permanent magnets are: parts in AC and DC engines production as well as synchronized motors, transformers, actuators, magnetic buffers, stationary fields [1, 2]. The most important advantages of polymer bonded magnets are low weight, can be produced in any shapes and high dimensional accuracy, i.e. near-net shape manufacture, good fracture toughness and corrosion resistance as well as isotropic magnetization [3]. Bonded magnets are produced by mixing a magnetic material with a binder and processed the mixture into a finished product. Magnets based on rare-earth-transition metal alloys like NdFeB alloy have excellent magnetic properties with high performance and maximum energy product value among other hard magnets [4, 5]. Nd-Fe-B magnets can be produced by sintering or bonding; in contrast to sintered magnets, bonded Nd-Fe-B magnets are produced by blending various polymer binders, such as epoxy or nylon, where melt-spun ribbons play the role of raw material. However, the magnetic properties of bonded magnets are inferior to those of sintered magnets. Commercial bonded magnets have a remanence (B_r) up to 0.7 T and an coercivity (H_{cl}) up to 1400 kA/m but sintered magnets possess magnetic properties as high as with a B_r up to 1.4 T and an H_{cl} up to 2500 kA/m^3 [6, 7]. The magnetic

IOP Conf. Series: Journal of Physics: Conf. Series **985** (2018) 012047 doi:10.1088/1742-6596/985/1/012047

properties of bonded magnets have been previously studied and reviewed by many researchers. In addition to magnetic properties, mechanical properties of bonded magnets also play critical roles for many advanced applications. However, the temperature-dependent mechanical strengths of bonded magnets are not readily available in open literature for magnet designers [9].

The type of binders and composition of binders are parameters which are important in making bonded permanent magnet. One would expect the remanence and energy product of a bonded magnet to be directly linked to the amount of binder used, typically 2 wt% in compression molded magnets and 8–15 wt% in injection molded magnets [9]. The purpose of this research is to make bonded magnetic NdFeB type MQP-B and adhesive polyvinyl alcohol with various composition. This research is expected to produce optimal composition for bonded permanent magnets.

2. Experimentals

Synthesis of composite magnet NdFeB was carried out using magnetic powder NdFeB type MQP-B and polyvinyl alcohol (PVA) as binder. The mixing compositions of the raw materials used are: 95 wt% NdFeB - 5 wt% PVA, 92.5 wt% NdFeB - 7.5 wt% PVA, 90 wt% NdFeB - 10 wt% PVA, and 87.5 wt% NdFeB - 12.5 wt% PVA. Both raw materials are weighed according to the composition, and then mixed until homogeneous. Furthermore, pellet forming was made using dry pressing at 50 kgf/cm^2 pressure. Then the pellets were dried at 100 °C and 10 mmbar for 4 hours. The characterization includes bulk density using Archimedes method, vickers hardness using Microhardness Tester, compressive strength measurements using Universal Testing Machine (UTM), and magnetic properties testing measurement by using Gauss meter and Vibration Sample Magnetometer (VSM).

3. Results and discussion

Bulk density of the dried pellets were measured to evaluate densification results as a function of binder composition (PVA). The results of bulk density measurement is shown at figure 1. Figure 1 shows that changes in bulk density values from composition of binder (PVA) of 5 wt% to 7.5 wt% is relatively small and constant. Nevertheless, density values tend to decrease sharply from 7.5 wt% PVA to 12.5 wt% PVA. The decreasing of bulk density value is due to the density value difference of NdFeB and polymer (PVA). PVA has density value of about 1.2–1.5 g/cm^3, but density value of NdFeB is higher, i.e. 6–7 g/cm^3 [9]. If the amount of polymer material is higher, the bulk density of mixed materials tends to decrease. Based on the reference, a commercial bonded magnet NdFeB has bulk density of about 5.6-6.8 g/cm^3 [10, 11]. Meanwhile, the results achieved from this research obtained the highest bulk density value of about 5.21-5.25 g/cm^3 on samples with 5% and 7.5% PVA. According to the result of bulk density measurement, the achieved densification is about 83.87 %. Figure 2 shows vicker hardness curve as a function of binder composition (PVA). The profile curve of vickers hardness is similar with profile curve of bulk density, where the highest value of hardness is about 304.32 Hv and 302.17 Hv for samples with 5 wt% PVA and 7.5 wt% PVA, respectively. The hardness value for sample with binder composition (PVA) of more than 7.5 wt% tends to decrease. The decreasing value of hardness value is due to the amount of PVA composition. It has similar correlation with the density.

Based on the results in figures 1 and 2, it is found that samples with 5% and 7.5% binder (PVA) have the highest bulk density and vickers hardness values. This is because the amount of PVA binder that it serves to fill the pores and binds NdFeB magnetic particles perfectly, so that it can obtain the maximum density. If the material has high bulk density value then the value of hardness will also increase as well. When the amount of PVA binder is more than 7.5%, the bulk density and hardness value drops sharply. This is because the density value of PVA material is much lower than that of the NdFeB particle. Therefore, with the addition of adhesive over the optimal limit there will be a decline in the value of bulk density and hardness. Compressive strength is one of many mechanical properties of a material and it is measured to determine the ability of PVA binder to bind magnetic particles.

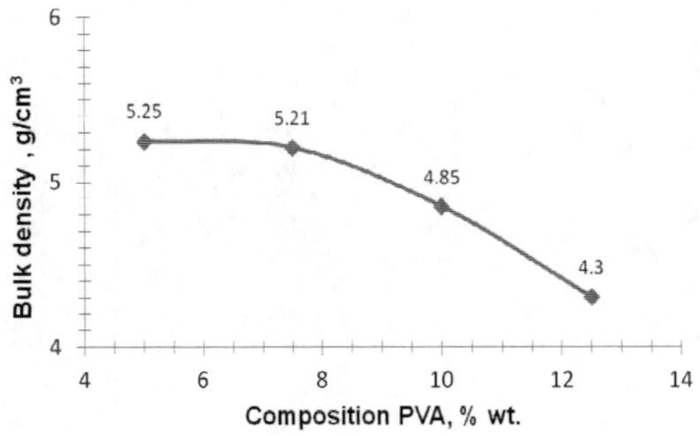

Figure 1. Bulk density curves as function of composition of binder (PVA).

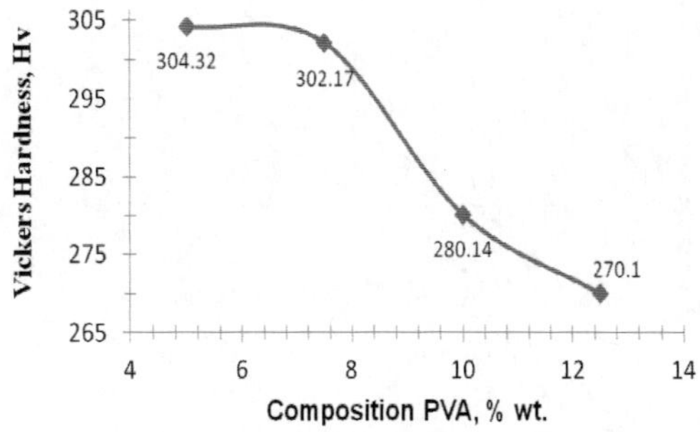

Figure 2. Vickers hardness curve as function of binder composition (PVA).

Figure 3 shows that compressive strength value tends to increase from 25.17 kgf/cm^2 to 33.75 kgf/cm^2 with increasing PVA binder composition from 5 % to 12.5 wt% PVA. Because of the higher volume fraction of binder, the sample exhibits higher compressive strength. It indicates that higher composition of PVA binder gives greater adhesion forces between powder particles. The highest compressive strength value of bonded magnet NdFeB using PVA adhesive is 33.75 kg/cm^2 whereas based on reference [10] bonded magnet NdFeB using epoxy resin adhesive has compressive strength value of about 55 kg/cm^2. The results achieved in this study indeed produced lower compressive strength, because different type of adhesive material can provide different mechanical strength. Therefore, the PVA adhesive produces lower adhesion when compared to the epoxy resin adhesive. More adhesive material will result in increasing compressive strength. However, it should also be considered that increasing amount of non-magnetic adhesive material will decrease its magnetic properties.

Magnetic properties of pellet samples were measured using Gaussmeter to obtain magnetic flux value. The result of magnetic flux measurement is shown in figure 4. Magnetic flux decreases slightly (1170 Gauss to 1150 Gauss) at samples with 5 wt% PVA to 7.5 wt% PVA and decreases sharply (1150 gauss to 810 Gauss) at samples with 7.5 wt% PVA to 12.5 wt% PVA. The decreasing of flux magnetic value is due to the increasing of binder composition and also the fact that the binder material (PVA) is a non magnetic materials. The optimum value of PVA composition is 5 wt% to 7.5 wt%.

Magnetic properties were investigated using VSM to observe the value of remanence, coercivity force, magnetic saturation, and energy product. The results are shown in figure 5. The hysteresis curve of sample bonded magnet with 5 wt% PVA and 7.5 wt% PVA is shown in figure 5, where value of magnetic properties were found: remanence, coercivity, and magnetic saturation (Ms) as shown at table 1. The hysteresis curve (figure 5) indicates that the hysteresis curve area of both samples do not look much different. The composition 5 wt% PVA and 7.5 wt% PVA do not affect significantly in the synthesis of bonded magnet NdFeB. The magnetic properties for both samples are relatively similar; the remanence value is 70.90–74.97 emu/g or 0.47–0.50 Tesla, coercivity value is 8.68–8.76 kOe and energy product 2.89–3.04 MgOe. When compared with other research, the remanence value of the results of this study (0.47-0.50 T) is still lower than that of Zhang Xiuhai and Xiong Weihao's study (0.6-0.7 T) [1, 9].

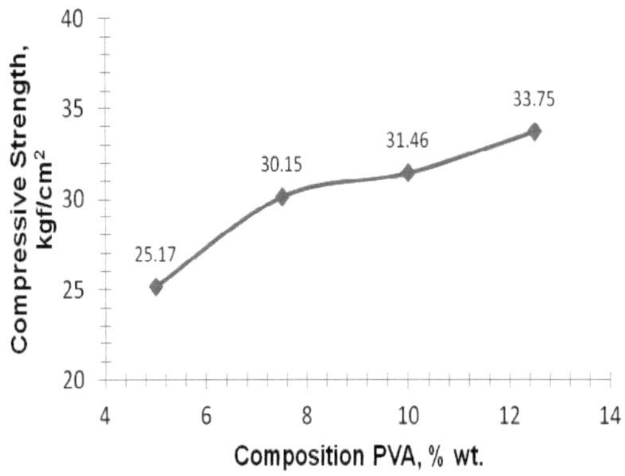

Figure 3. Compressive strength value curves as function of binder composition (PVA).

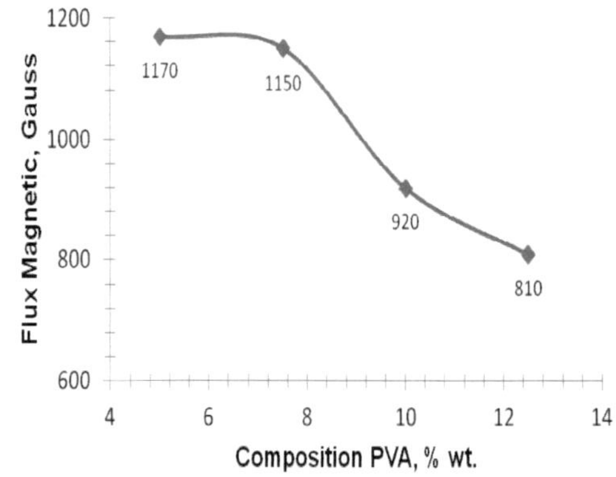

Figure 4. Magnetic flux value curves as a function of binder composition (PVA).

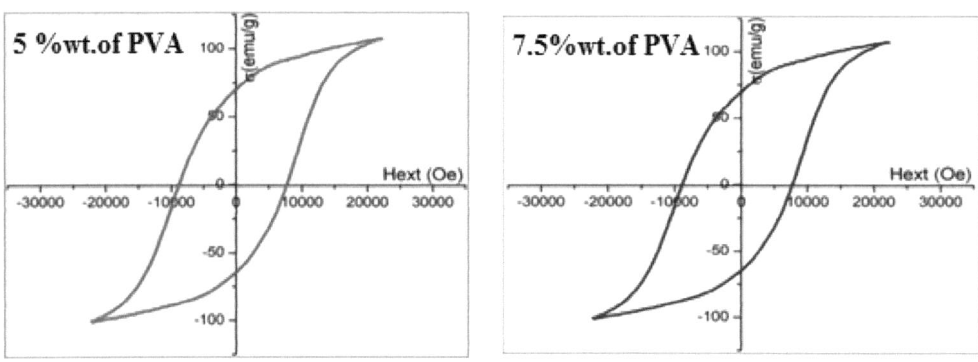

Figure 5. Hysteresis curve of sample bonded magnet with 5 %wt.of PVA and 7.5 %wt.of PVA.

It is possible that this difference is due to different types of different adhesive polymers, where this research used PVA adhesive while reference research using epoxy resin. Another reason regarding the low value of magnetic properties is due to the value of achieved densification. One of the most important parameters in fabrication of bonded magnet is hard and high densification [9, 12].

Table 1. Value of remanence, coercivity, magnetic saturation and energy product.

Composition PVA (%wt)	Remanence (emu/g) [T]	Coercivity (kOe)	Magnetic Saturation (emu/g)	Energy Product (MGOe)
5	74.97 [0.5]	8.76	119	3.04
7.5	70.90 [0.47]	8.68	109	2.89

4. Conclussions

A permanent bonded magnet NdFeB has been successfully made from raw materials NdFeB powder (MQP-B) and polyvinyl alcohol binder, with optimal composition about 5–7.5 wt% PVA. The obtained magnetic properties are bulk density = 5.21–5.25 g/cm^3, hardness = 302.17-304.32 Hv, compressive strength = 25.17–3.17 kgf/cm^2, magnetic flux = 1150-1170 Gauss, remanence = 70.90–74.97 emu/g (4.7–5.0 kGauss), coercivity = 8.68–8.76 kOe, and energy product = 2.89–3.04 MgOe.

Acknowledgments
The authors wish to acknowledge the work conducted by technical staff VSM from Research Center for Physics LIPI.

References

[1] Grujić A, Stajić-Trošić J, Stijepović M, Stevanović J and Aleksić R 2011 Magnetic and Dynamic Mechanical Properties of Nd-Fe-B Composite Materials with Polymer Matrix Metal *Ceramic and Polymeric Composites for Various Uses* Ed J Cuppoletti (London: InTech Open) chapter 25
[2] Haavisto M, Tuominen S, Santa-Nokki T, Kankaanpää H, Paju M and Ruuskanen P 2014 *Advances in Materials Science and Engineering* 760584
[3] Grujić A S, Lazić N L, Talijan N M, Spasojević V, Stajić-Trošić J T, Ćosović V R and Aleksić R 2010 *Acta Physica Polonica A* **117** 859
[4] Abhijit P. Jadhav, Ma H, Kim D S, Baek Y K, Choi C J and Kang Y S 2014 *Bull. Korean Chem. Soc* **35** 886
[5] Talijan N, Cosovic´ V, Grujic´ A, Stajic´-Tro´sic J and Zák T 2008 *Acta Physica Polonica A* **113** 525
[6] Sojer D, Kulj I, Kobe S, Kova J and McGuiness P J 2013 *Materials and Technology* **47** 223
[7] Kirchmayr H R 1996 *J. Phys. D* **29** 2763
[8] Garrell M G, Shih A J, Ma B M, Lara-Curzio E and Scattergood R O 2003 *J. Magn. Magn. Mater.* **257** 32
[9] Brown D, Ma B M and Chen Z 2002 *J. Magn. Magn. Mater.* **248** 432
[10] Bonded NdFeB magnet, http://www.e-magnet.cn/productsa2.html
[11] Zhang X and Xiong W 2009 *Rare Metals* **28** 248
[12] Paranthaman M P, Li L, Rio O, Elliott A M, Post B, Kunc V, Fredette R and Ormerod J 2016 *CRADA final report NFE-15-0576.1: Additive manufacturing of isotropic NdFeB bonded permanent magnets* (Oak Ridge: US Department of Energy)

Manufacture of barium hexaferrite ($BaO3.98Fe_2O_3$) from iron oxide waste of grinding process by using calcination process

N Idayanti, Dedi, T Kristiantoro, D Mulyadi, N Sudrajat and G F N Alam

Research Center for Electronics and Telecommunications, LIPI, Bandung 40135, Indonesia

E-mail: novr001@lipi.go.id/novrita89@yahoo.com

Abstract. The utilization of iron oxide waste of grinding process as raw materials for making barium hexaferrite has been completed by powder metallurgy method. The iron oxide waste was purified by roasting at 800 °C temperature for 3 hours. The method used varying calcination temperature at 1000, 1100, 1200, and 1250 °C for 3 hours. The starting iron oxide waste (Fe_2O_3) and barium carbonate ($BaCO_3$) were prepared by mol ratio of Fe_2O_3:$BaCO_3$ from the formula $BaO_3.98Fe_2O_3$. Some additives such as calcium oxide (CaO), silicon dioxide (SiO_2), and polyvinyl alcohol (PVA) were added after calcination process. The samples were formed at the pressure of 2 ton/cm^2 and sintered at the temperature of 1250 °C for 1 hour. The formation of barium hexaferrite compounds after calcination is determined by X-Ray diffraction. The magnetic properties were observed by Permagraph-Magnet Physik with the optimum characteristic at calcination temperature of 1250 °C with the induction of remanence (Br) = 1.38 kG, coercivity (HcJ) = 4.533 kOe, product energy maximum (BHmax) = 1.086 MGOe, and density = 4.33 g/cm^3.

1. Introduction

Barium hexaferrite is one of the ferrite permanent magnets with a complex hexagonal magnetoplumbite structure and general formula of $MFe_{12}O_{19}$, where M = Ba, Sr, Pb [1]. This magnet is the most extensively investigated material among the family of hard ferrites because it has high curie temperature, relatively large magnetization, excellent chemical stability, and corrosion resistivity, whereas its magnetic and electrical properties should be modulated to satisfy different applications [2]. There are many applications of barium hexaferrite in electronic devices such as automotive industry, magnetic recording media, microwave devices, electric generators and in many magnetically operated devices such as magnetic levitations, telephone ringers, receivers, etc [3].

The main raw material for making barium ferrite is iron oxide. There are many sources of iron oxide materials in Indonesia, such as iron sand and iron oxide waste from production process of steel industry. The use of local resources as materials for research is highly recommended by the government (in accordance with national development program) [4]. This is due to the high cost of imported chemicals and the entry restriction of these chemicals to Indonesia.

The waste material is often a problem because it can damage and pollute the environment. The problem needs research and technology to optimize it so that the waste can be utilized into a more useful product. One of the waste that can be processed into utilizable raw materials is iron powder waste from the grinding process. In grinding process, the iron is cut using a grinding eye and produce the residual iron in the form of coarse powder. There are plenty of iron and steel industries that produce tools and components from iron. They produce a lot of waste iron that has not been utilized. Iron waste taken from the rest of the grinding process can be seen in figure 1.

Figure 1. Pieces of waste iron from the grinding process.

Laurent et al reported that iron oxide commonly known in 3 forms, i.e. hematite (α-Fe_2O_3), magnetite (Fe_3O_4), and maghemite (γ-Fe_2O_3) [5]. There are many studies that have been developed to obtain iron oxide; powder metallurgy is one of the common technique used. The other methods that are mentioned by Teja et al. and Cheng et al. that often used are sol-gel, co-precipitation, microemulsion and thermal decomposition [6, 7]. Iron oxide can be applied in many area, such as on the development of sensors and actuators, material to produce inorganic pigments, main materials for making magnetic materials, adsorbent of wastewater treatment an so on [8]. Considering many uses of iron oxide, research to prepare the oxide as a main raw material to make magnetic materials is interesting.

This paper will introduce the manufacture of permanent magnet barium hexaferrite from iron waste of the grinding process by a powder metallurgy method. This research is a preliminary study to find out whether the iron waste from the grinding process can be used as a raw material for the manufacture of magnetic materials.

2. Experiment

The experiments were carried out in two stages: the preparation of iron oxide and the manufacture of barium hexaferrite magnets from iron oxide waste.

2.1 Preparation of iron oxide

The iron powder obtained from the rest of the grinding process must be prepared to increase the purity before being used. Characterization result of waste using X-Ray diffraction before purification process can be seen in figure 2. On the curve we can see the waste iron contains magnetite (Fe_3O_4) compound = 81.8%, silicone (Si) = 4.2%, Nickel (Ni) = 12.9%, and Zinc (Zn) = 11%.

This waste still contains many impurities that must be removed through several processes. To improve the purity of waste, we have done three purification processes: separation, roasting, and milling. The iron waste is separated from non-magnetic material by magnetic bars. Magnetic materials and metals stick on bars, while non-magnetic and non-metallic materials are left. Furthermore, the iron powders are washed with water to remove dust and other impurities. The iron powders are burned at 800 °C for 3 hours, then milled using a milling machine for 50 hours to produce a fine iron oxide powder (Fe_2O_3) that is ready for use as the main raw material for making barium hexaferrite permanent magnets.

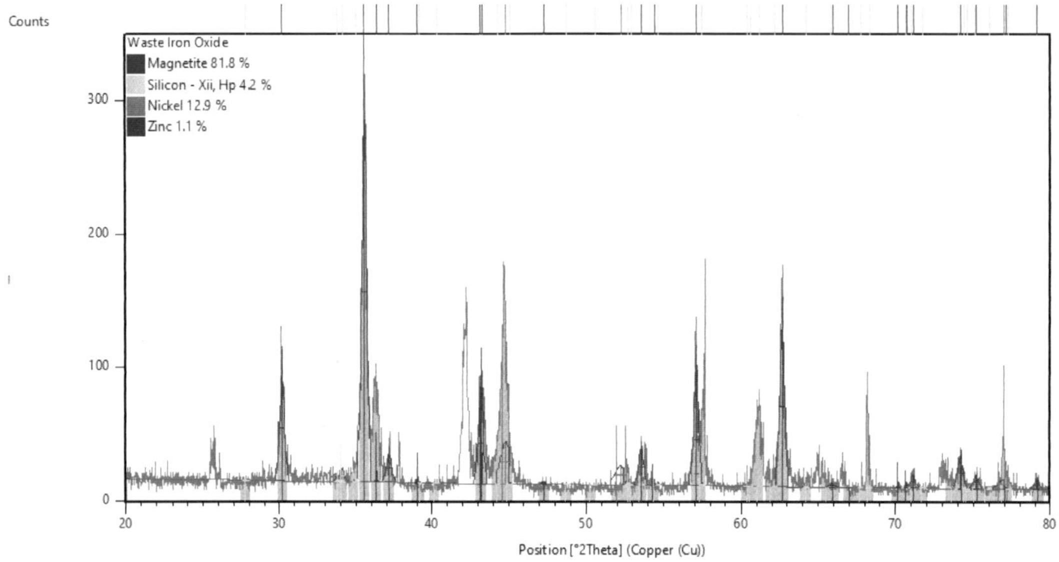

Figure 2. Characterization result of iron waste content.

The purity of iron oxide waste increased, as seen from the color change of the powder into a dark black color. The iron oxide powder of purification process result can be seen in figure 3. After being roasted and milled by grinding machine, the iron oxide waste contains hematite (Fe_2O_3) and maghemite (Fe_3O_4). There is no Si, Ni, and Zn anymore.

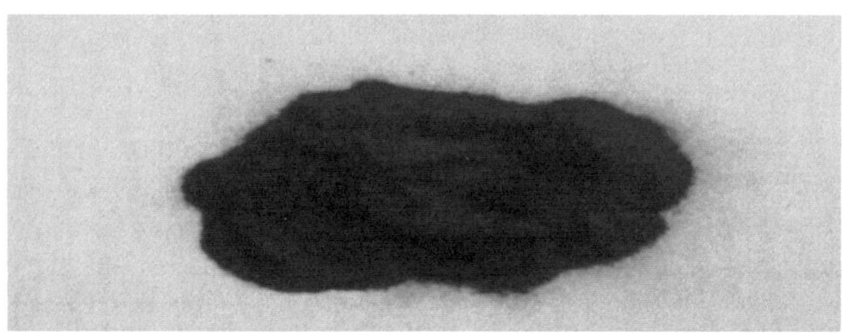

Figure 3. The iron oxide powder after purification process.

2.2 Manufacture of barium hexaferrite magnet

The raw main materials for making barium hexaferrite magnets are iron oxide and barium carbonate. Both materials are weighed on the weight percent Fe_2O_3: $BaCO_3$ = 85.00 wt%: 15.00 wt%. Next, the raw materials are mixed in a milling machine using steel balls in alcohol media for 6 hours. After that, the materials are calcined at varied temperatures of 1000, 1100, 1200, and 1250 °C for 3 hours of holding time. The variation of the calcination temperature was performed to determine the optimal temperature of initial formation of the barium hexaferrite compound. After the calcination process, the mixture is dried, weighed, and then some additives such as calcium oxide (CaO), silicon dioxide (SiO2), and polyvinyl alcohol (PVA) of about 0.75 to 1.5 wt% of the weight of calcine powder are added. Mixing process of calcine powders and additives are performed in a milling machine for 16 hours. The mixture is then dried at 100 °C temperature until it dries and is then filtered with 400 mesh sieve shaker to make the powder homogeneous and has the same particles size so it is ready to be pressed. The samples are pressed at 2 ton/cm^2 by using hydraulic press in a round shape. The samples that have been pressed are sintered at temperature of 1250 °C for 1 hour. The process of calcination

IOP Conf. Series: Journal of Physics: Conf. Series **985** (2018) 012048 doi:10.1088/1742-6596/985/1/012048

and sintering are done using a furnace Yamato 1700 °C. Magnetic properties are characterized using the Physik Magnetic Permagraph, and the formation of barium hexaferrite compounds using X-Ray Diffraction Maxima. The experimental details of preparation of iron oxide and manufacturing of barium hexaferrite can be seen in figure 4.

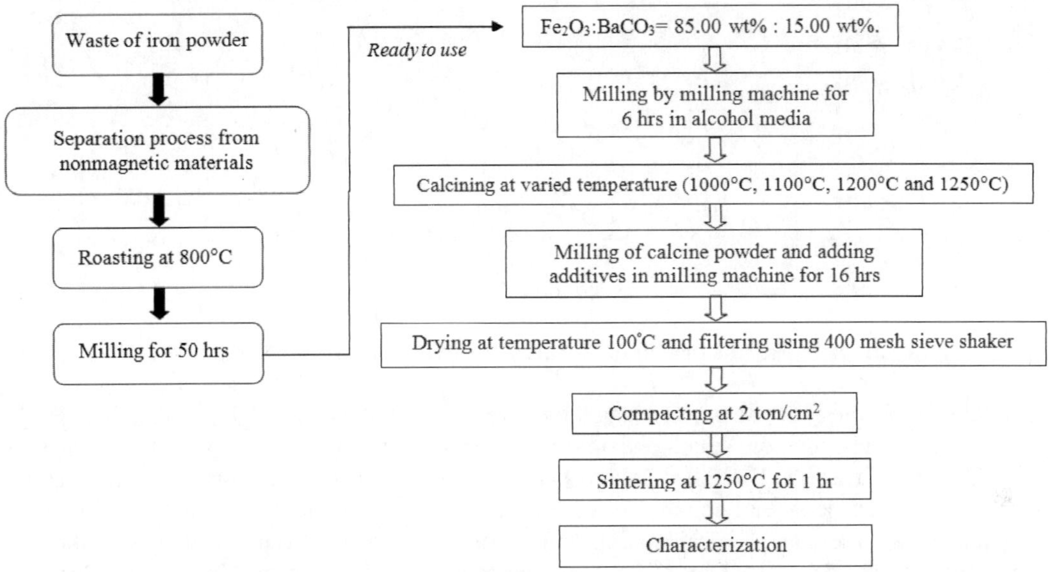

Figure 4. Flow chart of preparation of iron oxide and manufacturing of barium hexaferrite.

3. Results and discussion

The experimental utilization of iron waste from the grinding process to become the raw material for making permanent magnets has been completed. The characterization results shown by the test of barium hexaferrite compound formation using XRD, and magnetic properties test by using Permagraph. The characterization results using XRD shows that the calcination temperature influences the forming of barium hexaferrite compounds.

Curves 1-4 have similar pattern with different percentages of barium hexaferrite compound formation. On curve 1, formation percentage of barium hexaferrite ($BaFe_{12}O_{19}$) is 86.7% while hematite (α-Fe_2O_3) is 13.3%. Curve 2 showed 98.7 formation of $BaFe_{12}O_{19}$ and 1.3% magnetite(Fe_3O_4). On curve 3, $BaFe_{12}O_{19}$ formation is 99.6%, Fe_3O_4 = 0.4%, and the percentage of $BaFe_{12}O_{19}$ and Fe_3O_4 on curve 4 is almost the same with curve 3. The percentage of barium hexaferrite formation is increasing with the increase of calcination temperatures. Magnetic properties has reached the maximum at 1200 °C temperature, therefore that at temperature of 1250 °C it remains unchanged. The patterns of formation of barium hexaferrite can be seen in figure 5.

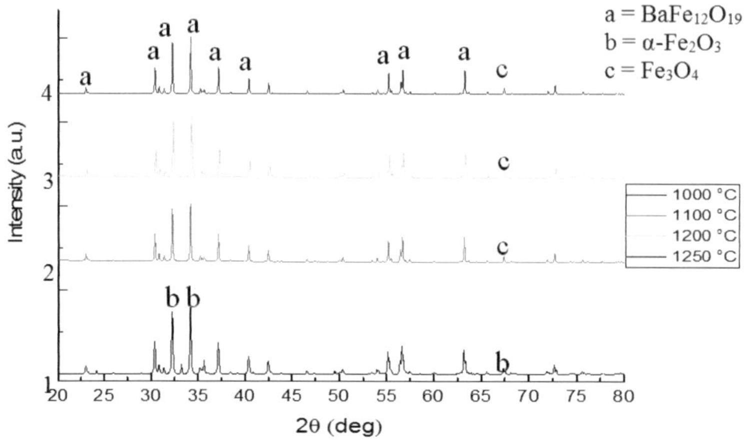

Figure 5. XRD patterns of barium hexaferrite at varying calcination temperatures.

Several other studies have also investigated by Sumangala et al. and Prengki et al. about the effect of calcination temperature on structures, microstructures, and magnetic properties of magnesium ferrite and nickel ferrite [9, 10]. Gharagozlou [11] observed the effect of calcination temperature on crystallinity and particle size of cobalt ferrite nanoparticle. It is known that crystallinity and particle size of ferrite increase at higher calcination temperature. The morphological changes to the cobalt ferrite magnet due to differences in calcination temperature have also been observed by Sezer [12]. The calcination process is performed by heating the solid material at a high temperature which aims to reduce impurities, removing volatile substances, and oxidizing an elements or compounds as reported by Rachel et al [13].

Hysteresis curve of barium ferrite at varying temperature were analyzed using Permagraph as shown in figure 6. It can be observed that the sample curve calcined at temperature of 1250 °C is larger than the other samples. This curve shows the magnitude of the maximum energy product of the sample. This magnetic properties are influenced by the level of sample crystallinity that are affected by the calcination temperature.

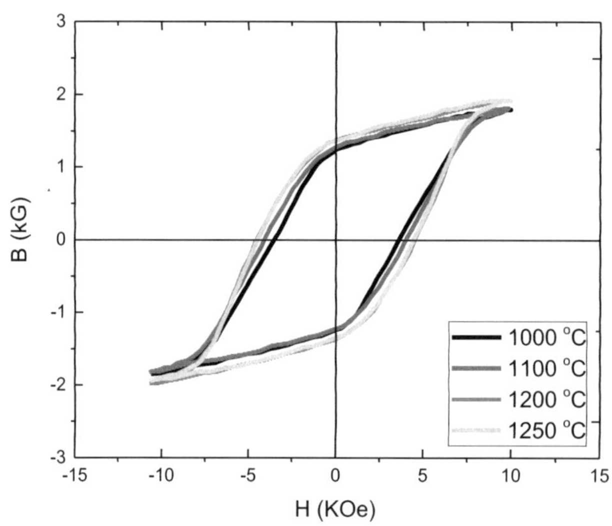

Figure 6. Hysteresis Curve.

The magnetic characteristic of samples such as induction of remanence (Br), coercivity (HcJ), maximum energy product (BHmax), and density are shown in table 1. The higher calcination temperature is, the Br value increases as well as the values of Hc, BHmax, and density.

Table 1. Magnetic properties and density of the barium hexaferrite with variation of calcination temperatures.

Magnetic Properties and Density	Calcination Temperature			
	1000 °C	1100 °C	1200 °C	1250 °C
Br (kG)	1.27	1.26	1.36	1.38
HcJ (kOe)	3.067	4.089	4.540	4.533
BHmax (MGOe)	0.35	0.34	0.39	1.086
ρ (g/cm^3)	4.24	4.29	4.31	4.33

The magnetic characteristics values increase because at higher calcination temperature the formation of ferrite compound and crystallinity are better. Grain growth at higher calcination temperature reduces porosity and increases density. Porosity is one factor that can disrupt the magnetization process. The higher the density, the magnetic properties will be better.

The magnetic characteristics obtained from these experimental results approached the previous experiment of making barium hexaferrite magnets from iron oxide waste by Novrita et al [14], where the remanent induction value is about 1.5 kG. Standard Specifications for Permanent Magnet Materials compares with the characteristics of commercial barium hexaferrite magnets on the market that range between 2.4 kG - 4 kG [15]. The magnetic strength value of this magnet still needs to be improved again, so further experiments are needed to improve the generated magnetic characteristics.

4. Conclusion

In summary, a preliminary research of the utilization of iron waste from the grinding process has been carried out with powder metallurgy technology. The results show that the magnetic characteristics of barium hexaferrite are almost similar with the previous experiments of ferrite magnets from other iron oxide wastes. The calcination temperature affects the magnetic characteristic of barium hexaferrite, where the optimum condition is obtained at calcination temperature of 1250 °C. This experiment still requires further research because there are still many conditions of process that will affect the characteristics of the magnet such as iron waste purification process, composition, grain size, and sintering temperature.

Acknowledgments

The authors are grateful to LIPI for providing support and funding under the Superior Research Program (Riset Unggulan) 2017 with the project title of Development of Functional Material Based on Local Natural Resources. The authors are also thankful for the research cooperation of the researchers and technicians of the magnetic group PPET-LIPI.

References

[1] Liu Y, Li Y, Liu Y, Yin H S, Wang L L, Shun K and Gao Y 2011 *Appl. Mech. Mater.* **69** 6
[2] Durmus Z, Kavas H, Sozeri H, Toprak M S, Aslan A and Baykal A 2012 *J. Supercond. Novel Magn.* **25** 1193
[3] Bhatia P G 2016 *International Journal of Chemical and Physical Sciences* **5** 28
[4] UU RI No. 25 Program pembangunan nasional 2000 compiled by Yayasan Titian https://luk.staff.ugm.ac.id/atur/UU25-2000Propenas.pdf
[5] Laurent S, Forge D, Port M, Roch A, Robic C, Elst L V and Muller R N 2008 *Chem. Rev.* **108** 2064
[6] Teja S A and Koh P Y 2009 *Prog. Cryst. Growth Charact. Mater.* **55** 22

[7] Cheng Z, Tan A L K, Tao Y, Shan D, Ting K E and Yin X J 2012 *International Journal of Photoenergy* 608298

[8] Campos E A, Pinto D V B S, Oliveira J I S, Mattos E C and Dutra R C L 2015 *Journal of Aerospace Technology and Management* **7** 267

[9] Sumangala T P, Mahender C and Venkataramani N 2013 *AIP Conf. Proc.* **1512** 382

[10] Pransisco P, Shafie A and Guan B H 2014 *AIP Conference Proceedings* **1621** 619

[11] Gharagozlou M 2011 *Chem. Cent. J.* **5** 19

[12] Erdem S, Erdem B, Öksüzoğlu R M and Çıtak A 2015 *J. Porous Mater.* **22** 689

[13] Rachel E T 2000 *High temperature processing of kaolinitic materials* Doctoral Thesis (Birmingham: University of Birmingham)

[14] Novrita I, Tony K, Ardita S and Ika K 2017 *Proc. Sriwijaya Int. Conf. on Engineering, Science and Technology* (Bangka Island) vol 101 (London: EDP Sciences)

[15] Magnetic materials producers association 2012 *Standard specifications for permanent magnet materials* (Chicago: Magnetic materials producers association) p 12

The influence of TiO_2 composition in LiBOB electrolyte polymer composite membrane characteristics for lithium ion batteries applications

T Lestariningsih, Q Sabrina and E M Wigayati

Research Center for Physics - Indonesian Institute of Sciences, Kawasan Puspiptek, Tangerang Selatan, Indonesia

E-mail: titi013@lipi.go.id

Abstract. Characterization of the composite membrane of LiBOB electrolyte polymers made from poly (vinylidene fluoride co-hexafluororopylene) (PVdF-HFP) as the polymer, LiBOB or $LiB(C_2O_4)_2$ as electrolyte salt and titanium dioxide (TiO_2) as ceramic filler of three different concentrations have been done. Sample of membrane was prepared using solution casting technique. Microstructural study by SEM shows non-uniform distribution of pore over the surface of the sample. X-ray structural analysis, impedance spectroscopy, and cyclic voltammetry (CV) studies were carried out. Membrane composite polymer of LiBOB electrolyte without additional ceramic filler with composition of 70% polymer, 30% LiBOB, and 0% TiO_2 has the greatest conductivity for forming amorphous phase and is compatible with material membrane composite. Meanwhile, sample with 70% polymer composition, 28% LiBOB and 2% TiO_2 shows oxidation reaction at the most perfect discharge despite very slow current speed.

1. Introduction

Nowadays, the development of electrolyte membrane/polymer electrolyte is very attractive and continues to grow. One practical application from the polymer electrolyte membrane is that it can effectively be used as an electrolysis medium in energy storage devices such as battery. The other potential applications for modification of polymer and lithium salt is for fuel cell, sensors, and super capacitors [1]. Various sources of lithium salt such as $LiFePO_4$ [2], $LiCF_3SO_3$ [1], $LiBF_4$ [3], LiBOB [4] become electrolyte sources that have been developed as a source of Li+ charges in polymer matrix.

Lithium bis(oxalato) borate with chemical formula $LiB(C_2O_4)_2$ is the active ingredient of electrolyte in lithium battery. LiBOB is the main candidates substituting $LiPF_6$ which is highly toxic and harmful to human health. LiBOB electrolyte is considered to be more environmentally friendly because it compounds do not contain halogen element which is potential to pollute the environment. LiBOB also has high thermal stability of 302 °C [5].

Performance of lithium ion battery is based on the ionic conductivity (Li^+ ions). Generally, conductivity is influenced by ion concentration as bearer charge and ion mobility [6] in accordance with equation:

$$\sigma = \sum n_i z_i \mu_i \quad (1)$$

where n_i is the amount of charge carrier, z_i the charge of ions, and μ_i the mobility of ions. At the same condition of ion mobility, larger amount of lithium ions in the membrane results in increasing conductivity. However, if the Li^+ ion mobility condition is different, for example, ion mobility is smaller; it is possible that conductivity value will not always increase. Ionic conductivity is related to the polymer parts. The mobility of carrier is determined by the carrier environment itself, the interaction between the ions and the polymer, the flexibility of polymer chains and characteristics of carrier, and so on.

Electrolyte membrane is one type of electrolyte polymer. Electrolyte polymer is defined as a solution of alkali metal salt present in the polymer matrix [7]. The electrolyte membrane is formed by a polymer matrix having alkali metal ions as a source of its charges. Ions with low atomic mass such as lithium ions (Li^+) will be easier to experience charge transport process. This transport is due to segmental relaxation in which the charge moves from one segment to another in the polymer matrix. The charge transport is more effective when polymer matrix is in the amorphous phase [8]. The simplest approach to grow amorphous phase is dispersed nanoparticle/ ceramic filler in polymer matrix. The existence of nano-particles/ceramic filler dispersed among the polymer chains impedes their reconstruction in crystalline form [9].

Meanwhile, various polymers have been used as polymer membranes such as poly(vinylidene fluoride) (PVdF), poly(vinyl alcohol) (PVA), acetate cellulose, and polyacrylonitrile-methylmethacrylate (P(AN-MMA)). Optimization of physical properties such as electrical conductivity becomes an important focus that still develops today. Poly (vinylidene fluoride-co-hexafluoropropylene) or PVdF-HFP is combined compound of poly vinylidene and hexafluropropylene, and both of them are often functioned as a polymer forming electrolyte composite.

In this paper, fabricating polymer composite membrane of LiBOB electrolyte using PVdF-HFP and TiO_2 as filler with various compositions has been conducted as preliminary study. The aim is to find the influence of TiO_2 composition on the characteristics of polymer composite membrane of LiBOB electrolyte for lithium battery cell application. The manufacture of polymer composite membrane of LiBOB electrolyte was prepared by solution casting technique. The characterization involves investigating crystal structure, morphological structure, electrical conductivity and cyclic voltammetry (CV) testing for battery performance.

2. Experimental method

2.1. Material

Material used to fabricate electrolyte membrane were poly(vinylidene fluoride)-hydro furan phosphate (PVdF-HFP) as polymer (Sigma Aldrich), lithium bis(oxalato) borate (LiBOB) as lithium salts (Sigma Aldrich), titanium oxide (TiO_2) as ceramic filler (Merck), and N,N-Dimethylacetamide (DMAc) solvents. Lithium ferro phosphate ($LiFePO_4$) from MTI and metal lithium were exploited as cathode and anode sheets. They were functioned for testing ion conductivity and cyclic voltammetry.

2.2. Sample preparation

Samples of LiBOB electrolyte membrane were made by solution cast technique. The first step was to make slurry (viscous solution) by dissolving LiBOB electrolytes salt and ceramic filler into DMAc solvent through mixing process using magnetic stirrer on a hot plate at temperature of 70-80 $^{\circ}$C and speed of 240 rpm. The next step was adding PVdF-HFP as polymer matrix into the mixed solution with the same process to get slurry solution. Then, the solution was cast on a glass substrate and dried at room temperature.

2.3. Sample characterization

The first characterization of LiBOB electrolyte membrane was measuring ionic conductivity by EIS (Electrochemical Impedance Spectroscopy) method using LCR meter instrument, and HIOKI 3532150 Chemical Impedance Meters. This measurement was carried out in the form of coin cell that was applied in the form of half-cell battery. The half-cell battery used $LiFePO_4$ as the cathode and Li metal as the anode. Characterization of LiBOB electrolyte membrane included crystal structure characterized by XRD and morphology observation by SEM. Regarding the application of lithium ion batteries, samples were tested by cyclic voltammetry (CV), using WBCS3000 tool, Automatic Battery Cycler. CV testing was performed in the form of half-cells using $LiFePO_4$ as cathode and Li metal as an anode, while LiBOB electrolyte membrane serves as separator and electrolyte. Variations of samples in this study were figured out by table 1.

Table 1. Sample composition for composite of LiBOB polymer electrolyte.

Sample	PVdF HFP (% Weight)	Filler: TiO_2 (% Weight)	Commercial LiBOB salt (%weight)	Solvent
CPE-1	70	0	0	DMAc
CPE-2	70	30	0	DMAc
CPE-3	70	0	30	DMAc
CPE-4	70	2	28	DMAc
CPE-5	70	5	25	DMAc

3. Results and discussion

EIS (electrochemical impedance spectroscopy) is one of the methods to investigate electrical characteristics of materials. In the present study, impedance measurement by LCR meter instrument presented by Cole-Cole graph is shown by figure 1. Ionic conductivity of the electrolyte polymer composite can be calculated by using equation

$$\tau = t / (Rb * A) \quad (2)$$

τ is ionic conductivity, t is the sample thickness (cm), Rb is measured resistance, and A is cross-sectional area of the sample (cm^2). Rb is obtained by determining semicircle line intersecting x axis. Measurement of ionic conductivity was only performed on CPE-3, CPE-4, and CPE-5 samples.

Conductivity measurement was carried out in the form of half-cells shown by Nyquist graph as figure 1. It shows that CPE-3 sample containing 70% polymer, 0% TiO_2, and 30% LiBOB salt has the smallest semi-circle shape. According to equation $\tau = t / (Rb * A)$, smaller semicircle line intersecting x-axis means greater conductivity value. Therefore, CPE-3 sample has the largest conductivity value. It can be seen from figure 1 that the addition of 2% TiO_2 will decrease conductivity value while adding 5% TiO_2 increases conductivity value. TiO_2 in polymer electrolyte promotes ionic conductivity. Adding TiO_2 is supposed to increase ionic conductivity. This phenomenon can be explained by the results of crystal structures and morphology analysis. XRD analysis was done to observe the change of crystalline properties due to the presence of Li^+ metal ionic in polymer matrix. The maximum intensity is at an angle (2θ) of 28.078°. This peak is characteristic of polymer membrane (PVdF-HFP) and LiBOB.

IOP Conf. Series: Journal of Physics: Conf. Series **985** (2018) 012049 doi:10.1088/1742-6596/985/1/012049

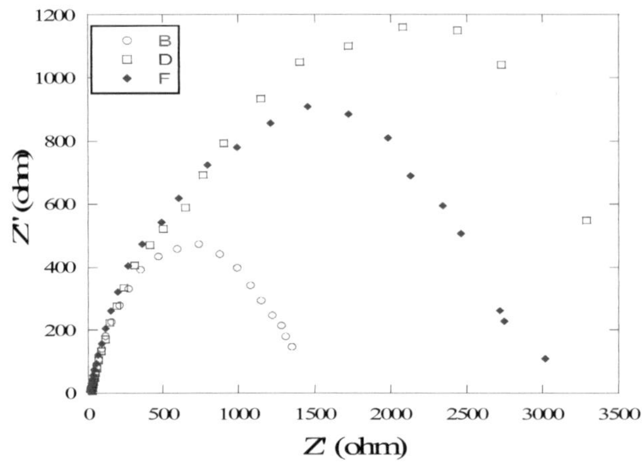

Figure 1. Nyquist graph of sample of LiBOB electrolyte polymer composite in the form of coin cell with variation of TiO_2 addition: CPE-3 (B), CPE-4 (D), and CPE-5 (F).

Figure 2 illustrates that adding 2% TiO_2 changes amorphous structure into crystalline structure. Meanwhile, an addition of 5% TiO_2 transforms crystal structures to amorphous structure. Therefore, membrane polymer has formed amorphous phase without adding TiO_2 as demonstrated by figure 2. The existence of amorphous phase in electrolyte polymer membrane will support the effective segmental relaxation of the polymer. In crystalline solid phase, arrangement of chains will regularly reduce segmental relaxation, so that ionic transport (conductivity) becomes very low [10]. Consequently, the presence of amorphous phase turns out to be an important factor for increasing ionic conductivity in electrolyte membrane. That is why CPE-3 sample has the highest conductivity in half-cells. In addition, this phenomenon can also be proven by SEM analysis. SEM photographs of polymer electrolyte indicate good compatibility among polymer, salt, and other contents as well as the amorphous phase formation in the manufacture of electrolyte polymer composite. SEM images of CPE-3, CPE-4, and CPE-5 samples are shown in figure 3.

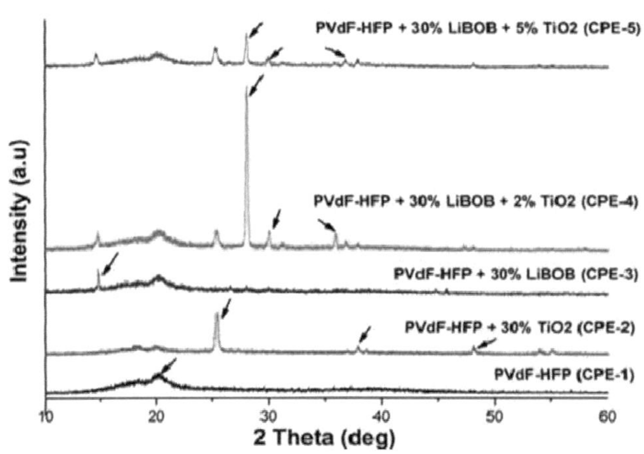

Figure 2. X-Ray diffractions patterns of composite polymer electrolyte.

IOP Conf. Series: Journal of Physics: Conf. Series **985** (2018) 012049 doi:10.1088/1742-6596/985/1/012049

Figure 3 demonstrates that CPE-3 sample with composition of 70% polymer, 0% TiO_2, and 30% LiBOB has good compatibility between polymers and salt, involving pore formation. CPE-4 sample with 2% TiO_2 addition does not show good compatibility among salt, polymer, and TiO_2, so amorphous phase transforms to crystalline phase. It is also evidenced by comparing the morphology of CPE-4 and the rutile TiO_2 sample [10] (figure 4).

(a) CPE-3 (b) CPE–4 (c) CPE–5

Figure 3. Morphology of electrolyte polymer composite membrane of CPE-3, CPE-4 and CPE-5 samples with 20 kV at 5000x.

TiO_2 does not blend with polymers and salt (figure 4). However, adding 5% TiO_2 converts crystalline phase to amorphous phase as shown on figure 3c. Based on XRD and SEM analysis, CPE-3 sample without additional addition of TiO_2 has the highest conductivity when applied in half-cell. Nevertheless, with addition of 5% TiO_2 can transform crystalline phase to amorphous phase and increasing amount of TiO_2 may grow greater amount of amorphous phase.

Figure 4. SEM microphotograph of composite polymer electrolyte for sample CPE-4.

Considering the results of CV analysis in the form of cells containing $LiFePO_4$ as cathode and Li metal as an anode, CPE-4 sample has a very slow speed current to reach oxidation reaction at discharge condition (figure 5). By the type of cell testing, it was found that 2% TiO_2 added to the functional electrolyte decomposed just before the decomposition of the basic electrolyte at about just over 4.5 V. It was also confirmed that the highest impedance among the two other samples due to the film produced on catode was negligibly small, as displayed by figure 1.

4. Conclusions

LiBOB electrolyte polymer composite membrane of CPE-3 sample without adding TiO_2 has an amorphous phase, therefore it has highest conductivity properties among the three samples. CPE-4 sample with addition of 2% TiO_2 forms crystalline phase and materials composing membrane cannot mix or not compatible resulting in declining conductivity, but with adding 5% TiO_2 crystalline phase starts to transform in amorphous phase resulting in increasing conductivity. However, with very slow current velocities, CPE-4 sample shows oxidation reaction at the most perfect discharge.

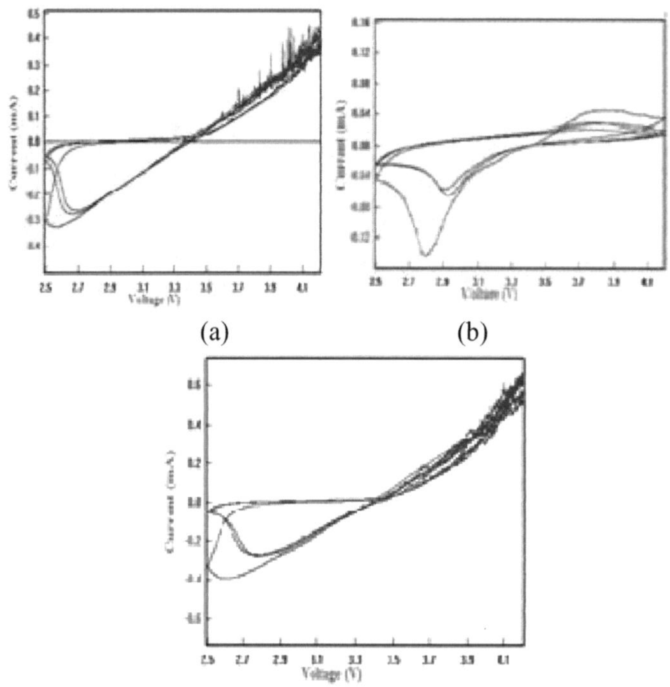

(a) (b)

(c)

Figure 5. Cyclic voltammograms of composit polimer electrolyte for sample (a) CPE-3, (b) CPE-4, and (c) CPE-5

References

[1] Ramesh S and Arof A K 2009 *J. Mater. Sci.* **44** 6404
[2] Mohan V M , Qiu W, Shen J and Chen W 2010 *J. Polym. Res.* **17** 143
[3] Akbulut O, Taniguchi I, Kumar S, Shao-Horn Y and Mayes A M 2007 *Electrochim. Acta* **52** 1983
[4] Appetecchi G B, Zane D and Scrosati B 2004 *J. Electrochem. Soc.* **151** A1369
[5] Azeez F and Fedkiw P S 2010 *J. Power Sources* **195** 7627
[6] Ratna D, Divekar S, Patchaiappan S, Samui A B and Chakraborty B C 2007 *Polym. Int.* **56** 900
[7] Meyer W H 1998 *Adv. Mater.* **10** 439
[8] Suthanthiraraj S A and Sheeba D J 2007 *Ionics* **13** 447
[9] Abdullah M, Lenggoro W and Okuyama K 2004 Polymer Electrolyte Nancomposites (*Encyclopedia of Nanoscience and Nanotechnology* vol 8) ed H S Nalwa (California: American Scientific Publishers) pp 731-762
[10] Swapna M V and Haridas K R 2015 *The Chemist* **88** 1

IOP Conf. Series: Journal of Physics: Conf. Series **985** (2018) 012050 doi:10.1088/1742-6596/985/1/012050

Effect of graphite addition into mill scale waste as a potential bipolar plates material of proton exchange membrane fuel cells

D S Khaerudini, G B Prakoso, D R Insiyanda, H Widodo, F Destyorini and N Indayaningsih

Fuel Cell Group, Research Center for Physics, Indonesian Institute of Sciences, Gd. 440 Kawasan Puspiptek Serpong, Tangerang Selatan 15314, Banten, Indonesia

E-mail: deni008@lipi.go.id

Abstract. Bipolar plates (BPP) is a vital component of proton exchange membrane fuel cells (PEMFC), which supplies fuel and oxidant to reactive sites, remove reaction products, collects produced current and provide mechanical support for the cells in the stack. This work concerns the utilization of mill scale, a by-product of iron and steel formed during the hot rolling of steel, as a potential material for use as BPP in PEMFC. On the other hand, mill scale is considered a very rich in iron source having characteristic required such as for current collector in BPP and would significantly contribute to lower the overall cost of PEMFC based fuel cell systems. In this study, the iron reach source of mill scale powder, after sieving of 150 mesh, was mechanically alloyed with the carbon source containing 5, 10, and 15 wt.% graphite using a shaker mill for 3 h. The mixed powders were then pressed at 300 MPa and sintered at 900 °C for 1 h under inert gas atmosphere. The structural changes of powder particles during mechanical alloying and after sintering were studied by X-ray diffractometry, optical microscopy, scanning electron microscopy, and microhardness measurement. The details of the presence of iron, carbon, and iron carbide (Fe-C) as the products of reactions as well as sufficient mechanical strength of the sintered materials were presented in this report.

1. Introduction

In recent years, due to the demand of energy consumption continuously increasing along with the depletion of resources, the development and research on renewable energy is essential to overcome this energy issue. Proton exchange membrane fuel cells, as one of the most promising renewable energy technology for stationary, portable and automotive applications, have attracted much attention mainly because of their light weight, low operating temperature and producing electricity from electrochemical reactions in an environmentally friendly manner [1]. PEMFC consists of various components, among which bipolar plates allocate about 60 to 80% of PEMFC weight and 30 to 45% of their cost [2], hence one of the most important components of PEMFC due to their volume and cost.

The bipolar plates are typically made from graphite, composite, and metal plates [3]. With respect to corrosion resistance, graphite is preferable. However, the conductivity of this material is still much lower than that of metallic one. The machining cost of graphite BPP is very high, which account for as much as 60% of the stack cost. The porous nature and brittleness of graphite also prevent people from utilizing it. Extensive works have been made to develop alternative BPP materials to replace it [4].

Alternatives fall into composite and metallic based BPP. It has been estimated that for composites and metal plates, the cost of BPP would be only 15 to 29% of the stack cost [5]. Due to their high strength, metallic BPP can be made of very thin sheets with a thickness of 0.1 mm or less [6-9]. Therefore, metallic BPP is considered as the most important substitute for graphitic ones, due to ability to mass production and decreasing PEMFC volume and weight, which lower the overall manufacturing cost [10]. Low cost materials are also essential for fruitful combination of PEMFC into the commercial energy sector [11]. As an alternative one, waste-based materials can also be utilized as raw materials, for instance mill scale. Mill scale is a waste material from the steelmaking industry and considered has rich iron content (about 70 - 75% Fe). Thus, the high content of Fe in mill scale is potentially used as raw material for manufacturing metallic-based BPP. In this study, in order to reduce the cost as well as to utilize the industrial mill scale waste, we report a study of the utilization of mill scale based BPP with some variation addition of graphite for evaluation and justification on their microstructural, physical, and mechanical properties.

2. Experimental

In the present study, the industrial mill scale (as received from PT Krakatau Steel, Cilegon, Indonesia) the chemical composition (wt.%) of Fe total: 74.24%, Fe metal: 0.20%, Fe_2O_3: 52.02%, SiO: 0.25%. CaO: 0.97%, MgO: balance (analyzed by XPS), was chosen as raw material. After sieving of 150 mesh and preparing the substrate sample, the graphitic material as a carbon source was added with variation of 5, 10, and 15 wt.%. The carbon was considered to lead the formation of Fe-C, specifically in their austenite phase. Thus, the blends was mechanically alloyed using a shaker mill in a discontinues milling manner with milling speed of 600 rpm, delay time of 20 minutes, and total milling time of 3 h. To prepare the green body, the milled powders were pressed with 300 MPa load at room temperature into the mould with diameter of 20 mm and targeted thickness of 3 mm. The green body was then sintered at 900 °C for 1 h under inert gas atmosphere. After grinding and polishing, the samples obtained were microstructure analyzed by optical microscopy (OM Bestscope BS6000AT) and scanning electron microscopy (SEM Hitachi SU 3500). Phase analysis was carried out by X-ray diffraction (XRD SmartLab, Rigaku) with a copper anode x-ray tube with 2θ from 10 to 90° using a 0.01^{o} step size and 0.25 second count time. To measure the surface hardness of sample, Vickers hardness was analyzed by Microhardness Tester (HV Leco LM100AT).

3. Results and discussion

3.1. Phase characterization

The X-ray diffraction analysis was used to identify the phase present in the material during the sintering process. Figure 1 shows the XRD peaks profile for sample in the powder (before sintering) and solid (after sintering) state. From the as received mill scale, it was known that iron oxides were the predominant substances in the material before mixing and the amount of metallic iron is not significant (0.20%). For the powder of pure mile scale (0 wt.% addition) as shown in figure 1a, the presence of wustite (FeO) and magnetite (Fe_3O_4) was recorded. At this condition the ferritic (Fe) phase was not detected, indicating insignificant of metallic iron or due to very low intensity of Fe peak. However, for the sintered sample it shows very high peaks of Fe species at 2θ value of 44.354^{o} with reflection index of (001). For all mixing powder with various addition of graphite after milling for 3 h (figure 1b, 1c, 1d), it has shown typical peaks profile that detected only wustite and magnetite, the difference is only wider peaks compared with the pure one (figure 1e). The wider peaks of mixing powder can be contributed due to the finer size of graphite powder; however the peak of graphite itself still cannot detected this due to the origin of requirement of much higher sinter temperature (>> 2000 °C) of graphite compared with the ferritic based material (mill scale). In this study, one can be considered the graphite also act like a "binder" or "matrix" for the mixing product of mill scale-graphite. This benefit for its homogenous bonding of Fe-C; yet it is still far from their ideal sinter condition. For the sintered mill scale-graphite (solid), one can see that the Fe species also could be detected, even though not the highest peak one, at 2θ about 70 to 80°. This showed that the flowing of inert gas and sintering process at 900 °C are suitable to converts the metal oxide (FeO or Fe_3O_4) that are present in mill scale into pure Fe, although still in the α-Fe (ferritic) phase; whereas, the desired phase in this study is the austenite (γ-Fe) phase.

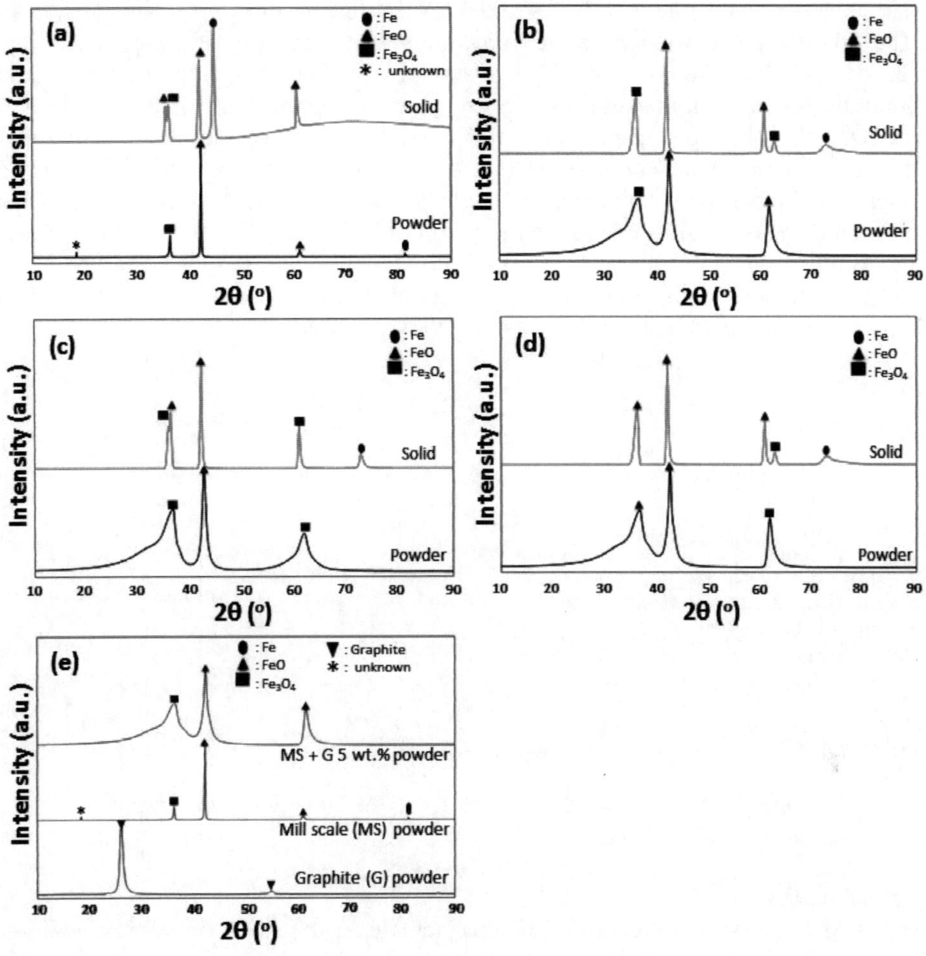

Figure 1. XRD patterns of milled powder and solid sample with graphite addition of (a) 0, (b) 5, (c) 10, (d) 15 wt.%, and (e) XRD patterns of original mill scale, graphite, and 5 wt.% of powder samples. The obtained phases are wustite (FeO), magnetite (Fe_3O_4) and ferritic (Fe).

The presence of ferritic is due to the origin of mill scale that starting emerged the iron species after sintering at 900 °C and flowing under inert atmosphere. From the evidence of XRD, there are two possibilities occurred. (i) The 'thermal energy' of sintering temperature causing the oxygen element in the metal oxide 'diffuse-out' of the matrix in a mill scale and allow the formation of new species, such as iron (Fe). Other one, (ii) it can be due to the nature of impurities that probably contained in the chemical composition of mill scale that will affect into their sintering temperature and process. Therefore, the presence of impurities makes graphite difficult to diffuse-out and react with the mill scale (Fe species) compound; for instance become Fe_3C, so that no atomic bonding reactions or alloy phase occurred between the elements on the mill scale-graphite at the condition of 900 °C.

3.2. Vickers hardness

Vickers hardness test was carried out on the surface of the sintered sample to evaluate Fe-C bonding with respect of graphite addition. The optical micrographs of the penetrated area after diamond indentation showed wider diamond flaking zone (not shown here) with increasing of graphite addition. Microhardness values of the samples are shown in figure 2. As clearly seen, the pure mill scale sample (0 wt.% graphite addition) have the highest microhardness (320.8 VHN) than the graphite added samples. The lowest microhardness value is 103.53 VHN belong to sample with 15 wt.% graphite addition. Thus, the sample shows a decrease in hardness with increasing of graphite addition. This is probably due to low interfacial bonding of Fe-C caused by inappropriate sintering temperature at 900 °C condition. One should consider that the desired sintering temperature of pure graphite must be higher than 2000 °C. Therefore, the initial stage for diffusion process of Fe-C shows incomplete bonding. However, when it was compared with the commercial pure graphite of BPP, i.e. 24 HVN (directly measured in our lab), the obtained hardness values are still much higher. One can be rationalised from the above results are due to the origin of low hardness value of graphite itself, when mixed with the mill scale (ferritic based materials) it will contribute into lower hardness value. This also inferred that the desired Fe-C phase (Fe-C bonding) at sinter condition of 900 °C is not fully achieved. Therefore, the samples are probably still in the composite state.

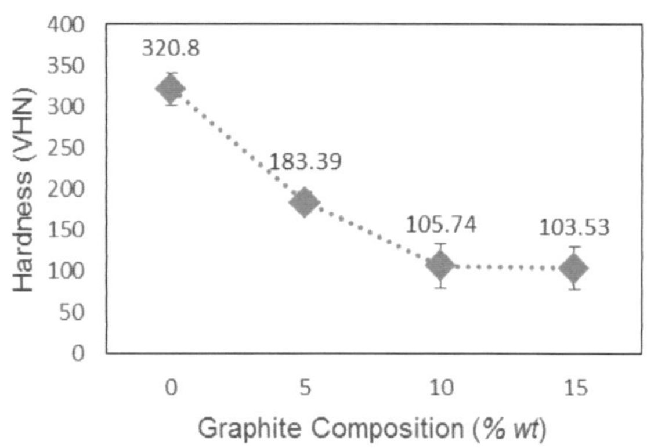

Figure 2. Vickers hardness result of the sample with graphite addition of 0 to 15 wt.%.

3.3. Micrograph analysis

Figure 3 shows SEM top view images of all sintered samples with respect of graphite addition of 0, 5, 10, and 15 wt.% after sintering at 900 °C for 1 h. It clearly shows a significant morphology different between with and without graphite addition. The pure mill scale based sample (figure 3a) shows inhomogeneous grain structure morphology while the graphite added samples (figures 3b, 3c, 3d) show relatively homogenous. At this condition, all samples still show some pore morphology. However, with increasing of graphite addition the pore morphology shows decrease. For the pure mill scale (0 wt.% graphite addition), the main reason of such pore morphology can be addressed due to the various of its original elemental/compound composition of mill scale that not totally react at sinter condition of 900 °C. Hence, the imperfect phase transformation process and diffusion between each compound causing such inhomogeneous morphology. This also probably can be explained due to different of diffusion rates and some different of lattice movement between each compound of mill scale so that causing such different of grain size and shape as well as its pore morphology existence.

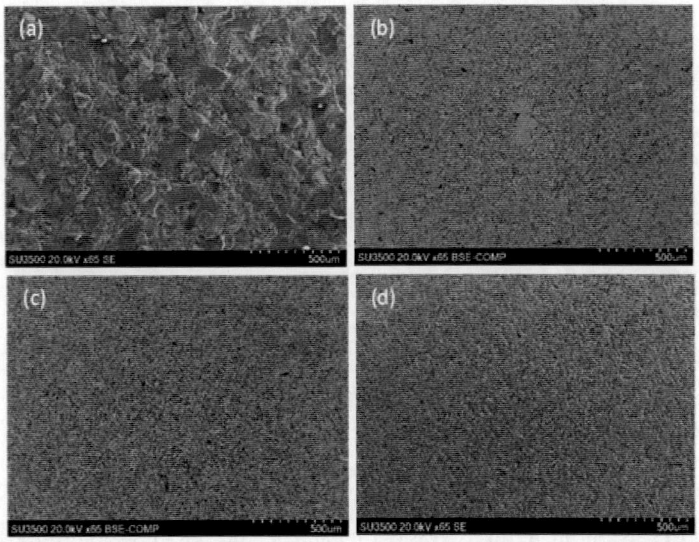

Figure 3. SEM micrograph of the sintered sample with graphite addition of (a) 0, (b) 5, (c) 10, and (d) 15 wt.%.

In the pure mill scale sample, indeed the pore structures are still exist; However, in term of hardness level, the pure one is the highest one. This can be explained due to its better compound bonding (as clearly showed in figure 4a) where the grain boundary is formed at this condition, even not fully yet homogenous. While the 15 wt.% graphite addition sample (figure 4b), the grain boundary is difficult to be identified. Therefore, this grain boundary state most probably contribute into its better hardness value of pure mill scale (0 wt.% graphite addition) than the graphite added samples. Figure 5 shows the optical micrograph of all samples. One can be seen that with increasing graphite addition the density of the 'white colour' is increasing. It means the graphite is distributing more with increase its addition number.

Figure 4. Comparison of SEM micrograph images of the sintered sample with (a) 0 and (b) 15 wt.% graphite addition at magnification of 5000x and 15000x.

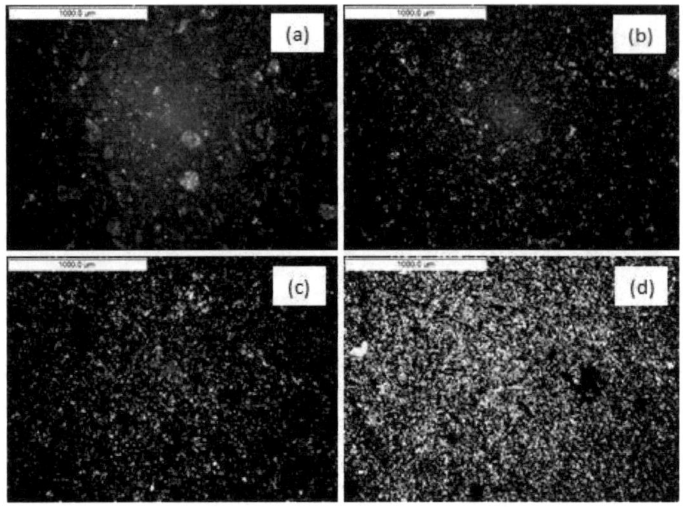

Figure 5. Optical micrograph images of the sintered sample with graphite addition of (a) 0, (b) 5, (c) 10, and (d) 15 wt.%.

4. Conclusions and future work

In this report, the elaboration of using mill scale-based bipolar plate material for PEMFC was studied. We fabricated with variation of graphite addition (0, 5, 10, 15 wt.%) and sintered at 900 °C for 1 h under inert gas atmosphere. The XRD analysis revealed that the obtained phases are wustite (FeO) and magnetite (Fe_3O_4), while for the sintered sample the ferritic (Fe) phase is also can be detected, inclusive for the pure mill scale. The hardness evaluation showed that the pure sample is much higher (320.8 VHN) than the graphite added samples. This is tentatively distributed to the not optimum sinter condition of mill scale-graphite. The micrograph analysis showed the grain boundary morphology of the pure mill scale sample. While the added graphite samples are hardly to be identified, however, their graphite distribution can be confirmed by its optical micrograph results. Those results suggest that the mill scale waste has some potential application as bipolar plate material; however it still need further research investigation and evaluation prior to their ideal bipolar plate component application.

Acknowledgments

The authors would like to thank Dr Bambang Prihandoko for raw material support, the Research Center for Physics – LIPI for technical and facilities support.

References

[1] Wee J H 2007 *Renew. Sustain. Energy Rev.* **11** 1720
[2] Feng K, Li Z, Sun H, Yu L, Cai X, Wu Y and Chu P K 2013 *J. Power Sources* **222** 351
[3] Karimi S, Fraser N, Roberts B and Foulkes F R 2012 *Adv. Mater. Sci. Eng.* **2012** 828070
[4] Oyarce A 2013 *Electrode degradation in proton exchange membrane fuel cells (Doctoral Dissertation)* (Stockholm: KTH Royal Institute of Technology) p 77
[5] Taherian R 2014 *J. Power Sources* **265** 370
[6] Rajasekar S, Chetty R and Neelakantan L 2015 *Int. J. Hydrog. Energy* **40** 12413
[7] Zhou T Y and Chen Y S 2015 *J. Fuel Cell Sci. Technol.* **12** 051001
[8] Jung S P, Lee C I, Chen C C, Chang W S and Yang C C 2015 *J. Power Sources* **283** 429
[9] Tawfik H, Hung Y and Mahajan D 2007 *J. Power Sources* **163** 755
[10] Lu C D, Mao P Z, Wen D H and Jia H 2013 *Key Engineering Materials* **546** 164
[11] Shimpalee S, Lilavivat V, McCrabb H, Khunatorn Y, Lee H K, Lee W K and Weidner J W 2016 *Int. J. Hydrog. Energy* **41** 13688

IOP Conf. Series: Journal of Physics: Conf. Series **985** (2018) 012051 doi:10.1088/1742-6596/985/1/012051

Frequency dependence of electrical properties of polyvinylidene fluoride/graphite electrode waste/natural carbon black composite

D R Insiyanda, N Indayaningsih, B Prihandoko, A Subhan, D S Khaerudini, H Widodo, F Destyorini and A Chaer

Research Centre for Physics, Indonesian Institute of Sciences, Gd. 440 Kawasan Puspiptek Serpong, Tangerang Selatan 15314, Banten, Indonesia

Email: dita002@lipi.go.id

Abstract. Polyvinylidene fluoride (PVdF) is a semi-crystalline thermoplastic material with remarkably high piezoelectric coefficient and an attractive polymer matrix for micro-composite with superior mechanical and electrical properties. The conductive filler is obtained from Graphite Electrode Waste (GEW) and Natural Carbon Black (NCB). The variation of composite content (%) of PVdF/NCB/GEW were 100/0/0, 95/5/0, 95/0/5, 95/2.5/2.5. This experiment employed dry dispersion method for material mixing. The materials were then moulded using hot press machine with compression parameters of P = 5.5 MPa, T = 150 °C, t = 60 minutes, $A = 5\times5\times(0.2 - 0.4)$ cm^3. The electrical conductivity properties of pure PVdF, as well as PVdF/GEW, PVdF/NCB, and PVdF/NCB/GEW composites were investigated in a frequency range of 100 to 100000 Hz. The PVdF/GEW sample obtained the highest electrical conductivity. It is concluded that GEW and NCB can be incorporated into PVdF as a conductive filler to increase the conductivity of conductive material composite without solvent.

1. Introduction

The Waste-to-Energy (WtE) technologies can be applied for various type of waste (e.g. semi-solid, liquid, and gaseous waste from the effluent of treatment industry or medium factory). These technologies deal with waste treatment process that generates energy in form of electricity, heat, transports fuels or composite material for energy devices [1].

The carbon black (CB) can be obtained from the decomposition of biomass waste, such as coconut fibers which are abundant and easy to obtain. Natural carbon black (NCB) decomposition occurs usually at a temperature corresponding to the thermodynamic conditions of the precursor to form charcoal, including by-products such as tar and gases [2]. Graphite is a carbonaceous material; expanded graphite (EG) with a high aspect ratio and excellent electrical conductivity can be produced by exfoliating graphite intercalation compound (GIC) through rapid heating in a furnace or microwave environment. Graphite electrode waste (GEW) from electric arc furnace (EAF) can substitute graphite synthetic as a reinforcement material for carbon composite.

Composites material has been used widely in industrial sectors as materials for antistatic, electromagnetic shielding, and electronic technology [3, 4]. In electronic technology, composites material has been used for a device with improved electromagnetic and energy storage system (ESS) [5] or electrochemical energy storage systems (EESS) [6].

Polyvinylidene fluoride (PVdF) is a highly non-reactive and pure thermoplastic fluoropolymer material. Below 150 °C, PVdF becomes ferroelectric. Thus, PVdF is electro-active and semi-crystalline polymer with pyro and piezoelectric properties at room temperature, which can be used for many applications [7]. It exhibits high mechanical strength, good chemical resistance and thermal stability, as well as excellent aging resistance [8]. Moreover, PVdF is an attractive polymer matrix for composite material with superior mechanical and electrical properties.

The performance of conducting polymer composite depends on filler concentration which determines the electrical properties. However, higher filler content can lead to poor process ability and low mechanical properties. EG reinforced-conducting polymer composites have shown substantial improvements in electrical conductivity and mechanical property when CB is incorporated [9].

In this study, the electrical properties of PVdF with addition of some additives (e.g. GEW from the EAF and NCB from coconut fiber) are investigated. In addition, the effect of solvent in the process is also studied.

2. Experimental

The main material is commercial-grade PVdF powder (182702-2506), supplied by Sigma-ALDRICH Inc., USA. Meanwhile, NCB powder from coconut fiber and GEW powder from iron or steel smelting factory are used as an additive. GEW has particle size of < 44 μm. GEW from the EAF was chosen as a reinforcement material for electrically conductive material due to its similarity with synthetic graphite. Structural examination of GEW from the EAF showed the crystalline carbon structure, instead of amorphous structure, which implies its good conductivity [10].

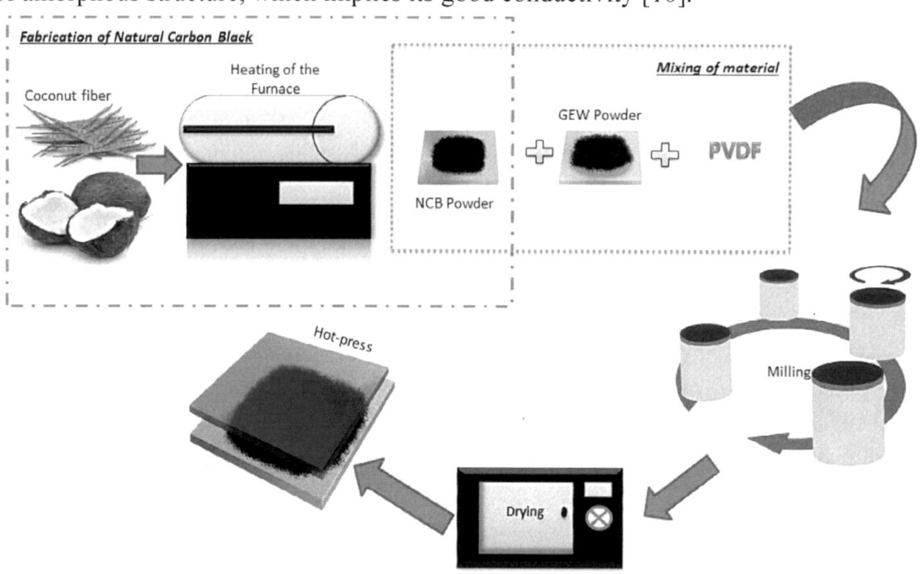

Figure 1. Schematic diagram of the dry process or non-solvent to shape as a bulk composite

Preparation of carbon materials, NCB powder was conducted in two main steps, namely carbonized and pyrolysis process. In the carbonization, process, coconut coir was carbonized at 500 °C temperature for an hour in a furnace with nitrogen atmosphere. Subsequently, the charcoal was pyrolyzed at 1300 °C temperature for an hour. NCB with particle size of < 44 μm was used as filler. Composite content (%) variation of PVdF/NCB/GEW were 100/0/0, 95/5/0, 95/0/5, and 95/2.5/2.5. Dry mixing or non-solvent method was used for material mixing. The waste of GEW/NCB/PVdF was mixed with a ball mill for 900 minutes, dried, and molded used compression hot press machine at 5.5 MPa pressure and 150 °C temperature for 60 minutes with 5×5×(0.2-0.4) cm^3 dimension. Total weight composition to make raw material of carbon composite polymer is 20 gram. HIOKI 3522-50 LCR

IOP Conf. Series: Journal of Physics: Conf. Series **985** (2018) 012051 doi:10.1088/1742-6596/985/1/012051

HiTESTER was used to measure AC electrical conductance of the composite materials sample with frequency ranged from 100 to 100000 Hz.

3. Results and discussion

This experiment was meant to study the behaviour and electrical properties of PVdF/GEW/NCB Composite. PVdF/GEW/NCB is designed from graphite electrode waste and carbon black material developed from natural resource, in which the content is abundant and the process feasible. Carbonization step of coconut coir was performed to degrade the chemical compound of the coconut coir and produce charcoal. In order to reach activated conductive carbon, the charcoal was then pyrolyzed. One of the advantages of this processed is the simplicity [11].

Figures 2 - 4 display the frequency dependence of electrical properties. For PVdF/5 wt.% GEW, there is a transition from an insulator to semiconductor. GEW was chosen as a reinforcement material for electrically conductive material because it has the characteristics and specifications similar to synthetic graphite. The insulator-conductor translation is also confirmed by the large dissipation factor (tan δ) for PVdF/5 wt.% GEW (figure 2).

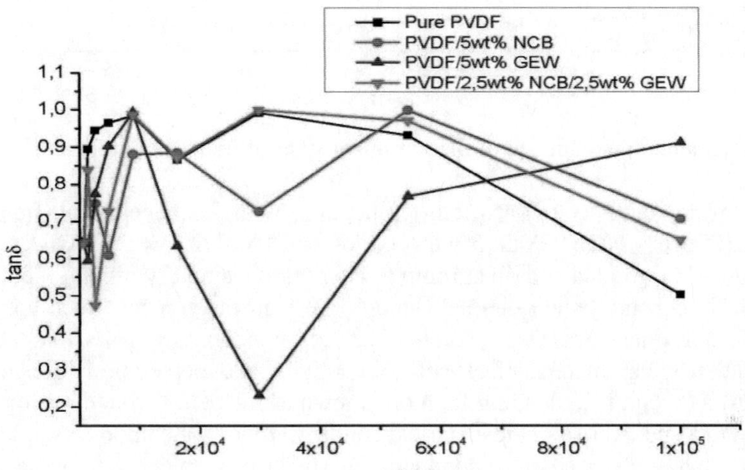

Figure 2. Frequency-dependence of the loss tangent of material composite.

The electric permittivity, energy loss, and dielectric relaxation strength are important parameters for material science because these parameters are used to decide on the suitability for given application. Measurement of AC electrical properties of the composite materials sample employed following formula.

The frequency dependence complex dielectric permittivity $\varepsilon^*(\omega)$ is given by

$$\varepsilon^*(\omega) = \varepsilon'(\omega) - \varepsilon''(\omega) \tag{1}$$

Here, $\varepsilon'(\omega)$ and $\varepsilon''(\omega)$ are the real and imaginary part of complex dielectric permittivity. PVdF/5 wt.% GEW as a function of frequencies at room temperature is shown in figure 3. As expected, the variation tendency of dielectric constant with frequency is inversely proportional to electrical conductivity (σ). The ε' attains high value at low frequency and decreases exponentially with increasing frequency.

Dielectric permittivity was measured with applied field at a frequency of 100 Hz to 100000 Hz. PVdF is dielectric material which stores charges. Dielectric properties also depend on the orientation of the structure in the material. Dielectric loss of permittivity can be calculated from the measured loss tangent and capacitance:

$$V_\theta = V_T \sin(\pi - \varnothing) \tag{2}$$

$$tan\delta = \frac{\left(V_T^{\,2} - V_\theta^{\,2}\right)^{\frac{1}{2}}}{V_\theta} \tag{3}$$

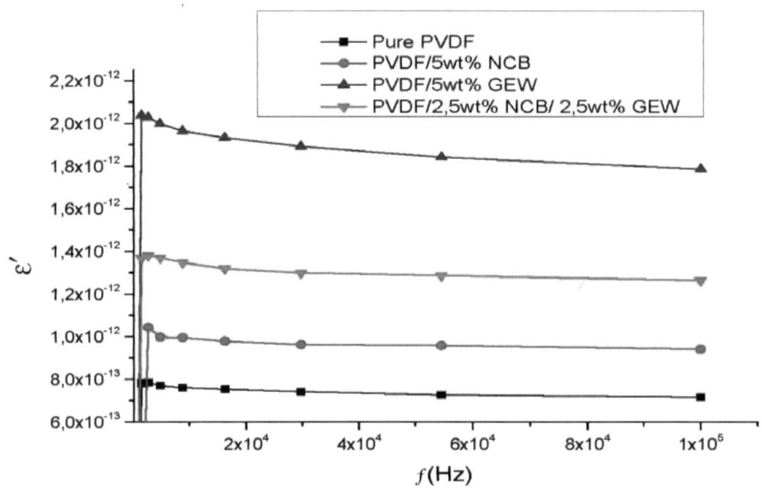

Figure 3. Frequency-dependence of dielectric constant of material composite.

The variation of conductivity (σ) at room temperature with frequency for different filler content (pure PVdF, PVdF/5 wt.% NCB, PVdF/5 wt.% GEW, and PVdF/2.5 wt.% NCB/2.5 wt.% GEW) of composites is shown in figure 4a and 4b at room temperature. Using dry mixing method, the electrical conductivity is fairly constant when the NCB and GEW load the content. Such composite materials generally exhibit non-linear increase of the electrical conductivity as a function of the filler concentration. The two parameters, electrical conductivity and percolation threshold, are together associated with. At a certain filler loading fraction, known as the percolation threshold, the fillers form a network leading to a sudden rise in the electrical conductivity of the composites.

The addition of a very low amount of conducting particles can make filler contact to form effective conducting paths and thus making the whole composite conductive. These specimens show a typical insulating behaviour with a frequency-dependent conductivity. However, the higher NCB loading in the polymer composite can cause serious NCB aggregation, and the electrical conductivity will lower or even decrease. Figure 4 shows electric conductivity of PVdF/2.5 wt.% NCB/2.5 wt.% GEW composites decreased due to the loading of NCB equal to the loading of GEW is in 2.5 wt.%. This occurs because the loading of NCB between the GEW particles does not act as a bridge for the mobility of electrons in any direction and does not provide gives electrical conductivity to the composite. Conductivity (σ) was calculated according to the relation:

$$\varepsilon'(\omega) = C\, d \,/\, \varepsilon_o A \tag{4}$$

$$\varepsilon''(\omega) = \varepsilon'\, tan\delta \tag{5}$$

Where, d is separation distance between two electrodes, A is electrodes area, ε_o is the permittivity of the free space, (ε_o =8.85×10^{-12} F/m).

$$\sigma = \varepsilon_o\, \omega\, \varepsilon'' \tag{6}$$

Where $\omega = 2\pi f$ is the angular frequency.

Figure 5 shows the electrical conductivity properties of pure PVdF, PVDF/5 wt.% NCB, PVdF/ 5 wt.% GEW, PVdF/2.5 wt.% NCB/2.5 wt.% GEW at frequency of 100 - 100000 Hz. The electrical

IOP Conf. Series: Journal of Physics: Conf. Series **985** (2018) 012051 doi:10.1088/1742-6596/985/1/012051

conductivity was enhanced the most in PVdF/5 wt.% GEW, i.e. at 1.26×10^{-7} S.cm^{-1}. However, graph exhibited decreasing value of electrical conductivity in PVdF/2.5 wt.% NCB/2.5 wt.% GEW, i.e. at 9.81×10^{-8} S.cm^{-1}.

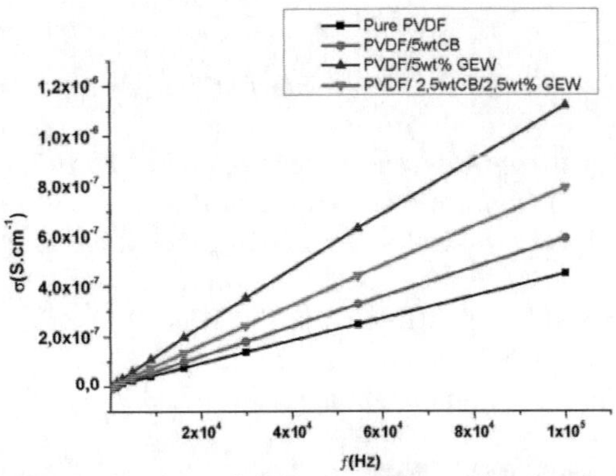

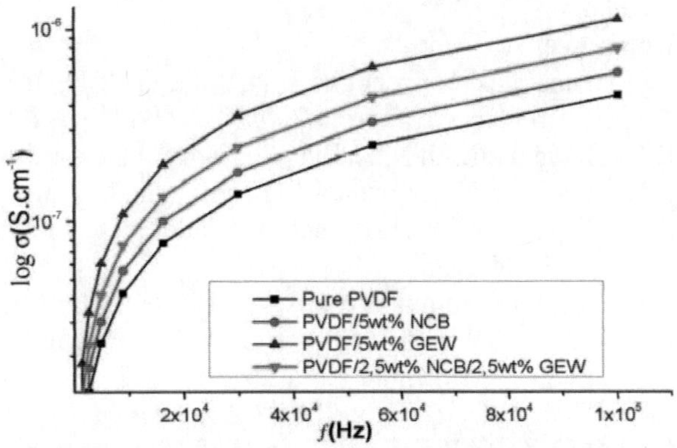

Figure 4. Plot of frequency-dependence of electrical conductivity of material composite (left) and frequency-dependence of log electrical conductivity (right).

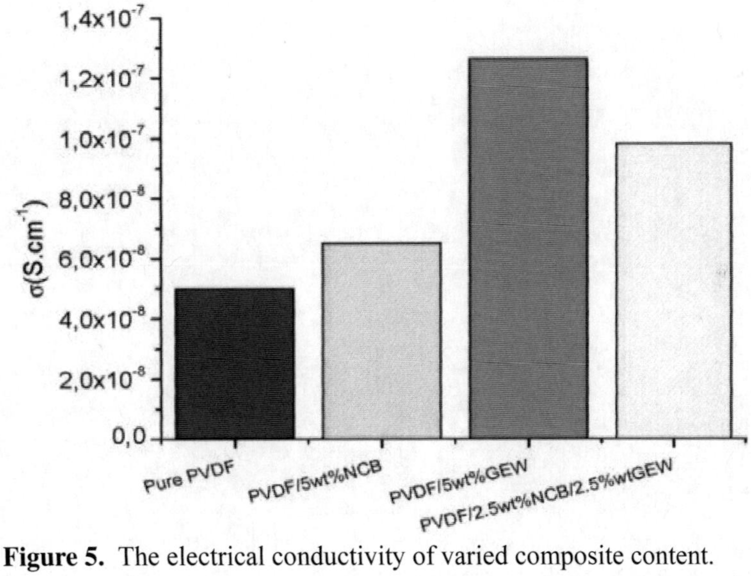

Figure 5. The electrical conductivity of varied composite content.

4. Conclusion and future work

Electrical properties of composite materials based on PVdF/5 wt.% NCB, PVdF/5 wt.% GEW, PVdF/2.5 wt.% NCB/2.5 wt.% GEW prepared by dry mixing were investigated depending on filler concentration, frequency, and deformation. The experimental results demonstrated that the investigated composites had some peculiarities of electrical properties in comparison with pure PVdF. The dielectric permittivity of PVdF/5 wt.% GEW composites was significantly higher than that of composites based NCB fillers. In addition, composites filled with GEW demonstrated the most stable electrical properties depending on such important exploitation parameters as deformation. This preparation method allows us to achieve a uniform distribution of filler in a polymer matrix and specific properties of filler particles. It makes GEW as alternative conductive filler for conductive material in electronic applications.

Acknowledgments
The authors gratefully acknowledge the financial support from Indonesian Institute of Sciences, PUI-Ministry of Research, Technology and Higher Education Republic of Indonesia, as well as the facilities provided by the Research Centre for Physics-Indonesian Institute of Sciences

References

[1] Omari A M, Kichonge B N, John G R, Njau K N and Mtui P L 2014 *International Journal of Renewable Energy Technology Research* **3** 1
[2] Indayaningsih N, Zulfia A, Priadi D and Kartini E 2013 *Jurnal Sains Materi Indonesia* **14** 147
[3] Ezquerra T A, Canalda J C, Sanz A and Linares A 2014 *Colloid Polym. Sci.* **292** 1989
[4] Hussain W A, Hussein A A, Khalaf J M, Al-Mowali A H and Sultan A A 2015 *Adv. Chem. Eng. Sci.* **5** 282
[5] Rajib M, Shuvo M A I, Karim H, Delfin D, Afrin S and Lin Y 2015 *Ceram. Int.* **41** 1807
[6] Wang Y, Wei H, Lu Y, Wei S, Wujcik E K and Guo Z 2015 *Nanomaterials* **5** 755
[7] Zhen Y, Arredondo J and Zhao G 2013 *Open Journal of Organic Polymer Materials* **3** 99
[8] Kang G D and Cao Y M 2014 *J. Membr. Sci.* **463** 145
[9] Li Y C, Li R K Y and Tjong S C 2010 *Journal of Nanomaterial* **2010** 261748
[10] Sadeli Y, Soedarsono J W, Prihandoko B and Harjanto S 2011 *Journal of Materials Science and Engineering B* **1** 178
[11] Noh Y J, Joh H I, Yu J, Hwang S H, Lee S, Lee C H, Kim S Y and Youn J R 2015 *Sci. Rep.* **5** 9141

Reliability performance of up-scaling DSSC into sub-module in series design using hermetic sealing

P N Anggraini, L Retnaningsih and J Hidayat

Research Center for Electronics and Telecommunications, Indonesian Institute of Sciences (PPET – LIPI), Kampus LIPI, Jl. Sangkuriang, Bandung 40135, Indonesia

E-mail: putr014@lipi.go.id

Abstract. Dye-sensitized solar sub-module has been prepared as a practical application of DSSC. TiO_2 paste was deposited onto the 3.3 cm × 5 cm FTO glass by screen printing method consisting of two active cells with a dimension of 1 cm × 5 cm for each cell. The sub-module was assembled using two parts (part A and part B) hermetic sealing compound as a sealant. The performance of the sub-module was then observed using two independent measurements in the span of six days using sun simulator. Photovoltaic effect was properly occurred as indicated by the I-V curve for both of sub-module. First test showed that the sub-module with hermetic sealant has good performance with the highest efficiency of 2.9%. Meanwhile, the second test showed significant decrease of the sub-module performance with the highest efficiency of 1.43%.

1. Introduction

Several alternative energy sources have been developed to replace fossil energy source. Solar energy, as one of renewable energy source which is abundant, clean, safe, and able to generate energy in remotes areas, become the main candidate energy source to replace fossil fuel [1]. To date, silicon solar cells are the main commercial photovoltaic technology. Unfortunately, this technology has high cost and is not environmentally friendly. To overcome these problems, researchers have been developing new photovoltaic energy based on dyed-TiO_2 semiconductor, known as dye-sensitized solar cells (DSSC). One study, conducted by Gratzel, et al [2], obtained cell efficiency of 11% [3].

Scaling up the cells into DSSC module is the most important part for practical application. Module designs in two different groups of series and parallel can be chosen according to its necessity. Serial connection consists of monolithic and W/Z interconnection schemes that has high output voltage and low current. Gratzel disclosed that performance of DSSC in module scale had reached efficiency of 7% [3].

In the fabrication of DSSC module, the performance stability is the main focus of investigation. Some factors such as materials, assembly process, and environmental condition can affect the performance of DSSC particularly on photoconversion efficiency and stability. One of the requirements to meet the criteria is good sealing between photoelectrode and counter electrode which could avoid the leakage of electrolyte liquid thermoplastic sealant is one of the best sealant materials for cells. It is not suitable for module due to its difficulty in the assembly process. One of many materials for sealant is a hermetic seal-based materials, which has semi-rigid seal for DSSC enclosure.

In this work, two-part hermetic sealing compound from Dyesol was employed as sealing material for DSSC sub-module using series interconnection. The hermetic compound can be applied easily by screen printing technique since it is in its paste form. In addition, the hermetic sealant does not require high temperature for drying which is suitable for dyed DSSC. Two sets of hermetic compound were prepared to obtain the optimum sealant based on its composition. The scope of investigation include performance optimization, as well as sealing endurance which affects its lifetime stability.

2. Experiments

2.1. Preparation of photoelectrode and counter electrode

The screen printing method was used in the fabrication process of DSSC sub-modules. Photoelectrode was prepared on a 3.3 cm × 5 cm area of FTO glass which was divided into two active areas of 1 cm × 5 cm for each area. Then, the pattern for sub-module screen (figure 1) consisted of active area, circuit contact, and sealant area was created using photolithography process. Meanwhile, FTO glass as a conductive substrate was prepared for photoelectrode and counter electrode by scribing the surface into two parts. These substrates were then cleaned with DI water using ultrasonic cleaner and immersed in ethanol solution for 1 hour. For photoelectrode preparation, $TiCl_4$ pretreatment was applied by dipping the cleaned FTO glass in $TiCl_4$ solution (0.04 M) for 30 minutes with temperatures of 70 °C. After the pretreatment process, TiO_2 paste (Dyesol) was deposited onto the FTO glass by screen-printing technique to make TiO_2 active layer. This film was dried in vacuum oven with temperatures of 100 °C for 10 minutes and sintered in infrared furnace at 450 °C for 45 minutes. $TiCl_4$ post-treatment was applied as well to the sintered film, using the same process as pretreatment. After rinsing in ethanol solution, the film was sintered once more with the same process as before and immersed in dye solution of N907 for 24 hours.

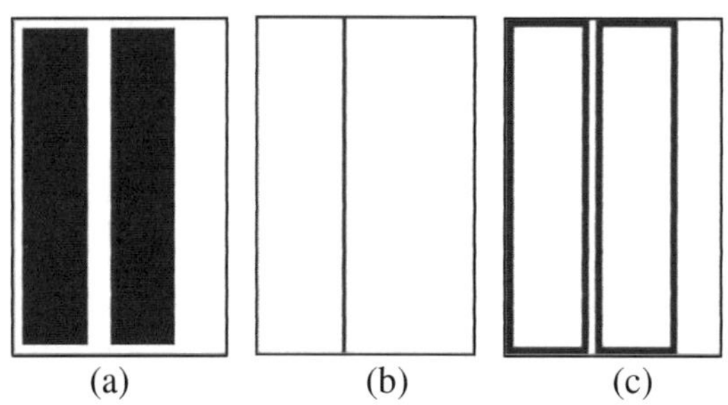

Figure 1. Pattern design for sub-module screen of (a) TiO_2 active area, (b) circuit contact, and (c) sealant area.

Counter electrode was prepared with an area equal to the photoelectrodes. Platinum (Pt) was deposited onto the substrate using sputtering process for 2 hours. The area of the electrode was divided into two parts by scribing the Pt layer as the area of photoelectrode substrate.

2.2. Assembly process and performance measurement

The assembly process of DSSC sub-module was preceded by the sealing preparation, which uses two parts hermetic sealing compound (part A and part B) from Dyesol. These parts were combined in two different compositions that can be seen in table 1. On sealant area of TiO_2 photoelectrode, hermetic compound was subsequently deposited by screen printing method. Silver conductive paste (Ag) was deposited as well on circuit contact area. Counter electrode was directly stacked on the photoelectrode to assemble them into sandwich construction in Z-type series interconnection. The sub-module was heated in temperature of 100 °C for 5 min to dry the sealant and Ag paste so that both electrodes attached to each other. The scheme of the interconnection was described in figure 2. Electrolyte liquid was injected to the assembled sub-module through a pre-drilled tiny hole on the counter electrode surface.

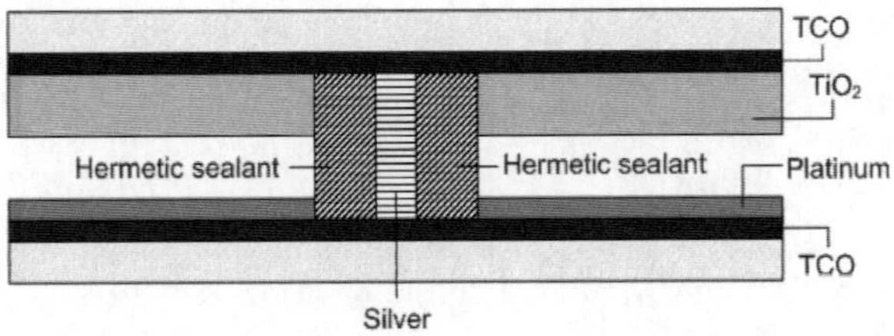

Figure 2. Interconnection scheme of DSSC sub-module.

Table 1. Composition ratio of hermetic sealing compound (Dyesol).

Sealant Compound	Part A	Part B
Hermetic A	1	0.75
Hermetic B	1	0.5

To investigate photovoltaic performance of the sub-module, I-V measurement was applied by illuminating the sub-module using sun simulator at 500 watt/m^2 intensity. Aside from measuring performance of the sub-module, this work also observed the quality of the compound of hermetic sealant in the stability aspect. Therefore, the I-V measurement was taken twice in the range of six days.

3. Results and discussion

In the fabrication of DSSC sub-modules, some aspects need to be observed including photoelectrode and counter electrode preparation, assembly process, and sealing material. In the preparation of TiO_2 photoelectrode, $TiCl_4$ treatment was applied to improve the inter-particle connectivity and make TiO_2 layer more compact [4]. Application of hermetic compound for the sub-modules becomes an attempt to find out the optimum sealing materials. With sufficient ratio between part A and part B, hermetic sealing compound should be able to carry out the high reliability of DSSC sub-module so that the sub-module can perform with long-term stability which is important for practical application.

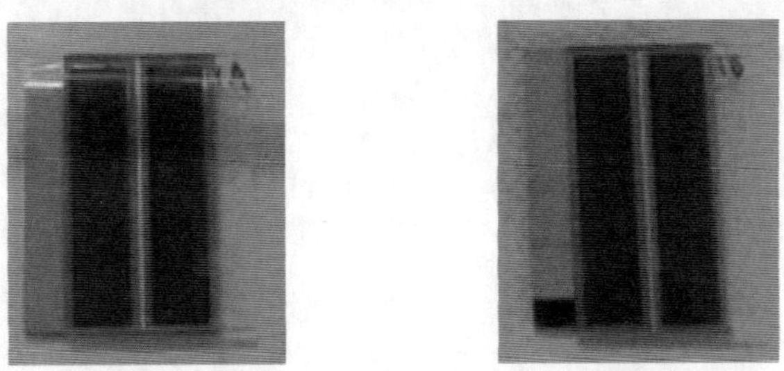

(a) (b)

Figure 3. Sample of DSSC sub-module (a) with hermetic A sealant compound and (b) with hermetic B sealant compound.

TiO_2 has been used as a semiconductor material for the DSSC sub-module. Characteristics of TiO_2 as main material is essential for photovoltaic performance which can be obtained by its morphology or crystalline phase. Figure 4 shows XRD pattern of TiO_2 layer deposited on FTO glass. It can be seen from the pattern that strong peak was in position of 2θ with intensity of 25.3° which is confirmed as anatase phase of TiO_2. Chemical and physical treatment in the fabrication process did not change the phase composition of TiO_2.

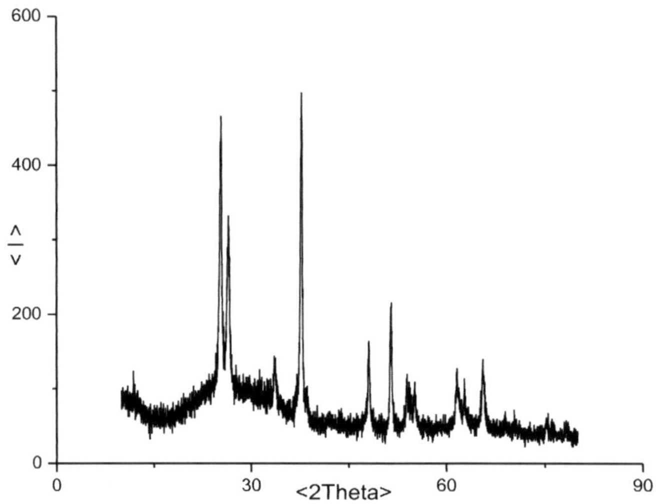

Figure 4. XRD pattern of TiO_2.

Performance of solar sub-module was investigated by I-V measurement using sun simulator. Figure 5 shows the I-V curve from the first test that describes photovoltaic performance of the sub-module. Both of sub-module using hermetic A and hermetic B sealant showed the photovoltaic effect confirms the proper work as a solar cells. Photoelectrodes had good performance which was affected by the treatment in the preparation process in spite of manual process in assembly.

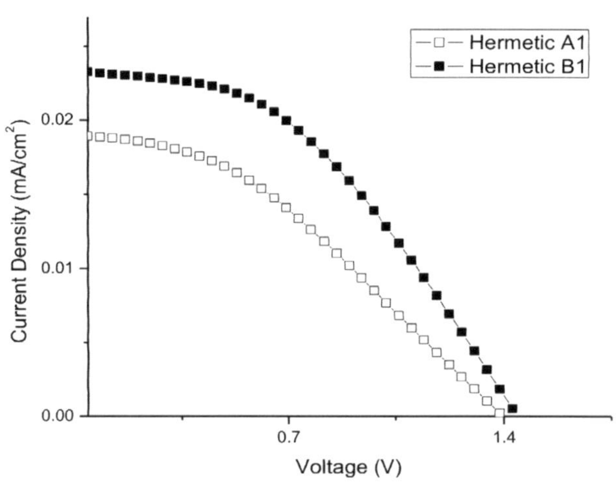

Figure 5. I-V curve of DSSC sub-module sample from the first I-V test.

Composition between two parts of hermetic sealant compound obviously affected the sub-modules performance as well. It can be seen from table 2 that higher performance of photovoltaic was achieved by DSSC sub-module that uses hermetic B as sealant, with the value of open-circuit voltage (Voc), short-circuit current density (Jsc), fill factor (FF), and photoconversion efficiency (η) of 1.427 V, 26.38 mA/cm^2, 0.38, and 2.9%, respectively. Hermetic B might have been a more effective compound which formed an effective hermetic seal that can make stronger bond with glass and minimize electrolyte leakage. The result also indicates that both cells of each sub-module had good series interconnection which could affect the value of open circuit voltage.

Table 2. Photovoltaic parameters of DSSC sub-module from the first I-V test.

Sealant Compound	Voc (V)	Jsc (mA/cm^2)	FF	η (%)
Hermetic A1	1.386	18.92	0.37	1.9
Hermetic B1	1.427	26.38	0.38	2.9

Second I-V test of sample was taken six days after the first test with same criteria. The result of the second test can be seen in figure 6. Both of I-V test of samples show curves that were less ideal compared with the curves from the first I-V test. Nevertheless, the effect of photovoltaic still appears which means that both hermetic sealants were able to insulate the interconnection of the cells and avoid unwanted mass transport between electrolytes of adjacent cells.

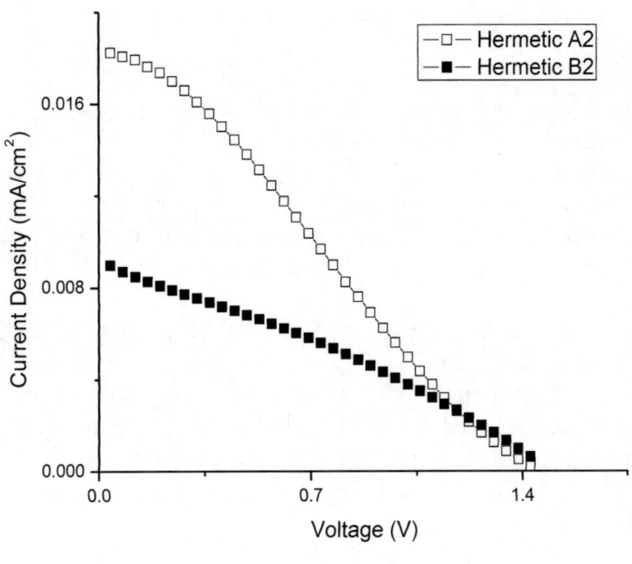

Figure 6. I-V curve of DSSC sub-module sample from the second I-V test.

Poor performance of the samples in the second test can also be seen from the quantity parameters, mainly for sub-module with hermetic B which yielded more significant decrease. The result shows that performance of sub-module with hermetic A as sealant was higher than sub-module with hermetic B, with best performance of Voc, Jsc, FF, and η respectively are 1.427 V, 18.29 mA/cm^2, 0.28, and 1.43%. In this case, sub-module with hermetic A was able to remain more stable after six days from the first measurement. Performance degradation of sub-module with hermetic A was 24% for the photoconversion efficiency while sub-module with hermetic B was 71.3%.

Hermetic A initially demonstrated good performance as sealant in terms of the ability to avoid the leakage and minimize the reaction between electrolytes. However, hermetic A failed to maintain its performance in long-term stability. Hermetic B, which had lower performance in the first test, became relatively stable according to the second test. Hermetic B was considered to have less reaction with electrolyte and less effect on surroundings in terms of temperature and humidity.

Table 3. Photovoltaic parameters of DSSC sub-module from the second I-V test.

Sealant Compound	Voc (V)	Jsc (mA/cm^2)	FF	η (%)
Hermetic A2	1.427	18.29	0.28	1.43
Hermetic B2	1.467	8.968	0.32	0.83

4. Conclusion

In summary, hermetic sealing compound was employed as sealant in the fabrication of DSSC sub-module. The optimized composition of sealant was hermetic compound with ratio part A : part B = 1 : 0.5 as the DSSC sub-module received the higher photoconversion efficiency of 2.9% in the first I-V test. On the contrary, in terms of obtaining the stability higher performance was received by sub-module with sealant containing hermetic compound part A : part B = 1 : 0.75 in the second I-V test. Initially, hermetic B sealant compound had better quality compared with hermetic A but hermetic A is more recommended for long term use.

Acknowledgment
This research is fully supported by DIPA LIPI funding program year of 2017.

References

[1] Longo C and De Paoli M A 2003 *J. Braz. Chem. Soc.* **14** 889
[2] O'Regan B and Grätzel M 1991 *Nature* **353** 737
[3] Grätzel M 2011 *Nature* **414** 338
[4] Park J and Lee M 2015 *Electron. Mater. Lett.* **11** 271
[5] Hagfeldt A, Boschloo G, Sun L, Kloo L and Pettersson H 2010 *Chem. Rev.* **110** 6595
[6] Chou J C, Huang C H, Liao Y H, Lin Y J, Chu C M and Nien Y H 2016 *Int. J. Photoenergy* **2016** 6595639
[7] Zhang Y D, Huang X M, Gao K Y, Yang Y Y, Luo Y H, Li D M and Meng Q B 2011 *Sol. Energy Mater. Sol. Cells* **95** 2564
[8] Yu H, Zhang S, Zhao H, Will G and Liu P 2009 *Electrochim. Acta* **54** 1319
[9] Chien C H, Tsai M L, Su T H, Hsieh C C, Li Y H, Chen L C and Gau H D 2010 Efficiency Enhancement of Dye-Sensitized Solar Cell *SEM Annual Conf.* **6** pp 137–144
[10] Huang Z, Liu X, Li K, Li D, Luo Y, Li H, Song W, Chen L Q and Meng Q 2007 *Electrochem. Commun.* **9** 596

Structural and morphological changes in supramolecular-structured polymer electrolyte membrane fuel cell on addition of phosphoric acid

S Hendrana[1], R F Pryliana[2], C L Natanael[2] and I Rahayu[2]

[1]Research Center for Physics, Indonesian Institute of Sciences (LIPI)
Kawasan Puspiptek No. 440-442, Kota Tangerang Selatan, Banten, Indonesia
[2]Departmen of Chemistry, Faculty of Mathematic and Natural Sciences, Padjadjaran University, Jl. Raya Bandung Sumedang Km 21, Jatinangor, Sumedang, Indonesia

E-mail: sunit.hendrana@lipi.go.id

Abstract. Phosphoric acid is one agents used in membrane fuel cell to modify ionic conductivity. Therefore, its distribution in membrane is a key parameter to gain expected conductivity. Efforts have been made to distribute phosphoric acid in a supramolecular-structured membrane prepared with a matrix. To achieve even distribution across bulk of the membrane, the inclusion of the polyacid is carried out under pressurized chamber. Image of scanning electron microscopy (SEM) shows better phosphoric acid distribution for one prepared in pressurized state. It also leads in better performing in ionic conductivity. Moreover, data from differential scanning calorimetry (DSC) indicate that the addition of phosphoric acid is prominent in the change of membrane structure, while morphological changes are captured in SEM images.

1. Introduction

Polymer membrane is an essential part in Polymer Electrolyte Membrane Fuel Cell (PEMFC). The membrane only permits ion and exhibit electrone. Therefore, the higher ionic conductivity of the membrane would end in higher performance of fuel cell. Efforts have been made to improve ionic conductivity of the membrane including inclusion of small molecules including polyacids [1] and phosphoric acid [2]. The latter is chosen from its ability on self-dissociation that produces proton ion. The produced proton ion can assist proton conduction in the membrane. From this respect it is expected that the membrane prepared with inclusion of phosphoric acid will have characteristic of high conductivity in higher temperature and lower humidity [2].

However, it is well known that in PEMFC, sulfonate groups in the membrane are clustered [3] in order to be able to facilitate ion movement from anode to cathode. The inclusion of small molecules in such clustered-structure could change the structure itself.

Membrane developed ini LIPI is not constructed in such cluster-structured but rather aligned to form a rod-structured in order to arranged of sulfonate groups in sulfonated polystyrene. Sulfonate groups in sulfonated polystyrene are difficult to form cluster-like structure because the sulfonate groups are attached in benzene pendant group in polystyrene. A rod-like structure is likely to be one possibility and is constructed from hydrogen bond between succinic/anhydride group of polyethylene-*graft*-maleic anhydride (PE-*g*-MA) and sulfonate groups in sulfonated polystyrene. The stucture form thereafter called as supramolecular structure. SANS (Small Angle Neutron Scattering) analysis has been performed for the evident of the supramolecular structure [4].

As also occurs in cluster-like structured polymer electrolyte membrane fuel cell, the addition of phosphoric acid in the supramolecular-structured polymer electrolyte membrane fuel cell could also promote changes in structural arrangement and morphology. Therefore, this work will emphasize on the above mentioned problems. The inclusion of phosphoric acid will depend on the preparation Method, in order to study the effect of the inclusion of phosphoric acid to the structural and morphological changes, the acid is added with and without pressure to the base of the membrane.

2. Experimental

2.1. Chemicals
Toluene and phosphoric acid are p.a. grade and they were supplied from E-Meck. The chemicals were used without any purification. Water used was demineralized water. Polyethylene-graft-maleic anhydride or PE-*g*-MA was from Aldrich (lot# MKBG0507V). Sulfonated polystyrene (SPS) was prepared by using method developed in our laboratorium.

2.2. Preparation of membrane
Solution of sulfonated polystyrene in toluene:methanol (9:1, v/v) and solution of PE-g-MA and benzimidazol in toluene were prepared separately before mixing at 40 °C. The solution thereafter called as supramolecular solution. Template for membrane was prepared by mixing SPS powder and PE-g-MA powder at 1:1 (w/w). Further, the mixed powder was hot-pressed at 105 °C at 10 kgf.cm^{-2} for 20 seconds. Phosphoric acid is added in two ways, i.e. for pressurized membrane samples, the template immersed in phosphoric acid in a closed vessel and heated at 50 °C for one hour, while non-pressurized membrane samples, the immersion is in open space and no heat is applied.

Table 1. Membrane samples used in this work.

Code	Components				Pressurized
	SPS	PE-*g*-MAH	H_3PO_4	Supramolecular Solution*)	
(a)	(b)	(c)	(d)	(e)	(f)
dHdS	V	V	V	V	-
dHtS	V	V	V	-	-
dPdS	V	V	V	V	V
dPtS	V	V	V	-	V
tHdS	V	V	-	V	V

*) Supramolecular solution is prepared by method described above and have to be in very dilute solution.

The supramolecular solution was added dropwise into the template according to the variation listed in table 1. The resulted products were hot-rolled at 150 °C using laminating machine prior to hot-pressed at 150 °C, 50 kgf cm^{-2} for 10 seconds, and immediately followed by compression at 150 °C, 100 kgf cm^{-2} for 20 seconds.

2.3. Scanning electron microscope
Samples were immersed in liquid nitrogen for couple minutes and suddenly broken into two parts. The cross-section area were gold-coated for SEM observation.

2.4. Fourier Transform Infra Red (FT-IR)
FTIR data were obtained using FTIR 4300 of Shimadzu. Samples were scanned at wavenumber of 400 to 5000 cm^{-1} at resolution of 2 cm^{-1}. The data were collected after 25 scans for each samples.

2.5. Differential Scanning Calorimetry (DSC)

IOP Conf. Series: Journal of Physics: Conf. Series **985** (2018) 012053 doi:10.1088/1742-6596/985/1/012053

The samples were contained in closed chamber and applied into heat scan from 40 °C to 300 °C at 10 °C per minutes under nitrogen atmosphere. The presentations are taken from the second-run or second-heating of each sample.

3. Result and discussion

Figure 1 shows the thermal scan of differential scanning calorimetry of the membranes. It shows melting temperature which related to their crystalline region.

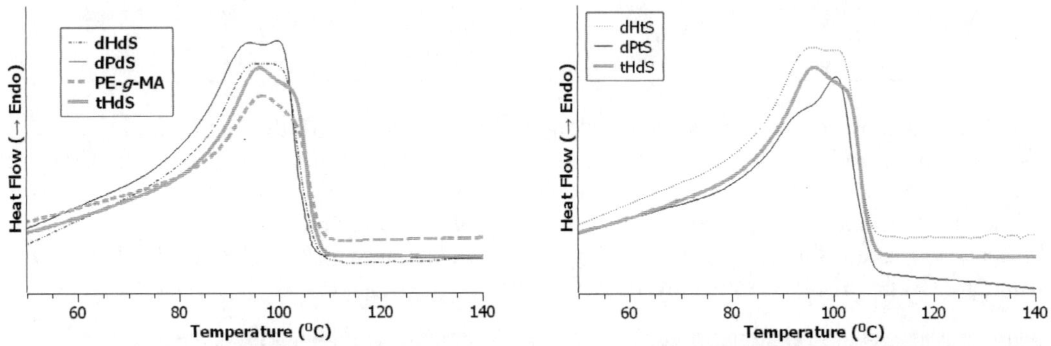

(a) **(b)**

Figure 1. Differential scanning calorimetry (DSC) scan of membrane prepared with phosphoric acid, prepared with and without pressure: (a) comparison of melting profile of membranes with supramolecular structure at variation of preparation methods; (b) comparison of melting profile of membranes to observe the effect of pressure application in membrane preparation

Figure 1 shows that each treatment produces different melting temperature profile, indicating different crystallite structure. PE-*g*-MA shows two melting temperature peaks and the lower melting temperature peak is higher (or dominant) than the higher melting temperature peak. Each melting temperature peak represents a different crystallite size or domain of the sample [5].

An addition of phosphoric acid (sample dHdS) leads to profile changes where two peaks have similar high. When an addition of phosphoric acid is added in pressurized condition (sample dPdS) the higher melting peaks is dominant. As the peak is related to the crystallite structure [6], therefore the DSC scans in figure 1 above indicate changes of crystalline structure within membranes, e.g. the sample has more part which has a crystalline domain that melts at slightly higher temperature indicated by the melting peak temperature.

Table 2 shows melting points of the membrane. The onset temperature of all membranes are only have small differences or within experimental error as well as the first melting temperature peaks and the first melting temperature peaks.

Eventhough the peak temperatures are not significantly differ, the DSC scans shown in figure 1 exhibit an importance of phophoric acid addition and method of the addition to the changes discussed above. The melting profile of membrane without addition of phosphoric acid (tHdS membrane, figure 1) is similar to melting profile of PE-*g*-MA. Addition of phosphoric acid changes the melting profile (dHdS membrane in figure 1a), where the two melting peaks have similar height. The application of pressure during addition of phosphoric acid produces membrane with higher portion of higher melting peak temperature (dPdS membrane in figure 1a). This is the evidence that method of preparation of membranes could affect the structure of the membrane produced.

Table 2. Melting point of the membranes.

Membrane	Melting Points (°C)		
	Onset	1st Peak	2nd Peak
(a)	(b)	(c)	(d)
PE-*g*-MA	85.2	96.6	102.4
dHdS	83.5	94.9	99.3
dPdS	83.5	94.0	99.8
dHtS	83.5	95.1	101.1
dPtS	83.2	95.9	101.7
tHdS	84.0	95.9	100.6

Moreover, membrane prepared without supramolecular solution (dPtS membrane in figure 1b) shows that the second melting peak temperature is much higher than membrane prepared with supramolecular solution and with pressure (dPdS membrane in figure 1a). This indicates that the phosphoric acid have more prominent effect in their crystalline-structural change.

To achieve goals of this works, morphology of the membranes were investigated using scanning electron microscope (SEM) as shown in figure 2 below.

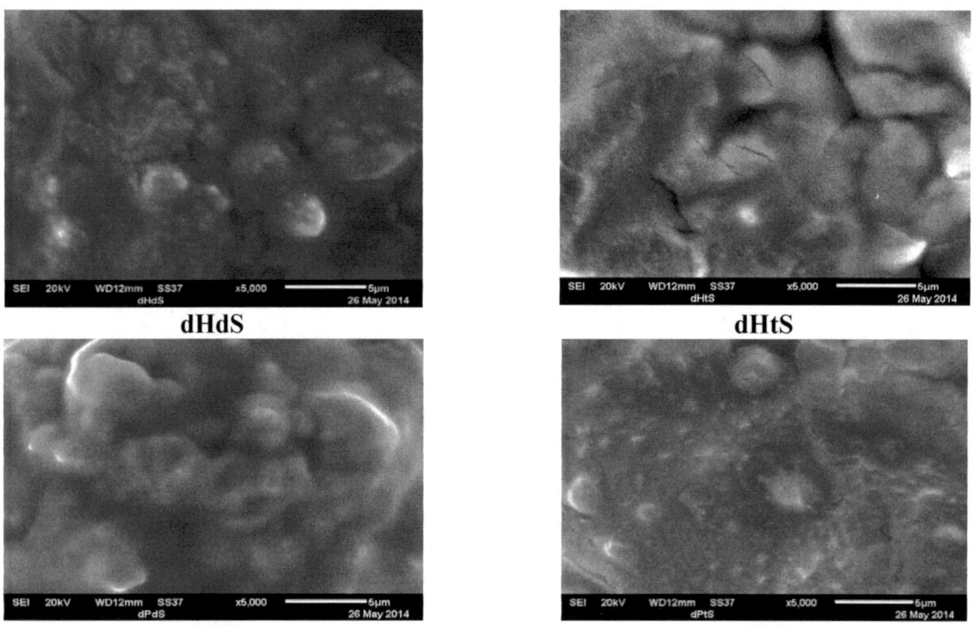

Figure 2. SEM images of PEMFC membrane prepared in this work. Each sample is distinguished by sample code written below each picture.

Figure 2 shows the various morphologies with type of membrane preparation, especially on inclusion of phosphoric acid. According to the previous investigation a small dot represent a part of phosphoric acid [7]. Membrane prepared with pressurized (dPdS membrane) shows no small dots indicating even distribution of phosphoric acid. Meanwhile, membrane prepared without pressure (dHdS) shows some small dots indicating uneven distribution of phosphoric acid.

Moreover, membranes prepared without supramolecular structure (dHtS and dPtS membranes) always showing cracks. The cracks are not observed in dHdS membrane. The data imply the importance of supramolecular solution in the rigidity or robustness of the membrane. Those facts are evidence that application of pressure in the preparation of the membranes could affect into morphological changes. In addition, the role of supramolecular structure can be seen by investigation of Fourier Transform Infra Red (FTIR) spectra, as shown in figure 3 below. Both membranes in the figure contain phosphoric acid but have no addition of supramolecular structure.

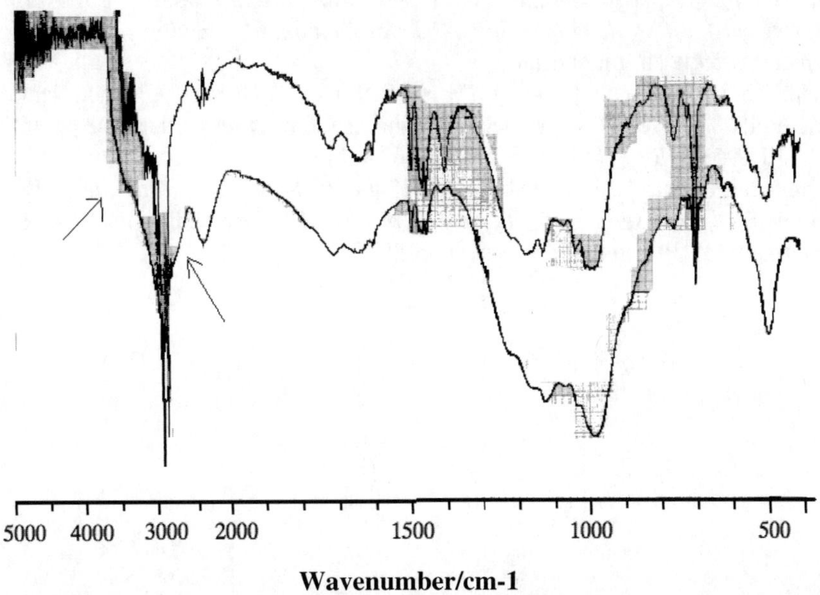

Figure 3. Fourier Transform Infra Red (FTIR) spectra of dHtS membrane (top) and dHdS membrane (bottom)

The figure shows that with addition of supramolecular structure (dHdS membrane) promotes intermolecular hydrogen bonding which can be seen in broader hydrogen bond absorption in wave number 3300 cm^{-1}. Broader absorption can be seen by the shape of the spectra [8], e.g. shown as convex shape for dHdS membrane or concave shape for dHtS membrane (see the arrows in the figure). Therefore, the FTIR data emphasize the presence of supramolecular structured membrane introduced in this paper.

The supramolecular structure is built from hydrogen bonding betweeen anhydride or succinic group with sulfonate group along every polymer molecule of PE-*g*-MA and SPS, respectively. One polymer molecule of SPS may build hydrogen bonding with more than one PE-g-MA molecules, and vice versa. As the interaction occurs in the very dilute solution, it can be understood that the the bonding could also occurs in the crystalline part of PE-g-MA, when it is in solid form. Consequently, when the solution is removed during the process of membrane preparation, the crystalline portions are affected.

4. Conclusion

The work has shown clearly that supramolecular structure contribute to the robustness of the membrane. Moreover the inclusion of phosphoric acid would lead to the change of its structure, especially in crystallite region. The important features of this work is that application of pressure in the sample preparation would lead to improved phosphoric acid distribution in the bulk of membrane, and the the addition of phosphoric acid would provide more prominent effect in their crystalline-structural changes.

Acknowledgments
Authors would like to acknowledge to Indonesian Institute of Sciences (LIPI) for funding through Program Unggulan–LIPI 2016-2017.

References

[1] Lade H, Kumar V, Arthanareeswaran G and Ismail A F 2017 *Int. J. of Hydrog. Energy* **42** 1063
[2] Li W, Shen C, Gao S, Yin S and Li H 2016 *Solid State Ion.* **287** 1
[3] Mauritz K A and Moore R B 2004 *Chem. Rev.* **104** 4535
[4] Sunit Hendrana 2010 *Laporan Akhir Program Insentif KNRT 2010 Aplikasi polistiren tersulfonasi (sPS) dengan sulfonasi heterogen untuk membran PEMFC generator listrik kapasitas 5 kW* (In Indonesian)
[5] Wunderlich B 1997 *The Basic of Thermal Analysis* (Thermal Characterization of Polymeric Materials Vol 1) ed E.A. Turi (San Diego: Academic Press) chapter 2 pp 366-379
[6] Weeks J J 1963 *J. Res. Natl. Bur. Stand. A* **67A** 441-451
[7] Hendrana S, Hartanti F T and Rahayu I 2013 *Jurnal Sains Materi Indonesia* **14** 241
[8] Kalsi P S *Spectroscopy of Organic Compounds* sixth edition (New Delhi: New Age International (P) Limited Publisher) pp 117

Synthesis of lithium mangan dioxide ($LiMn_2O_4$) for lithium-ion battery cathode from various lithium sources

S Priyono[1], N R Ginting[2], S Humaidi[2], A Subhan[1] and B Prihandoko[1]

[1]Research Center for Physics – Indonesian Institute of Sciences Kawasan PUSPIPTEK Gedung 442 Serpong, Tangerang Selatan, Banten, Indonesia
[2]Department of Physics- Faculty of Mathematics and Natural Sciences-University of North Sumatera Jl. Bioteknologi I Kampus USU Medan, 20155, Indonesia

E-mail: slam013@lipi.go.id

Abstract. $LiMn_2O_4$ as a cathode material has been synthesized via solid state reaction. The synthesis has been done by varying lithium sources such as $LiOH.H_2O$ and Li_2CO_3 while MnO_2 was used as Mn sources. All raw materials were mixed stoichiometrically to be the precursors of $LiMn_2O_4$. The precursors were sintered using high temperature furnace at 800 °C for 4 hours in atmospheric condition to form final product. The final products were sieved to separate the finer and smoother particles from the coarse ones. The products were characterized by X-Ray Diffractometer (XRD) to identify phases and crystal structure. The peak wave number was also determined using Fourier Transform Infra Red (FTIR) to find functional group. $LiMn_2O_4$ sheets were prepared by mixing active material with polyvinylidene fluoride (PVdF) and acetylene black (AB) in mass ratio of 85:10:5 wt.% in N,N-Dimethylacetamide (DMAc) solvents to form slurry. The slurry was then coated onto Al foil with thickness of about 0.15 mm and dried in an oven. $LiMn_2O_4$ sheet was cut into circular discs and arranged with separator, metallic lithium, and electrolyte in a coin cell. Automatic battery cycler was used to measure electrochemical performance and specific capacity of the cell. XRD analysis showed that sample synthesized with Li_2CO_3 has higher crystallinity and more pristine than sample synthesized with $LiOH.H_2O$. FTIR analysis revealed that both of samples have identical functional group but sample with Li_2CO_3 source tend to degrade. Cyclic voltammetry data gave information that sample with $LiOH.H_2O$ source has better electrochemical performance. It showed double oxidation/reduction peaks more clearly but sample with Li_2CO_3 source has higher specific capacity (64.78 mAh/g) than sample with $LiOH.H_2O$ (50 mAh/g).

1. Introduction

Increasing dependence on fossil fuels has negative impact on the environment, so alternative energy becomes very important. Alternative energy systems such as solar, wind, and air often require batteries to store the energy. Batteries are devices that convert chemical energy into electrical energy. At present, batteries have become a part of everyday life, for today's modern life batteries have become a necessity inherenty in every human activity especially related to electronic devices.

Batteries are used in almost every aspect of modern life. Although the primary (non-rechargeable) battery has a great usability, the secondary battery (rechargeable) looks more popular. In the secondary battery, the reversible electrode reaction and cells can be recharged [1]. Among the types of secondary batteries, lithium ion batteries have received special attention [2]. The main reason for using Li ion battery technology is lithium is the lightest and most electropositive metal, thus providing high energy density. Li-ion batteries show a stable life cycle (over 500 cycles); those can be made in various sizes and also require less maintenance when compared to other batteries [3].

Lithium-ion battery consist of four main parts: cathode, anode, electrolyte, and separator. Cathode has higher voltage than anode and plays a role as the place where oxidation process take place, while anode has low voltage and acts as a place where lithium ion inserted into this during charge process. So, cathode material will determine the capacity of a battery and we will focus on this material including how to synthesize and characterize for lithium ion battery.

Some of the cathode materials in lithium ion batteries that have been synthesized are lithium manganese oxide ($LiMn_2O_4$), lithium cobalt oxide ($LiCoO_2$), and lithium ferro phosphate (LFP) [4] or mixed metal oxides which include cobalt (Co), nickel (Ni), aluminum (Al), and manganese oxides such as cobalt nickel aluminate (NCA) [5]. The cathode material used is lithium mangan oxide ($LiMn_2O_4$) which has theoretical specific capacity of 148 mAh/g with an electrical conductivity of about 10^{-6} S/cm, and higher voltage of about 3.9 Volt [6]. $LiMn_2O_4$ also has higher energy density than others because of high working voltage. $LiMn_2O_4$ has advantages over $LiCoO_2$ and Ni due to its abundant availability, cheap, environmentally friendly, and greater thermal stability [7]. Those are reasons why we will choose and synthesize $LiMn_2O_4$ for cathode material in lithium ion battery.

The simplest route to produce $LiMn_2O_4$ is via solid state reaction because it simply combines Li and Mn sources stochiometrically and then sinters with appropriate temperature in atmospheric condition. There are two common raw materials that can be used as lithium sources: Li_2CO_3 and $LiOH.H_2O$. Li_2CO_3 is conventional raw material which has high purity while $LiOH.H_2O$ has low purity commonly used for paint. In this research cathode materials will be prepared with different source material of lithium such as $LiOH.H_2O$ and Li_2CO_3 while MnO_2 is used as Mn sources. The goals of this research are to synthesize and characterize $LiMn_2O_4$ from different raw materials as well as to study the electrochemical performance of this material. This study is very important to subtitute costly material like Li_2CO_3 with lower one like $LiOH.H_2O$.

2. Experimental method

The synthesis process was done with two different sources of lithium. The first synthesis of $LiMn_2O_4$ used Li_2CO_3 and MnO_2 (sample A) while the second synthesis used $LiOH.H_2O$ and MnO_2 (sample B). Synthesis on this research followed solid state reaction method, by mixing each of these materials based on the calculation of stoichiometry. The reaction can be explained by following formulas:

Sample A

$$2Li_2CO_3 + 8MnO_2 \rightarrow 4LiMn_2O_4 + O_2 + 2CO_2 \quad (1)$$

Sample B

$$4LiOH.H_2O + 8MnO_2 \rightarrow 4LiMn_2O_4 + 6H_2O + O_2 \quad (2)$$

Raw materials were mixed and homogenized using mortar and pestle. Subsequently, the mixtures were sintered at 800 °C with high temperature furnace for 4 hours in air atmosphere.$LiMn_2O_4$ products were grinded and sieved to produce finer particle, which were then characterized by X-Ray diffraction (RIGAKU Japan) using Cu-Kα as radiation source to know the phase and crystal structure. The angle used in diffraction measurement is 10 - 70°, with a scanning speed of 50 min^{-1}. Characterization using FTIR with iTR method was performed at wavenumber range of 400 - 4000 cm^{-1} to know the existence of functional groups contained in the sample and to confirm the existence of phases.

Cathode sheets were made by mixing active materials (sample A and B) with binder (PVdF) and additive material (AB/acetylene black) in ratio 85:10:5 wt.% and diluted in N,N-DMAc solution until slurry was obtained. After the slurry is formed, it was coated on Al foil with 0.15 um in thickness and then dried using an oven at 80 °C for 60 minutes.

In order to know the electrochemical performance, electrode sheet must be assembled into half-cell battery using coin cell in the glove box. Electrode sheets were cut into a circular shape with 14 mm in diameter and the separator was also cut with 18 mm in diameter. $LiPF_6$ solution was used as an electrolyte. Battery performance test includes Cyclic Voltammetry (CV) and Charge/ Discharge (CD) performed with WBCS 3000, Automatic Battery Cycler Ver. 3.2. For performance test, two tests were conducted: cyclic voltammetry (CV) and charge discharge. For the CV test, the potential range is 3 V to 4.6 V with 0.01 mV/s while for the charge discharge test was done with a constant current of 0.05 mA (0.1 C) with a potential range of 3.3 V- 4.6 V.

3. Results and discussion

Figure 1 shows XRD pattern of $LiMn_2O_4$ samples prepared by solid state process. The ICDD (No 01-070-3120) card data of $LiMn_2O_4$ is also shown in the figure as the standard. It can be seen that both pattern have similarity with $LiMn_2O_4$ phase. Both samples were identified to have $LiMn_2O_4$ crystals. Sample A has purer phase and higher crystallinity with higher and sharper peaks than sample B. Sample B shows some impurities such as Li_2O and MnO_2 are existing denoted by triangle and star shape in the figure 1 respectively. The existence of these impurities can be caused by inhomogeneous precursors because $LiOH.H_2O$ has different particle size. $LiOH.H_2O$ is technical grade material with coarse grains so milling process must be done to homogenize the precusors.

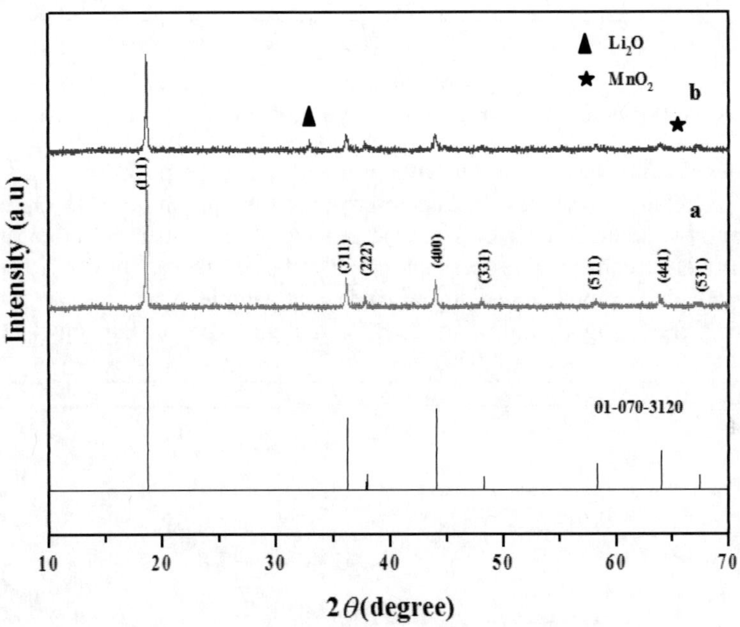

Figure 1. XRD pattern of $LiMn_2O_4$ synthesized with (a) Li_2CO_3 and (b) $LiOH.H_2O$ as raw materials.

Infrared test has been done to study the metal-oxygen and metal-metal bond in the synthesized lithium manganate sample. Figure 2 shows IR spectrum of the synthesized $LiMn_2O_4$ sample. The vibrational frequencies at wavenumber 514.80 cm^{-1} and 526.73 cm^{-1} are associated with vibration of stretching group symmetry of Mn-O derived from element MnO_6 [8]. The Li-Mn vibration is located at the wavenumber of 310 cm^{-1} [9] so it does not appear at the spectrum. A peak below 1000 cm^{-1} indicates the frequency of metal-oxide vibration. The sample also shows an absorption band below 1000 cm^{-1} wavenumber [9]. In the infrared spectrum of $LiMn_2O_4$ also appears C=O group with vibration stretching at wave numbers 1630.42 cm^{-1} and 1630.47 cm^{-1} with low intensity which might come from CO_2 in air [10]. In the sample A peak at wave number of 1047.70 cm^{-1} is present, indicating the characteristic peak of CO-CO stretching C-O vibration [11]. This peak is not possessed by sample

B which was likely to have degraded to Li_2CO_3. At wave number 3437.78 cm^{-1} and 3443.53 cm^{-1} there is an absorption band of the OH functional group of water molecules [12].

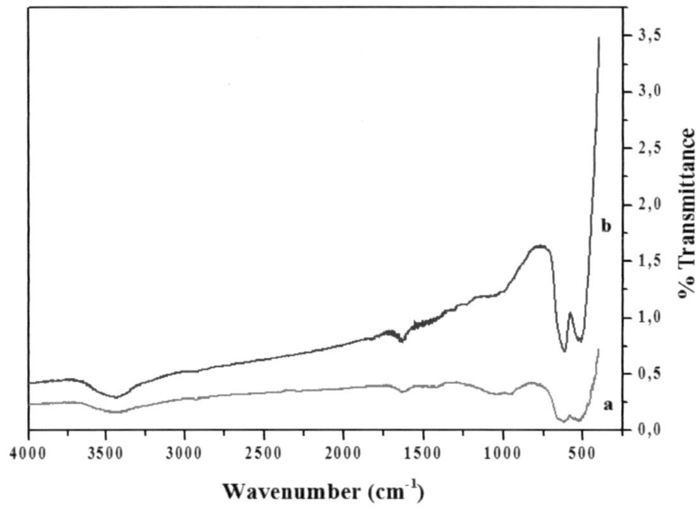

Figure 2. The FTIR spectra of $LiMn_2O_4$ which synthesized with (a) Li_2CO_3 and (b) $LiOH.H_2O$.

Figure 3 shows double peaks of a pair reduction and oxidation peak for each sample. Reduction-oxidation peaks are related to insertion and de-insertion of lithium ion in $LiMn_2O_4$. Every peak has two oxidation peak and two reduction peaks. The peaks in sample B is sharper and clearer than sample A, so sample B shows electrochemical process taking place faster than sample A. However, sample A has higher current peak than sample B which indicated that sample A has higher current response and gives higher ionic diffuse. Higher ionic diffuse will give higher specific capacity [13].

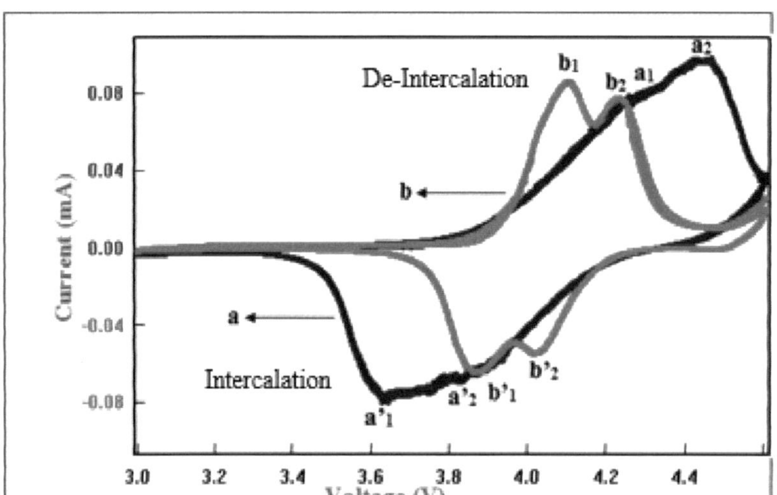

Figure 3. Cyclic Voltamogram of $LiMn_2O_4$ which synthesized with (a) Li_2CO_3 and (b) $LiOH.H_2O$.

Figure 4 shows that specific capacity of $LiMn_2O_4$ synthesized with Li_2CO_3 and $LiOH.H_2O$. As stated in table 3, Sample A has higher specific capacity than sample B with 65 mAh/g and 51 mAh/g, respectively. Sample A having higher capacity can be caused by no existence of impurities phases in the sample, unlike with sample B which has other impurities such as Li_2O and MnO_2. Both samples show high coulombic efficiency indicating that both materials are good for battery but sample A can perform charge-discharge process with higher speed than sample B.

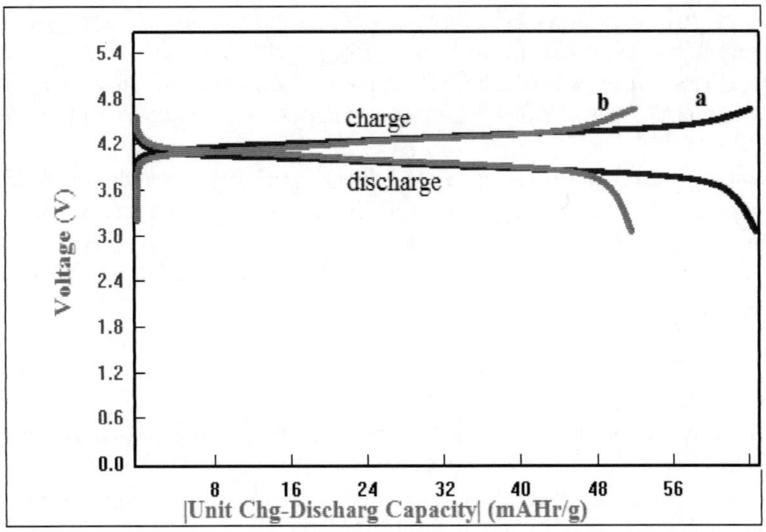

Figure 4. Charge-discharge of $LiMn_2O_4$ synthesized with (a) Li_2CO_3 and (b) $LiOH.H_2O$.

Table 1. Specific Capacity and coulombic efficiency of $LiMn_2O_4$.

Sealant Compound	Voc (V)	Jsc (mA/cm^2)	FF	η (%)
Hermetic A1	1.386	18.92	0.37	1.9
Hermetic B1	1.427	26.38	0.38	2.9

4. Conclusion

$LiMn_2O_4$ had been successfully synthesized via solid state reaction with MnO_2 and Li_2CO_3/$LiOH.H_2O$ as starting materials. Sample A has better purity and crystallinity than that of sample B. However, FTIR spectrum shows CO_2 bonding within sample A, which is expected from Li_2CO_3 as lithium sources. Cyclic voltammograms show that both samples give oxidation and reduction peak as an indication that the sample has reversibility characteristic. Charge-discharge testing shows that sample A has higher specific capacity (65 mAh/g) than that of sample B (54 mAh/g).

Acknowledgment

The author would like to acknowledge the financial and experimental tools support from Minister Research, Technology and Higher Education (Menristek-Dikti); and Research Center for Physics – Indonesian Institute of Sciences (LIPI). The author also would like to thank the Battery Research Group in Research Center for Physics – Indonesian Institute of Sciences (LIPI) for helping the battery fabrication and analysis.

References

[1] Julien C M, Mauger A, Zaghib K and Groult H 2014 *Inorganics* **2** 132
[2] Lange J 2012 *Improving lithium-ion battery power and energy densities using novel cathode architectures and materials* Master Thesis (Champaign:University of Illinois at Urbana-Champaign)
[3] Philippe B 2013 *Insights in Li-ion Battery Interfaces through Photoelectron Spectroscopy Depth*

Profiling PhD Thesis (Uppsala: Uppsala University)
[4] Eriksson T 2001 *$LiMn_2O_4$ as a Li-Ion Battery Cathode: From Bulk to Electrolyte Interface* PhD Thesis (Uppsala: Uppsala University)
[5] Kasvayee K A 2011 *Synthesis of Li-ion battery cathode materials via freeze granulation* (Gothenburg: Chalmers University of Technology) p 8
[6] Paravasthu R 2012 *Synthesis and Characterization of Lithium-Ion Catahode Materials in The System $(1\text{-}x\text{-}y)LiNi_{1/3}Mn_{1/3}Co_{1/3}O_2.xLi_2MnO_3.yLiCoO_2$* (Colorado: Colorado State University) pp 1-5
[7] Kiani M A, Mousavi M F and Rahmanifar M S 2011 *Int. J. Electrochem. Sci.* **6** 2581
[8] Chitra S, Kalyani P, Mohan T, Gangadharan R, Yebka B, Castro-Garcia S, Massot M, Julien C and Eddrief M 1999 *J. Electroceramics* **3** 433
[9] Lagashetty A, Havanoor V, Basavaraja S and Venkataraman A 2008 *Indian Journal of Chemical Technology* **15** 41
[10] Matsui M, Dokko K and Kanamura K 2010 *J. Electrochem. Soc.* **157** A121
[11] Xiang M W, Zhou X Y, Zhang Z F, Chen M M, Bai H L and Guo J M 2013 *Advanced Materials Research* **652-654** 825
[12] Seyedahmadian M, Houshyarazar S and Amirshaghaghi A 2013 *Bull. Korean Chem. Soc.* **34** 622
[13] Wang G 2000 *Investigations on Electrode Material for Lithium-Ion Batteries* PhD Thesis (Wollongong: University of Wollongong) pp 13-20

Enhancing the efficiency of dye-sensitized solar cells by hydrothermal post-treatment in acidic environment

A Nathania[1], N M Nursam[2], Shobih[2], J Hidayat[2] and N Prastomo[1]

[1]Department of Physics Engineering, Faculty of Engineering, Surya University, Jl. Boulevard Gading Serpong, Tangerang, Banten 15810
[2]Research Centre for Electronics and Telecommunication – Indonesian Institute of Science, Building 20 level 4 Kampus LIPI Jl. Sangkuriang, Cisitu, Bandung 40135, Indonesia

E-mail: natalita.maulani.nursam@lipi.go.id

Abstract. Dye-sensitized solar cell (DSSC) have been extensively studied due to its low production cost and simple production process. In this research, DSSC with improved performance is acquired by modification of TiO_2 layer through hydrothermal post-treatment with different hydrochloric acid (HCl) concentrations to obtain various particles and pore sizes. Qualitative and quantitative characterizations of the TiO_2 film were conducted using thickness measurement, scanning electron microscope (SEM), and X-ray diffraction (XRD), while the solar cell performances were characterized using current-voltage (I-V) measurement under 0.5 Sun. When hydrothermally treated with 1 mol/L HCl at 180 °C for 3 h, the DSSC showed the most optimum photo-electricity conversion performance of 3.60%, which improved the efficiency of the non-treated DSSC by a factor of 1.2. As the HCl concentration increased, the treated TiO_2 film became thinner with smaller particle size and denser structure. It was suspected that the modification in the TiO_2 film morphology has led to better light absorption, which consequently resulted in the improvement of DSSC performance.

1. Introduction

Since its first invention by Gräetzel and coworkers in 1991, power conversion efficiency of DSSC up to 12% has been reported, using co-sensitizer of carboxy-anchor organic dye of LEG4 as the collaborative sensitizer to the silyl-anchor dye of ADEKA-1[1]. Typically, DSSC is a sandwich structure of photo anode with dye-sensitizer adsorbed on its surface, iodine based redox-coupled electrolyte and a counter electrode. The entire assembly is embedded between two conducting glass supports.

The photo anode is a key component of DSSCs that plays important roles in dye loading, electron injection, transportation and collection, and thus has significant influence on the photocurrent, photo voltage and the power conversion efficiency [2]. The photo anode materials should meet several conditions: first, the semiconductor bandgap must match with the dye energy level to ensure an effective injection of the photo-generated electrons from the dyes to the semiconductor; second, it must have a high surface area to adsorb more dye molecules, so that more photon from the solar energy can be harvested; third, the porosity of the film must be high enough to ensure infiltration of electrolyte into the interior of the film and form a good contact between them; fourth, the electron mobility of the semiconductor should be high enough to transport the injected electrons quickly to the outer circuit.

Among numerous material candidates for photo anode, titanium dioxide (TiO_2) is a typical material used as the photoelectrode in DSSC. TiO_2 has wide band gap (~3.2 eV), high energy level that match with the energy level of Ru-based dye, and good chemical stability [3]. Nanoparticle and microporous TiO_2 are expected to have high surface area for better dye particles adsorption [4]. However, smaller pore size may lead to poor electrolyte infiltration [5]. It is also reported that high porosity of TiO_2 may lead to complex electron transport pathways, hence decreasing the electron transport rate [4], [6].

Many techniques have been developed to synthesize suitable porous TiO_2 structure for obtaining better efficiency in DSSC. Various particle sizes, pore size distributions and crystallinity can be achieved by different TiO_2 growing technique or structure modification. Hydrothermal process has been widely used to produce nanostructured TiO_2. Hydrothermal treatment could improve the crystallinity of a powder at relatively low temperature [7]. Hydrothermal can also be used as post treatment to modify TiO_2 structure after fabrication. Water vapor exposure in the post-hydrothermal treatment can successfully enhance the nano-crystallinity of TiO_2 produced from sol-gel method [8]. Hydrothermally prepared TiO_2 films that were treated by hydrochloric acid (HCl) via hydrothermal method was reported to reform the nanostructure of TiO_2 and improve the efficiency of DSSC by 18% from the initial non-treated condition [9]. It has been demonstrated that the electron recombination between conduction band and I_3^- in the electrolyte was reduced and electron injection efficiency was increased when the nano-porous TiO_2 film was modified with 0.1 M hydrochloric acid compared to other acid [10]. Nevertheless, effect of hydrothermal post-treatment in acidic environment for commercial TiO_2 paste with anatase phase has never been discussed. In this research, different hydrochloric acid (HCl) concentrations were used in hydrothermal post-treatment to obtain various particles and pore sizes of TiO_2.

2. Experiment

2.1. TiO_2 photoelectrode preparation

Precut (2 × 2 cm^2) fluorine-doped tin oxide (FTO) coated glass TEC™ 15 (15 Ω/sq, Dyesol) were used as substrate. The FTO glasses were cleaned with Teepol, water, and ethanol by ultrasonic cleaning, respectively. The cleaned FTO glasses were immersed in 40 mmol L^{-1} $TiCl_4$ solution at 70 °C for 30 min followed by sintering process at 450 °C for 15 min. TiO_2 films were prepared by screen printing of the TiO_2 commercial paste (DSL 18 NR-AO opaque titanium paste, Dyesol) onto the pre-treatment FTO glass in size of 0.5 × 0.5 cm^2 and 8 μm thickness. After drying, the films were sintered at 500 °C for 45 min.

The prepared TiO_2 photoelectrodes were placed against the inside wall of teflon-liner hydrothermal chamber with the TiO_2 film facing down. The electrodes were immersed in different concentrations of hydrochloric acid, 0.5 mol L^{-1}, 1 mol L^{-1}, 2 mol L^{-1}, and 4 mol L^{-1}. The chamber was put in an electric oven for 3 hours at 180 °C. The samples were taken out when the temperature down to ± 25 °C, dried in the ambient air, and subsequently sintered at 500 °C for 45 min.

2.2. DSSC fabrication

The post-treatment TiO_2 electrodes were put into dye solution of Z-907 (cis-Bis(isothiocyanato)(2,2'-bipyridyl-4,4'-dicarboxylato)(4,4'-di-nonyl-2'-bipyridyl)ruthenium(II), Dyesol) dissolving in ethanol, and kept for 24 hours at room temperature. The films were further rinsed with ethanol and dried in the air. Counter electrodes were prepared using DC magnetron sputtering on pre-cut 2×2 cm^2 of FTO coated glass TEC™ 15 (15 Ω/sq, Dyesol). The DSSC was fabricated by sandwiching the working and counter electrodes. Thermoplastic Surlyn (50 μm thickness, Dyesol) was used as spacer between the electrodes. The I^-/I_3^- based liquid electrolyte (Dyesol, EL-HSE) was added into the gap between the electrodes by an injector.

2.3. Characterizations

Film thickness was measured with digital thickness measurer Mitutoyo, Model ID-C112B. The field emission scanning electron microscopy (FE-SEM) was performed on a STEM JEOL JB-4610F using 15 kV beam to characterize the morphology, size and the structure of TiO_2. The phase purity and crystallite structure of TiO_2 samples were characterized by X-ray diffraction (XRD) on Rigaku, RINT 2200 VK/PC diffractometer and radiation of copper Kα (λ=1.54056 Å), 40 kV 30 mA. The crystallite sizes of TiO_2 nanoparticles were estimated using Scherrer's equation [11]:

$$L = \frac{0.9}{B\cos\theta} \tag{1}$$

where L is the average crystallite size, λ is the X-ray wavelength, B is the full-width at half maximum (FWHM) in radians and θ is the Bragg's angle. The photocurrent-voltage (I-V) characteristics were measured with National instrument I-V measurement system under simulated AM 1.5 illumination with constant light intensity of 500 W/m^2 provided by a solar light simulator (Oriel, Model: 91192). The size of the active area of the cells used for this measurement was 0.25 cm^2.

3. Results and discussion

3.1. Morphology analysis of TiO_2 film

The thickness of TiO_2 film as the result of hydrothermal treatment were 5.0±1.1 µm for sample with 0.5 M HCl treatment, 3.7±0.9 µm for 1 M HCl, 1.9±0.6 µm for 2 M HCl, and 1.1±0.3 µm for 4 M HCl, while sample without hydrothermal post treatment had thickness of 5.0±1.1 µm. The thickness of the film decreased when the HCl concentration increased. TiO_2 is known as chemically stable, even in acidic environment [12]. However, hydrothermal treatment in high pressure and temperature had caused TiO_2 particles to dissolve in HCl solution and reduced the thickness of the film. Higher concentration of HCl had a greater effect in dissolution of TiO_2 particles, which resulted in TiO_2 film that was treated with 4 M HCl having the lowest thickness.

The acidic environment also induced some changes in the film morphology. The transformation of the TiO_2 films morphology as the result of hydrothermal post-treatment was observed in SEM images in figure 1. With the increase in the HCl molar concentration, the particles started to agglomerate and stick to each other so the film become denser with smaller gap between the particles. However, the particle size was smaller and distributed evenly when the acid concentration increased. The particle size was 38.9±9.6 nm for the control sample and it was reduced for hydrothermal treated sample, i.e. 35.0±6.7 nm, 34.9±6.6 µm, 75.9±23.9 µm, and 34.0±5.1 µm for 0.5 M, 1 M, 2 M, and 4 M samples, respectively.

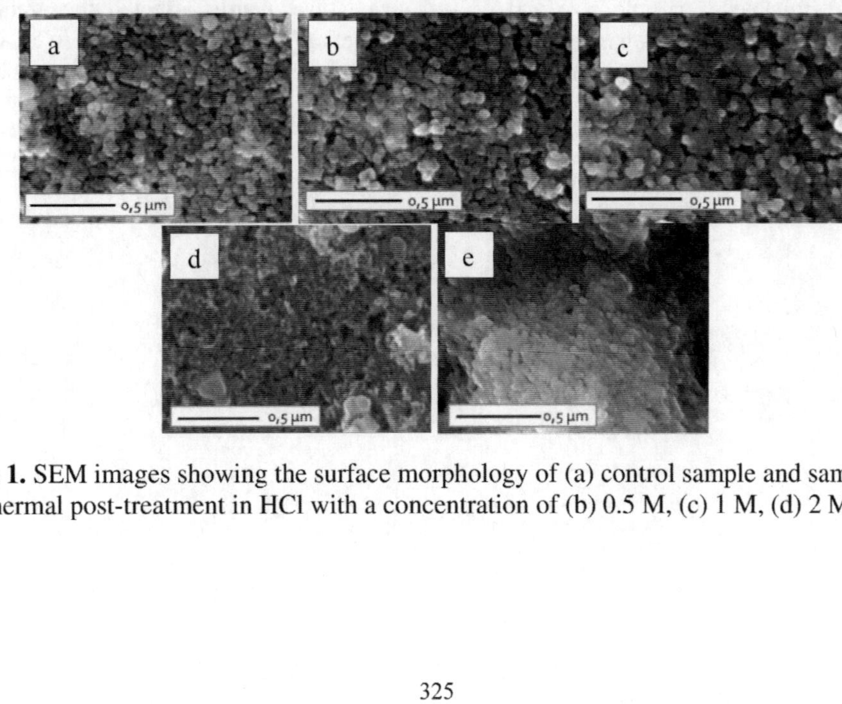

Figure 1. SEM images showing the surface morphology of (a) control sample and samples with hydrothermal post-treatment in HCl with a concentration of (b) 0.5 M, (c) 1 M, (d) 2 M, (e) 4 M.

Better photovoltaic performance is normally obtained for anatase TiO_2 compared to the rutile structure as the result of higher dye adsorption and faster electron transport rate [13]. XRD analysis was carried out to study the crystallinity of the TiO_2 film (figure 2). The diffraction peaks of the TiO_2 nanoparticles on FTO glass match with the anatase phase (ICSD 98-015-4604), while the other peaks come from tin oxide glass substrate match with ICSD 98-009-1517 for SnO_2. No other phases of TiO_2 were observed, suggesting that the hydrothermal post-treatment have no effect on the phase of TiO_2 crystal.

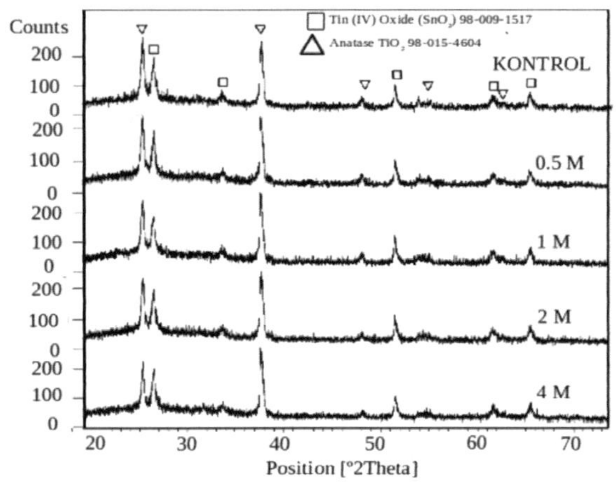

Figure 2. XRD patterns of TiO_2 films sample.

Hydrothermal post-treatment in acidic environment also caused the increase in the grain size of TiO_2. The estimated crystallite diameter from observed XRD pattern (figure 2) were 283.32 Å, 286.62 Å, 292.67 Å, 334.07 Å, and 341.66 Å for control, 0.5 M, 1 M, 2 M, and 4 M samples, respectively. Bigger crystallite size was observed when the HCl concentration was increased. The growth of grain size for synthesizing TiO_2 using hydrothermal method in acidic environment had also been reported [11]. Even though the particle size was smaller at higher acid concentrations, the crystallite size was increased. TiO_2 typically has a great chemical stability. However, when hydrothermally treated with high pressure and temperature, it was possible for TiO_2 particles to be slightly dissolved and reformed with bigger crystallite structure [14]. The acidic condition may also induce the grinding of the particles surface, hence the diameter of TiO_2 particles became smaller. The changes in the morphology of the TiO_2 film, i.e. thickness, particle size and crystal size, consequently affected the amount of dye adsorbed on the surface of TiO_2 particles. From visual observation (as shown in figure 3), it was clear that the 4 M sample showed the brightest red colour, while the other samples had dark red colour. The dark red colour of TiO_2 film indicated that the dye is adsorbed well by the film.

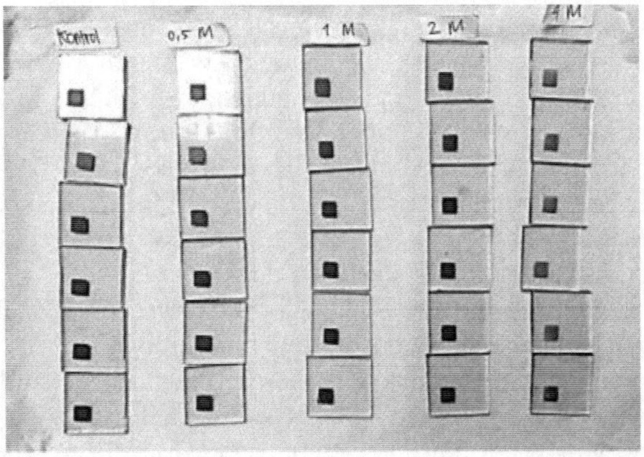

Figure 3. Photograph showing the TiO_2 photoelectrodes after the sensitizing process (from left to right: control, 0.5 M, 1 M, 2 M, and 4 M).

3.2. Performance analysis of DSSC

Hydrothermal post-treatment has caused the transformation in the TiO_2 film morphology, which subsequently affect the DSSC performance. The performance of DSSC can be described with several parameters, such as efficiency, fill factor, and short circuit current. Current-voltage characteristics of DSSC with different photoelectrodes is shown by figure 4. Highest efficiency, short circuit current density (J_{sc}) and open circuit voltage (V_{oc}) were exhibited by 1 M sample (table 1). The efficiency of 1 M sample was 3.30%, which was 1.2 times higher than the sample without treatment. Furthermore, the lowest resistance value also shown by 1 M sample. Low value of series resistance indicated high conductivity of electron in the DSSC system, which contributed to higher current [15].

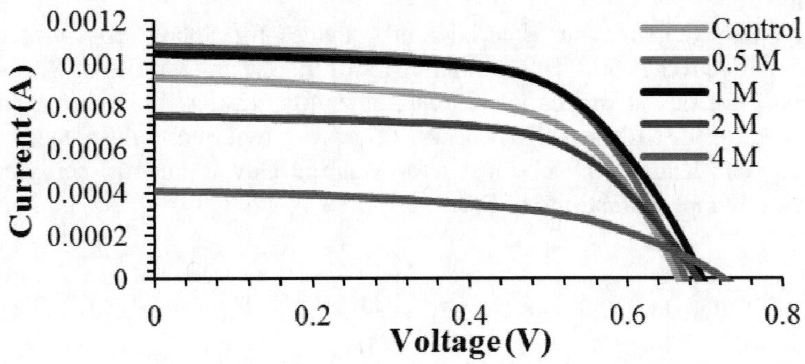

Figure 4. J-V Curve of the DSSC with different photoelectrodes.

Table 1. Photovoltaic characteristics of the DSSC with different photoelectrodes.

	J_{sc} (mA cm^{-2})	V_{oc} (Volt)	R_s (ohm)	Efficiency (%)	FF
Control	3.32	0.66	175.15	2.67	0.61
0,5 M	3.30	0.67	188.20	2.75	0.62
1 M	3.91	0.68	166.68	3.30	0.62
2 M	1.67	0,66	497.24	1.34	0.58
4 M	1.26	0,66	726.87	0.88	0.50

The DSSC performance was affected by key structural properties of the film, such as thickness, phase composition, particle size and porosity [3]. Film thickness served as electron pathways. Film with higher thickness would have longer electron pathways, and thus had higher possibility for electron recombination. Fewer defects and smaller particle size are known to inhibit the recombination of electrons and holes [16]. Materials with higher crystallite size tend to have fewer defects and thus decreasing the recombination rate, which led to greater current flow. Smaller particle size attributed to larger surface area, permitting a greater amount of dye molecules to be adsorbed to the film surface [2]. However, further reduction of film thickness could decrease the available surface of TiO_2 particles for adsorbing dye molecules. Although 4 M sample has the smallest particle size, and biggest crystallite size, the efficiency was the lowest due to its thickness.

The pores of the dye-coated TiO_2 film are filled with electrolyte for regeneration of the dye through reduction-oxidation process and transport of positive charges through the cell. The pore size influences the electrolyte infiltration and dye adsorption. It is more difficult for electrolyte to permeate into the film with smaller pore size. Smaller particles tend to have smaller gaps between them which cause poor pore-penetration. This will lead to low electron transport from electrolyte to the TiO_2 film.

After being hydrothermally treated, the film thickness was reduced and the particle was smaller and had better interconnection; thus, more electron can be transferred from dye to external load through TiO_2 film as photoelectrode. Smaller chance of electron recombination was also expected at 1 M sample, therefore leading to higher current density.

4. Conclusion

TiO_2 film treated with hydrothermal post-treatment in hydrochloric acid solution become thinner with denser structure and the particle is distributed evenly as the HCl molar concentration increased. The particle size of treated TiO_2 decreases and the crystallite size increase in the higher concentration of HCl compared to TiO_2 without treatment. Hydrothermal post-treatment with 1 mol/L HCl produces the optimum TiO_2 film, with thickness of 3.7 μm, average particle size of 34.9 nm and crystallite size of 292.7 Å, which resulted in efficiency of 3.3% and J_{sc} of 3.9 A/cm^2. The efficiency increases 1.2 times compared to TiO_2 film without treatment.

Acknowledgment

The authors would like to thank the Materials and Devices for Solar Cells research group at the Research Center for Electronics and Telecommunication, Indonesia Institute of Science (P2ET LIPI) and Surya University for the assistance. Research Center for Physics, Indonesia Institute of Science (P2F LIPI) is thanked for the SEM and XRD access. The work reported in this paper was supported by research grant "Kegiatan Unggulan" sub-cluster for Material Development, Energy and Manufacture Engineering from the Indonesia Institute of Science.

References

[1] Kakiage K, Aoyama Y, Yano T, Oya K, Fujisawa J and Hanaya M 2015 *Chem. Commun.* **51** 15894
[2] Chen H Y, Kuang D B and Su C Y 2012 *J. Mater. Chem.* **22** 15475
[3] Tricoli A, Wallerand A S and Righettoni M 2012 *J. Mater. Chem.* **22** 14254
[4] Yu H, Zhang S, Zhao H, Xue B, Liu P and Will G 2009 *J. Phys. Chem. C* 113 16277
[5] Ferber J and Luther J 1998 *Sol. Energy Mater. Sol. Cells* **54** 265
[6] Benkstein K D, Kopidakis N, van de Lagemaat J and Frank A J 2003 *J. Phys. Chem. B* **107** 7759
[7] Byrappa K and Yoshimura M 2001 Handbook of Hydrothermal Technology *Ceramic and Other Materials—Processing and Technology Advanced* (New Jersey: Noyes Publications)
[8] Yuwono A H, Sofyan N, Kartini I, Ferdiansyah A and Pujianto T H 2011 *Advanced Materials Research* **277** 90
[9] Wen Q, Yu J, Sun X, Zhuang J, He Q, You X, Guo J and Tao L 2016 *New J. Chem.* **40** 3233
[10] Wang Z S, Yamaguchi T, Sugihara H and Arakawa H 2015 *Langmuir* **21** 4272
[11] Speakman S 2014 Estimating Crystallite Size using XRD

http://prism.mit.edu/xray/oldsite/CrystalSizeAnalysis.pdf (last accessed on 1 May, 2017)
[12] Zhou J, Song B, Zhao G and Han G 2012 *Nanoscale Res. Lett.* **7** 217
[13] Park N G, van de Lagemaat J and Frank A J 2000 *J. Phys. Chem. B* **104** 8989
[14] Wen Q, Yu J, Sun X, Zhuang J, He Q, You X, Guo J and Tao L 2016 *New J. Chem.* **40** 3233
[15] Liu Y C, Lu Y F, Zeng Y Z, Liao C H, Chung J C and Wei T Y 2011 *International Journal of Photoenergy* **2011** 619069
[16] Jafari T, Moharreri E, Amin A S, Miao R, Song W and Suib S L 2016 *Molecules* **21** E900

IOP Conf. Series: Journal of Physics: Conf. Series **985** (2018) 012056 doi:10.1088/1742-6596/985/1/012056

The influence of bamboo-packed configuration to mixing characteristics in a fixed-bed reactor

M Detalina[1], S A Pradanawati[1], Widyarani[2], Mamat[2], D Nilawati[2] and N Sintawardani[2]

[1]Department of Physics-Energy Engineering, Surya University, Jl. Boulevard Gading Serpong Kav. M5 No. 21, Tangerang 15810, Indonesia
[2]Research Unit for Clean Technology, Indonesian Institute of Sciences (LIPI) Jl. Cisitu 21/154D, Bandung 40135, Indonesia

E-mail: widyarani@lipi.go.id

Abstract. Fixed-bed reactors are commonly used as bioreactors for various applications, including chemicals production and organic wastewater treatment. Bioreactors are fixed with packing materials for attaching microorganisms. Packing materials should have high surface area and enable sufficient fluid flow in the reactor. Natural materials e.g. rocks and fibres are often used as packing materials. Commercially, packing materials are also produced from polymer with the advantage of customizable shapes. The objective of this research was to study the mixing pattern in a packed-bed reactor using bamboo as packing material. Bamboo was selected for its pipe-like and porous form, as well as its abundant availability in Indonesia. The cut bamboo sticks were installed in a reactor in different configurations namely vertical, horizontal, and random. Textile dye was used as a tracer. Our results show that the vertical configuration gave the least liquid resistant flow. Yet, the random configuration was the best configuration during mixing process.

1. Introduction

Reactor type and design are determining factors in chemical and biological processes. Fixed-bed reactors are high performance reactors where the reactions occur in packing (support) material. Fixed-bed reactors are commonly used as bioreactors for various applications, including chemicals production and organic wastewater treatment [1].

Fixed-bed reactors use materials such as plastics, rocks, or ceramics as support material and locus of microbial attachment [2]. Performance of a fixed-bed reactor depends on the availability of surface area; the higher the surface area, more microorganisms can attach and form biofilm. Even though a higher surface area can be achieved by reducing the size of packing materials, smaller materials can increase the risk of pressure drop in the reactor. The ideal characteristic of packing material should have enough surface area, high porosity level, stable at various pH condition, not easily degraded or oxidized, and not toxic [3]. On the other hand, the support materials should be easy to process, low cost, and preferably available using local material.

Toda et al. [4] compared vertical and horizontal configuration of rectangular tapered packed-bed reactors. They found horizontal packed-bed reactor produced higher yield of alcohol from glucose fermentation compared with vertical packed-bed reactor. Horizontal packed-bed reactor showed high sedimentation and accumulation of free cells, which was not occurred in the vertical packed-bed reactor.

Reactor performance is also influenced by fluid flow inside the reactor. In an up-flow fixed-bed reactor, the fluid flows from the bottom to the outlet at the upper part of the reactor. This reactor type is often used to reduce the high concentration of organic content in the agro-industrial waste [5]. This wastewater must be treated properly before being discharged into the environment.

Bamboo can grow easily and abundantly in Indonesia. The tubular form and the structural strength of bamboo can be used as an alternative for packed bed material. Bamboo has been used as packing materials for treating domestic, leather industry, milk processing, and tapioca wastewater [6 10]. Organic matter removal on bamboo media takes place through two mechanisms: absorption and decomposition by microbes [10,11]. Both of these mechanisms are supported by a specific surface area of bamboo that can reach 2100 m^2/m^3, which is comparable to some synthetic media of polypropylene and polyethylene [7,10].

As packing material, bamboo has the advantage of high lignocellulosic content. Experiments with various acid, alkali, and solvents showed that bamboo fibre was insoluble in 10% sulphuric acid, 15% hydrochloric acid, 17% nitric acid, 5% sodium hydroxide, ethyl acetate, carbon tetrachloride, and 13% sodium hypochlorite [12]. For application in bioreactors, such extreme condition rarely occurs. The degradation of bamboo as filter material in an anaerobic process treating wastewater from cassava starch production showed no degradation of the bamboo pieces at the end of the experiment. The lignin and cellulose remaining should not be subjected to further degradation [13].

The flow inside a fixed-bed reactor is often assumed as homogeneous. However, the structure and configuration of packing materials have significant influences on the flow [14]. For bamboo, this includes the hollow cylinder dimension and shape that influence pressure inside the reactor. The configuration of cutting bamboo as a support material must be designed to minimize disturbance of fluid flow, but still enable efficient mixing process.

In this research, the fluid flow and mixing process in a lab-scale fixed-bed reactor were investigated. Three bamboo configurations: random, vertical, and horizontal configurations were compared using water as the working fluid. Textile dye was used as the tracer.

2. Material and methods

2.1. TiO_2 photoelectrode preparation

To enable observation of fluid flow in the reactor, the experiments were performed in a transparent acrylic reactor with 18.5 cm diameter and 50 cm height (figure 1(a)). A plate was placed at 10 cm from the bottom to support the bamboo. The empty volume of the reactor (without bamboo) was 12.6 L, while the volume between bottom and top plates was 6.6 L.

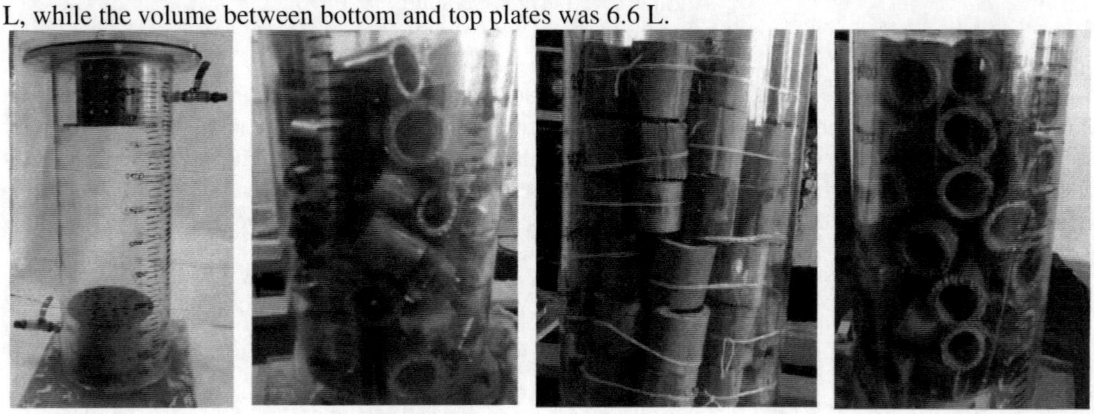

(a) (b) (c) (d)

Figure 1. Reactor with different bamboo configuration: (a) Empty (control), (b) Random, (c) Vertical, (d) Horizontal.

Bamboos with diameter of 3 - 4.5 cm were obtained from local vendor. The bamboo pieces were cut into 4.5 cm length. Bamboo pieces were placed in the reactor in random, vertical, or horizontal configuration (figures 1b - (d)). The numbers of pieces were 84 for random, 105 for vertical, and 86 for horizontal configurations. The bamboos were soaked for seven days before used for the experiment. Compared with bamboo before soaking, the soaked bamboo absorbed 0.9 L of water during the soaking process.

Water was used as the working fluid for the experiment. Red textile dye at the concentration of 0.005% (w/v) was used as the tracer.

2.2. Experimental setup

The experimental setup is presented in figure 2. The reactor was first filled with water without tracer. Colored water as the tracer was flowed into the feeding tank and let overflow to keep the water level in the feeding tank. The colored water was subsequently flowed by gravity from the feeding tank to the reactor and mixed with the water inside the reactor. The tracer needed about 60 minutes to replace the water in the reactor until they are completely mixed.

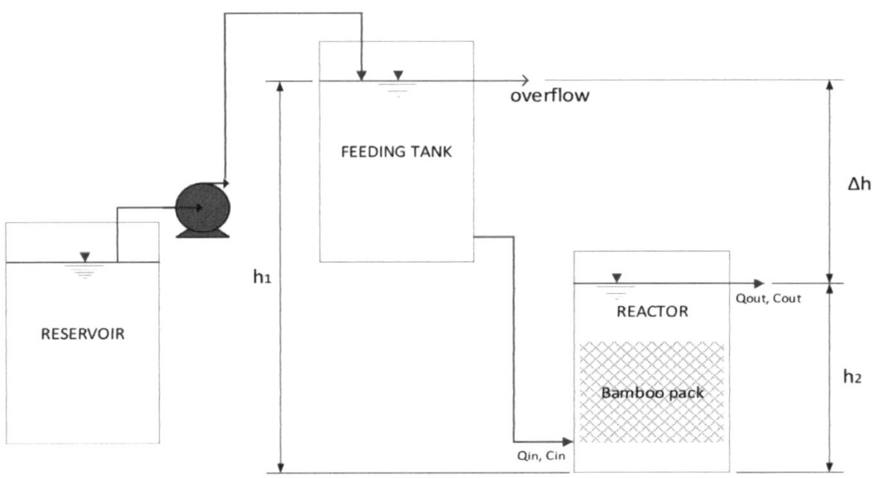

Figure 2. Experimental setup.

2.3. Analysis

The measured parameters during this experiment were flow rate and tracer concentration. The input flow rate was constant due to the use of pump. The output flow rate from the reactor was calculated by measuring the accumulated volume of fluid in the outlet with a measuring cylinder. Flow acceleration was analysed by comparing flow rate gradient between the three configurations during the unsteady state. To determine the flow regime inside the reactor, Reynold number (Re) was calculated as follow [15]:

$$\Re = \frac{Inertialforce}{Viscousforce} = \frac{\rho v L}{\mu} \tag{1}$$

ρ = fluid density
v = velocity along fluid trajectory
L = Length of fluid trajectory
μ = Dynamic viscosity

Tracer concentration was measured with a spectrophotometer at 496 nm. The change of relative concentration between the output and input of the reactor was used as an indication of mixing process inside the reactor.

3. Results and discussion

3.1. Fluid flow profile

The flow rate from the reactor was measured and the results for the three bamboos configurations, i.e. random, vertical, and horizontal, are presented in figure 3. Experiment with a reactor with no bamboo was used as a control.

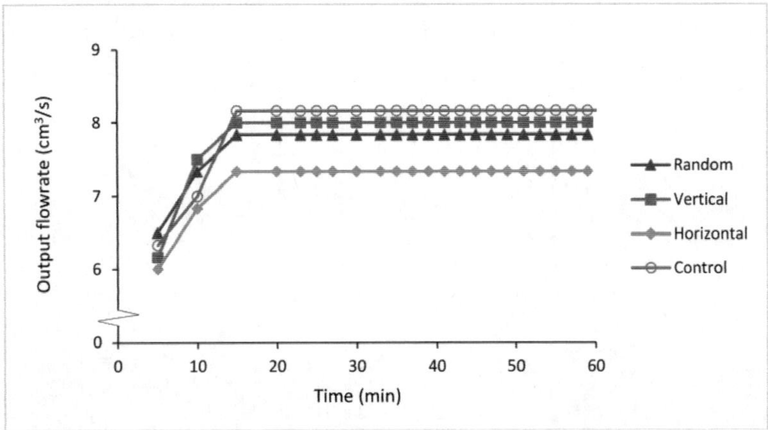

Figure 3. Output flow rate from the reactor.

Figure 3 shows that all configurations required 15 minutes to reach a steady state. The flow rates for random, vertical, and horizontal configurations were 7.8, 8.0, and 7.3 cm^3/s, respectively. Compared with the flow rate of the control experiment (8.2 cm^3/s), vertical configuration had the closest and highest flow rate. As shown in figure 4, vertical configuration also showed similar trend in reaching steady state compared with control.

Due to the limitation of experimental set up, the flow rate inside the reactor could not be measured. Assuming there was no significant difference between output flow rate and flow rate inside the reactor; Reynold numbers for all experiments were in the range of 126 - 141, which means the flow inside the reactor was laminar [15]. The presence of bamboo should create local turbulences; however, the overall flow was still laminar.

In this experiment, the ratio between bamboo length and diameter was 1:1 to 1.5:1. Higher ratio (longer bamboo cut) may cause a higher flow resistance in random and horizontal configurations. On the other hand, a higher length:diameter resulted in a lower flow resistance in vertical configuration [16]. In a laboratory, lab-scale bioreactor with the same dimension using cut-bamboo with length: diameter 2:1 in vertical configuration resulted in no significant flow reduction between inlet and outlet (unpublished result).

3.2. Mixing profile

Mixing characteristic of the fluid in the reactors could be approached by comparing the color intensity of the colored water in the outlet with the inlet. Bamboo could influence mixing process in the reactors. Figure 5 shows the fluctuation of tracer concentration presented as relative concentration to inlet.

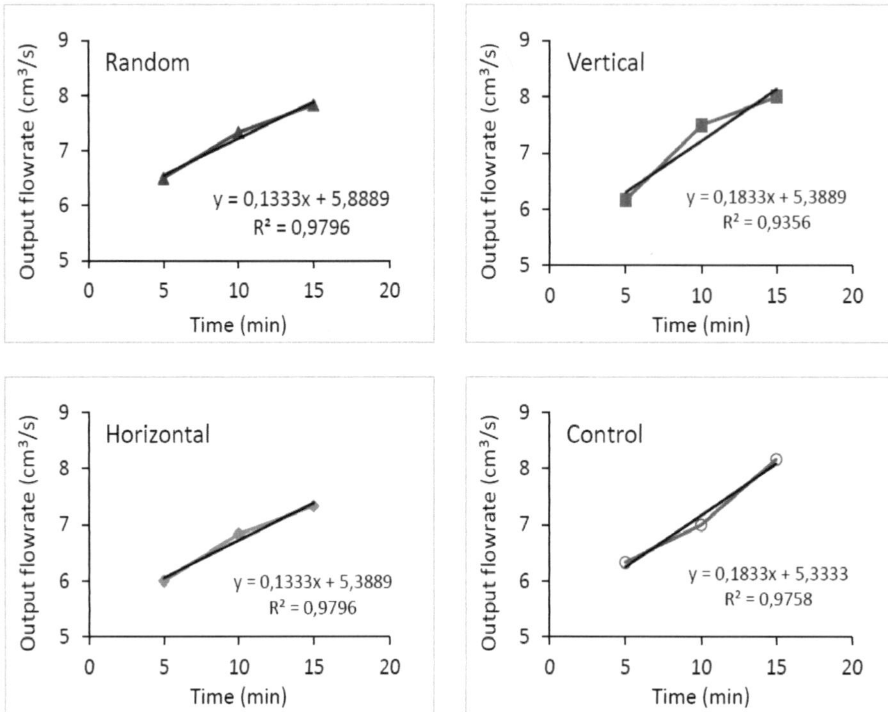

Figure 4. Flowrate changes during unsteady condition.

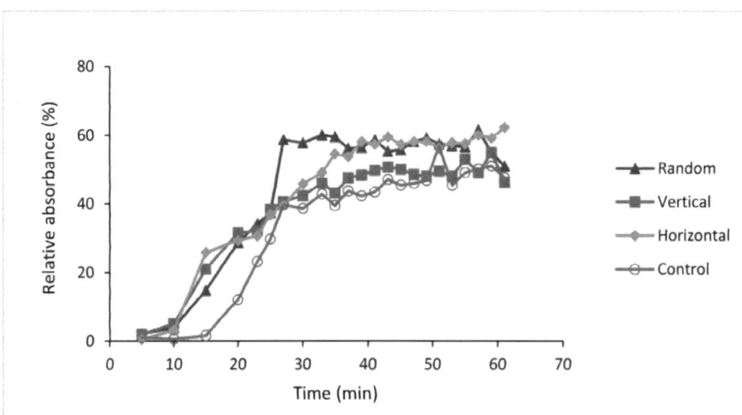

Figure 5. Tracer concentration in the reactor outlet.

Experiment without bamboo reached a steady state after 27 minutes, after which the steady state concentration still fluctuating at 46±5%. Experiment with configuration took 33 minutes to reach the steady state concentration of 49±3%. The concentration gradient for both experiments were quite similar at 1.7 (figure 6). Random configuration showed a distinct transfer from unsteady state to steady state. The steady state concentration of 57±2% was reached after 27 minutes. Similar steady state concentration of 58±2% was occurred in horizontal configuration. However, the time to reach steady state was 33 minutes. The results suggest that random and horizontal configuration is more optimum for mixing.

Compared to this experiment, a higher length:diameter ratio of bamboo cut may decrease the Reynolds number in random and horizontal configurations [17]. It resulted in less mixing. On the other hand, a higher length:diameter may increase Reynolds number in vertical configuration that resulted in better mixing [16].

The number of bamboo pieces in vertical configuration (105) was higher than horizontal (86) and random (84). The dye seemed to attach on the surface area of bamboo, as observed in the change of bamboo's color. Figure 5 shows that tracer concentration in vertical configuration was lower than horizontal and random. With the higher surface area in vertical configuration, more dye in the liquid could be attached compared with the other configurations. The attachment could reduce the color intensity in the reactor outlet.

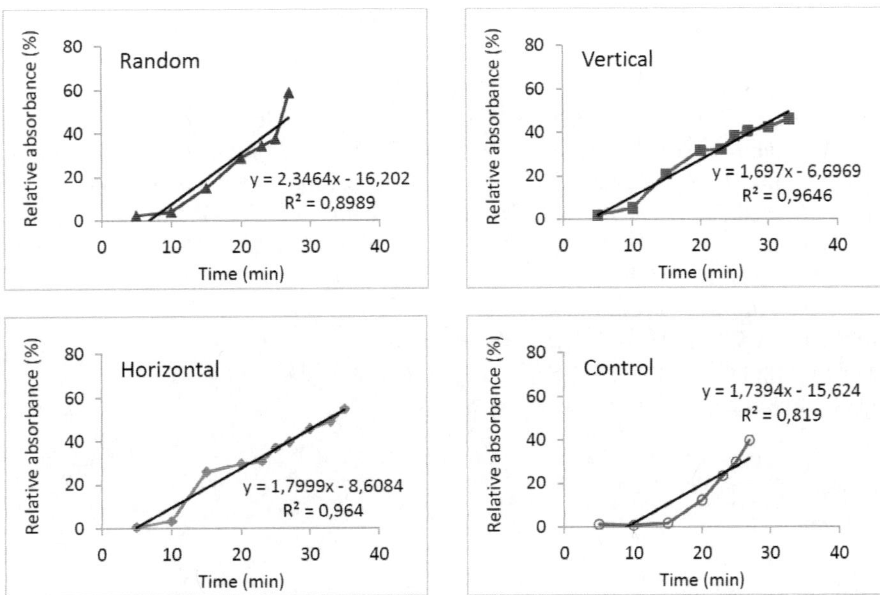

Figure 6. Concentration changes during unsteady condition.

A study conducted by Swaine and Daugulis [18] confirmed the difficulties of getting a determined certainty of liquid flow characteristics in the reactor. Aspect ratio of system, packing size, reactor size, and selection of suitable tracer could influence the result of liquid flow. On the other hand, tensile stress on the position parallel to the bamboo fibre direction was about 18 times higher than to the perpendicular direction of the fibre. This characteristic can be used for the application of bamboo cutting in the bigger scale anaerobic reactor [19].

4. Conclusions

Bamboo configurations in packed/fixed bed system influence the flow rate and mixing process in the reactor. Vertical configuration in fixed bed reactors showed the least liquid resistant flow in the reactor that results in a fastest flow rate. Random configuration showed the best configuration in mixing process.

Acknowledgment

This research was funded by Indonesian Institute of Sciences (LIPI) through the Deputy of Engineering Sciences priority program and LEF Award NTU-NEWRI Comm.Dev. Singapore.

References

[1] Zhao H, Li J, Li J, Yuan X, Piao R, Zhu W, Li H, Wang X and Cui Z 2013 *Bioresour. Technology* **140** 211
[2] Parlina I and Widodo L 2013 *Jurnal Teknologi Lingkungan* **14** 7
[3] Gong W, Liang H, Li W and Wang Z 2011 *Energy* **36** 3572
[4] Toda K, Ohtake H and Asakura T 1986 *Appl. Microbiol. Biotechnol.* **24** 97
[5] Switzenbaum M S 1983 *Enzyme Microb. Technol.* **5** 242
[6] Cahyaningsih S and Widyarani 2005 *Indonesian Journal on Environmental Chemistry and Toxicology* **4** 21
[7] Feng H, Hu L, Mahmood Q, Qiu C, Fang C and Shen D 2008 *Int. Biodeterior. Biodegradation* **62** 232
[8] Qazi J I, Nadeem M, Baig S S, Baig S and Syed Q 2011 *Middle-East Journal of Scientific Research* **8** 590
[9] Torres D G B, Lucas S D M, Andreani C L, Carvalho K Q, Coelho S R M and Gomes S D 2017 *Eng. Agric.* **37** 160
[10] Xiao J and Chu S 2015 *Desalin. Water Treat.* **56** 574
[11] Cao W, Zhang H, Wang Y and Pan J Z 2012 *Ecol. Eng.* **42** 146
[12] Yuan X H, Chen D S, Yin W and Yang M X 2015 *Chemical Engineering Transaction* **46** 25
[13] Colin X, Farinet J L, Rojas O and Alazard D 2007 *Bioresour. Technol.* **98** 1602
[14] McGreavy C, Foumeny E A and Javed K H 1986 *Chem. Eng. Sci.* **41** 787
[15] Cengel Y A and Cimbala J M 2014 *Fluid Mechanics Fundamentals and Applications* (New Delhi: Mc Graw Hill India)
[16] Tritton D 1959 *J. Fluid Mech.* **6** 547
[17] Nishioka M and Sato H 1974 *J. Fluid Mech.* **65** 97
[18] Swaine D E and Daugulis A J 1988 *Biotechnol. Prog.* **4** 134
[19] Chand N, Shukla M and Sharma M K 2008 *J. Nat. Fibers* **5** 127

IOP Conf. Series: Journal of Physics: Conf. Series **985** (2018) 012057 doi:10.1088/1742-6596/985/1/012057

Synthesis and properties of platinum on multiwall carbon nanotube modified by chitosan

A K Fikriyyah, E R Chaldun and Indriyati

Research Unit for Clean Technology, Indonesian Institute of Sciences (LIPI), Jalan Sangkuriang Komplek LIPI Gedung 50 Bandung, Indonesia 40135

E-mail: indriyati@lipi.go.id

Abstract. Platinum nanoparticles on multiwall carbon nanotubes (Pt/MWCNT) play an important role in fuel cell to convert the chemical energy from a fuel into electricity. In this study, Pt/MWCNT electrocatalysts were prepared by chemical reduction of the metal salts in chitosan as the support. Firstly, commercial MWCNTs were functionalized by oxidative process using a mixture of nitric acid and sulfuric acid. Then, functionalized MWCNTs were mixed with chitosan-acetic acid solution to conduct grafting reaction with NH_2 groups in chitosan by solution polymerization method. Platinum nanoparticles were loaded onto the surface of the MWCNTs after hexachloroplatinic acid was reduced by sodium hydroxide solution. The result showed that Pt was attached on MWCNT based on analysis from EDS, XRD, and UV Vis Spectroscopy. UV Vis analysis indicates the plasmon absorbance band of Pt nanoparticles in Pt/MWCNT, while XRD analysis confirmed the size of Pt particle in nanometer. This elucidates the potential procedure to synthesize Pt/MWCNT using chitosan.

1. Introduction

Platinum (Pt) as catalyst in fuel cell devices has a significant influence in the process of converting chemical energy into electricity. However, since platinum prices are quite expensive [1] and the inherent lack of stability of Pt when supported on carbon-black (Pt/C) [2], the reduction of Pt is required by making platinum on a nanometer scale. This makes platinum require catalyst support that can help increase platinum capacity and utilization [3, 4] A suitable support for an electrocatalyst should possess high surface area, good electrical properties, and high electrochemical stability under PEM Fuel Cell operating conditions [5]. Carbon Nanotubes (CNTs) potentially match these requirements since it has high surface area, high platinum dispersion, and smaller particle size, so it can reduce production costs by reducing platinum use [6].

To improve the catalytic activity of the platinum catalysts, many polymers have been used as support for metal nanoparticles, such as nafion, polyaniline, and chitosan [7]. Chitosan is a polymer derived from shrimp and crab shells waste that has cation polyelectrolyte properties so it can bind to metal [8]. The amine and two hydroxyl groups on each glucosamine monomer act as adsorption sites, especially the amine groups which are strongly reactive with metal ion [9]. Honda Wu [10] had been using chitosan to modify CNTs by solution polymerization method that adsorption of platinum was better for carbon nanotube modified by chitosan. The oxidation performance for carbon nanotube grafted by chitosan trigger the chelation reaction between chitosan and platinum ion [10]. Another study reported preparation of Pt, Fe and Co nanoparticles by chemical reduction of the metal salts in chitosan as support [7].

In this study, we report a novel process to synthesize Pt/MWCNT by modifying functionalized CNT with chitosan that conducted at 'mild' condition. The presence of Pt in Pt/MWCNT was carried out using UV-Vis spectroscopy analysis. Chemical composition and crystallographic structure of the obtained Pt/MWCNT will be confirmed by energy dispersive X-ray spectroscopy (EDS) and by X-ray diffraction (XRD) measurements.

2. Materials and method

2.1. Materials and chemicals

The MWCNTs were purchased from He Ji Company (95% purity, outer diameter 8-15 nm, Hongkong). The MWCNTs were firstly functionalized by surface oxidation method using a mixture of nitric acid and sulfuric acid as previous procedure [11]. Chitosan was obtained from local industry in Bogor, West Java. Chloroplatinic acid hexahydrate ($H_2PtCl_6.6H_2O$) was purchased from Sigma-Aldrich Company and was used as received. Nafion solution 5% was purchased from Fuelcell Earth. All other chemicals such as sulphuric acid, nitric acid, acetic acid glacial, sodium hydroxide, 2-propanol, and ethanol were purchased from Merck and used without further purification. Deionized water was used in all aqueous solutions and rinsing procedures.

2.2. Preparation of Pt/MWCNT electrocatalyst

The preparation of Pt/MWCNT electrocatalyst has been done according to the following procedures. A quantity of 100 mg of functionalized MWCNTs were dispersed in 20 mL of chitosan solution (1 g of chitosan was dissolved in 100 mL of 2% (v/v) acetic acid solution), followed by 1 h of sonication to achieve good dispersion. Afterward, the transparent yellow solution of $H_2PtCl_6.6H_2O$/etanol (15 mL, 0.007 M) was added slowly into the mixture of MWCNT-chitosan at the rate of 5 s/drop under stirring condition of 450 rpm. This solution was then stirred for 1 h. Subsequently, NaOH solution (2 M) as the reducing agent was added drop-wise into the mixture to increase the pH value until it reached 11.5. The reduction process of Pt precursors into Pt nanoparticles was carried out by refluxing the mixture under 'mild' condition, i.e. at 80 °C for 1 h and then cooled down to room temperature. The obtained material was filtered and washed with deionized water several times until neutral condition was achieved and finally dried in the vacuum oven at 40 °C.

2.3 Characterizations

Samples for transmission electron microscopy (TEM) were prepared by dispersing powder in ethanol, placing it in an ultrasonic bath, then putting droplets onto the grids and drying it in air at room temperature. TEM images were recorded using a FEI Tecnai G2 20 S-Twin operated at 200 kV. The surface morphology and elemental identification of the samples were characterized by Scanning Electron Microscope (SEM) equipped with EDS (Hitachi SU3500) under operating voltage of 20 kV. The crystal structure of the Pt nanoparticles supported on MWCNTs was determined by powder XRD (Shimadzu XRD-7000) with CuKα radiation. The average crystallite size of the Pt particles in the catalyst powder was calculated from full width at half maximum (FWHM) values using Debye-Scherrer equation [12]. For UV Visible characterization, a mixture of nafion solution and 2-propanol was used for dispersing Pt/MWCNT, while *f*-MWCNT and H_2PtCl_6 were dispersed in ethanol. About 20 mL of 2-propanol/nafion solution (ratio of 35:65) was mixed with 0.001 g of Pt/MWCNT and dispersed using ultrasonic bath for 3 h. UV-visible absorption spectra were recorded by spectrophotometer HITACHI U-2800 operated between 200 up to 300 nm at a scan speed of 100 nm/min.

3. Results and discussion

It is well known that functionalization of CNT by oxidised acid (acidic mixture of nitric acid and sulphuric acid) destroy and shorten the CNT to small segments by the breakage of the chain [6]. Figure 1 shows TEM images of MWCNT before and after oxidized treatment. The arrows in figure 1b serves as evidence that the outer walls of the functionalized MWCNT has been destroyed, which leads to the breakdown of the C-C bond in the CNT backbones. However, oxidized treatment of MWCNT also introduces some new functional groups, such as hydroxyl and carboxylic groups, which are formed on the MWCNT surfaces. Figure 2 illustrates our concept on synthesis method of Pt/MWCNT modified by chitosan. Introduction of chitosan into the functionalized MWCNT creates link between amine groups and some part of carboxylic groups. The reduction process was completed by adding NaOH solution to reduce Pt^{2-} into Pt (IV) metal nanoparticles. In this reaction, deposition of Pt nanoparticles on the surface of MWCNT occurred through amine groups and carboxylic groups.

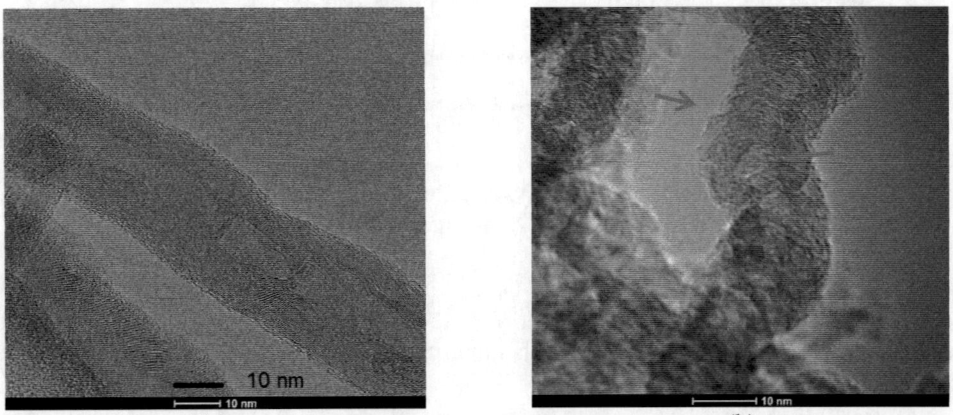

(a) (b)

Figure 1. TEM images of MWCNT (a) before, (b) after surface oxidized treatment.

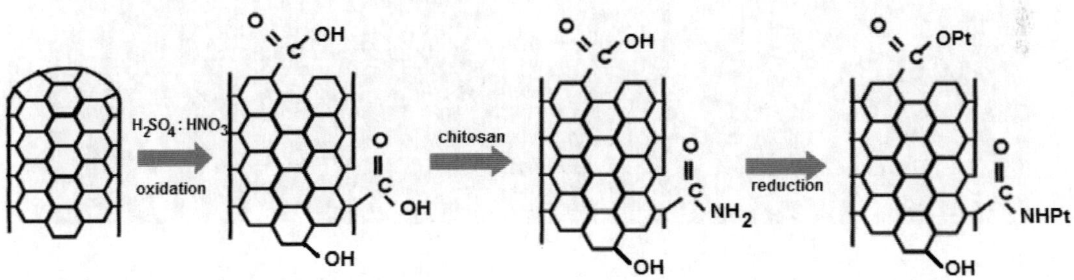

Figure 2. A schematic diagram of synthesis of Pt/MWCNT modified chitosan.

The XRD pattern of Pt/MWCNTs is shown in figure 3. The diffraction peaks in the XRD pattern indicate good crystallinity of the supported nanoparticles. The peak at ~26° is attributed to graphite crystallographic planes (002) of MWCNT. Pt/MWCNT shows the presence of diffraction peaks at nearly 40°, 46°, 68°, and 82° corresponding to (111), (200), (220), and (311) planes of platinum, respectively. These are analogous to the face-centered cubic (fcc) structure of the noble metal. The average crystallite size of platinum calculated from the Debye-Scherrer equation is less than 10 nm, which proves that Pt is nanoparticles.

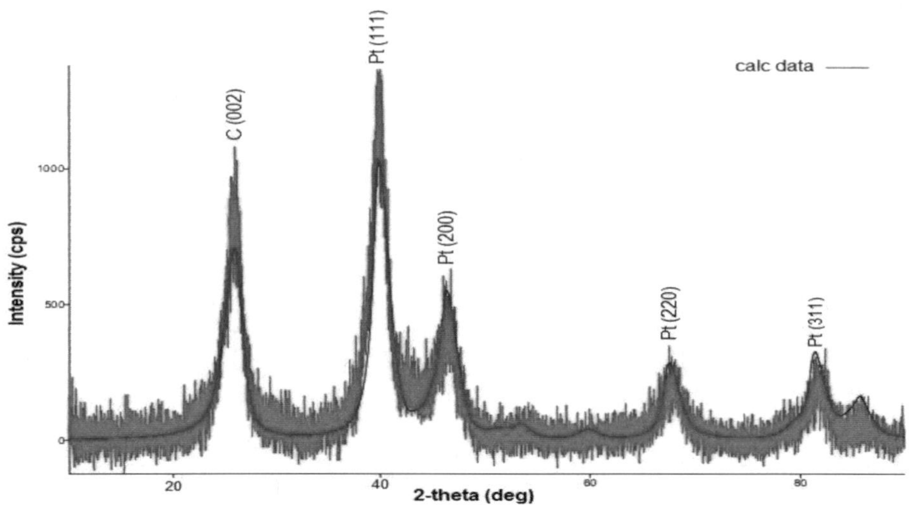

Figure 3. XRD pattern of Pt/MWCNT-chitosan.

SEM images and elemental analysis using EDS are shown in figures 4 and 5, respectively. It can be seen that Pt nanoparticles have been successfully deposited on the surface of MWCNT. The Pt phase appearance is convincingly confirmed by EDS analysis (figure 5b). Besides C and Pt signals, EDS spectrum includes also signals connected with O, Cl, and Ca elements. The peak of O element is connected with the functionalization process of CNTs, while the peaks of Cl and Ca element are connected with the presence of remained catalysts impurities. Cl signal suggested that H_2PtCl_6 was not reduced completely to Pt by NaOH.

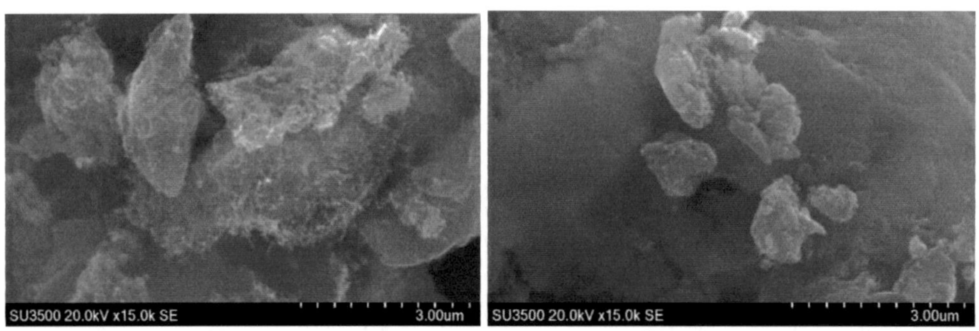

(a) (b)

Figure 4. SEM images of the (a) surface oxidised MWCNT and (b) Pt/MWCNT-chitosan.

UV-visible absorption can generally be used to analyze the content of a material. The testable material should be in liquid phase. The results can be seen in figure 6. UV-Vis characterization performed at 200-300 nm that was based on a study by Yang et al (2004) showing in the wavelength of 300-800 nm there would not be a plasmon absorbance band of metal Pt nanoparticles [13]. The curves formed from Pt/MWCNT have plasmon absorbance band at around 235 nm. The peak indicates the presence of Pt nanoparticles dispersed into 2-propanol/nafion solution. Meanwhile, the solution of $PtCl_6^{2-}$ compound has plasmon absorbance band at 265 nm indicating the presence of Pt (IV) metal content in $PtCl_6^{2-}$ compounds. This is consistent with the study of Ekrami-Kakhki et al. [7] showing that the plasmon absorbance band of Pt (IV) in $PtCl_6^{2-}$ compound is about 265 nm. When $PtCl_6^{2-}$ reacts with chitosan and NaOH, the particle of Pt (IV) will be discharged and will form colloid Pt [7].

Meanwhile, the functionalization of MWCNT (*f*-MWCNT) did not form a plasmon absorbance band. It could be due to quantum dot size effect as described by other [14].

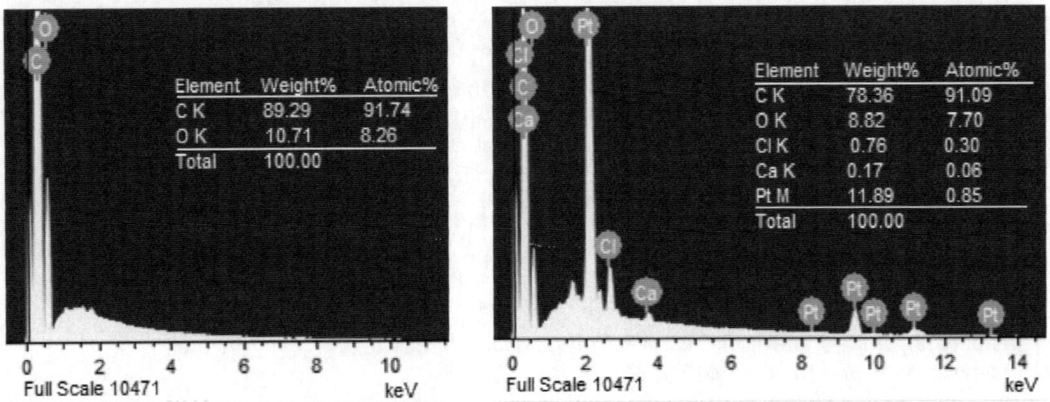

(a) (b)

Figure 5. Elemental analysis of the (a) surface oxidised MWCNT and (b) Pt/CNT-chitosan.

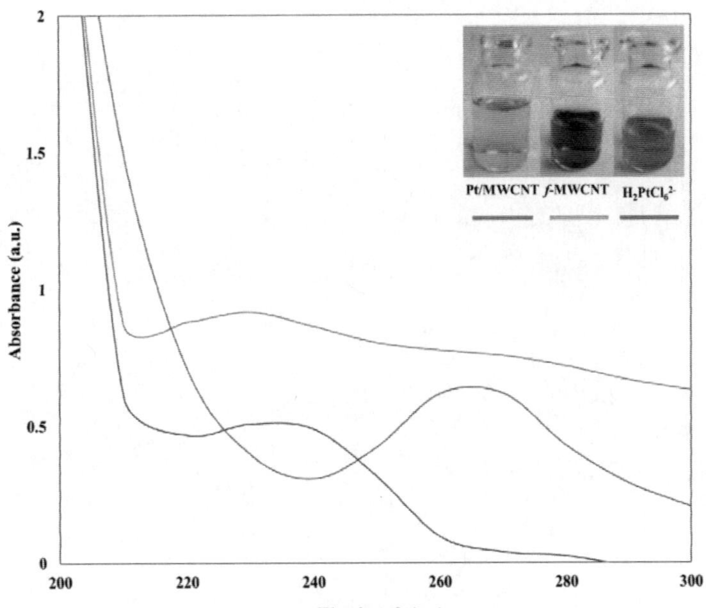

Figure 6. UV-visible absorption spectra of Pt/MWCNT, functionalized MWCNT and $PtCl_6^{2-}$.

4. Conclusion

Pt/MWCNT had been successfully synthesized in the presence of chitosan under mild condition, i.e. 80 °C. The presence of Pt nanoparticles on the surface of f-MWCNT was confirmed by XRD, EDS, and UV Vis analysis. In addition, XRD diffractogram indicated a good crystallinity of the supported nanoparticles. Calculation of average crystallite size of Pt using the Debye-Scherrer formula proved that Pt is nanoparticle. Obtained material can be employed in constructing catalyst for fuel cell.

References

[1] Jayakumar K, Pandiyan S, Rajalakshmi N and Dhathathreyan K S 2006 *J. Power Sources* **161**

454
[2] Zhai Y, Zhang H, Xing D and Shao Z 2007 *J. Power Sources* **164** 126
[3] Zhang W, Sherrell P, Minett A I, Razal J M and Chen J 2010 *Energy Environ. Sci.* **3** 1286
[4] Gong J et al. 2012 *Appl. Catal., B* **117–118** 185
[5] Zhang X, Zhu H, Guo Z, Wei Y and Wang F 2010 *Int. J. Hydrog. Energy* **35** 8841
[6] Luo C, Xie H, Wang Q, Luo G and Liu C 2015 *Journal of Nanomaterials* **2015** 560392
[7] Ekrami-Kakhki M S, Yavari Z, Saffari J and Abbasi S 2016 *Journal of Nanostructures* **6** 221
[8] Rinaudo M 2006 *Prog. Polym. Sci.* **31** 603
[9] Hiramatsu M and Hori M 2010 *Materials* **3** 1559
[10] Wu H 2014 *Journal of Materials Science and Engineering A* **4** 83
[11] Yudianti R, Onggo H, Sudirman, Saito Y, Iwata T and Azuma J 2011 *The Open Materials Science Journal* **5** 242
[12] Alexander L and Klug H P 1950 *J. Appl. Phys.* **21** 137
[13] Huang H, Yuan Q and Yang X 2004 *Colloids Surf. B* **39** 31
[14] Daoush W M and Imae T 2012 *J. Mater. Res.* **27** 1680